W9-BMN-731

Lial/Miller/Greenwell

Student's Solution Manual
to accompany
Calculus with Applications, Fifth Edition, and Calculus with Applications, Brief Version, Fifth Edition

Prepared with the assistance of

Jane Brandsma
Suffolk County Community College

Kathleen Pellissier

August Zarcone
College of DuPage

Richard Zylstra

HarperCollinsCollegePublishers

Student's Solution Manual to accompany *Lial /Miller/Greenwell: CALCULUS WITH APPLICA-TIONS, FIFTH EDITION, and CALCULUS WITH APPLICATIONS, BRIEF VERSION, FIFTH EDITION*

Copyright © 1993 by HarperCollins*CollegePublishers*

All rights reserved. Printed in the United States of America. No part of this book may be reproduced in any manner whatsoever without written permission. For information, address HarperCollins*CollegePublishers*, 10 E. 53rd Street, New York, NY 10022.

ISBN 0-673-46755-4

92 93 94 95 9 8 7 6 5 4 3 2 1

PREFACE

This book provides complete solutions for many of the exercises in <u>Calculus with</u> <u>Applications</u>, fifth edition, by Margaret L. Lial, Charles D. Miller, and Raymond N. Greenwell. Solutions are included for odd–numbered exercises. Solutions are not provided for exercises with open–response answers or for those that require the use of a computer. Sample tests are provided to help you determine if you have mastered the concepts in a given chapter.

This book should be used as an aid as you work to master your course work. Try to solve the exercises that your instructor assigns before you refer to the solutions in this book. Then, if you have difficulty, read these solutions to guide you in solving the exercises. The solutions have been written so that they are consistent with the methods used in the textbook.

You may find that some of the solutions are presented in greater detail than others. Thus, if you cannot find an explanation for a difficulty that you encountered in one exercise, you may find the explanation in the solution for a similar exercise elsewhere in the exercise set.

Solutions that require graphs will refer you to the answer section of the textbook. These graphs are not included in this book.

In addition to solutions, you will find a list of suggestions on how to be successful in mathematics. A careful reading will be helpful for many students.

The following people have made valuable contributions to the production of this <u>Student's</u> <u>Solution</u> <u>Manual</u>: Marjorie Seachrist, editor; Judy Martinez, typist; Therese Brown and Charles Sullivan, artists; and Carmen Eldersveld, proofreader.

We also want to thank Tommy Thompson of Seminole Community College for his suggestions for the essay "To the Student: Success in Mathematics" that follows this preface.

TO THE STUDENT: SUCCESS IN MATHEMATICS

The main reason students have difficulty with mathematics is that they don't know how to study it. Studying mathematics *is* different from studying subjects like English or history. The key to success is regular practice.

This should not be surprising. After all, can you learn to play the piano or to ski well without a lot of regular practice? The same thing is true for learning mathematics. Working problems nearly every day is the key to becoming successful. Here is a list of things you can do to help you succeed in studying mathematics.

1. *Attend class regularly.* Pay attention in class to what your instructor says and does, and make careful notes. In particular, note the problems the instructor works on the board and copy the complete solutions. Keep these notes separate from your homework to avoid confusion when you read them over later.

2. Don't hesitate to ask questions in class. It is not a sign of weakness, but of strength. There are always other students with the same question who are too shy to ask.

3. *Read your text carefully.* Many students read only enough to get by, usually only the examples. Reading the complete section will help you to be successful with the homework problems. Most exercises are keyed to specific examples or objectives that will explain the procedures for working them.

4. Before you start on your homework assignment, rework the problems the instructor worked in class. This will reinforce what you have learned. Many students say, "I understand it perfectly when you do it, but I get stuck when I try to work the problem myself."

5. Do your homework assignment only *after* reading the text and reviewing your notes from class. Check your work with the answers in the back of the book. If you get a problem wrong and are unable to see why, mark that problem and ask your instructor about it. Then practice working additional problems of the same type to reinforce what you have learned.

6. Work as neatly as you can. Write your symbols clearly, and make sure the problems are clearly separated from each other. Working neatly will help you to think clearly and also make it easier to review the homework before a test.

7. After you have completed a homework assignment, look over the text again. Try to decide what the main ideas are in the lesson. Often they are clearly highlighted or boxed in the text.

8. Use the chapter test at the end of each chapter as a practice test. Work through the problems under test conditions, without referring to the text or the answers until you are finished. You may want to time yourself to see how long it takes you. When you have finished, check your answers against those in the back of the book and study those problems that you missed. Answers are referenced to the appropriate sections of the text.

9. Keep any quizzes and tests that are returned to you and use them when you study for future tests and the final exam. These quizzes and tests indicate what your instructor considers most important. Be sure to correct any problems on these tests that you missed, so you will have the corrected work to study.

10. Don't worry if you do not understand a new topic right away. As you read more about it and work through the problems, you will gain understanding. Each time you look back at a topic you will understand it a little better. No one understands each topic completely right from the start.

CONTENTS

Student's Solution Manual
to accompany
Calculus with Applications, Fifth Edition, and Calculus with Applications, Brief Version, Fifth Edition

ALGEBRA REFERENCE

Section R.1

1. $(2x^2 - 6x + 11) + (-3x^2 + 7x - 2)$

   $= 2x^2 - 6x + 11 - 3x^2 + 7x - 2$

   $= (2 - 3)x^2 + (7 - 6)x + (11 - 2)$

   $= -x^2 + x + 9$

3. $-3(4q^2 - 3q + 2) + 2(-q^2 + q - 4)$

   $= -12q^2 + 9q - 6 - 2q^2 + 2q - 8$

   $= -14q^2 + 11q - 14$

5. $(.613x^2 - 4.215x + .892)$

   $- .47(2x^2 - 3x + 5)$

   $= .613x^2 - 4.215x + .892$

   $\quad - .94x^2 + 1.41x - 2.35$

   $= -.327x^2 - 2.805x - 1.458$

7. $-9m(2m^2 + 3m - 1)$

   $= -9m(2m^2) - 9m(3m) - 9m(-1)$

   $= -18m^3 - 27m^2 + 9m$

9. Use the FOIL method to find

   $(5r - 3s)(5r + 4s)$

   $= (5r)(5r) + (5r)(4s) + (-3s)(5r)$

   $\quad + (-3s)(4s)$

   $= 25r^2 + 20rs - 15rs - 12s^2$

   $= 25r^2 + 5rs - 12s^2.$

11. $\left(\frac{2}{5}y + \frac{1}{8}z\right)\left(\frac{3}{5}y + \frac{1}{2}z\right)$

    $= \left(\frac{2}{5}y\right)\left(\frac{3}{5}y\right) + \left(\frac{2}{5}y\right)\left(\frac{1}{2}z\right) + \left(\frac{1}{8}z\right)\left(\frac{3}{5}y\right)$

    $\quad + \left(\frac{1}{8}z\right)\left(\frac{1}{2}z\right)$

    $= \frac{6}{25}y^2 + \frac{1}{5}yz + \frac{3}{40}yz + \frac{1}{16}z^2$

    $= \frac{6}{25}y^2 + \left(\frac{8}{40} + \frac{3}{40}\right)yz + \frac{1}{16}z^2$

    $= \frac{6}{25}y^2 + \frac{11}{40}yz + \frac{1}{16}z^2$

13. $(12x - 1)(12x + 1)$

    $= (12x)(12x) + (12x)(1)$

    $\quad - 1(12x) - 1(1)$

    $= 144x^2 + 12x - 12x - 1$

    $= 144x^2 - 1$

15. $(3p - 1)(9p^2 + 3p + 1)$

    $= (3p - 1)(9p^2) + (3p - 1)(3p)$

    $\quad + (3p - 1)(1)$

    $= 3p(9p^2) - 1(9p^2) + 3p(3p)$

    $\quad - 1(3p) + 3p(1) - 1(1)$

    $= 27p^3 - 9p^2 + 9p^2 - 3p + 3p - 1$

    $= 27p^3 - 1$

17. $(2m + 1)(4m^2 - 2m + 1)$

    $= 2m(4m^2 - 2m + 1) + 1(4m^2 - 2m + 1)$

    $= 8m^3 - 4m^2 + 2m + 4m^2 - 2m + 1$

    $= 8m^3 + 1$

19. $(m - n + k)(m + 2n - 3k)$

    $= m(m + 2n - 3k) - n(m + 2n - 3k)$

    $\quad + k(m + 2n - 3k)$

    $= m^2 + 2mn - 3km - mn - 2n^2 + 3kn$

    $\quad + km + 2kn - 3k^2$

    $= m^2 + mn - 2km - 2n^2 + 5kn - 3k^2$

Section R.2

1. $8a^3 - 16a^2 + 24a$

   $= 8a \cdot a^2 - 8a \cdot 2a + 8a \cdot 3$

   $= 8a(a^2 - 2a + 3)$

3. $25p^4 - 20p^3q + 100p^2q^2$

   $= 5p^2 \cdot 5p^2 - 5p^2 \cdot 4pq + 5p^2 \cdot 20q^2$

   $= 5p^2(5p^2 - 4pq + 20q^2)$

5. $m^2 + 9m + 14 = (m + 2)(m + 7)$

since $2 \cdot 7 = 14$ and $2 + 7 = 9$.

7. $z^2 + 9z + 20 = (z + 4)(z + 5)$

since $4 \cdot 5 = 20$ and $4 + 5 = 9$.

9. $a^2 - 6ab + 5b^2 = (a - b)(a - 5b)$

since $(-b)(-5b) = 5b^2$ and
$-b + (-5b) = -6b$.

11. $y^2 - 4yz - 21z^2$

$\quad = (y + 3z)(y - 7z)$

since $(3z)(-7z) = -21z^2$ and
$3z + (-7z) = -4z$.

13. $3m^3 + 12m^2 + 9m$

$\quad = 3m(m^2 + 4m + 3)$

$\quad = 3m(m + 1)(m + 3)$

15. $3a^2 + 10a + 7$

The possible factors of $3a^2$ are $3a$ and a and the possible factors of 7 are 7 and 1. Try various combinations until one works.

$\quad 3a^2 + 10a + 7 = (a + 1)(3a + 7)$

17. $15y^2 + y - 2 = (5y + 2)(3y - 1)$

19. $24a^4 + 10a^3b - 4a^2b^2$

$\quad = 2a^2(12a^2 + 5ab - 2b^2)$

$\quad = 2a^2(4a - b)(3a + 2b)$

21. $x^2 - 64 = x^2 - 8^2$

$\quad\quad = (x + 8)(x - 8)$

23. $121a^2 - 100$

$\quad = (11a)^2 - 10^2$

$\quad = (11a - 10)(11a + 10)$

25. $z^2 + 14zy + 49y^2$

$\quad = z^2 + 2 \cdot 7zy + 7^2y^2$

$\quad = (z + 7y)^2$

27. $9p^2 - 24p + 16$

$\quad = (3p)^2 - 2 \cdot 3p \cdot 4 + 4^2$

$\quad = (3p - 4)^2$

29. $8r^3 - 27s^3$

$\quad = (2r)^3 - (3s)^3$

$\quad = (2r - 3s)(4r^2 + 6rs + 9s^2)$

Section R.3

When writing rational expressions in lowest terms in the following exercises, first write numerators and denominators in factored form. Then use the fundamental property of rational expressions.

1. $\dfrac{7z^2}{14z} = \dfrac{7z \cdot z}{2(7)z} = \dfrac{z}{2}$

3. $\dfrac{8k + 16}{9k + 18} = \dfrac{8(k + 2)}{9(k + 2)} = \dfrac{8}{9}$

5. $\dfrac{8x^2 + 16x}{4x^2} = \dfrac{8x(x + 2)}{4x^2}$

$\quad\quad = \dfrac{2(x + 2)}{x} \quad \text{or} \quad \dfrac{2x + 4}{x}$

7. $\dfrac{m^2 - 4m + 4}{m^2 + m - 6}$

$\quad = \dfrac{(m - 2)(m - 2)}{(m - 2)(m + 3)}$

$\quad = \dfrac{m - 2}{m + 3}$

9. $\dfrac{x^2 + 3x - 4}{x^2 - 1}$

$= \dfrac{(x - 1)(x + 4)}{(x - 1)(x + 1)}$

$= \dfrac{x + 4}{x + 1}$

11. $\dfrac{8m^2 + 6m - 9}{16m^2 - 9}$

$= \dfrac{(4m - 3)(2m + 3)}{(4m - 3)(4m + 3)}$

$= \dfrac{2m + 3}{4m + 3}$

13. $\dfrac{9k^2}{25} \cdot \dfrac{5}{3k} = \dfrac{3 \cdot 3 \cdot 5k^2}{5 \cdot 5 \cdot 3k} = \dfrac{3k^2}{5k} = \dfrac{3k}{5}$

15. $\dfrac{a + b}{2p} \cdot \dfrac{12}{5(a + b)}$

$= \dfrac{(a + b)12}{2p(5)(a + b)} = \dfrac{12}{10p}$

$= \dfrac{6}{5p}$

17. $\dfrac{2k + 8}{6} \div \dfrac{3k + 12}{2}$

$= \dfrac{2(k + 4)}{6} \cdot \dfrac{2}{3(k + 4)}$

$= \dfrac{2(k + 4)(2)}{2 \cdot 3 \cdot 3(k + 4)}$

$= \dfrac{2}{9}$

19. $\dfrac{4a + 12}{2a - 10} \div \dfrac{a^2 - 9}{a^2 - a - 20}$

$= \dfrac{4(a + 3)}{2(a - 5)} \cdot \dfrac{(a - 5)(a + 4)}{(a - 3)(a + 3)}$

$= \dfrac{2(a + 4)}{a - 3}$

21. $\dfrac{k^2 - k - 6}{k^2 + k - 12} \cdot \dfrac{k^2 + 3k - 4}{k^2 + 2k - 3}$

$= \dfrac{(k - 3)(k + 2)(k - 1)(k + 4)}{(k + 4)(k - 3)(k + 3)(k - 1)}$

$= \dfrac{k + 2}{k + 3}$

23. $\dfrac{2m^2 - 5m - 12}{m^2 - 10m + 24} \div \dfrac{4m^2 - 9}{m^2 - 9m + 18}$

$= \dfrac{2m^2 - 5m - 12}{m^2 - 10m + 24} \cdot \dfrac{m^2 - 9m + 18}{4m^2 - 9}$

$= \dfrac{(2m + 3)(m - 4)(m - 6)(m - 3)}{(m - 6)(m - 4)(2m - 3)(2m + 3)}$

$= \dfrac{m - 3}{2m - 3}$

25. $\dfrac{a + 1}{2} - \dfrac{a - 1}{2}$

$= \dfrac{(a + 1) - (a - 1)}{2}$

$= \dfrac{a + 1 - a + 1}{2}$

$= \dfrac{2}{2}$

$= 1$

27. $\dfrac{2}{y} - \dfrac{1}{4} = \left(\dfrac{4}{4}\right)\dfrac{2}{y} - \left(\dfrac{y}{y}\right)\dfrac{1}{4}$

$= \dfrac{8 - y}{4y}$

29. $\dfrac{1}{m - 1} + \dfrac{2}{m}$

$= \dfrac{m}{m}\left(\dfrac{1}{m - 1}\right) + \dfrac{m - 1}{m - 1}\left(\dfrac{2}{m}\right)$

$= \dfrac{m + 2m - 2}{m(m - 1)}$

$= \dfrac{3m - 2}{m(m - 1)}$

31. $\dfrac{8}{3(a - 1)} + \dfrac{2}{a - 1}$

$= \dfrac{8}{3(a - 1)} + \dfrac{3}{3}\left(\dfrac{2}{a + 1}\right)$

$= \dfrac{8 + 6}{3(a - 1)}$

$= \dfrac{14}{3(a - 1)}$

33. $\dfrac{2}{x^2 - 2x - 3} + \dfrac{5}{x^2 - x - 6}$

$$= \dfrac{2}{(x - 3)(x + 1)} + \dfrac{5}{(x - 3)(x + 2)}$$

$$= \left(\dfrac{x + 2}{x + 2}\right)\dfrac{2}{(x - 3)(x + 1)}$$

$$\quad + \left(\dfrac{x + 1}{x + 1}\right)\dfrac{5}{(x - 3)(x + 2)}$$

$$= \dfrac{2x + 4 + 5x + 5}{(x + 2)(x - 3)(x + 1)}$$

$$= \dfrac{7x + 9}{(x + 2)(x - 3)(x + 1)}$$

35. $\dfrac{3k}{2k^2 + 3k - 2} - \dfrac{2k}{2k^2 - 7k + 3}$

$$= \dfrac{3k}{(2k - 1)(k + 2)} - \dfrac{2k}{(2k - 1)(k - 3)}$$

$$= \left(\dfrac{k - 3}{k - 3}\right)\dfrac{3k}{(2k - 1)(k + 2)}$$

$$\quad - \left(\dfrac{k + 2}{k + 2}\right)\dfrac{2k}{(2k - 1)(k - 3)}$$

$$= \dfrac{(3k^2 - 9k) - (2k^2 + 4k)}{(2k - 1)(k + 2)(k - 3)}$$

$$= \dfrac{k^2 - 13k}{(2k - 1)(k + 2)(k - 3)}$$

$$= \dfrac{k(k - 13)}{(2k - 1)(k + 2)(k - 3)}$$

Section R.4

1. $\quad .2m - .5 = .1m + .7$

$$10(.2m - .5) = 10(.1m + .7)$$

$$2m - 5 = m + 7$$

$$m - 5 = 7$$

$$m = 12$$

The solution is 12.

3. $3r + 2 - 5(r + 1) = 6r + 4$

$$3r + 2 - 5r - 5 = 6r + 4$$

$$-3 - 2r = 6r + 4$$

$$-3 = 8r + 4$$

$$-7 = 8r$$

$$-\dfrac{7}{8} = r$$

The solution is $-7/8$.

5. $|3x + 2| = 9$

$3x + 2 = 9$ or $-(3x + 2) = 9$

$\quad 3x = 7 \qquad\qquad -3x - 2 = 9$

$\qquad\qquad\qquad\qquad\qquad -3x = 11$

$\quad x = \dfrac{7}{3} \qquad\qquad\quad x = -\dfrac{11}{3}$

Substituting each solution in the original equation shows that both $7/3$ and $-11/3$ are solutions.

7. $|2x + 8| = |x - 4|$

$$2x + 8 = x - 4$$

$$x = -12$$

or

$$-(2x + 8) = x - 4$$

$$-2x - 8 = x - 4$$

$$-3x = 4$$

$$x = -\dfrac{4}{3}$$

The solutions are -12 and $-4/3$.

9. $\quad x^2 + 5x + 6 = 0$

$$(x + 3)(x + 2) = 0$$

$$x + 3 = 0$$

$$x + 3 - 3 = 0 - 3$$

$$x = -3$$

or

$$x + 2 = 0$$

$$x + 2 - 2 = 0 - 2$$

$$x = -2$$

The solutions are -3 and -2.

11.
$$m^2 + 16 = 8m$$
$$m^2 + 16 - 8m = 8m - 8m$$
$$m^2 + 16 - 8m = 0$$
$$m^2 - 8m + 16 = 0$$
$$(m)^2 - 2(4m) + (4)^2 = 0$$
$$(m - 4)^2 = 0$$
$$m - 4 = 0$$
$$m - 4 + 4 = 0 + 4$$
$$m = 4$$

The solution is 4.

13.
$$6x^2 - 5x = 4$$
$$6x^2 - 5x - 4 = 0$$
$$(3x - 4)(2x + 1) = 0$$

$$3x - 4 = 0$$
$$3x - 4 + 4 = 4$$
$$3x = 4$$
$$\frac{1}{3}(3x) = \frac{1}{3}(4)$$
$$x = \frac{4}{3}$$

or

$$2x + 1 = 0$$
$$2x + 1 - 1 = 0 - 1$$
$$2x = -1$$
$$\frac{1}{2}(2x) = \frac{1}{2}(-1)$$
$$x = -\frac{1}{2}$$

The solutions are 4/3 and -1/2.

15.
$$9x^2 - 16 = 0$$
$$(3x)^2 - (4)^2 = 0$$
$$(3x + 4)(3x - 4) = 0$$

$$3x + 4 = 0$$
$$3x + 4 - 4 = 0 - 4$$
$$3x = -4$$
$$\frac{1}{3}(3x) = \frac{1}{3}(-4)$$
$$x = -\frac{4}{3}$$

or

$$3x - 4 = 0$$
$$3x - 4 + 4 = 0 + 4$$
$$3x = 4$$
$$\frac{1}{3}(3x) = \frac{1}{3}(4)$$
$$x = \frac{4}{3}$$

The solutions are -4/3 and 4/3.

17.
$$12y^2 - 48y = 0$$
$$12y(y) - 12y(4) = 0$$
$$12y(y - 4) = 0$$
$$12y = 0 \quad \text{or} \quad y - 4 = 0$$
$$\frac{1}{12}(12y) = \frac{1}{12}(0) \quad y - 4 + 4 = 0 + 4$$
$$y = 0 \qquad\qquad y = 4$$

The solutions are 0 and 4.

19.
$$2m^2 = m + 4$$
$$2m^2 - m - 4 = 0$$

Use the quadratic formula.

$$x = \frac{-(-1) \pm \sqrt{(-1)^2 - 4(2)(-4)}}{2(2)}$$
$$x = \frac{1 \pm \sqrt{1 + 32}}{4}$$
$$x = \frac{1 \pm \sqrt{33}}{4}$$

The solutions are $(1 + \sqrt{33})/4 \approx 1.686$ and $(1 - \sqrt{33})/4 \approx -1.186$.

21.
$$k^2 - 10k = -20$$
$$k^2 - 10k + 20 = 0$$
$$k = \frac{-(-10) \pm \sqrt{(-10)^2 - 4(1)(20)}}{2(1)}$$
$$k = \frac{10 \pm \sqrt{100 - 80}}{2}$$
$$k = \frac{10 \pm \sqrt{20}}{2}$$
$$k = \frac{10 \pm \sqrt{4}\sqrt{5}}{2}$$
$$k = \frac{10 \pm 2\sqrt{5}}{2}$$
$$k = \frac{2(5 \pm \sqrt{5})}{2}$$
$$k = 5 \pm \sqrt{5}$$

The solutions are $5 + \sqrt{5} \approx 7.236$ and $5 - \sqrt{5} \approx 2.764$.

23.
$$2r^2 - 7r + 5 = 0$$
$$(2r - 5)(r - 1) = 0$$
$$2r - 5 = 0 \quad \text{or} \quad r - 1 = 0$$
$$2r = 5 \qquad\qquad r = 1$$
$$r = \frac{5}{2}$$

The solutions are 5/2 and 1.

25.
$$3k^2 + k = 6$$
$$3k^2 + k - 6 = 0$$
$$k = \frac{-1 \pm \sqrt{1 - 4(3)(-6)}}{2(3)}$$
$$= \frac{-1 \pm \sqrt{73}}{6}$$

The solutions are $(-1 + \sqrt{73})/6 \approx 1.257$ and $(-1 - \sqrt{73})/6 \approx -1.591$.

27.
$$\frac{3x - 2}{7} = \frac{x + 2}{5}$$
$$35\left(\frac{3x - 2}{7}\right) = 35\left(\frac{x + 2}{5}\right)$$
$$5(3x - 2) = 7(x + 2)$$
$$15x - 10 = 7x + 14$$
$$8x = 24$$
$$x = 3$$

29.
$$\frac{4}{x - 3} - \frac{8}{2x + 5} + \frac{3}{x - 3} = 0$$
$$\frac{4}{x - 3} + \frac{3}{x - 3} - \frac{8}{2x + 5} = 0$$
$$\frac{7}{x - 3} - \frac{8}{2x + 5} = 0$$

Multiply both sides by $(x - 3)(2x + 5)$. Note that $x \neq 3$ and $x \neq -5/2$.

$$(x - 3)(2x + 5)\left(\frac{7}{x - 3} - \frac{8}{2x + 5}\right) = (x - 3)(2x + 5)(0)$$
$$7(2x + 5) - 8(x - 3) = 0$$
$$14x + 35 - 8x + 24 = 0$$
$$6x + 59 = 0$$
$$6x = -59$$
$$x = -\frac{59}{6}$$

Note: It is especially important to check solutions of equations that involve rational expressions. Here, a check shows that $-59/6$ is a solution.

31.
$$\frac{2}{m} + \frac{m}{m + 3} = \frac{3m}{m^2 + 3m}$$
$$\frac{2}{m} + \frac{m}{m + 3} = \frac{3m}{m(m + 3)}$$

Multiply both sides by $m(m + 3)$. Note that $m \approx 0$, and $m \neq -3$.

$$m(m + 3)\left(\frac{2}{m} + \frac{m}{m + 3}\right) = m(m + 3)\left(\frac{3m}{m^2 + 3m}\right)$$

$$2(m + 3) + m(m) = 3m$$

$$2m + 6 + m^2 = 3m$$

$$m^2 - m + 6 = 0$$

$$m = \frac{-(-1) \pm \sqrt{(-1)^2 - 4(1)(6)}}{2(1)}$$

$$= \frac{1 \pm \sqrt{1 - 24}}{2}$$

$$= \frac{1 \pm \sqrt{-23}}{2}$$

There are no real-number solutions.

33. $\dfrac{1}{x - 2} - \dfrac{3x}{x - 1} = \dfrac{2x + 1}{x^2 - 3x + 2}$

$$\frac{1}{x - 2} - \frac{3x}{x - 1} = \frac{2x + 1}{(x - 2)(x - 1)}$$

Multiply both sides by $(x - 2)(x - 1)$.
Note that $x \neq 2$ and $x \neq 1$.

$$(x - 2)(x - 1)\left(\frac{1}{x - 2} - \frac{3x}{x - 1}\right)$$

$$= (x - 2)(x - 1)$$

$$\cdot \left[\frac{2x + 1}{(x - 2)(x - 1)}\right]$$

$$(x - 2)(x - 1)\left(\frac{1}{x - 2}\right) - (x - 2)(x - 1) \cdot \left(\frac{3x}{x - 1}\right)$$

$$= \frac{(x - 2)(x - 2)(2x + 1)}{(x - 2)(x - 1)}$$

$$(x - 1) - (x - 2)(3x) = (2x + 1)$$

$$x - 1 - 3x^2 + 6x = 2x + 1$$

$$-3x^2 + 7x - 1 = 2x + 1$$

$$-3x^2 + 5x - 2 = 0$$

$$3x^2 - 5x + 2 = 0$$

$$(3x - 2)(x - 1) = 0$$

$$3x - 2 = 0 \quad \text{or} \quad x - 1 = 0$$

$$x = \frac{2}{3} \qquad\qquad x = 1$$

1 is not a solution since $x \neq 1$.
The solution is 2/3.

35. $\dfrac{2b^2 + 5b - 8}{b^2 + 2b} + \dfrac{5}{b + 2} = -\dfrac{3}{b}$

$$\frac{2b^2 + 5b - 8}{b(b + 2)} + \frac{5}{(b + 2)} = \frac{-3}{b}$$

Multiply both sides by $b(b + 2)$.
Note that $b \neq 0$ and $b \neq -2$.

$$b(b + 2)\left[\frac{2b^2 + 5b - 8}{b^2 + 2b}\right] + b(b + 2)\left[\frac{5}{b + 2}\right]$$

$$= b(b + 2)\left(-\frac{3}{b}\right)$$

$$2b^2 + 5b - 8 + 5b = (b + 2)(-3)$$

$$2b^2 + 10b - 8 = -3b - 6$$

$$2b^2 + 13b - 2 = 0$$

$$b = \frac{-(13) \pm \sqrt{(13)^2 - 4(2)(-2)}}{2(2)}$$

$$= \frac{-13 \pm \sqrt{169 + 16}}{4}$$

$$b = \frac{-13 \pm \sqrt{185}}{4}$$

The solutions are $(-13 + \sqrt{185})/4 \approx$
.150 and $(-13 - \sqrt{185})/4 \approx -6.650$.

Section R.5

For Exercises 1–23, see the answer graphs
in the back of the textbook.

1. $-3p - 2 \geq 1$

$$-3p \geq 3$$

$$\left(-\frac{1}{3}\right)(-3p) \leq \left(-\frac{1}{3}\right)(3)$$

$$p \leq -1$$

The solution in interval notation is
$(-\infty, -1]$.

3. $m - (4 + 2m) + 3 < 2m + 2$

$m - 4 - 2m + 3 < 2m + 2$

$-m - 1 < 2m + 2$

$-3m - 1 < 2$

$-3m < 3$

$-\frac{1}{3}(-3m) > -\frac{1}{3}(3)$

$m > -1$

The solution is $(-1, \infty)$.

5. $3p - 1 < 6p + 2(p - 1)$

$3p - 1 < 6p + 2p - 2$

$3p - 1 < 8p - 2$

$-5p - 1 < -2$

$-5p < -1$

$-\frac{1}{5}(-5p) > -\frac{1}{5}(-1)$

$p > \frac{1}{5}$

The solution is $(1/5, \infty)$.

7. $-7 < y - 2 < 4$

$-7 + 2 < y - 2 + 2 < 4 + 2$

$-5 < y < 6$

The solution is $(-5, 6)$.

9. $-4 \leq \frac{2k - 1}{3} \leq 2$

$3(-4) \leq 3\left(\frac{2k - 1}{3}\right) \leq 3(2)$

$-12 \leq 2k - 1 \leq 6$

$-12 + 1 \leq 2k - 1 + 1 \leq 6 + 1$

$-11 \leq 2k \leq 7$

$\frac{1}{2}(-11) \leq \frac{1}{2}(2k) \leq \frac{1}{2}(7)$

$-\frac{11}{2} \leq k \leq \frac{7}{2}$

The solution is $[-11/2, 7/2]$.

11. $\frac{3}{5}(2p + 3) \geq \frac{1}{10}(5p + 1)$

$10\left(\frac{3}{5}\right)(2p + 3) \geq 10\left(\frac{1}{10}\right)(5p + 1)$

$6(2p + 3) \geq 5p + 1$

$12p + 18 \geq 5p + 1$

$7p \geq -17$

$p \geq -\frac{17}{7}$

The solution is $[-17/7, \infty)$.

13. $(m + 2)(m - 4) < 0$

Solve $(m + 2)(m - 4) = 0$.

$m = -2$ or $m = 4$

Intervals: $(-\infty, -2), (-2, 4), (4, \infty)$

For $(-\infty, -2)$, choose -3 to test   fo
m.

$(-3 + 2)(-3 - 4) = -1(-7) = 8 \not< 0$

For $(-2, 4)$, choose 0.

$(0 + 2)(0 - 4) = 2(-4) = -8 < 0$

For $(4, \infty)$, choose 5.

$(5 + 2)(5 - 4) = 7(1) = 7 \not< 0$

The solution is $(-2, 4)$.

15. $y^2 - 3y + 2 < 0$

$(y - 2)(y - 1) < 0$

Solve $(y - 2)(y - 1) = 0$.

$y = 2$ or $y = 1$

Intervals: $(-\infty, 1), (1, 2), (2, \infty)$

For $(-\infty, 1)$, choose $y = 0$.

$0^2 - 3(0) + 2 = 2 \not< 0$

For (1, 2), choose $y = 3/2$.

$$\left(\frac{3}{2}\right)^2 - 3\left(\frac{3}{2}\right) + 2 = \frac{9}{4} - \frac{9}{2} + 2$$

$$= \frac{9 - 18 + 8}{4}$$

$$= -\frac{1}{4} < 0$$

For (2, ∞), choose 3.

$$3^2 - 3(3) + 2 = 2 \not< 0$$

The solution is (1, 2).

17. $q^2 - 7q + 6 \le 0$

Solve $q^2 - 7q + 6 = 0$.

$(q - 1)(q - 6) = 0$

$q = 1$ or $q = 6$

These solutions are also solutions of the given inequality because the symb $\le$ indicates that the end- points are included.

Intervals $(-\infty, 1)$, $(1, 6)$, $(6, \infty)$

For $(-\infty, 1)$, choose 0.

$$0^2 - 7(0) + 6 = 6 \not< 0$$

For (1, 6), choose 2.

$$2^2 - 7(2) + 6 = -4 \le 0$$

For (6, ∞), choose 7.

$$7^2 - 7(7) + 6 = 6 \not< 0$$

Since 1 and 6 are solutions, the solution is [1, 6].

19. $6m^2 + m > 1$

Solve $6m^2 + m = 1$.

$$6m^2 + m - 1 = 0$$

$(2m + 1)(3m - 1) = 0$

$m = -\frac{1}{2}$ or $m = \frac{1}{3}$

Intervals: $(-\infty, -1/2)$, $(-1/2, 1/3)$, $(1/3, \infty)$

For $(-\infty, -1/2)$, choose -1.

$$6(-1)^2 + (-1) = 5 > 1$$

For $(-1/2, 1/3)$, choose 0.

$$6(0)^2 + 0 = 0 \not> 1$$

For $(1/3, \infty)$, choose 1.

$$6(1)^2 + 1 = 7 > 1$$

The solution is

$$(-\infty, -1/2) \cup (1/3, \infty).$$

21. $2y^2 + 5y \le 3$

Solve $2y^2 + 5y = 3$.

$$2y^2 + 5y - 3 = 0$$

$(y + 3)(2y - 1) = 0$

$y = -3$ or $y = \frac{1}{2}$

Intervals: $(-\infty, -3)$, $(-3, 1/2)$, $(1/2, \infty)$

For $(-\infty, -3)$, choose -4.

$$2(-4)^2 + 5(-4) = 12 \not< 3$$

For $(-3, 1/2)$, choose 0.

$$2(0)^2 + 5(0) = 0 \le 3$$

For $(1/2, \infty)$, choose 1.

$$2(1)^2 + 5(1) = 7 \not< 3$$

The solution is $[-3, 1/2]$.

23. $x^2 \le 25$

Solve $x^2 = 25$.

$x = -5$ or $x = 5$

Intervals: $(-\infty, -5)$, $(-5, 5)$, $(5, \infty)$

For $(-\infty, -5)$, choose -6.

$$(-6)^2 = 36 \not\le 25$$

For $(-5, 5)$, choose 0.

$$0^2 = 0 \le 25$$

For $(5, \infty)$, choose 6.

$$6^2 = 36 \not\le 25$$

The solution is $[-5, 5]$.

25. $\dfrac{m - 3}{m + 5} \le 0$

Solve $\dfrac{m - 3}{m + 5} = 0$.

$$(m + 5)\dfrac{m - 3}{m + 5} = (m + 5)(0)$$

$$m - 3 = 0$$

$$m = 3$$

Set the denominator equal to 0 and solve.

$$m + 5 = 0$$

$$m = -5$$

Intervals: $(-\infty, -5)$, $(-5, 3)$, $(3, \infty)$

For $(-\infty, -5)$, choose -6.

$$\dfrac{-6 - 3}{-6 + 5} = 9 \not< 0$$

For $(-5, 3)$, choose 0.

$$\dfrac{0 - 3}{0 + 5} = -\dfrac{3}{5} \le 0$$

For $(3, \infty)$, choose 4.

$$\dfrac{4 - 3}{4 + 5} = \dfrac{1}{9} \not< 0$$

Although the $\le$ symbol is used, in-cluding -5 in the solution would caus the denominator to be zero.
The solution is $(-5, 3]$.

27. $\dfrac{k - 1}{k + 2} > 1$

Solve $\dfrac{k - 1}{k + 2} = 1$.

$$k - 1 = k + 2$$

$$-1 \ne 2$$

The equation has no solution.
Solve $k + 2 = 0$.

$$k = -2$$

Intervals: $(-\infty, -2)$, $(-2, \infty)$
For $(-\infty, -2)$, choose -3.

$$\dfrac{-3 - 1}{-3 + 2} = 4 > 1$$

For $(-2, \infty)$, choose 0.

$$\dfrac{0 - 1}{0 + 2} = -\dfrac{1}{2} \not> 1$$

The solution is $(-\infty, -2)$.

29. $\dfrac{2y + 3}{y - 5} \le 1$

Solve $\dfrac{2y + 3}{y - 5} = 1$.

$$2y + 3 = y - 5$$

$$y = -8$$

Solve $y - 5 = 0$.

$$y = 5$$

Intervals: $(-\infty, -8)$, $(-8, 5)$, $(5, \infty)$
For $(-\infty, -8)$, choose $y = -10$.

$$\dfrac{2(-10) + 3}{-10 - 5} = \dfrac{17}{15} \not< 1$$

For $(-8, 5)$, choose $y = 0$.

$$\dfrac{2(0) + 3}{0 - 5} = -\dfrac{3}{5} \le 1$$

For $(5, \infty)$, choose $y = 6$.

$$\frac{2(6) + 3}{6 - 5} = \frac{15}{1} \not< 1$$

The solution is $[-8, 5)$.

31. $\dfrac{7}{k + 2} \geq \dfrac{1}{k + 2}$

Solve $\dfrac{7}{k + 2} = \dfrac{1}{k + 2}$

$$\frac{7}{k + 2} - \frac{1}{k + 2} = 0$$

$$\frac{6}{k + 2} = 0$$

The equation has no solution.

Solve $k + 2 = 0$.

$$k = -2$$

Intervals: $(-\infty, -2)$, $(-2, \infty)$

For $(-\infty, -2)$, choose $k = -3$.

$$\frac{6}{-3 + 2} = -6 \not\geq 0$$

For $(-2, \infty)$, choose $k = 0$.

$$\frac{6}{0 + 2} = 3 \geq 0$$

The solution is $(-2, \infty)$.

33. $\dfrac{3x}{x^2 - 1} < 2$

Solve $\dfrac{3x}{x^2 - 1} = 2$.

$$3x = 2x^2 - 2$$
$$-2x^2 + 3x + 2 = 0$$
$$(2x + 1)(-x + 2) = 0$$
$$x = -\frac{1}{2} \quad \text{or} \quad x = 2$$

Set $x^2 - 1 = 0$.

$$x = 1 \quad \text{or} \quad x = -1$$

Intervals: $(-\infty, -1)$, $(-1, -1/2)$, $(-1/2, 1)$, $(1, 2)$, $(2, \infty)$

For $(-\infty, -1)$, choose $x = -2$.

$$\frac{3(-2)}{(-2)^2 - 1} = -\frac{6}{3} = -2 < 2$$

For $(-1, -1/2)$, choose $x = -3/4$.

$$\frac{3(-3/4)}{(-3/4)^2 - 1} = \frac{-9/4}{9/16 - 1} = \frac{36}{7} \not< 2$$

For $(-1/2, 1)$, choose $x = 0$.

$$\frac{3(0)}{0^2 - 1} = 0 < 2$$

For $(1, 2)$, choose $x = 3/2$.

$$\frac{3(3/2)}{(3/2)^2 - 1} = \frac{9/2}{5/4} = \frac{18}{5} \not< 2$$

For $(2, \infty)$, choose $x = 3$.

$$\frac{3(3)}{3^2 - 1} = \frac{9}{8} < 2$$

The solution is

$(-\infty, -1) \cup (-1/2, 1) \cup (2, \infty)$.

35. $\dfrac{z^2 + z}{z^2 - 1} \geq 3$

Solve $\dfrac{z^2 + z}{z^2 - 1} = 3$.

$$z^2 + z = 3z^2 - 3$$
$$-2z^2 + z + 3 = 0$$
$$(-z - 1)(2z - 3) = 0$$
$$z = -1 \quad \text{or} \quad z = \frac{3}{2}$$

Set $z^2 - 1 = 0$.

$$z^2 = 1$$
$$z = -1 \quad \text{or} \quad z = 1$$

Intervals: $(-\infty, -1)$, $(-1, 1)$, $(1, 3/2)$, $(3/2, \infty)$

For $(-\infty, -1)$, choose $x = -2$.

$$\frac{(-2)^2 + 3}{(-2)^2 - 1} = \frac{7}{3} \not\geq 3$$

For $(-1, 1)$, choose $x = 0$.

$$\frac{0^2 + 3}{0^2 - 1} = -3 \not\geq 3$$

For $(1, 3/2)$, choose $x = 3/2$.

$$\frac{(3/2)^2 + 3}{(3/2)^2 - 1} = \frac{21}{5} \geq 3$$

For $(3/2, \infty)$, choose $x = 2$.

$$\frac{2^2 + 3}{2^2 - 1} = \frac{7}{3} \not\geq 3$$

The solution is $(1, 3/2]$.

Section R.6

1. $8^{-2} = \frac{1}{8^2} = \frac{1}{64}$

3. $6^{-3} = \frac{1}{6^3} = \frac{1}{216}$

5. $(-12)^0 = 1$, by definition.

7. $-2^{-4} = -\frac{1}{2^4} = -\frac{1}{16}$

9. $-(-3)^{-2} = -\frac{1}{(-3)^2} = -\frac{1}{9}$

11. $\left(\frac{5}{8}\right)^2 = \frac{5^2}{8^2} = \frac{25}{64}$

13. $\left(\frac{1}{2}\right)^{-3} = \frac{1^{-3}}{2^{-3}} = \frac{\frac{1}{1^3}}{\frac{1}{2^3}} = \frac{1}{\frac{1}{8}}$

$$= 1 \div \frac{1}{8} = 1 \cdot \frac{8}{1} = 8$$

15. $\left(\frac{2}{7}\right)^{-2} = \frac{2^{-2}}{7^{-2}} = \frac{\frac{1}{2^2}}{\frac{1}{7^2}} = \frac{\frac{1}{4}}{\frac{1}{49}}$

$$= \frac{1}{4} \div \frac{1}{49} = \frac{1}{4} \cdot \frac{49}{1} = \frac{49}{4}$$

17. $\frac{7^5}{7^9} = 7^{5-9} = 7^{-4} = \frac{1}{7^4}$

19. $\frac{2^{-5}}{2^{-2}} = 2^{-5-(-2)} = 2^{-5+2} = 2^{-3} = \frac{1}{2^3}$

21. $4^{-3} \cdot 4^6 = 4^{-3+6} = 4^3$

23. $\frac{10^8 \cdot 10^{-10}}{10^4 \cdot 10^2}$

$$= \frac{10^{8+(-10)}}{10^{4+2}} = \frac{10^{-2}}{10^6}$$

$$= 10^{-2-6} = 10^{-8}$$

$$= \frac{1}{10^8}$$

25. $\frac{x^4 \cdot x^3}{x^5} = \frac{x^{4+3}}{x^5} = \frac{x^7}{x^5} = x^{7-5} = x^2$

27. $\frac{(4k^{-1})^2}{2k^{-5}} = \frac{4^2 k^{-2}}{2k^{-5}} = \frac{16k^{-2-(-5)}}{2}$

$$= 8k^{-2+5} = 8k^3$$

$$= 2^3 k^3$$

29. $\frac{2^{-1}x^3 y^{-3}}{xy^{-2}} = 2^{-1}x^{3-1}y^{-3-(-2)}$

$$= 2^{-1}x^2 y^{-3+2} = 2^{-1}x^2 y^{-1}$$

$$= \frac{1}{2}x^2 \cdot \frac{1}{y} = \frac{x^2}{2y}$$

31. $\left(\frac{a^{-1}}{b^2}\right)^{-3} = \frac{(a^{-1})^{-3}}{(b^2)^{-3}} = \frac{a^{(-1)(-3)}}{b^{2(-3)}}$

$$= \frac{a^3}{b^{-6}} = \frac{a^3}{\frac{1}{b^6}} = a^3 \cdot \frac{b^6}{1}$$

$$= a^3 b^6$$

For Exercises 33–37, $a = 2$, $b = -3$.

33. $a^{-1} + b^{-1} = 2^{-1} + (-3)^{-1}$

$$= \frac{1}{2} + \frac{1}{-3}$$

$$= \frac{1}{2}\left(\frac{3}{3}\right) + \frac{1}{-3}\left(\frac{2}{2}\right)$$

$$= \frac{3}{6} + \left(-\frac{2}{6}\right)$$

$$= \frac{1}{6}$$

35. $\dfrac{2b^{-1} - 3a^{-1}}{a + b^2} = \dfrac{2(-3)^{-1} - 3(2)^{-1}}{2 + (-3)^2}$

$$= \frac{\frac{2}{-3} - 3\left(\frac{1}{2}\right)}{2 + (-3)^2}$$

$$= \frac{-\frac{2}{3}\left(\frac{2}{2}\right) - \frac{3}{2}\left(\frac{3}{3}\right)}{2 + 9}$$

$$= \frac{-\frac{4}{6} - \frac{9}{6}}{11} = \frac{-\frac{13}{6}}{\frac{11}{1}}$$

$$= -\frac{13}{6} \cdot \frac{1}{11} = -\frac{13}{66}$$

37. $\left(\dfrac{a}{3}\right)^{-1} + \left(\dfrac{b}{2}\right)^{-2} = \left(\dfrac{2}{3}\right)^{-1} + \left(-\dfrac{3}{2}\right)^{-2}$

$$= \frac{2^{-1}}{3^{-1}} + \frac{(-3)^{-2}}{2^{-2}}$$

$$= \frac{\frac{1}{2}}{\frac{1}{3}} + \frac{\frac{1}{(-3)^2}}{\frac{1}{2^2}}$$

$$= \frac{1}{2} \div \frac{1}{3} + \frac{1}{9} \div \frac{1}{4}$$

$$= \frac{1}{2} \cdot \frac{3}{1} + \frac{1}{9} \cdot \frac{4}{1}$$

$$= \frac{3}{2} + \frac{4}{9}$$

$$= \frac{9}{9} \cdot \frac{3}{2} + \frac{2}{2} \cdot \frac{4}{9}$$

$$= \frac{27}{18} + \frac{8}{18} = \frac{35}{18}$$

39. $81^{1/2} = (9^2)^{1/2} = 9^{2(1/2)} = 9^1 = 9$

41. $8^{2/3} = (8^{1/3})^2 = (2)^2 = 4$

We could also write $8^{2/3} = (8^2)^{1/3}$, but this procedure involves taking a cube root of a larger number, 64. Whenever possible, take the root first.

43. $32^{2/5} = (32^{1/5})^2 = 2^2 = 4$

45. $\left(\dfrac{4}{9}\right)^{1/2} = \dfrac{4^{1/2}}{9^{1/2}} = \dfrac{2}{3}$

47. $16^{-5/4} = (16^{1/4})^{-5} = 2^{-5}$

$$= \frac{1}{2^5} \quad \text{or} \quad \frac{1}{32}$$

49. $\left(\dfrac{27}{64}\right)^{-1/3} = \dfrac{27^{-1/3}}{64^{-1/3}} = \dfrac{\frac{1}{27^{1/3}}}{\frac{1}{64^{1/3}}} = \dfrac{\frac{1}{3}}{\frac{1}{4}}$

$$= \frac{4}{3}$$

51. $2^{1/2} \cdot 2^{3/2} = 2^{1/2 + 3/2} = 2^{4/2}$

$$= 2^2 \quad \text{or} \quad 4$$

53. $\dfrac{4^{2/3} \cdot 4^{5/3}}{4^{1/3}} = \dfrac{4^{2/3 + 5/3}}{4^{1/3}}$

$$= 4^{7/3 - 1/3} = 4^{6/3}$$

$$= 4^2 \quad \text{or} \quad 16$$

55. $\dfrac{7^{-1/3} \cdot 7r^{-3}}{7^{2/3} \cdot (r^{-2})^2}$

$$= \frac{7^{-1/3 + 1}r^{-3}}{7^{2/3} \cdot r^{-4}}$$

$$= 7^{-1/3 + 3/3 - 2/3}r^{-3-(-4)}$$

$$= 7^0 r^{-3+4} = 1 \cdot r^1 = r$$

57. $\dfrac{6k^{-4} \cdot (3k^{-1})^{-2}}{2^3 \cdot k^{1/2}}$

$\quad = \dfrac{2 \cdot 3k^{-4}(3^{-2})(k^2)}{2^3 k^{1/2}}$

$\quad = 2^{1-3}3^{1+(-2)}k^{-4+2-1/2}$

$\quad = 2^{-2}3^{-1}k^{-5/2}$

$\quad = \dfrac{1}{2^2} \cdot \dfrac{1}{3} \cdot \dfrac{1}{k^{5/2}} = \dfrac{1}{2^2 3 k^{5/2}}$

$\quad = \dfrac{1}{12k^{5/2}}$

59. $\dfrac{a^{4/3}}{a^{2/3}} \cdot \dfrac{b^{1/2}}{b^{-3/2}} = a^{4/3 \,-\, 2/3}\, b^{1/2 \,-\,(-3/2)}$

$\quad\quad\quad\quad\quad\quad = a^{2/3}\, b^2$

61. $\dfrac{k^{-3/5} \cdot h^{-1/3} \cdot t^{2/5}}{k^{-1/5} \cdot h^{-2/3} \cdot t^{1/5}}$

$\quad = k^{-3/5 \,-\,(-1/5)} h^{-1/3 \,-\,(-2/3)} t^{2/5 \,-\, 1/5}$

$\quad = k^{-3/5 \,+\, 1/5} h^{-1/3 \,+\, 2/3} t^{2/5 \,-\, 1/5}$

$\quad = k^{-2/5} h^{1/3} t^{1/5}$

$\quad = \dfrac{h^{1/3} t^{1/5}}{k^{2/5}}$

63. $(x^2 + 2)(x^2 - 1)^{-1/2}(x) + (x^2 - 1)^{1/2}(2x)$

$\quad = (x^2 + 2)(x^2 - 1)^{-1/2}(x)$

$\quad\quad + (x^2 - 1)^1(x^2 - 1)^{-1/2}(2x)$

$\quad = x(x^2 - 1)^{-1/2}[(x^2 + 2) + (x^2 - 1)(2)]$

$\quad = x(x^2 - 1)^{-1/2}(x^2 + 2 + 2x^2 - 2)$

$\quad = x(x^2 - 1)^{-1/2}(3x^2)$

$\quad = 3x^3(x^2 - 1)^{-1/2}$

65. $(2x + 5)^2\left(\dfrac{1}{2}\right)(x^2 - 4)^{-1/2}(2x)$

$\quad + (x^2 - 4)^{1/2}(2)(2x + 5)$

$\quad = (2x + 5)^2(x^2 - 4)^{-1/2}(x)$

$\quad\quad + (x^2 - 4)^{1/2}(2)(2x + 5)$

$\quad = (2x + 5)^2(x^2 - 4)^{-1/2}(x)$

$\quad\quad + (x^2 - 4)^1(x^2 - 4)^{-1/2}(2)(2x + 5)$

$\quad = (2x + 5)(x^2 - 4)^{-1/2}$

$\quad\quad \cdot [(2x + 5)(x) + (x^2 - 4)(2)]$

$= (2x + 5)(x^2 - 4)^{-1/2}$

$\quad \cdot (2x^2 + 5x + 2x^2 - 8)$

$= (2x + 5)(x^2 - 4)^{-1/2}(4x^2 + 5x - 8)$

Section R.7

1. $\sqrt[3]{125} = 5$ because $5^3 = 125$.

3. $\sqrt[5]{-3125} = -5$ because $(-5)^5 = -3125$.

5. $\sqrt{2000} = \sqrt{4 \cdot 100 \cdot 5} = 2 \cdot 10\sqrt{5}$

$\quad\quad\quad = 20\sqrt{5}$

7. $7\sqrt{2} - 8\sqrt{18} + 4\sqrt{72}$

$\quad = 7\sqrt{2} - 8\sqrt{9 \cdot 2} + 4\sqrt{36 \cdot 2}$

$\quad = 7\sqrt{2} - 8(3)\sqrt{2} + 4(6)\sqrt{2}$

$\quad = 7\sqrt{2} - 24\sqrt{2} + 24\sqrt{2}$

$\quad = 7\sqrt{2}$

9. $2\sqrt{5} - 3\sqrt{20} + 2\sqrt{45}$

$\quad = 2\sqrt{5} - 3\sqrt{4 \cdot 5} + 2\sqrt{9 \cdot 5}$

$\quad = 2\sqrt{5} - 3(2)\sqrt{5} + 2(3)\sqrt{5}$

$\quad = 2\sqrt{5} - 6\sqrt{5} + 6\sqrt{5}$

$\quad = 2\sqrt{5}$

11. $\sqrt[3]{2} - \sqrt[3]{16} + 2\sqrt[3]{54}$

$\quad = \sqrt[3]{2} - (\sqrt[3]{8 \cdot 2}) + 2(\sqrt[3]{27 \cdot 2})$

$\quad = \sqrt[3]{2} - \sqrt[3]{8}\sqrt[3]{2} + 2(\sqrt[3]{27}\sqrt[3]{2})$

$\quad = \sqrt[3]{2} - 2\sqrt[3]{2} + 2(3\sqrt[3]{2})$

$\quad = \sqrt[3]{2} - 2\sqrt[3]{2} + 6\sqrt[3]{2}$

$\quad = 5\sqrt[3]{2}$

13. $\sqrt[3]{32} - 5\sqrt[3]{4} + 2\sqrt[3]{108}$

$= \sqrt[3]{8 \cdot 4} - 5\sqrt[3]{4} + 2\sqrt[3]{27 \cdot 4}$

$= \sqrt[3]{8}\sqrt[3]{4} - 5\sqrt[3]{4} + 2\sqrt[3]{27}\sqrt[3]{4}$

$= 2\sqrt[3]{4} - 5\sqrt[3]{4} + 2(3\sqrt[3]{4})$

$= 2\sqrt[3]{4} - 5\sqrt[3]{4} + 6\sqrt[3]{4}$

$= 3\sqrt[3]{4}$

15. $\sqrt{98r^3s^4t^{10}}$

$= \sqrt{(49 \cdot 2)(r^2 \cdot r)(s^4)(t^{10})}$

$= \sqrt{(49r^2s^4t^{10})(2r)}$

$= \sqrt{49r^2s^4t^{10}}\sqrt{2r}$

$= 7rs^2t^5\sqrt{2r}$

17. $\sqrt[4]{x^8y^7z^{11}} = \sqrt[4]{(x^8)(y^4 \cdot y^3)(z^8z^3)}$

$= \sqrt[4]{(x^8y^4z^8)(y^3z^3)}$

$= \sqrt[4]{x^8y^4z^8} \sqrt[4]{y^3z^3}$

$= x^2yz^2 \sqrt[4]{y^3z^3}$

19. $\sqrt{p^7q^3} - \sqrt{p^5q^9} + \sqrt{p^9q}$

$= \sqrt{(p^6p)(q^2q)} - \sqrt{(p^4p)(q^8q)}$

$\quad + \sqrt{(p^8p)q}$

$= \sqrt{(p^6q^2)(pq)} - \sqrt{(p^4q^8)(pq)}$

$\quad + \sqrt{(p^8)(pq)}$

$= \sqrt{p^6q^2}\sqrt{pq} - \sqrt{p^4q^8}\sqrt{pq} + \sqrt{p^8}\sqrt{pq}$

$= p^3q\sqrt{pq} - p^2q^4\sqrt{pq} + p^4\sqrt{pq}$

$= p^2pq\sqrt{pq} - p^2q^4\sqrt{pq} + p^2p^2\sqrt{pq}$

$= p^2\sqrt{pq}(pq - q^4 + p^2)$

21. $\dfrac{-2}{\sqrt{3}} = \dfrac{-2}{\sqrt{3}} \cdot \dfrac{\sqrt{3}}{\sqrt{3}} = \dfrac{-2\sqrt{3}}{\sqrt{9}} = -\dfrac{2\sqrt{3}}{3}$

23. $\dfrac{4}{\sqrt{8}} = \dfrac{4}{\sqrt{8}} \cdot \dfrac{\sqrt{8}}{\sqrt{8}} = \dfrac{4\sqrt{8}}{\sqrt{64}} = \dfrac{4\sqrt{8}}{8} = \dfrac{4(\sqrt{4}\sqrt{2})}{8}$

$= \dfrac{4(2\sqrt{2})}{8} = \dfrac{8\sqrt{2}}{8} = \sqrt{2}$

25. $\dfrac{5}{2 - \sqrt{6}} = \dfrac{5}{2 - \sqrt{6}} \cdot \dfrac{2 + \sqrt{6}}{2 + \sqrt{6}}$

$= \dfrac{5(2 + \sqrt{6})}{4 + 2\sqrt{6} - 2\sqrt{6} - \sqrt{36}}$

$= \dfrac{5(2 + \sqrt{6})}{4 - \sqrt{36}} = \dfrac{5(2 + \sqrt{6})}{4 - 6}$

$= \dfrac{5(2 + \sqrt{6})}{-2}$

$= -\dfrac{5(2 + \sqrt{6})}{2}$

27. $\dfrac{1}{\sqrt{10} + \sqrt{3}} = \dfrac{1}{\sqrt{10} + \sqrt{3}} \cdot \dfrac{\sqrt{10} - \sqrt{3}}{\sqrt{10} - \sqrt{3}}$

$= \dfrac{\sqrt{10} - \sqrt{3}}{\sqrt{100} - \sqrt{30} + \sqrt{30} - \sqrt{9}}$

$= \dfrac{\sqrt{10} - \sqrt{3}}{\sqrt{100} - \sqrt{9}} = \dfrac{\sqrt{10} - \sqrt{3}}{10 - 3}$

$= \dfrac{\sqrt{10} - \sqrt{3}}{7}$

29. $\dfrac{5}{\sqrt{m} - \sqrt{5}} = \dfrac{5}{\sqrt{m} - \sqrt{5}} \cdot \dfrac{\sqrt{m} + \sqrt{5}}{\sqrt{m} + \sqrt{5}}$

$= \dfrac{5(\sqrt{m} + \sqrt{5})}{\sqrt{m^2} + \sqrt{5m} - \sqrt{5m} - \sqrt{25}}$

$= \dfrac{5(\sqrt{m} + \sqrt{5})}{\sqrt{m^2} - \sqrt{25}} = \dfrac{5(\sqrt{m} + \sqrt{5})}{m - 5}$

31. $\dfrac{z - 11}{\sqrt{z} - \sqrt{11}} = \dfrac{z - 11}{\sqrt{z} - \sqrt{11}} \cdot \dfrac{\sqrt{z} + \sqrt{11}}{\sqrt{z} + \sqrt{11}}$

$= \dfrac{(z - 11)(\sqrt{z} + \sqrt{11})}{\sqrt{z^2} + \sqrt{11z} - \sqrt{11z} - \sqrt{121}}$

$= \dfrac{(z - 11)(\sqrt{z} + \sqrt{11})}{\sqrt{z^2} - \sqrt{121}}$

$= \dfrac{(z - 11)(\sqrt{z} + \sqrt{11})}{(z - 11)}$

$= \sqrt{z} + \sqrt{11}$

33. $\dfrac{\sqrt{p} + \sqrt{p^2 - 1}}{\sqrt{p} - \sqrt{p^2 - 1}}$

$= \dfrac{\sqrt{p} + \sqrt{p^2 - 1}}{\sqrt{p} - \sqrt{p^2 - 1}} \cdot \dfrac{\sqrt{p} + \sqrt{p^2 - 1}}{\sqrt{p} + \sqrt{p^2 - 1}}$

$= \dfrac{\sqrt{p}^2 + \sqrt{p}\sqrt{p^2 - 1} + \sqrt{p}\sqrt{p^2 - 1} + (\sqrt{p^2 - 1})^2}{\sqrt{p}^2 + \sqrt{p}\sqrt{p^2 - 1} - \sqrt{p}\sqrt{p^2 - 1} - (\sqrt{p^2 - 1})^2}$

$= \dfrac{p + 2\sqrt{p}\sqrt{p^2 - 1} + (p^2 - 1)}{p - (p^2 - 1)}$

$= \dfrac{p^2 + p + 2\sqrt{p(p^2 - 1)} - 1}{-p^2 + p + 1}$

35. $\dfrac{1 - \sqrt{3}}{3} = \dfrac{1 - \sqrt{3}}{3} \cdot \dfrac{1 + \sqrt{3}}{1 + \sqrt{3}}$

$= \dfrac{1 + \sqrt{3} - \sqrt{3} - \sqrt{9}}{3(1 + \sqrt{3})}$

$= \dfrac{1 - \sqrt{9}}{3(1 + \sqrt{3})} = \dfrac{1 - 3}{3(1 + \sqrt{3})}$

$= -\dfrac{2}{3(1 + \sqrt{3})}$

37. $\dfrac{\sqrt{p} + \sqrt{p^2 - 1}}{\sqrt{p} - \sqrt{p^2 - 1}}$

$= \dfrac{\sqrt{p} + \sqrt{p^2 - 1}}{\sqrt{p} - \sqrt{p^2 - 1}} \cdot \dfrac{\sqrt{p} - \sqrt{p^2 - 1}}{\sqrt{p} - \sqrt{p^2 - 1}}$

$= \dfrac{\sqrt{p}^2 - \sqrt{p}\sqrt{p^2 - 1} + \sqrt{p}\sqrt{p^2 - 1} - (\sqrt{p^2 - 1})^2}{\sqrt{p}^2 - \sqrt{p}\sqrt{p^2 - 1} - \sqrt{p}\sqrt{p^2 - 1} + (\sqrt{p^2 - 1})^2}$

$= \dfrac{p - (p^2 - 1)}{p - 2\sqrt{p(p^2 - 1)} + p^2 - 1}$

$= \dfrac{p - p^2 + 1}{p - 2\sqrt{p(p^2 - 1)} + p^2 - 1}$

$= \dfrac{-p^2 + p + 1}{p^2 + p - 2\sqrt{p(p^2 - 1)} - 1}$

39. $\sqrt{4y^2 + 4y + 1}$

$\quad = \sqrt{(2y + 1)(2y + 1)}$

$\quad = \sqrt{(2y + 1)^2}$

$\quad = |2y + 1|$

Since $\sqrt{}$ denotes the nonnegative root, we must have $2y + 1 \geq 0$.

41. $\sqrt{9k^2 + h^2}$

The expression $9k^2 + h^2$ is the sum o two squares and cannot be fac‑ tored Therefore, $\sqrt{9k^2 + h^2}$ cannot be simplified.

FUNCTIONS AND GRAPHS

Section 1.1

1. The x-value of 82 corresponds to two
    y-values, 93 and 14. In a function,
    each x must correspond to exactly
    one y.
    The rule is not a function.

3. Each x-value corresponds to exactly
    one y-value.
    The rule is a function.

5. $y = x^3$

    Each x-value corresponds to exactly
    one y-value.
    The rule is a function.

7. $x = |y|$

    Each value of x (except 0) corre-
    sponds to two y-values.
    The rule is not a function.

In Exercises 9-23, the domain is
$\{-2, -1, 0, 1, 2, 3\}$. See the
answer graphs in the back of your
textbook.

9. $y = x - 1$

x	-2	-1	0	1	2	3
y	-3	-2	-1	0	1	2

    Pairs: (-2, -3), (-1, -2), (0, -1),
       (1, 0), (2, 1), (3, 2)
    Range: $\{-3, -2, -1, 0, 1, 2\}$

11. $y = -4x + 9$

x	-2	-1	0	1	2	3
y	17	13	9	5	1	-3

     Pairs: (-2, 17), (-1, 13), (0, 9),
        (1, 5), (2, 1), (3, -3)
     Range: $\{-3, 1, 5, 9, 13, 17\}$

13. $2x + y = 9$
       $y = -2x + 9$

x	-2	-1	0	1	2	3
y	13	11	9	7	5	3

     Pairs: (-2, 13), (-1, 11), (0, 9),
        (1, 7), (2, 5), (3, 3)
     Range: $\{3, 5, 7, 9, 11, 13\}$

15. $2y - x = 5$
       $2y = 5 + x$
       $y = \frac{1}{2}x + \frac{5}{2}$

x	-2	-1	0	1	2	3
y	3/2	2	5/2	3	7/2	4

     Pairs: (-2, 3/2), (-1, 2), (0, 5/2),
        (1, 3), (2, 7/2), (3, 4)
     Range: $\{3/2, 2, 5/2, 3, 7/2, 4\}$

17. $y = x(x + 1)$
       $y = x^2 + x$

x	-2	-1	0	1	2	3
y	2	0	0	2	6	12

     Pairs: (-2, 2), (-1, 0), (0, 0),
        (1, 2), (2, 6), (3, 12)
     Range: $\{0, 2, 6, 12\}$

19. $y = x^2$

x	-2	-1	0	1	2	3
y	4	1	0	1	4	9

Pairs: (-2, 4), (-1, 1), (0, 0),
       (1, 1), (2, 4), (3, 9)

Range: $\{0, 1, 4, 9\}$

21. $y = \dfrac{1}{x + 3}$

x	-2	-1	0	1	2	3
y	1	1/2	1/3	1/4	1/5	1/6

Pairs: (-2, 1), (-1, 1/2), (0, 1/3),
       (1, 1/4), (2, 1/5), (3, 1/6)

Range: $\{1, 1/2, 1/3, 1/4, 1/5, 1/6\}$

23. $y = \dfrac{3x - 3}{x + 5}$

x	-2	-1	0	1	2	3
y	-3	-3/2	-3/5	0	3/7	6/8

Pairs: (-2, -3), (-1, -3/2),
       (0, -3/5), (1, 0),
       (2, 3/7), (3, 3/4)

Range: $\{-3, -3/2, -3/5, 0, 3/7, 3/4\}$

For Exercises 25–29, see the answer graphs in the back of the textbook.

25. $x < 0$

This expression represents all numbers less than 0, or $(-\infty, 0)$, The number 0 is not included in the interval.

27. $1 \le x < 2$

This expression represents all numbers greater than or equal to 1 and less than 2, or [1, 2). The number 1 is included in the interval, but the number 2 is not included.

29. $-9 > x$

This expression represents all numbers less than -9, or $(-\infty, -9)$. The number -9 is not included in the interval.

31. (-4, 3)

This interval represents all numbers greater than -4 and less than 3. Neither -4 nor 3 is included. As an inequality, this is represented as

$$-4 < x < 3.$$

33. $(-\infty, -1]$

This interval represents all numbers less than or equal to -1. -1 is included in the interval. As an inequality, this is represented as

$$x \le -1.$$

35. This interval represents all numbers greater than or equal to -2 and less than 6. -2 is included but 6 is not. As an inequality, this is represented by

$$-2 \le x < 6.$$

37. This interval represents all numbers less than or equal to −4 or greater than or equal to 4. As an inequality, this is represented as

$$x \le -4 \quad \text{or} \quad x \ge 4.$$

39. $f(x) = 2x$

x can take on any value, so the domain is the set of real numbers, $(-\infty, \infty)$.

41. $f(x) = x^4$

x can take on any value, so the domain is the set of real numbers, $(-\infty, \infty)$.

43. $f(x) = \sqrt{16 - x^2}$

For $f(x)$ to be a real number, $16 - x^2 \ge 0$.
Solve $16 - x^2 = 0$.

$$(4 - x)(4 + x) = 0$$
$$x = 4 \quad \text{or} \quad x = -4$$

The numbers form the intervals $(-\infty, -4)$, $(-4, 4)$ and $(4, \infty)$. Only values in the interval $(-4, 4)$ satisfy the inequality. The domain is $[-4, 4]$.

45. $f(x) = (x - 3)^{1/2} = \sqrt{x - 3}$

For $f(x)$ to be a real number,

$$x - 3 \ge 0$$
$$x \ge 3.$$

The domain is $[3, \infty)$.

47. $f(x) = \dfrac{2}{x^2 - 4}$

$$= \dfrac{2}{(x - 2)(x + 2)}$$

Since division by zero is not definded, $(x - 2)(x + 2) \ne 0$.
When $(x - 2)(x + 2) = 0$,

$$x - 2 = 0 \quad \text{or} \quad x + 2 = 0$$
$$x = 2 \quad \text{or} \quad x = -2$$

Thus, x can be any real number except ±2.
The domain is

$$(-\infty, -2) \cup (-2, 2) \cup (2, \infty).$$

49. $f(x) = -\sqrt{\dfrac{2}{x^2 + 9}}$

x can take on any value. No choice for x will cause the denominator to be zero. Also, no choice for x will produce a negative number under the radical. The domain is $(-\infty, \infty)$.

51. $f(x) = \sqrt{x^2 - 4x - 5}$
$$= \sqrt{(x - 5)(x + 1)}$$

See the method used in Exercise 43.

$$(x - 5)(x + 1) \ge 0$$

when $x \ge 5$ and when $x \le -1$.
The domain is $(-\infty, -1] \cup [5, \infty)$.

53. $f(x) = \dfrac{1}{\sqrt{x^2 - 6x + 8}}$

$$= \dfrac{1}{\sqrt{(x - 4)(x - 2)}}$$

$(x - 4)(x - 2) > 0$, since the radicand cannot be negative and the denominator of the function cannot be zero.

Solve $(x - 4)(x - 2) = 0$

$$x - 4 = 0 \quad \text{or} \quad x - 2 = 0$$
$$x = 4 \quad \text{or} \quad x = 2$$

Use the values 2 and 4 to divide the number line into 3 intervals, $(-\infty, 2)$, $(2, 4)$ and $(4, \infty)$. Only the values in the intervals $(-\infty, -2)$ and $(4, \infty)$ satisfy the inequality.

The domain is $(-\infty, -2) \cup (4, \infty)$.

55. By reading the graph, the domain is all numbers greater than or equal to -5 and less than or equal to 4. The range is all numbers greater than or equal to -2 and less than or equal to 6.

Domain: $[-5, 4]$; range: $[-2, 6]$

57. By reading the graph, x can take on any value, but y is less than or equal to 12.

Domain: $(-\infty, \infty)$; range: $(-\infty, 12]$

59. $f(x) = 3x + 2$

    (a) $f(4) = 3(4) + 2 = 12 + 2 = 14$

    (b) $f(-3) = 3(-3) + 2 = -9 + 2$
    $$= -7$$

    (c) $f\left(-\frac{1}{2}\right) = 3\left(-\frac{1}{2}\right) + 2$
    $$= -\frac{3}{2} + \frac{4}{2}$$
    $$= \frac{1}{2}$$

    (d) $f(a) = 3(a) + 2 = 3a + 2$

    (e) $f\left(\frac{2}{m}\right) = 3\left(\frac{2}{m}\right) + 2$
    $$= \frac{6}{m} + 2$$

61. $f(x) = -x^2 + 5x + 1$

    (a) $f(4) = -(4)^2 + 5(4) + 1$
    $$= -16 + 20 + 1$$
    $$= 5$$

    (b) $f(-3) = -(-3)^2 + 5(-3) + 1$
    $$= -9 - 15 + 1$$
    $$= -23$$

    (c) $f\left(-\frac{1}{2}\right) = -\left(-\frac{1}{2}\right)^2 + 5\left(-\frac{1}{2}\right) + 1$
    $$= -\frac{1}{4} - \frac{5}{2} + 1$$
    $$= -\frac{7}{4}$$

    (d) $f(a) = -(a)^2 + 5(a) + 1$
    $$= -a^2 + 5a + 1$$

    (e) $f\left(\frac{2}{m}\right) = -\left(\frac{2}{m}\right)^2 + 5\left(\frac{2}{m}\right) + 1$
    $$= -\frac{4}{m^2} + \frac{10}{m} + 1$$
    $$\text{or} \quad \frac{-4 + 10m + m^2}{m^2}$$

63. $f(x) = \dfrac{2x + 1}{x - 2}$

    (a) $f(4) = \dfrac{2(4) + 1}{4 - 2} = \dfrac{9}{2}$

    (b) $f(-3) = \dfrac{2(-3) + 1}{-3 - 2} = \dfrac{-6 + 1}{-5}$
    $$= \frac{-5}{-5} = 1$$

    (c) $f\left(-\frac{1}{2}\right) = \dfrac{2(-1/2) + 1}{-1/2 - 2}$
    $$= \frac{-1 + 1}{5/2}$$
    $$= \frac{0}{5/2} = 0$$

(d) $f(a) = \dfrac{2(a) + 1}{(a) - 2} = \dfrac{2a + 1}{a - 1}$

(e) $f\left(\dfrac{2}{m}\right) = \dfrac{2\left(\dfrac{2}{m}\right) + 1}{\dfrac{2}{m} - 2}$

$= \dfrac{\dfrac{4}{m} + \dfrac{m}{m}}{\dfrac{2}{m} - \dfrac{2m}{m}}$

$= \dfrac{\dfrac{4 + m}{m}}{\dfrac{2 - 2m}{m}}$

$= \dfrac{4 + m}{m} \cdot \dfrac{m}{2 - 2m}$

$= \dfrac{4 + m}{2 - 2m}$

In Exercises 65–67, count squares on the graph paper. On the horizontal axis, note that two squares correspond to one unit.

65. **(a)** $f(-2) = 0$

(b) $f(0) = 4$

(c) $f\left(\dfrac{1}{2}\right) = 3$

(d) $f(4) = 4$

67. **(a)** $f(-2) = -3$

(b) $f(0) = -2$

(c) $f\left(\dfrac{1}{2}\right) = -1$

(d) $f(4) = 2$

In Exercises 69–73, $f(x) = 6x - 2$ and $g(x) = x^2 - 2x + 5$.

69. $f(m - 3) = 6(m - 3) - 2$

$= 6m - 18 - 2$

$= 6m - 20$

71. $g(r + h)$

$= (r + h)^2 - 2(r + h) + 5$

$= r^2 + 2hr + h^2 - 2r - 2h + 5$

73. $g\left(\dfrac{3}{q}\right) = \left(\dfrac{3}{q}\right)^2 - 2\left(\dfrac{3}{q}\right) + 5$

$= \dfrac{9}{q^2} - \dfrac{6}{q} + 5$

or $\dfrac{9 - 6q + 5q^2}{q^2}$

75. A vertical line drawn anywhere through the graph will intersect the graph in only one place. The graph represents a function.

77. A vertical line drawn through the graph may intersect the graph in two places. The graph does not represent a function.

79. A vertical line drawn anywhere through the graph will intersect the graph in only one place. The graph represent a function.

81. $f(x) = x^2 - 4$

(a) $f(x + h) = (x + h)^2 - 4$

$= x^2 + 2hx + h^2 - 4$

(b) $f(x + h) - f(x)$

$= [(x + h)^2 - 4] - (x^2 - 4)$

$= x^2 + 2hx + h^2 - 4 - x^2 + 4$

$= 2hx + h^2$

(c) $\dfrac{f(x + h) - f(x)}{h}$

$= \dfrac{[(x + h)^2 - 4] - (x^2 - 4)}{h}$

$= \dfrac{x^2 + 2hx + h^2 - 4 - x^2 + 4}{h}$

$= \dfrac{2hx + h^2}{h}$

$= 2x + h$

83. $f(x) = 6x + 2$

(a) $f(x + h) = 6(x + h) + 2$

$\qquad\qquad = 6x + 6h + 2$

(b) $f(x + h) - f(x)$

$\qquad = [6(x + h) + 2] - (6x + 2)$

$\qquad = 6x + 6h + 2 - 6x - 2$

$\qquad = 6h$

(c) $\dfrac{f(x + h) - f(x)}{h}$

$= \dfrac{[6(x + h) + 2] - (6x + 2)}{h}$

$= \dfrac{6x + 6h + 2 - 6x - 2}{h} = \dfrac{6h}{h}$

$= 6$

85. $f(x) = \dfrac{1}{x}$

(a) $f(x + h) = \dfrac{1}{x + h}$

(b) $f(x + h) - f(x)$

$= \dfrac{1}{x + h} - \dfrac{1}{x}$

$= \left(\dfrac{x}{x}\right)\dfrac{1}{x + h} - \dfrac{1}{x}\left(\dfrac{x + h}{x + h}\right)$

$= \dfrac{x - (x + h)}{x(x + h)}$

$= \dfrac{-h}{x(x + h)}$

(c) $\dfrac{f(x + h) - f(x)}{h}$

$= \dfrac{\dfrac{1}{x + h} - \dfrac{1}{x}}{h}$

$= \dfrac{\dfrac{1}{x + h}\left(\dfrac{x}{x}\right) - \dfrac{1}{x}\left(\dfrac{x + h}{x + h}\right)}{h}$

$= \dfrac{\dfrac{x - (x + h)}{(x + h)x}}{h}$

$= \dfrac{1}{h}\left[\dfrac{x - x - h}{(x + h)x}\right]$

$= \dfrac{1}{h}\left[\dfrac{-h}{(x + h)x}\right]$

$= \dfrac{-1}{(x + h)x}$

87. (a) To determine I(1976), locate 1976 on the horizontal axis, find the corresponding point on the energy intensity graph and read across (to the left) to 26,300 BTU per dollar.

(b) To determine S(1984), locate 1984 on the horizontal axis, find the corresponding point on the savings graph and read across (to the right) to $120 billion.

(c) When the two graphs cross, I(t) = 23,500 BTU per dollar and S(t) = $85 billion. These values occur in 1980.

(d) There is no significance to the point at which the two graphs cross. They each measure different, in-dependent outcomes.

89. If x is a whole ounce, the cost of mailing a letter in cents is $C(x) = 29 + 23(x - 1)$. For x in whole ounces and a fraction of an ounce, substitute the next whole number for x in $C(x) = 29 + 23(x - 1)$ because a fraction of an ounce is billed as a whole ounce.

(a) $C\left(\frac{2}{3}\right) = C(1) = 29 + 23(1 - 1)$

$= 29 + 23(0) = 29¢$

(b) $C\left(1\frac{1}{3}\right) = C(2) = 29 + 23(2 - 1)$

$= 29 + 23 = 52¢$

(c) $C(2) = 29 + 23(2 - 1) = 29 + 23$

$= 52¢$

(d) $C\left(3\frac{1}{8}\right) = 29 + 23(4 - 1)$

$= 29 + 23(3) = 98¢$

(e) To determine the postage on a letter that weighs 2 5/8 ounces, calculate $C(2\ 5/8)$.

$C\left(2\frac{5}{8}\right) = 29 + 23(3 - 1)$

$= 29 + 23(2) = 75¢$

It costs 75¢ to send a letter that weighs 2 5/8 ounces.

(g) See the answer graph in the back of the textbook.

(h) The independent variable is x, the weight of the letter.

(i) The dependent variable is $C(x)$, the cost to mail the letter.

Section 1.2

1. Find the slope of the line through $(4, 5)$ and $(-1, 2)$.

$$m = \frac{5 - 2}{4 - (-1)}$$

$$= \frac{3}{5}$$

3. Find the slope of the line through $(8, 4)$ and $(8, -7)$.

$$m = \frac{4 - (-7)}{8 - 8}$$

$$= \frac{11}{0}$$

The slope is not defined; the line is vertical.

5. $y = 2x$

Using the slope-intercept form, $y = mx + b$, we see that the slope is 2.

7. $5x - 9y = 11$

Rewrite the equation in slope-intercept form.

$$9y = 5x - 11$$

$$y = \frac{5}{9}x - \frac{11}{9}$$

The slope is 5/9.

9. $x = -6$

This is a vertical line and the slope is not defined.

11. Find the slope of the line parallel to $2y - 4x = 7$. Rewrite the equation in slope–intercept form.

$$2y = 4x + 7$$
$$y = 2x + \frac{7}{2}$$

This slope is 2, so a parallel line will also have slope 2.

13. Find the slope of the line through $(-1.978, 4.806)$ and $(3.759, 8.125)$. (Note that there are four digits in each number.)

$$m = \frac{8.125 - 4.806}{3.759 - (-1.978)}$$
$$= \frac{3.319}{5.737}$$
$$= .5785$$

(We give the answer correct to four significant digits.)

15. As shown on the graph, the line goes through the points $(0, 2)$ and $(-5, 0)$.

$$m = \frac{0 - 2}{-5 - 0} = \frac{2}{5}$$

17. As shown on the graph, the line goes through the points $(4, 0)$ and $(0, 1)$.

$$m = \frac{1 - 0}{0 - 4} = -\frac{1}{4}$$

19. The line goes through $(1, 3)$, with slope $m = -2$.
Use point–slope form.

$$y - 3 = -2(x - 1)$$
$$y = -2x + 2 + 3$$
$$2x + y = 5$$

21. The line goes through $(6, 1)$, with slope $m = 0$.
Use point–slope form.

$$y - 1 = 0(x - 6)$$
$$y - 1 = 0$$
$$y = 1$$

23. The line goes through $(4, 2)$ and $(1, 3)$.
Find the slope, then use point–slope form with either of the two given points.

$$m = \frac{3 - 2}{1 - 4}$$
$$= -\frac{1}{3}$$

$$y - 3 = -\frac{1}{3}(x - 1)$$
$$y = -\frac{1}{3}x + \frac{1}{3} + 3$$
$$y = -\frac{1}{3}x + \frac{10}{3}$$
$$3y = -x + 10$$
$$x + 3y = 10$$

25. The line goes through $(1/2, 5/3)$ and $(3, 1/6)$.

$$m = \frac{\frac{1}{6} - \frac{5}{3}}{3 - \frac{1}{2}} = \frac{\frac{1}{6} - \frac{10}{6}}{\frac{6}{2} - \frac{1}{2}}$$

$$= \frac{-\frac{9}{6}}{\frac{5}{2}} = -\frac{18}{30} = -\frac{3}{5}$$

$$y - \frac{5}{3} = -\frac{3}{5}\left(x - \frac{1}{2}\right)$$

$$y - \frac{5}{3} = -\frac{3}{5}x + \frac{3}{10}$$

$$30\left(y - \frac{5}{3}\right) = 30\left(-\frac{3}{5}x + \frac{3}{10}\right)$$

$$30y - 50 = -18x + 9$$

$$18x + 30y = 59$$

27. The line has x-intercept 3 and y-intercept −2.

Two points on the line are (3, 0) and (0, −2). Find the slope, then use slope-intercept form.

$$m = \frac{0 - (-2)}{3 - 0}$$

$$= \frac{2}{3}$$

$$b = -2$$

$$y = \frac{2}{3}x - 2$$

$$3y = 2x - 6$$

$$2x - 3y = 6$$

29. The vertical line through (−6, 5) goes through the point (−6, 0), so the equation is

$$x = -6.$$

31. The line goes through (−1.76, 4.25) and has slope −5.081.
Use point-slope form.

$$y - 4.25 = -5.081[x - (-1.76)]$$
$$y = -5.081x - 8.94256$$
$$+ 4.25$$
$$y = -5.081x - 4.69256$$
$$5.081x + y = -4.69$$

For Exercises 33–45, see the answer graphs in the back of your textbook.

33. Graph the line through (−1, 3), m = 3/2. The slope is $\frac{3}{2}$, so $m = \frac{\Delta y}{\Delta x} = \frac{3}{2}$. Thus, $\Delta y = 3$ and $\Delta x = 2$. Another point on the line has coordinates
[(−1 + 2), (3 + 3)] = (1, 6).

Graph the line through (−1, 3) and (1, 6).

35. Graph the line through (3, −4), m = −1/3.
Since m = −1/3, $\Delta y = -1$ and $\Delta x = 3$. Another point on the line has coordinates
[(3 + 3), (−4 − 1)] = (6, −5).

Graph the line through (3, −4) and (6, −5).

37. Graph the line 3x + 5y = 15.
Let x = 0.

$$3(0) + 5y = 15$$
$$y = 3$$

One point is (0, 3).
Let y = 0.

$$3x + 5(0) = 15$$
$$x = 5$$

A second point is (5, 0).
Draw the graph through (0, 3) and (5, 0).

39. Graph line $4x - y = 8$.

Let $x = 0$.

$$4(0) - y = 8$$
$$y = -8$$

One point on the line is $(0, -8)$.

Let $y = 0$.

$$4x - 0 = 8$$
$$4x = 8$$
$$x = 2$$

Another point on the line is $(2, 0)$.
Draw the graph through $(0, -8)$ and $(2, 0)$.

41. Graph the line $x + 2y = 0$.

Let $x = 0$.

$$0 + 2y = 0$$
$$y = 0$$

One point on the line is $(0, 0)$.

Let $x = -2$.

$$-2 + 2y = 0$$
$$2y = 2$$
$$y = 1$$

Another point on the line is $(-2, 1)$.
Draw the graph through $(0, 0)$ and $(-2, 1)$.

43. Graph the line $x = -1$. Since this line has the form $x = a$, it is a vertical line through the point $(-1, 0)$.

45. Graph the line $y = -3$.
Since $y = -3$ for all values of x, the line is horizontal and passes through the point $(0, -3)$.

47. Write the equation of the line through $(-1, 4)$, parallel to $x + 3y = 5$. Rewrite the equation in slope-intercept form.

$$x + 3y = 5$$
$$3y = -x + 5$$
$$y = -\frac{1}{3}x + \frac{5}{3}$$

The slope is $-1/3$.
Use $m = -1/3$ and the point $(-1, 4)$ in the point-slope form.

$$y - 4 = -\frac{1}{3}[x - (-1)]$$
$$y = -\frac{1}{3}(x + 1) + 4$$
$$y = -\frac{1}{3}x - \frac{1}{3} + 4$$
$$= -\frac{1}{3}x + \frac{11}{3}$$
$$x + 3y = 11$$

49. Write the equation of the line through $(3, -4)$, perpendicular to $x + y = 4$.
Rewrite the equation as

$$y = -x + 4.$$

The slope of this line is -1. To find the slope of a perpendicular line, solve

$$-1m = -1.$$
$$m = 1$$

Use $m = 1$ and $(3, -4)$ in the point-slope form.

$$y - (-4) = 1(x - 3)$$
$$y = x - 3 - 4$$
$$y = x - 7$$
$$x - y = 7$$

51. Write the equation of the line with
x-intercept −2, parallel to y = 2x.
The given line has slope 2. A
parallel line will also have m = 2.
Since the x-intercept is −2, the
point (−2, 0) is on the required
line. Using point-slope form, we
have

$$y - 0 = 2[x - (-2)]$$
$$y = 2x + 4$$
$$-2x + y = 4.$$

53. Write the equation of the line with
y-intercept 2, perpendicular to
3x + 2y = 6.
Find slope of given line.

$$3x + 2y = 6$$
$$2y = -3x + 6$$
$$y = -\frac{3}{2}x + 3$$

The slope is −3/2, so the slope of
the perpendicular line will be 2/3.
If the y-intercept is 2, then the
point (0, 2) lies on the line.
Using point-slope form, we have

$$y - 2 = \frac{2}{3}(x - 0)$$
$$y = \frac{2}{3}x + 2$$
$$2x - 3y = -6.$$

55. Do (4, 3), (2, 0), (−18, −12) lie on
the same line?
The line containing (4, 3) and
(2, 0) has slope

$$m = \frac{3 - 0}{4 - 2}$$
$$= \frac{3}{2}.$$

The line containing (2, 0) and
(−18, −12) has slope

$$m = \frac{0 - (-12)}{2 - (-18)}$$
$$= \frac{12}{20}$$
$$= \frac{3}{5}.$$

The two lines do not have the same
slope. Therefore, the three points
do not lie on the same line.

57. Show that the quadrilateral with
vertices (1, 3), (−5/2, 2),
(−7/2, 4), (2, 1) is a parallelogram. A parallelogram is a four-
sided plane figure which has
opposite sides parallel. A sketch
will show which pairs of sides are
opposite.

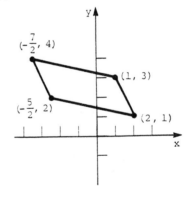

Show that the line containing
(−7/2, 4) and (−5/2, 2) is parallel
to the line containing (1, 3) and
(2, 1) and that the line containing
(−7/2, 4) and (1, 3) is parallel to
the line containing (−5/2, 2) and
(2, 1).

For $(-7/2, 4)$ and $(-5/2, 2)$, the
slope is

$$m = \frac{2 - 4}{\left(-\frac{5}{2}\right) - \left(-\frac{7}{2}\right)}$$

$$= \frac{-2}{1}$$

$$= -2$$

For $(1, 3)$ and $(2, 1)$, the slope is

$$m = \frac{1 - 3}{2 - 1} = \frac{-2}{1} = -2.$$

Since the slopes are equal, these
lines are parallel.
For $(-7/2, 4)$ and $(1, 3)$, the slope
is

$$m = \frac{3 - 4}{1 - \left(-\frac{7}{2}\right)}$$

$$= \frac{-1}{\frac{9}{2}}$$

$$= -\frac{2}{9}.$$

For $(-5/2, 2)$ and $(2, 1)$, the slope
is

$$m = \frac{1 - 2}{2 - \left(-\frac{5}{2}\right)}$$

$$= \frac{-1}{\frac{9}{2}}$$

$$= -\frac{2}{9}.$$

Since the slopes are equal, these
lines are parallel.
Both pairs of opposite sides are
parallel so the quadrilateral is
a parallelogram.

59. Let P, P_1, and P_2 be points on a
nonvertical line L.
Draw a line N through point P parallel to the x-axis. Draw vertical
lines through P_1 and P_2 that intersect N at points Q_1 and Q_2. Angle P
is common to triangles PP_1Q_1 and
PP_2Q_2.

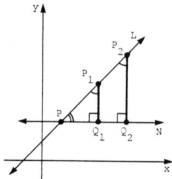

Angle Q_1 = angle Q_2 because they are
right angles.
Angle P_1 = angle P_2 because they are
corresponding angles formed by
parallel lines intersected by a
transversal. Since corresponding
angles of the triangles are equal,
the triangles are similar.
Let m_1 be the slope determined by P
and P_1 and m_2 be the slope determined by P and P_2. In geometry,
recall that P_1Q_1 is the symbol for
the length of segment P_1Q_1.

$$m_1 = \frac{P_1Q_1}{PQ_1} \quad \text{and} \quad m_2 = \frac{P_2Q_2}{PQ_2}$$

Since the triangles are similar,
$P_2Q_2 = k(P_1Q_1)$ and $PQ_2 = k(PQ_1)$

so $m_2 = \dfrac{k(P_1Q_1)}{k(PQ_1)} = \dfrac{P_1Q_1}{PQ_1}$

$$= m_1.$$

61. **(a)** See the figure in your textbook. Segment MN is drawn perpendicular to segment PQ. Recall that MQ is the length of segment MQ.

$$m_1 = \frac{\Delta y}{\Delta x} = \frac{MQ}{PQ}$$

From the diagram, we know that PQ = 1. Thus, $m_1 = \frac{MQ}{1}$ so MQ has length m_1.

(b) $m_2 = \frac{\Delta y}{\Delta x} = \frac{-QN}{PQ} = \frac{-QN}{1}$

$QN = -m_2$

(c) Triangles MPQ, PNQ, and MNP are right triangles by construction. In triangles MPQ and MNP

angle M = angle M

and in right triangles PNQ and MNP

angle N = angle N.

Since all right angles are equal and since triangles with two equal angles are similar,

triangle MPQ is similar to triangle MNP

triangle PNQ is similar to t iangle MNP.

Therefore, triangles MPQ and PNQ are similar to each other.

(d) Since corresponding sides in similar triangles are proportional,

$MQ = k \cdot PQ$ and $PQ = k \cdot QN$.

$$\frac{MQ}{PQ} = \frac{k \cdot PQ}{k \cdot QN}$$

$$\frac{MQ}{PQ} = \frac{PQ}{QN}$$

From the diagram, we know that PQ = 1.

$$MQ = \frac{1}{QN}$$

From (a) and (b), m_1 = MQ and $-m_2$ = QN.

Substituting, we get

$$m_1 = \frac{1}{-m_2}.$$

Multiplying both sides by m_2, we have

$$m_1 m_2 = -1.$$

63. **(a)** Graph $S(q) = \frac{2}{5}q$ and

$$D(q) = 100 - \frac{2}{5}q.$$

See the answer graph in the back your textbook.

(b) The equilibrium quantity is the number of units, q, of wool sweaters for which the price on the demand curve equals the price on the supply curve.

Set $\frac{2}{5}q = 100 - \frac{2}{5}q$ and solve for q.

$$\frac{2}{5}q + \frac{2}{5}q = 100$$

$$\frac{4}{5}q = 100$$

$$q = \frac{5}{4}(100)$$

$$= 125 \text{ units}$$

(c) The equilibrium price is the price corresponding to q = 125 units.

Use $S(q) = \frac{2}{5}q$ and $q = 125$.

$$S(125) = \frac{2}{5}(125) = \$50$$

Or use

$$D(q) = 100 - \frac{2}{5}q \text{ and } q = 125.$$

$$D(q) = 100 - \frac{2(125)}{5}$$

$$= \$50$$

65. $C(x) = 27(x) + 25$, C in dollars, where x is a whole number of days. If x includes a part of a day, then the next whole number is used to find $C(x)$ in the formula.

(a) $C\left(\frac{3}{4}\right) = C(1)$

$$= 27(1) + 25$$

$$= \$52$$

(b) $C\left(\frac{9}{10}\right) = C(1)$

$$= 27(1) + 25$$

$$= \$52$$

(c) $C(1) = 27(1) + 25 = \$52$

(d) $C\left(1\frac{5}{8}\right) = C(2)$

$$= 27(2) + 25$$

$$= \$79$$

(e) $C(2.4) = C(3)$

$$= 27(3) + 25$$

$$= \$106$$

(f) Graph $y = C(x)$.

See the answer graph in the back of your textbook.

(g) Yes.

C is function because a vertical line cuts the graph in no more than one point.

(h) No.

C is not a linear function because its graph is not a straight line.

67. **(a)** To find the equation, find the slope using the points (40, 177) and (43, 185).

$$m = \frac{185 - 177}{43 - 40} = \frac{8}{3}$$

Then use the point-slope form with (40, 177).

$$h - 177 = \frac{8}{3}(t - 40)$$

$$h = \frac{8}{3}t - \frac{8}{3}(40) + 177$$

$$h = \frac{8}{3}t + \frac{211}{3}$$

(b) $t = 38$ cm

$$h = \frac{8}{3}t + \frac{211}{3}$$

$$= \frac{8}{3}(38) + \frac{211}{3}$$

$$= \frac{304 + 211}{3}$$

$$\approx 172$$

$t = 45$ cm

$$h = \frac{8}{3}(45) + \frac{211}{3}$$

$$= \frac{360 + 211}{3}$$

$$\approx 190$$

Males with tibias having lengths of 38 cm to 45 cm would have been about 172 cm to 190 cm tall.

(c) h = 190 cm

$$190 = \frac{8}{3}t + \frac{211}{3}$$

$$\frac{8}{3}t = 190 - \frac{211}{3}$$

$$\frac{8}{3}t = \frac{570 - 211}{3}$$

$$t = \frac{359}{8}$$

$$\approx 45$$

A male whose height is 190 cm would have a tibia bone about 45 cm long.

69. **(a)** Find the slope using (45, 42.5) and (55, 67.5).

$$m = \frac{67.5 - 42.5}{55 - 45} = \frac{25}{10} = 2.5 \text{ or } \frac{5}{2}$$

Use point-slope form with (45, 42.5).

$$(y - 42.5) = \frac{5}{2}(x - 45)$$

$$y = \frac{5}{2}x - \frac{5}{2}(45) + 42.5$$

$$y = \frac{5}{2}x - 70$$

$$= 2.5x - 70$$

(b) Let y = 60.

$$60 = 2.5x - 70$$
$$130 = 2.5x$$
$$52 = x$$

The Democrats would need 52% of the vote.

71. **(a)** If the temperature rises .3°C per decade, and a decade is 10 years, then the temperature rises $\frac{.3}{10} = .03°C$ per year. This is the slope of the line, the change in

temperature per year. When t = 0, in 1970, the temperature was 15°C. This is the intercept. The equation can be written in slope-intercept form as

$$T = .03t + 15.$$

(b) Let T = 19.

$$19 = .03t + 15$$
$$4 = .03t$$
$$133 \approx t$$

Since t = 0 corresponds to 1970, t = 133 corresponds to 1970 + 133 = 2103. The sea level will increase by 65 cm in about the year 2103.

Section 1.3

1. $12 is the fixed cost and $1 is the cost per hour.

    Let x = number of hours

    and

    C(x) = cost of renting a saw for x hours.

    So,

    C(x) = fixed cost + (cost per hour) · (number of hours)

    C(x) = 12 + 1x

    = 12 + x.

3. 50¢ is the fixed cost and 35¢ is the cost per half-hour.

    Let x = the number of half-hours

    and

    C(x) = the cost of parking a car for x half-hours.

So,

$$C(x) = 50 + 35x$$
$$= 35x + 50.$$

5. Fixed cost, $100; 50 items cost $1600 to produce.

$C(x)$ = cost of producing x items.

$C(x) = mx + b$ where b is the fixed cost.

$$C(x) = mx + 100$$

Now,

$C(x) = 1600$ when $x = 50$.

$$1600 = m(50) + 100$$
$$1500 = 50m$$
$$30 = m$$

Thus, $C(x) = 30x + 100$.

7. Fixed cost, $1000; 40 items cost $2000 to produce.

$$C(x) = mx + 1000$$

Now, $C(x) = 2000$ when $x = 40$.

$$2000 = m(40) + 1000$$
$$1000 = 40m$$
$$25 = m$$

Thus, $C(x) = 25x + 1000$.

9. Marginal cost, $50; 80 items cost $4500 to produce.

Let x = number of items produced and

$C(x)$ = cost of producing x items.

$C(x) = mx + b$, where m is the marginal cost.

$$C(x) = 50x + b$$

Now, $C(x) = 4500$ when $x = 80$.

$$4500 = 50(80) + b$$
$$4500 = 4000 + b$$
$$500 = b$$

Thus, $C(x) = 50x + 500$.

11. Marginal cost, $90; 150 items cost $16,000 to produce.

$$C(x) = 90x + b$$

Now $C(x) = 16,000$ when $x = 150$.

$$16,000 = 90(150) + b$$
$$16,000 = 13,500 + b$$
$$2500 = b$$

Thus, $C(x) = 90x + 2500$.

13. $S(x) = 300x + 2000$; 1987 is year 0, so $x = 0$ for 1987.

(a) 1987: $x = 0$

$$S(0) = 300(0) + 200$$
$$= 2000$$

(b) 1990: $x = 3$

$$S(3) = 300(3) + 2000$$
$$= 900 + 2000$$
$$= 2900$$

(c) 1991: $x = 4$

$$S(4) = 300(4) + 2000$$
$$= 1200 + 2000$$
$$= 3200$$

(d) 1996: $x = 9$

$$S(9) = 300(9) + 2000$$
$$= 2700 + 2000$$
$$= 4700$$

Yes, they will reach their goal of 4000.

(e) Annual rate of change of sales is the slope of the linear function, or 300.

15. **(a)** Let $S(x)$ represent sales. $x = 0$ represents 1981, so

$$S(0) = 200,000.$$

$x = 7$ represents 1988, so

$$S(7) = 1,000,000.$$

$$m = \frac{1,000,000 - 200,000}{7 - 0}$$

$$= \frac{800,000}{7}$$

The value of $S(x)$ when $x = 0$ is the y-intercept of the line, b.
So,

$$b = 200,000.$$

Thus,

$$S(x) = \frac{800,000}{7}x + 200,000.$$

(b) 1992: $x = 11$

$$S(11) = \frac{800,000}{7}(11) + 200,000$$

$$= 1,257,142.8 + 200,000$$

$$= 1,457,142.8 \text{ or about}$$
$$\$1,457,000$$

(c) When sales reach $2,000,000,

$$2,000,000 = \frac{800,000}{7}x + 200,000$$

$$1,800,000 = \frac{800,000}{7}x$$

$$12,600,000 = 800,000x$$

$$15.75 = x.$$

They will want to negotiate a contract in about 16 years, or in 1997.

17. $C(x) = .097x$

(a) $C(1000) = .097(1000)$
$$= \$97$$

(b) $C(1001) = .097(1001)$
$$= \$97.097$$

(c) Marginal cost of 1001st cup
$$= C(1001) - C(1000)$$

Use parts (a) and (c).
$$= 97.097 - 97$$
$$= .097 \text{ or } 9.7¢$$

(d) This value is the slope so the marginal cost of any cup is .097 or 9.7¢.

19. $\bar{C}(x) = \dfrac{C(x)}{x} = \dfrac{800 + 20x}{x}$

$$= \frac{800}{x} + \frac{20x}{x}$$

$$= \frac{800}{x} + 20$$

(a) $\bar{C}(10) = \dfrac{800}{10} + 20$
$$= 80 + 20$$
$$\bar{C}(10) = \$100$$

(b) $\bar{C}(50) = \dfrac{800}{50} + 20$
$$= 16 + 20$$
$$\bar{C}(50) = \$36$$

(c) $\bar{C}(200) = \dfrac{800}{200} + 20$
$$= 4 + 20$$
$$\bar{C}(200) = \$24$$

21. $C(x) = 50x + 5000$

$R(x) = 60x$

Let $C(x) = R(x)$ to find the number of units at the break-even point.

$$50x + 5000 = 60x$$
$$5000 = 10x$$
$$x = 500 \text{ units}$$

Revenue for 500 units:

$$R(500) = 60(500)$$
$$= \$30,000$$

23. $C(x) = 85x + 900$

$R(x) = 105x$

Set $C(x) = R(x)$ to find the number of units at the break-even point.

$$85x + 900 = 105x$$
$$900 = 20x$$
$$45 = x$$

The break-even point would be 45 units, so you decide not to produce since not more than 38 units can be sold.

25. $C(x) = 70x + 500$

$R(x) = 60x$

$$70x + 500 = 60x$$
$$10x = -500$$
$$x = -50$$

It is impossible to make a profit when the break-even point is negative. Cost will always be greater than revenue.

27. From the graph, the break-even point was about \$140 billion, which occurred in the middle of 1981.

29. Let $x = 0$ correspond to 1900. Then the "life expectancy from birth" line contains the points (0, 46) and (75, 75).

$$m = \frac{75 - 46}{75 - 0} = \frac{29}{75}$$

Use point slope form.

$$y - 46 = \frac{29}{75}(x - 0)$$
$$y = \frac{29}{75}x + 46$$

The "life expectancy from age 65" line contains the points (0, 76) and (75, 80).

$$m = \frac{80 - 76}{75 - 0} = \frac{4}{75}$$

$$y - 76 = \frac{4}{75}(x - 0)$$
$$y = \frac{4}{75}x + 76$$

Set the two equations equal to determine where the lines intersect. At this point life expectancy should increase no further.

$$\frac{29}{75}x + 46 = \frac{4}{75}x + 76$$
$$29x + 3450 = 4x + 5700$$
$$25x = 2250$$
$$x = 90$$

Determine the y-value when x = 90.

$$y = \frac{4}{75}(90) + 76$$

$$y = 4.8 + 76$$

$$= 80.8$$

Thus, the maximum life expectancy for humans is about 81 years.

31. Let x represent the force and y represent the speed.
The linear function contains the points (.75, 2) and (.93, 3).

$$m = \frac{3 - 2}{.93 - .75} = \frac{1}{.18} = \frac{1}{\frac{18}{100}}$$

$$= \frac{100}{18} = \frac{50}{9}$$

Use point-slope form to write the equation.

$$y - 2 = \frac{50}{9}(x - .75)$$

$$y - 2 = \frac{50}{9}x - \frac{50}{9}(.75)$$

$$y = \frac{50}{9}x - \frac{75}{18} + 2$$

$$= \frac{50}{9}x - \frac{39}{18}$$

Now determine y, the speed, when x, the force, is 1.16.

$$y = \frac{50}{9}(1.16) - \frac{39}{18}$$

$$= \frac{58}{9} - \frac{39}{18}$$

$$= \frac{77}{18} \approx 4.3$$

The pony switches from a trot to a gallop at approximately 4.3 m/sec.

33. $y = mx + b$
$y = 1.25x - 5$

(a) $y = 1.25(30) - 5$
$y = 37.5 - 5$
$y = 32.5$ min

(b) $y = 1.25(60) - 5$
$y = 75 - 5$
$y = 70$ min

(c) $y = 1.25(120) - 5$
$y = 150 - 5$
$y = 145$ min

(d) $y = 1.25(180) - 5$
$y = 225 - 5$
$y = 220$ min

35. (a) $C = \frac{5}{9}(F - 32)$

$$C = \frac{5}{9}(58 - 32)$$

$$C = \frac{5}{9}(26)$$

$$C = 14.4°$$

The temperature is 14.4° C.

(b) $F = \frac{9}{5}C + 32$

$$F = \frac{9}{5}(50) + 32$$

$$F = 90 + 32$$

$$F = 122°$$

The temperature is 122°F.

37. Water freezes at 32° F and 0° C. It boils at 212° F and 100° C. The ordered pairs (32, 0) and (212, 100) lie on the graph giving the Celsius temperature (C) as a linear function of the Fahrenheit temperature.

$$m = \frac{100 - 0}{212 - 32} = \frac{100}{180} = \frac{5}{9}$$

$$C - 0 = \frac{5}{9}(F - 32)$$

$$C = \frac{5}{9}(F - 32)$$

39. $R(x) = -8x + 240$

(a) $R(0) = -8(0) + 240$

$= 0 + 240$

$= 240$

(b) $R(5) = -8(5) + 240$

$= -40 + 240$

$= 200$

(c) $R(10) = -8(10) + 240$

$= -80 + 240$

$= 160$

(d) The rate of change is the slope of the line, or −8 students per hour of study. The negative sign indicates that student enrollment decreases if the hours of required study increases.

(e) $R(x) = 8x + 240$

$16 = -8x + 240$

$16 - 240 = -8x$

$-224 = -8x$

$28 = x$

28 hr of study are required.

Section 1.4

For Exercises 1–21, see the answer graphs in the back of the textbook.

1. $f(x) = kx^2$ (k is a constant)

(e) All the curves have the same vertex: (0, 0). As the coefficient k increases in absolute value, the parabola becomes narrower.

3. $f(x) = x^2 + k$ (k is a constant)

(e) All the curves have the same shape. The coordinates of the vertices are (0, k) so the graphs are shifted upward or downward.

5. $y = x^2 + 6x + 5$

$y = (x + 5)(x + 1)$

Set $y = 0$ to find the x−intercepts.

$0 = (x + 5)(x + 1)$

$x = -5, \; x = -1$

The x−intercepts are −5 and −1.
Set $x = 0$ to find the y−intercept.

$y = 0^2 + 6(0) + 5$

$y = 5$

The y−intercept is 5.
The x−coordinate of the vertex is

$x = \frac{-b}{2a} = \frac{-6}{2} = -3.$

Substitute to find the y−coordinate.

$y = (-3)^2 + 6(-3) + 5$

$= 9 - 18 + 5$

$= -4$

The vertex is $(-3, -4)$.

The axis is $x = -3$, the vertical line through the vertex.

7. $y = x^2 - 4x + 4$

$\quad = (x - 2)(x - 2)$

Let $y = 0$.

$0 = (x - 2)(x - 2)$

2 is the x-intercept.

Let $x = 0$.

$y = 0^2 - 4(0) + 4$

4 is the y-intercept.

Vertex: $x = \dfrac{-b}{2a} = \dfrac{4}{2} = 2$

$\quad\quad\quad y = 2^2 - 4(2) + 4$

$\quad\quad\quad\quad = 4 - 8 + 4 = 0$

The vertex is $(2, 0)$.

The axis is $x = 2$, the vertical line through the vertex.

9. $y = -2x^2 - 12x - 16$

$\quad = -2(x^2 + 6x + 8)$

$\quad = -2(x + 4)(x + 2)$

Let $y = 0$.

$0 = -2(x + 4)(x + 2)$

$x = -4, \; x = -2$

-4 and -2 are the x-intercepts.

Let $x = 0$.

$y = -2(0)^2 + 12(0) - 16$

-16 is the y-intercept.

Vertex: $x = \dfrac{-b}{2a} = \dfrac{12}{-4} = -3$

$y = -2(-3)^2 - 12(-3) - 16$

$\quad = -18 + 36 - 16$

$\quad = 2$

The vertex is $(-3, 2)$.

The axis is $x = -3$, the vertical line through the vertex.

11. $y = 2x^2 + 12x - 16$

$\quad = 2(x^2 + 6x - 8)$

Let $y = 0$.

$0 = 2(x^2 + 6x - 8)$

Divide by 2.

$0 = x^2 + 6x - 8$

Use the quadratic formula.

$x = \dfrac{-6 \pm \sqrt{6^2 - 4(1)(-8)}}{2(1)}$

$\quad = \dfrac{-6 \pm \sqrt{36 - (-32)}}{2}$

$\quad = \dfrac{-6 \pm \sqrt{68}}{2}$

$\quad = \dfrac{-6 \pm 2\sqrt{17}}{2}$

$\quad = -3 \pm \sqrt{17}$

The x-intercepts are $-3 + \sqrt{17} \approx 1.12$ and $-3 - \sqrt{17} \approx -7.12$.

Let $x = 0$.

$y = 2(0)^2 + 12(0) - 16$

-16 is the y-intercept.

Vertex: $x = \dfrac{-b}{2a} = \dfrac{-12}{2(2)} = -\dfrac{12}{4} = -3$

$y = 2(-3)^2 + 12(-3) - 16$

$\quad = 18 - 36 - 16$

$\quad = -34$

The vertex is $(-3, -34)$.

The axis is $x = -3$, the vertical line through the vertex.

13. $y = 3x^2 + 6x + 12$

Let $y = 0$.

$0 = 3x^2 + 6x + 2$

$$x = \frac{-6 \pm \sqrt{6^2 - 4(3)(2)}}{2(3)}$$

$$= \frac{-6 \pm \sqrt{36 - 24}}{6} = \frac{-6 \pm \sqrt{12}}{6}$$

$$= \frac{-6 \pm 2\sqrt{3}}{6} = -1 \pm \frac{\sqrt{3}}{3}$$

The x-intercepts are $-1 + \frac{\sqrt{3}}{3} \approx -.42$

and $-1 - \frac{\sqrt{3}}{3} \approx -1.58$.

Let $x = 0$.

$y = 3(0)^2 + 6(0) + 2$

2 is the y-intercept.

Vertex: $x = \frac{-b}{2a} = \frac{-6}{2(3)} = -\frac{6}{6} = -1$

$y = 3(-1)^2 + 6(-1) + 2$

$= 3 - 6 + 2$

$= -1$

The vertex is $(-1, -1)$.
The axis is $x = -1$.

15. $y = -x^2 + 6x - 6$

Let $y = 0$.

$0 = -x^2 + 6x - 6$

$$x = \frac{-6 \pm \sqrt{6^2 - 4(-1)(-6)}}{2(-1)}$$

$$= \frac{-6 \pm \sqrt{36 - 24}}{-2} = \frac{-6 \pm \sqrt{12}}{-2}$$

$$= \frac{-6 \pm 2\sqrt{3}}{-2}$$

$$= 3 \pm \sqrt{3}$$

The x-intercepts are $3 + \sqrt{3} \approx 4.73$
and $3 - \sqrt{3} \approx 1.27$.

Let $x = 0$.

$y = -0^2 + 6(0) - 6$

-6 is the y-intercept.

Vertex: $x = \frac{-b}{2a} = \frac{-6}{2(-1)} = \frac{-6}{-2} = 3$

$y = -3^2 + 6(3) - 6$

$= -9 + 18 - 6$

$= 3$

The vertex is $(3, 3)$.

The axis is $x = 3$.

17. $y = -3x^2 + 24x - 36$

$= -3(x^2 - 8x + 12)$

$= -3(x - 6)(x - 2)$

Let $y = 0$.

$0 = -3(x - 6)(x - 2)$

Divide by -3.

$0 = (x - 6)(x - 2)$

$x = 6, \ x = 2$

6 and 2 are the x-intercepts.

Let $x = 0$.

$y = -3(0)^2 + 24(0) - 36$

-36 is the y-intercept.

Vertex: $x = \frac{-b}{2a} = \frac{-24}{2(-3)} = \frac{-24}{-6} = 4$

$y = -3(4)^2 + 24(4) - 36$

$= -48 + 96 - 36$

$= 12$

The vertex is $(4, 12)$.

The axis is $x = 4$.

19. $y = \frac{5}{2}x^2 + 10x + 8$

Let $y = 0$.

$0 = \frac{5}{2}x^2 + 10x + 8$

$x = \dfrac{-10 \pm \sqrt{10^2 - 4\left(\frac{5}{2}\right)(8)}}{2\left(\frac{5}{2}\right)}$

$ = \dfrac{-10 \pm \sqrt{100 - 80}}{5}$

$ = \dfrac{-10 \pm \sqrt{20}}{5}$

$ = -2 \pm \dfrac{2\sqrt{5}}{5}$

The x-intercepts $-2 + 2\sqrt{5}/5 \approx -1.11$ and $-2 - 2\sqrt{5}/5 \approx -2.89$.
Let $x = 0$.

$y = \frac{5}{2}(0)^2 + 10(0) + 8$

8 is the y-intercept.

Vertex: $x = \dfrac{-b}{2a} = \dfrac{-10}{2\left(\frac{5}{2}\right)} = \dfrac{-10}{5} = -2$

$y = \frac{5}{2}(-2)^2 + 10(-2) + 8$

$ = \frac{5}{2}(4) - 20 + 8$

$ = 10 - 20 + 8$

$ = -2$

The vertex is $(-2, -2)$.
The axis is $x = -2$.

21. $y = \frac{2}{3}x^2 - \frac{8}{3}x + \frac{5}{3}$

Let $y = 0$.

$0 = \frac{2}{3}x^2 - \frac{8}{3}x + \frac{5}{3}$

Multiply by 3.

$0 = 2x^2 - 8x + 5$

$x = \dfrac{-(-8) \pm \sqrt{(-8)^2 - 4(2)(5)}}{2(2)}$

$ = \dfrac{8 \pm \sqrt{64 - 40}}{4} = \dfrac{8 \pm \sqrt{24}}{4}$

$ = \dfrac{8 \pm 2\sqrt{6}}{4} = 2 \pm \dfrac{\sqrt{6}}{2}$

The x-intercepts are $2 + \sqrt{6}/2 \approx$ 3.22 and $2 - \sqrt{6}/2 \approx .78$.
Let $x = 0$.

$y = \frac{2}{3}(0)^2 - \frac{8}{3}(0) + \frac{5}{3}$

5/3 is the y-intercept.

Vertex: $x = \dfrac{-b}{2a} = \dfrac{-\left(-\frac{8}{3}\right)}{2\left(\frac{2}{3}\right)} = \dfrac{\frac{8}{3}}{\frac{4}{3}} = 2$

$y = \frac{2}{3}(2)^2 - \frac{8}{3}(2) + \frac{5}{3}$

$ = \frac{8}{3} - \frac{16}{3} + \frac{5}{3}$

$ = -\frac{3}{3} = -1$

The vertex is $(2, -1)$.
The axis is $x = 2$.

23. Let $y = x(1 - x)$.

$y = x - x^2$

$ = -x^2 + x$

Let $y = 0$.

$0 = x(1 - x)$

$x = 0 \quad \text{or} \quad x = 1$

0 and 1 are the x-intercepts.
Let $x = 0$.

$y = 0(1 - 0)$

0 is the y-intercept.

Vertex: $x = \dfrac{-b}{2a} = \dfrac{-1}{2(-1)} = \dfrac{-1}{-2} = \dfrac{1}{2}$

$$y = -\left(\frac{1}{2}\right)^2 + \frac{1}{2}$$

$$= -\frac{1}{4} + \frac{1}{2}$$

$$= \frac{1}{4}$$

The vertex is at $\left(\frac{1}{2}, \frac{1}{4}\right)$.

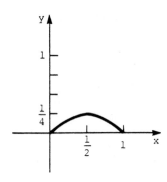

This parabola never reaches a height greater than $\frac{1}{4}$, thus the product $x(1 - x)$ is always less than or equal to $\frac{1}{4}$. The product equals $\frac{1}{4}$ at the vertex, when $x = \frac{1}{2}$.

25. $R(x) = 5000 + 50x - x^2$
$$= -x^2 + 50x + 5000$$

The maximum revenue occurs at the vertex.

$$x = \frac{-b}{2a} = \frac{-50}{2(-1)} = 25$$

$$y = 5000 + 50(25) - (25)^2$$
$$= 5000 + 1250 - 625$$
$$= 5625$$

The vertex is (25, 5625).
The maximum revenue of $5625 is realized when 25 seats are left unsold.

27. $p = 500 - x$

(a) Revenue is $R(x) = px$
$$= (500 - x)(x)$$
$$= 500x - x^2$$

(b) See the answer graph in your textbook.

(c) From the graph, the vertex is halfway between $x = 0$ and $x = 500$ so $x = 250$ units corresponds to maximum revenue. Then the price is

$$p = 500 - x$$
$$= 500 - 250 = \$250.$$

Note that price, p, cannot be read directly from the graph of

$$R(x) = 500x - x^2.$$

(d) $R(x) = 500x - x^2$
$$= -x^2 + 500x$$

Find the vertex.

$$x = \frac{-b}{2a} = \frac{-500}{-2} = 250$$

$$y = -(250)^2 + 500(250)$$
$$= 62,500$$

The vertex is (250, 62,500).
The maximum revenue is $62,500.

29. Let x = the number of $20 increases.

(a) Rent per apartment: $200 + 20x$

(b) Number of apartments rented:
     $80 - x$

(c) Revenue:

R(x) = (number of apartments
      rented)(rent per apart-
      ment)

    = (80 − x)(200 + 20x)

    = 16,000 + 1400x − 20x²

(d) Find the vertex.

$$x = \frac{-b}{2a} = \frac{-1400}{2(-20)} = \frac{-1400}{-40} = 35$$

y = 16,000 + 1400(35) − 20(35)²

  = 40,500

The vertex is (35, 40,500).
The maximum revenue occurs when
x = 35.

(e) The maximum revenue is the y-
coordinate of the vertex or $40,500.

31. C(x) = 10x + 50   (June–September)
C(x) = −20(x − 5)² + 100
               (October–December)

(a) June: x = 1

C(x) = 10 + 50
    = 60

(b) July: x = 2

C(x) = 10(2) + 50
    = 70

(c) September: x = 4

C(x) = 10(4) + 50
    = 90

(d) October: x = 5

C(x) = −20(5 − 5)² + 100
    = 100

(e) November: x = 6

C(x) = −20(6 − 5)² + 100
    = −20 + 100
    = 80

(f) December: x = 7

C(x) = −20(7 − 5)² + 100
    = −80 + 100
    = 20

33. h = 32t − 16t²
   = −16t² + 32t

Find the vertex.

$$x = \frac{-b}{2a} = \frac{-32}{-32} = 1$$

y = −16(1)² + 32(1)
  = 16

The vertex is (1, 16), so the maximum height is 16 ft. When the
object hits the ground, h = 0, so

32t − 16t² = 0
16t(2 − t) = 0
t = 0 or t = 2

When t = 0, the object is about to
be thrown. When t = 2, the object
hits the ground; that is, after 2
sec.

35. Let x = the width.
Then 320 − 2x = the length.

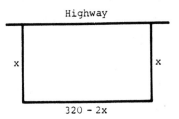

Area = x(320 − 2x)

$\qquad$ = −2x² + 320x

Find the vertex:

x = $\dfrac{-b}{2a}$ = $\dfrac{-320}{-4}$ = 80

y = −2(80)² + 320(80)

$\quad$ = 12,800

The graph of the area function is a parabola with vertex (80, 12,800). The maximum area of 12,800 sq ft occurs when the width is 80 ft and the length is

$$320 − 2x = 320 − 2(80)$$
$$= 160 \text{ ft.}$$

37. Let x = the one number.

Then 20 − x = the other number.

Product = x(20 − x)

$\qquad$ = 20x − x²

$\qquad$ = −x² + 20x

Find the vertex.

x = $\dfrac{-b}{2a}$ = $\dfrac{-20}{-2}$ = 10

y = −10² + 20(10)

$\quad$ = 100

The vertex is (10, 100), so the product is a maximum when

$$x = 10.$$

The other number is

$$20 − 10 = 10.$$

39. Draw a sketch of the arch with the base on the x-axis and the vertex at (0, 15).

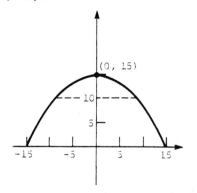

The equation of the parabola is in the form

$$y = −ax² + 15.$$

The points (−15, 0) and (15, 0) are on the parabola. Use (15, 0) as one point on the parabola.

$$0 = −a(15)² + 15$$
$$−15 = −a(225)$$
$$a = \frac{1}{15}$$

The equation is

$$y = −\frac{1}{15}x² + 15.$$

When y = 10,

$$10 = −\frac{x²}{15} + 15$$
$$−5 = −\frac{x²}{15}$$
$$x² = 75$$
$$x = ±\sqrt{75}$$
$$= ±5\sqrt{3}.$$

The width of the arch is

$$5\sqrt{3} + |-5\sqrt{3}|$$
$$= 10\sqrt{3} \text{ m} \approx 17.32 \text{ m}.$$

Section 1.5

1. The range is $[-.5, \infty)$, which is of the form $[k, \infty)$. Thus, the function is of even degree. Since there are 3 turning points the degree is 4 or greater. Possible values for the degree are 4, 6, and so on. The shape is similar to the first graph of Figure 34 so the sign of the x^n term is +.

3. Range: $(-\infty, \infty)$

   The function is of odd degree.
   Turing points: 4
   The function has degree 5 or greater. Possible values for degree: 5, 7, and so on. Shape: similar to the first graph of Figure 33. The sign of the x^n term is +.

5. Range: $(-\infty, \infty)$

   The function is of odd degree.
   Turning points: 4
   The function is of degree 5 or greater. Possible values for degree: 5, 7, and so on. Shape: similar to the first graph of Figure 33. The sign of the x^n term is +.

7. Range: $(-\infty, \infty)$

   The function is of odd degree.
   Turing points: 6
   The function is of degree 7 or greater. Possible values for degree: 7, 9, and so on. Shape: similar to the second graph of Figure 33. The sign of the x^n term is -.

For Exercises 9–55, see the answer graphs in the back of your textbook.

9. $y = \dfrac{-4}{x - 3}$

   The function is undefined for $x = 3$, so the line $x = 3$ is a vertical asymptote.

x	-97	-7	-2	0	2	4	13	103
$x - 3$	-100	-10	-5	-3	-1	1	10	100
y	$.04$	$.4$	$.8$	1.3	4	-4	$-.4$	$-.04$

   The graph approaches $y = 0$, so the line $y = 0$ (the x-axis) is a horizontal asymptote.
   Asymptotes: $y = 0$, $x = 3$
   x-intercept:
   none, because the x-axis is an asymptote

   y-intercept:
   4/3, the value when $x = 0$

11. $y = \dfrac{2}{3 + 2x}$

    $3 + 2x = 0$ when $2x = -3$ or $x = -\dfrac{3}{2}$, so the line $x = -3/2$ is a vertical asymptote.

x	-51.5	-6.5	-2	-1	3.5	48.5
3 + 2x	-100	-10	-1	1	10	100
y	-.02	-.2	-2	2	.2	.02

The graph approaches y = 0, so the line y = 0 (the x-axis) is a horizontal asymptote.

Asymptote: y = 0, x = -3/2

x-intercept:

none, since the x-axis is an asymptote

y-intercept:

2/3, the value when x = 0

13. $y = \dfrac{3x}{x - 1}$

x - 1 = 0 when x = 1, so the line x = 1 is a vertical asymptote.

x	-99	-9	-1
3x	-297	-27	-3
x - 1	-100	-10	-2
y	2.97	2.7	1.5

0	.5	1.5	12	11	101
0	1.5	4.5	6	33	303
-1	-.5	.5	1	10	100
0	-3	9	6	3.3	3.03

As x gets larger, $\dfrac{3x}{x - 1} \approx \dfrac{3x}{x} = 3$.

Thus, y = 3 is a horizontal asymptote.

Asymptotes: y = 3, x = 1

x-intercept:

0, the value when y = 0

y-intercept:

0, the value when x = 0

15. $y = \dfrac{x + 1}{x - 4}$

x - 4 = 0 when x = 4, so x = 4 is a vertical asymptote.

x	-96	-6	-1	0	3	3.5	4.5
x + 1	-95	-5	0	1	4	4.5	5.5
x - 4	-100	-10	-5	-4	-1	-.5	.5
y	.95	.5	0	-.25	-4	-9	11

5	14	104
6	15	105
1	10	100
6	1.5	1.05

As x gets larger, $\dfrac{x + 1}{x - 4} \approx \dfrac{x}{x} = 1$.

Thus y = 1 is a horizontal asymptote.

Asymptotes: y = 1, x = 4

x-intercept:

-1, the value when y = 0

y-intercept:

-.25 = -1/4, the value when x = 0

17. $y = \dfrac{1 - 2x}{5x + 20}$

5x + 20 = 0 when 5x = -20 or x = -4, so the line x = -4 is a vertical asymptote.

x	−7	−6	−5	−3	−2	−1	0
1 − 2x	15	13	11	7	5	3	1
5x + 20	−15	−10	−5	5	10	15	20
y	−1	−1.3	−2.2	1.4	.5	.2	.05

As x gets larger, $\dfrac{1-2x}{5x+20} \approx \dfrac{-2x}{5x} = \dfrac{-2}{5}$.

Thus, the line y = −2/5 is a horizontal asymptote.

Asymptotes: x = −4, y = −2/5

x−intercept:

1/2, the value when y = 0

y−intercept:

.05 = 1/20, the value when x = 0

19. $y = \dfrac{-x-4}{3x+6}$

3x + 6 = 0 when 3x = −6 or x = −2, so the line x = −2 is a vertical asymptote.

x	−5	−4	−3	−1	0	1
−x − 4	1	0	−1	−3	−4	−5
3x + 6	−9	−6	−3	3	6	9
y	−.11	0	.33	−1	−.67	−.56

As x gets larger, $\dfrac{-x-4}{3x+6} \approx \dfrac{-x}{3x} = -\dfrac{1}{3}$.

The line y = −1/3 is a horizontal asymptote.

Asymptotes: y = −1/3, x = −2

x−intercept:

−4, the value when y = 0

y−intercept:

−.67 = −2/3, the value when x = 0

21. $C(x) = \dfrac{500}{x+30}$

(a) $C(10) = \dfrac{500}{10+30}$

$= \dfrac{500}{40}$

$= \$12.50$

$C(20) = \dfrac{500}{20+30}$

$= \dfrac{500}{50}$

$= \$10$

$C(50) = \dfrac{500}{50+30}$

$= \dfrac{500}{80}$

$= \$6.25$

$C(75) = \dfrac{500}{75+30}$

$= \dfrac{500}{105}$

$\approx \$4.76$

$C(100) = \dfrac{500}{100+30}$

$= \dfrac{500}{130}$

$= 3.8461538$

$\approx \$3.85$

(b) (0, ∞) would be a more reasonable domain for average cost than [0, ∞). If zero were included in the domain, there would be no units produced. It is not reasonable to discuss the average cost per unit of zero units.

(c) See the answer graph in the back of your textbook.

23. $y = x(100 - x)(x^2 + 500)$,

y = tax revenue in hundreds of thousands of dollars

x = tax rate.

(a) $x = 10$

$y = 10(100 - 10)(10^2 + 500)$

$= 10(90)(600)$

$= \$54$ billion

(b) $x = 40$

$y = 40(100 - 40)(40^2 + 500)$

$= 40(60)(2100)$

$= \$504$ billion

(c) $x = 50$

$y = 50(100 - 50)(50^2 + 500)$

$= 50(50)(3000)$

$= \$750$ billion

(d) $x = 80$

$y = 80(100 - 80)(80^2 + 500)$

$= 80(20)(6900)$

$= \$1104$ billion

(e) See the answer graph in the back of the textbook.

25. $y = \dfrac{60x - 6000}{x - 120}$,

y = revenue in millions

x = tax rate.

(a) $x = 50$

$y = \dfrac{60(50) - 6000}{50 - 120}$

$= \dfrac{-3000}{-70}$

$= \$42.9$ million

(b) $x = 60$

$y = \dfrac{60(60) - 6000}{60 - 120}$

$= \dfrac{-2400}{-60}$

$= \$40$ million

(c) $x = 80$

$y = \dfrac{60(80) - 6000}{80 - 120}$

$= \dfrac{-1200}{-40}$

$= \$30$ million

(d) $x = \dfrac{60(100) - 6000}{100 - 120}$

$= \$0$

(e) See the answer graph in the back of the textbook.

27. $y = \dfrac{6.7x}{100 - x}$,

y = cost in thousands

x = percent of pollutant.

(a) $x = 50$

$y = \dfrac{6.7(50)}{100 - 50}$

$= 6.7$

The cost is $\$6700$.

(b) $x = 70$

$y = \dfrac{6.7(70)}{100 - 70}$

≈ 15.6

The cost is $\$15,600$.

(c) $x = 80$

$y = \dfrac{6.7(80)}{100 - 80}$

$= 26.8$

The cost is $\$26,800$.

(d) x = 90

$$y = \frac{6.7(90)}{100 - 90}$$

$$= 60.3$$

The cost is $60,300.

(e) x = 95

$$y = \frac{6.7(95)}{100 - 95}$$

The cost is $127,300.

(f) x = 98

$$y = \frac{6.7(98)}{100 - 98}$$

$$= 328.3$$

The cost is $328,300.

(g) x = 99

$$y = \frac{6.7(99)}{100 - 99}$$

$$= 663.3$$

The cost is $663,300.

(h) No, because x = 100% makes the denominator zero so x = 100% is a vertical asymptote.

(i) See the answer graph in the back of your textbook.

29. **(a)** See the answer graph in the back of your textbook.

(b) The function defined by

g(x) = −.006x⁴ + .140x³ − .053x² + 1.79x,

a fourth-degree polynomial, will have a shape similar to the graphs in Figure 34. The negative coefficient on the x⁴ term indicates that both "ends" of the graph will

eventually head toward −∞, as shown in the second graph in Figure 34. So, the function will not keep increasing but will turn and begin to decrease.

31. A(x) = −.015x² + 1.058x

(a) See the answer graph at the back of your textbook.

(b) Reading the graph, we find that concentration is maximum between x = 4 hr and x = 5 hr, but closer to 5 hr.

(c) Concentration exceeds .08% from less than 1 hr to about 8.4 hr.

33. $f(x) = \dfrac{\lambda x}{1 + (ax)^b}$

(a) A reasonable domain for the function is [0, ∞). Populations are not measured using negative numbers and they may get extremely large.

(b) See the answer graph in the back of the textbook.

(c) See the answer graph in the back of the textbook.

(d) As seen from the graphs, when b increases, the population of the next generation, f(x), gets smaller when the current generation, x, is larger.

35. $G(R) = \dfrac{r}{r + R}$

R is resistance; r = 1000 ohms

(a) $G(R) = \dfrac{R}{1000 + R}$

G is undefined if R = −1000 ohms, so R = −1000 is a vertical asymptote.

(b) $G(R) = \dfrac{R}{1000 + R}$

As R gets larger,

$$\dfrac{R}{1000 + R} \approx \dfrac{R}{R} = 1.$$

Thus G(R) = 1 is a horizontal asymptote.

(c) See the answer graph at the back of your textbook.

37. $f_t(u) = u^3 + tu$

(a) t = 1

$f_1(u) = u^3 + u$

u	−2	−1	0	1	2
f(u)	−10	−2	0	2	10

See the answer graph in the back of the textbook.

(b) t = −1

$\quad f_{-1}(u) = u^3 + (-1)u$
$\qquad\quad = u^3 - u$

u	−2	−1	−.5	0	.5	1	2
f(u)	−6	0	.375	0	−.375	0	6

See the answer graph in the back of the textbook.

39. $C(x) = \dfrac{10x}{49(101 - x)}$

C(x) is in thousands of dollars; x represents points.

(a) $C(99) = \dfrac{10(99)}{49(101 - 99)}$

$\qquad\qquad = \dfrac{990}{49(2)}$

$\qquad\qquad = \dfrac{990}{98}$

$\qquad\qquad = 10.102040$

The cost is about \$10,000.

(c) $C(100) = \dfrac{10(100)}{49(101 - 100)}$

$\qquad\qquad = \dfrac{1000}{49(1)}$

$\qquad\qquad = \dfrac{1000}{49}$

$\qquad\qquad = 20.408163$

The cost is about \$20,000.

41. $f(x) = -x^3 + 4x^2 + 3x - 8$

x	−3	−2	−1	0	1	2	3	4	5
f(x)	46	10	−6	−8	−2	6	10	4	−18

See the answer graph in the textbook.

43. $f(x) = 2x^3 + 4x - 1$

x	−2	−1	0	1	2
f(x)	−25	−7	−1	5	23

See the answer graph in the textbook.

45. $f(x) = x^4 + x^3 - 2$

x	-2	-1	0	1	2
f(x)	6	-2	-2	0	22

See the answer graph in the textbook.

47. $f(x) = -x^4 - 2x^3 + 3x^2 + 3x + 5$

x	-3	-2	-1	0	1	2
f(x)	-4	11	6	5	8	-9

See the answer graph in the textbook.

49. $f(x) = -x^5 + 6x^4 - 11x^3 + 6x^2 + 5$

x	-2	-1	-.5	0	.5
f(x)	245	29	5.1	5	5.2

x	1	1.5	2	2.5	3
f(x)	5	4.2	5	7.3	5

See the answer graph in the textbook.

51. $f(x) = \dfrac{-2x^2 + x - 1}{2x + 3}$

Vertical asymptote:

$$2x + 3 = 0$$
$$2x = -3$$
$$x = -\frac{3}{2}$$

Horizontal asymptote:
None, because the degree of the numerator is greater than the degree of the denominator.

x	-5	-4	-3	-2	-1	0	1	2
f(x)	8	7.4	7.3	12	-4	-.3	-.4	-1

See the answer graph in the textbook.

53. $f(x) = \dfrac{2x^2 - 5}{x^2 - 1}$

Vertical asymptotes:

$$x^2 - 1 = 0$$
$$(x + 1)(x - 1) = 0$$
$$x = -1, \; x = 1$$

Horizontal asymptote:
$y = 2$, since $f(x)$ approaches 2 as x gets larger.

x	-4	-3	-2	0	2	3	4
f(x)	1.8	1.6	3	5	3	1.6	1.8

See the answer graph in the textbook.

55. $f(x) = \dfrac{-2x^2}{x^2 - 10}$

Vertical asymptotes:

$$x^2 - 10 = 0$$
$$x^2 = 10$$
$$x = \pm\sqrt{10} \approx \pm 3.16$$

Horizontal asymptote:
$y = -2$, since $f(x)$ approaches -2 as x gets larger.

x	-5	-4	-3	-2	-1	0
f(x)	-3.3	-5.3	18	1.3	.22	0

x	1	2	3	4	5
f(x)	.22	1.3	18	5.3	3.3

See the answer graph in the textbook.

Section 1.6

For Exercises 1–15, see the answer
graphs in the back of the textbook.

1. $f(x) = (x + 2)^3$

   Translate the graph of $f(x) = x^2$
   2 units left.

3. $f(x) = (4 - x)^3 + 1$
   $= -(x - 4)^3 + 1$

   Translate the graph of $f(x) = x^2$
   4 units right and 1 unit up.
   Reflect it vertically.

5. $f(x) = -2(x + 1)^3 + 5$

   Translate the graph of $f(x) = x^3$
   1 unit left and 5 units up. Reflect
   the graph vertically and stretch it
   vertically by a factor of 2.

7. $f(x) = -2x^5 + 5$

   Translate the graph of $f(x) = x^5$
   5 units up. Reflect it vertically.
   Then stretch it vertically by a
   factor of 2.

9. $f(x) = \dfrac{2}{x + 3} - 2$

   Translate the graph of $f(x) = \dfrac{1}{x}$
   3 units left and 2 units down.
   Stretch it vertically by a factor
   of 2.

11. $f(x) = \dfrac{-1}{x - 1} + 2$

    Translate the graph of $f(x) = \dfrac{1}{x}$
    1 unit right and 2 units up.
    Reflect it vertically.

13. $f(x) = \sqrt{x - 1} + 3$

    Translate the graph of $f(x) = \sqrt{x}$
    1 unit right and 3 units up.

15. $f(x) = -\sqrt{-4 - x} - 2$
    $= -\sqrt{-(x + 4)} - 2$

    Translate the graph of $f(x) = \sqrt{x}$
    4 units left and 2 units down.
    Reflect vertically and horizontally.

For Exercises 17–23, see the answer
graphs in the textbook.

17. $f(x) = x^2 - 2x - 2$
    $= x^2 - 2x + \left(\dfrac{-2}{2}\right)^2 - \left(\dfrac{-2}{2}\right)^2 - 2$
    $= (x^2 - 2x + 1) - 1 - 2$
    $= (x - 1)^2 - 3$

    Translate: 1 unit right and 3 units
    down

19. $f(x) = x^2 - 4x - 2$
    $= x^2 - 4x + \left(\dfrac{-4}{2}\right)^2 - \left(\dfrac{-4}{2}\right)^2 - 2$
    $= (x^2 - 4x + 4) - 4 - 2$
    $= (x - 2)^2 - 6$

    Translate: 2 units right and 6 units
    down

21. $f(x) = 2x^2 + 4x - 1$

$\qquad = 2(x^2 + 2x) - 1$

$\qquad = 2\left[x^2 + 2x + \left(\frac{2}{2}\right)^2 - \left(\frac{2}{2}\right)^2\right] - 1$

$\qquad = 2(x^2 + 2x + 1) - 2(1) - 1$

$\qquad = (x + 2)^2 - 3$

Translate: 2 units left and 3 units down

Stretch vertically by a factor of 2.

23. $f(x)$

$\qquad = -3x^2 + 24x - 55$

$\qquad = -3(x^2 - 8x) - 55$

$\qquad = -3\left[x^2 - 8x + \left(-\frac{8}{2}\right)^2 - \left(-\frac{8}{2}\right)^2\right] - 55$

$\qquad = -3(x^2 - 8x + 16) - 3(-16) - 55$

$\qquad = -3(x - 4)^2 - 7$

Translate: 4 units right and 7 units down

Reflect: vertically

Stretch vertically by a factor of 3.

For Exercises 25–33, see the answer graphs in the textbook.

25. If $0 < a < 1$, the graph of $f(ax)$ will be flatter and wider than the graph of $f(x)$.
Multiplying x by a fraction makes the y-values less than the original y-values.

27. If $-1 < a < 0$, the graph of $f(ax)$ will be reflected horizontally, since a is negative. It will be flatter because multiplying x by a fraction decreases the corresponding y-values.

29. If $0 < a < 1$, the graph of af(x) will be flatter and wider than the graph of f(x). Each y-value is only a fraction of the height of the original y-values.

31. If $-1 < a < 0$, the graph will be reflected vertically, since a will be negative. Also, because a is a fraction, the graph will be flatter because each y-value will only be a fraction of its original height.

33. **(a)** If $f(x) = \frac{c}{x}$, then

$$af(x - h) + k = a\left[\frac{c}{x - h}\right] + k$$

$$= \frac{ac}{x - h} + k$$

$$= \frac{ac + kx - hk}{x - h}$$

(c) and

(d) See the answer graphs in the textbook.

35. $S(x) = \frac{1}{25}(x - 10)^3 + 40$

(a) $S(0) = \frac{1}{25}(0 - 10)^3 + 40$

$\qquad = \frac{1}{25}(-1000) + 40$

$\qquad = -40 + 40$

$\qquad = \$0$

(b) $S(10) = \frac{1}{25}(10 - 10)^3 + 40$

$\qquad = \frac{1}{25} \cdot 0 + 40$

$\qquad = 40$ or $\$40,000$

(c) See the answer graph in the textbook.

(d) $A(x) = \bar{S}(x)$

$$= \frac{S(x)}{x} = \frac{\frac{1}{25}(x-10)^3 + 40}{x}$$

(e) $A(10) = \dfrac{\frac{1}{25}(10-10)^3 + 40}{x}$

$$= \frac{0+40}{10} = \$4$$

(f) $A(50) = \dfrac{\frac{1}{25}(50-10)^3 + 40}{50}$

$$= \frac{2560 + 40}{50} = \$52$$

Chapter 1 Review Exercises

For Exercises 5–13, the domain is $\{-3, -2, -1, 0, 1, 2, 3\}$. See the graphs in the answer section of your textbook.

5. $2x - 5y = 10$

$\qquad -5y = -2x + 10$

$\qquad\quad y = \dfrac{2}{5}x - 2$

x	-3	-2	-1	0
y	-16/5	-14/5	-12/5	-2

x	1	2	3
y	-8/5	-6/5	-4/5

Pairs: $(-3, -16/5)$, $(-2, -14/5)$, $(-1, -12/5)$, $(0, -2)$, $(1, -8/5)$, $(2, -6/5)$, $(3, -4/5)$

Range: $\{-16/5, -14/5, -12/5, -2, -8/5, -6/5, -4/5\}$

7. $y = (2x + 1)(x - 1)$

$\qquad = 2x^2 - x - 1$

x	-3	-2	-1	0	1	2	3
y	20	9	2	-1	0	5	14

Pairs: $(-3, 20)$, $(-2, 9)$, $(-1, 2)$, $(0, -1)$, $(1, 0)$, $(2, 5)$, $(3, 14)$

Range: $\{-1, 0, 2, 5, 9, 14, 20\}$

9. $y = -2 + x^2$

x	-3	-2	-1	0	1	2	3
y	7	2	-1	-2	-1	2	7

Pairs: $(-3, 7)$, $(-2, 2)$, $(-1, -1)$, $(0, -2)$, $(1, -1)$, $(2, 2)$, $(3, 7)$

Range: $\{-2, -1, 2, 7\}$

11. $y = \dfrac{2}{x^2 + 1}$

x	-3	-2	-1	0	1	2	3
y	1/5	2/5	1	2	1	2/5	1/5

Pairs: $(-3, 1/5)$, $(-2, 2/5)$, $(-1, 1)$, $(0, 2)$, $(1, 1)$, $(2, 2/5)$, $(3, 1/5)$

Range: $\{1/5, 2/5, 1, 2\}$

13. $y + 1 = 0$

$\qquad\quad y = -1$

x	-3	-2	-1	0	1	2	3
y	-1	-1	-1	-1	-1	-1	-1

Pairs: (-3, -1), (-2, -1), (-1, -1),
 (0, -1), (1, -1), (2, -1),
 (3, -1)

Range: $\{-1\}$

15. $f(x) = 4x - 1$

   (a) $f(6) = 4(6) - 1$
            $= 24 - 1$
            $= 23$

   (b) $f(-2) = 4(-2) - 1$
            $= -8 - 1$
            $= -9$

   (c) $f(-4) = 4(-4) - 1$
            $= -16 - 1$
            $= -17$

   (d) $f(r + 1) = 4(r + 1) - 1$
              $= 4r + 4 - 1$
              $= 4r + 3$

17. $f(x) = -x^2 + 2x - 4$

   (a) $f(6) = -(6)^2 + 2(6) - 4$
           $= -36 + 12 - 4$
           $= -28$

   (b) $f(-2) = -(-2)^2 + 2(-2) - 4$
            $= -4 - 4 - 4$
            $= -12$

   (c) $f(-4) = -(-4)^2 + 2(-4) - 4$
           $= -16 - 8 - 4$
           $= -28$

   (d) $f(r + 1)$
      $= -(r + 1)^2 + 2(r + 1) - 4$
      $= -(r^2 + 2r + 1) + 2r + 2 - 4$
      $= -r^2 - 3$

19. $f(x) = 5x - 3$
    $g(x) = -x^2 + 4x$

   (a) $f(-2) = 5(-2) - 3$
            $= -10 - 3$
            $= -13$

   (b) $g(3) = -(3)^2 + 4(3)$
           $= -9 + 12$
           $= 3$

   (c) $g(-4) = -(-4)^2 + 4(-4)$
           $= -16 - 16$
           $= -32$

   (d) $f(5) = 5(5) - 3$
           $= 25 - 3$
           $= 22$

   (e) $g(-k) = -(-k)^2 + 4(-k)$
           $= -k^2 - 4k$

   (f) $g(3m) = -(3m)^2 + 4(3m)$
           $= -9m^2 + 12m$

   (g) $g(k - 5)$
      $= -(k - 5)^2 + 4(k - 5)$
      $= -(k^2 - 10k + 25) + 4k - 20$
      $= -k^2 + 10k - 25 + 4k - 20$
      $= -k^2 + 14k - 45$

   (h) $f(3 - p) = 5(3 - p) - 3$
              $= 15 - 5p - 3$
              $= 12 - 5p$

For Exercises 21–27, see the answer
graphs in your textbook.

21. $y = 4x + 3$

   Let $x = 0$.

        $y = 4(0) + 3$
        $y = 3$

Let y = 0.

$$0 = 4x + 3$$
$$-3 = 4x$$
$$-\frac{3}{4} = x$$

Draw the line through (0, 3) and (−3/4, 0).

23. 3x − 5y = 15

$$-5y = -3x + 15$$
$$y = \frac{3}{5}x + 3$$

When x = 0, y = −3.
When y = 0, x = 5.
Draw the line through (0, −3) and (5, 0).

25. x + 2 = 0

x = −2

This is the vertical line through (−2, 0).

27. y = 2x

When x = 0, y = 0.
When x = 1, y = 2.
Draw the line through (0, 0) and (1, 2).

29. Through (−2, 5) and (4, 7)

$$m = \frac{5 - 7}{-2 - 4}$$
$$= \frac{-2}{-6}$$
$$= \frac{1}{3}$$

31. Through the origin and (11, −2)

$$m = \frac{-2 - 0}{11 - 0} = -\frac{2}{11}$$

33. 2x + 3y = 15

$$3y = -2x + 15$$
$$y = -\frac{2}{3}x + 15$$
$$m = -\frac{2}{3}$$

35. x + 4 = 9

$$x = 9 - 4$$
$$x = 5$$

The line is vertical. The slope is not defined.

37. Through (5, −1), with slope $\frac{2}{3}$.

Use point−slope form.

$$y - (-1) = \frac{2}{3}(x - 5)$$
$$y + 1 = \frac{2}{3}(x - 5)$$
$$y = \frac{2}{3}x - \frac{10}{3} - 1$$
$$3y = 2x - 10 - 3$$
$$2x - 3y = 13$$

39. Through (5, −2) and (1, 3)

$$m = \frac{-2 - 3}{5 - 1} = \frac{-5}{4}$$

Use point−slope form.

$$y - (-2) = \frac{-5}{4}(x - 5)$$
$$y + 2 = \frac{-5}{4}x + \frac{25}{4}$$
$$4y + 8 = -5x + 25$$
$$5x + 4y = 17$$

41. Undefined slope, through (-1, 4)

Undefined slope means the line is vertical. The vertical line through (-1, 4) is x = -1.

43. Through (0, 5) and perpendicular to 8x + 5y = 3 Find the slope of the given line first.

$$8x + 5y = 3$$
$$5y = -8x + 3$$
$$y = \frac{-8}{5}x + \frac{3}{5}$$
$$m = -\frac{8}{5}$$

The perpendicular line has m = 5/8. Use point-slope form.

$$y - 5 = \frac{5}{8}(x - 0)$$
$$y = \frac{5}{8}x + 5$$
$$5x - 8y = -40$$

45. Through (3, -5), parallel to y = 4

Find the slope of the given line. y = 0x + 4, so m = 0, and the required line will also have slope 0.

Use the point-slope form.

$$y - (-5) = 0(x - 3)$$
$$y + 5 = 0$$
$$y = -5$$

Another method: Note y = 4 is a horizontal line. The required line then will be horizontal, and have an equation of the form y = b, that is, all points on the line have the same y-coordinate. Since (3, -5) is on the line, the y-coordinate is -5, and the equation of the line is y = -5.

For Exercises 47-69, see the answer graphs in the back of your textbook.

47. Through (2, -4), $m = \frac{3}{4}$

One point on the line is (2, -4)

Slope $= \frac{3}{4} = \frac{\Delta y}{\Delta x}$

$x_2 = x_1 + \Delta x = 2 + 4 = 6$
$y_2 = y_1 + \Delta y = -4 + 3 = -1$

A second point on the line is (6, -1).

Draw the line through these points.

49. Through (-4, -1), m = 3

One point on the line is (-4, 1).

Slope $= \frac{3}{1} = \frac{\Delta y}{\Delta x}$

$x_2 = x_1 + \Delta x = -4 + 1 = -3$
$y_2 = y_1 + \Delta y = 1 + 3 = 4$

A second point is (-3, 4).

Draw the line through these points.

51. $y = x^2 - 4$
$\quad = (x + 2)(x - 2)$

Set y = 0 to find the x-intercepts.

$0 = (x + 2)(x - 2)$
$x = -2$ or $x = 2$

Set x = 0 to find the y—intercept.

y = 0² − 4

y = −4

The graph is a parabola.

Vertex: $x = \dfrac{-b}{2a} = \dfrac{-0}{2} = 0$

y = 0² − 4 = −4

The vertex is (0, −4).

53. y = 3x² + 6x − 2

The graph is a parabola.
Let y = 0.

0 = 3x² + 6x − 2

$x = \dfrac{-6 \pm \sqrt{6^2 - 4(3)(-2)}}{2(3)}$

$= \dfrac{-6 \pm \sqrt{36 + 24}}{6} = \dfrac{-6 \pm \sqrt{60}}{6}$

$= \dfrac{-6 \pm 2\sqrt{15}}{6} = -1 \pm \dfrac{\sqrt{15}}{3}$

The x—intercepts are −1 + √15/3 ≈
.29 and −1 − √15/3 ≈ −2.29.
Let x = 0.

y = 3(0)² + 6(0) − 2

−2 is the y—intercept.

Vertex: $x = \dfrac{-b}{2a} = \dfrac{-6}{2(3)} = -1$

y = 3(−1)² + 6(−1) − 2

= 3 − 6 − 2

= −5

The vertex is (−1, −5).

55. y = x² − 4x + 2

Let y = 0.

0 = x² − 4x + 2

$x = \dfrac{-(-4) \pm \sqrt{(-4)^2 - 4(1)(2)}}{2(1)}$

$= \dfrac{4 \pm \sqrt{8}}{2}$

$= 2 \pm \sqrt{2}$

The x—intercepts are 2 + √2 ≈ 3.41
and 2 − √2 ≈ −1.41.
Let x = 0.

y = 0² − 4(0) + 2

2 is the y—intercept.

Vertex: $x = \dfrac{-b}{2a} = \dfrac{-(-4)}{2(1)} = \dfrac{4}{2} = 2$

y = 2² − 4(2) + 2 = −2

The vertex is (2, −2).

57. f(x) = x³ + 5

Translate the graph of f(x) = x³
5 units up.

59. y = −(x − 1)³ + 4

Translate the graph of y = x³
1 unit right and 4 units up.
Reflect vertically.

61. y = 2√x + 3 + 1

Translate the graph of y = √x
3 units left and 1 unit up. Stretch
the graph vertically by a factor
of 2.

63. $f(x) = \dfrac{8}{x}$

Vertical asymptote:

$x = 0$

Horizontal asymptote:

$\dfrac{8}{x}$ approaches zero as x gets larger.

$y = 0$ is an asymptote.

x	−4	−3	−2	−1	1	2	3	4
y	−2	−2.7	−4	−8	8	4	2.7	2

65. $f(x) = \dfrac{4x - 2}{3x + 1}$

Vertical asymptote:

$$3x + 1 = 0$$
$$x = -\frac{1}{3}$$

Horizontal asymptote:

As x get larger,

$$\frac{4x - 2}{3x - 1} \approx \frac{4x}{3x} = \frac{4}{3}.$$

$y = \dfrac{4}{3}$ is an asymptote.

x	−3	−2	−1	0	1	2	3
y	1.75	2	3	−2	.5	.86	1

67. Exercise 53 is

$$y = 3x^2 + 6x - 2.$$

Completing the square produces

$$y = 3(x + 1)^2 - 5.$$

Translate $y = x^2$ 1 unit left and 5 units down. Stretch vertically by a factor of 3.

69. Exercise 55 is

$$y = x^2 - 4x + 3.$$

Completing the square produces

$$y = (x - 2)^2 - 2.$$

Translate $y = x^2$ 2 units right and 2 units down.

71. Supply: $p = q + 3$
Demand: $p = 19 - 2q$

(a)

$10 = 6q + 3$	$10 = 19 - 2q$
$7 = 6q$	$2q = 19 - 10$
	$2q = 9$
$\dfrac{7}{6} = q$	$q = \dfrac{9}{2}$

Supply: 7/6   Demand: 9/2

(b)

$15 = 6q + 3$	$15 = 19 - 2q$
$12 = 6q$	$2q = 19 - 15$
$2 = q$	$2q = 4$
	$q = 2$

Supply: 2   Demand: 2

(c)

$18 = 6q + 3$	$18 = 19 - 2q$
$15 = 6q$	$2q = 19 - 18$
$\dfrac{15}{6} = q$	$2q = 1$
$\dfrac{5}{2} = q$	$q = \dfrac{1}{2}$

Supply: 5/2   Demand: 1/2

(d) See the answer graphs in the back of your textbook.

(e) The equilibrium price is 15 (refer to (b) above or read graph).

(f) The equilibrium quantity is 2 (refer to (b) above or read graph).

73. Eight units cost $300; fixed cost is $60.

The fixed cost is the cost if zero units are made. (8, 300) and (0, 60) are points on the line.

$$m = \frac{60 - 300}{0 - 8} = 30$$

Use slope–intercept form.

$$y = 30x + 60$$
$$C(x) = 30x + 60$$

$$A(x) = \frac{C(x)}{x} = \frac{30x + 60}{x} = 30 + \frac{60}{x}$$

Thirty units cost $1500; 120 units cost $5640.

75. Thirty units cost $1500; 120 units cost $5640.

Points on the line are (30, 1500), (120, 5640).

$$m = \frac{5640 - 1500}{120 - 30}$$

$$= \frac{4140}{90}$$

$$= 46$$

Use point–slope form.

$$y - 1500 = 46(x - 30)$$
$$y = 46x - 1380 + 1500$$
$$y = 46x + 120$$
$$C(x) = 46x + 120$$

$$A(x) = \frac{C(x)}{x} = \frac{46x + 120}{x}$$

$$= 46 + \frac{120}{x}$$

77. (a) For x in the interval $0 < x \le 1$, the renter is charged the fixed cost of $40 and 1 days rent of $40 so

$$C\left(\tfrac{3}{4}\right) = \$40 + \$40(1)$$

$$= \$40 + \$40$$

$$= \$80.$$

(b) $C\left(\tfrac{9}{10}\right) = \$40 + \$40(1)$

$$= \$40 + \$40$$

$$= \$80$$

(c) $C(1) = \$40 + \$40(1)$

$$= \$40 + \$40$$

$$= \$80$$

(d) For x in the interval $1 < x \le 2$, the renter is charged the fixed cost of $40 and 2 days rent of $80. So

$$C\left(1\tfrac{5}{8}\right) = \$40 + \$40(2)$$

$$= \$40 + \$80$$

$$= \$120.$$

(e) For x in the interval $2 < x \le 3$ the renter is charged the fixed cost of $40 and 3 days rent of $120. So

$$C\left(2\tfrac{1}{9}\right) = \$40 + \$40(3)$$

$$= \$40 + \$120$$

$$= \$160.$$

(f) See the answer graph in text-book.

(g) The independent variable is the number of days, or x.

(h) The dependent variable is the cost, or C(x).

79. **(a)** As seen on the graph, the rate was the highest halfway through 1984.

(b) As seen on the graph, the maximum rate was 160,000.

(c) The range of the function is the interval of possible y–values or [0, 160,000].

81. **(a)** The dosages are equal when

$$\frac{12}{A + 12} = \frac{A + 1}{24}$$

$$24A = (A + 12)(A + 1)$$

$$= A^2 + 13A + 12$$

$$0 = A^2 - 11A + 12$$

$$A = \frac{11 \pm \sqrt{121 - 48}}{2}$$

$$A = \frac{11 - \sqrt{72}}{2} \quad \text{or} \quad A = \frac{11 + \sqrt{72}}{2}$$

$$\approx 1.2 \qquad\qquad \approx 9.8.$$

The two dosages are at approximately 1.2 years and 9.8 years.

(b) See the answer graph in the back of your textbook.

$f(A) > g(A)$ for $1.2 < A < 9.8$

$g(A) > f(A)$ for $A < 1.2$ or $A > 9.8$.

In the interval shown in the graph,

$f(A) > g(A)$ for $2 < A < 9.8$

$g(A) > f(A)$ for $9.8 < A < 13$.

(c) The functions f and g seem to differ most at 5 yr and 13 yr.

83. **(a)** From the graph, the maximum horsepower occurs at 5750 rpm.

(b) From the graph, the maximum horsepower is about 310.

(c) When the horsepower is at its maximum, the graph shows the corresponding torque as 280 ft–lb.

85. $C(x) = \dfrac{5x + 3}{x + 1}$

(a) See the answer graph in the textbook.

(b)

$C(x + 1) - C(x)$

$$= \frac{5(x + 1) + 3}{(x + 1) + 1} - \frac{5x + 3}{x + 1}$$

$$= \frac{5x + 8}{x + 2} - \frac{5x + 3}{x + 1}$$

$$= \frac{(5x + 8)(x + 1) - (5x + 3)(x + 2)}{(x + 2)(x + 1)}$$

$$= \frac{5x^2 + 13x + 8 - 5x^2 - 13x - 6}{(x + 2)(x + 1)}$$

$$= \frac{2}{(x + 2)(x + 1)}$$

(c) $A(x) = \dfrac{C(x)}{x} = \dfrac{\dfrac{5x + 3}{x + 1}}{x}$

$$= \frac{5x + 3}{x(x + 1)}$$

(d)

$A(x + 1) - A(x)$

$$= \frac{5(x + 1) + 3}{(x + 1)[(x + 1) + 1]} - \frac{5x + 3}{x(x + 1)}$$

$$= \frac{5x + 8}{(x + 1)(x + 2)} - \frac{5x + 3}{x(x + 1)}$$

$$= \frac{x(5x + 8) - (5x + 3)(x + 2)}{x(x + 1)(x + 2)}$$

$$= \frac{5x^2 + 8x - 5x^2 - 13x - 6}{x(x + 1)(x + 2)}$$

$$= \frac{-5x - 6}{x(x + 1)(x + 2)}$$

Extended Application

1. $y = .133x + 10.09$, x in millions of units. Selling price is $10.73. Then

    $$10.73 = .133x + 10.09$$
    $$.64 = .133x$$
    $$4.8 \approx x.$$

    So the company has marginal revenue = marginal cost when it makes 4.8 million units.

2. For product B,

    $$y = .0667x + 10.29.$$

    When $x = 3.1$, $y = 10.50$.
    When $x = 5.7$, $y = 10.67$.
    The graph is the line segment connecting the points

    $$(3.1, 10.5) \quad \text{and} \quad (5.7, 10.67).$$

    See the graph in the answer section at the back of your textbook.

3. For product B,

    $$y = .0667x + 10.29,$$

    which is always ≥ 10.29 for $x \geq 0$. The marginal revenue is only 9.65. Therefore, there is no production level possible where the marginal cost = marginal revenue.

4. For product C, the marginal cost y is

    $$y = .133x + 9.46.$$

    (a) At 3.1 million units,

    $$y = .133(3.1) + 9.46$$
    $$= \$9.87.$$

    At 5.7 million units,

    $$y = .133(5.7) + 9.46$$
    $$= \$10.22.$$

    (b) The marginal cost graph is the portion of a line connecting the points (3.1, 9.87) and (57, 10.22).

    See the graph in the answer section at the back of your textbook.

    (c) At a selling price of $9.57, the cost equals the selling price when

    $$9.57 = .133x + 9.46$$
    $$x = .83 \quad \text{million units,}$$

    which is not in the interval under discussion.

CHAPTER 1 TEST

[1.1]   **1.** List the ordered pairs obtained from each of the following if the domain of x for each exercise is $\{-3, 2, -1, 0, 1, 2, 3\}$. Graph each set of ordered pairs. Give the range.

   (a) $2x + 3y = 6$ **(b)** $y = \dfrac{1}{x^2 - 2}$

[1.1]   **2.** Let $f(x) = 3x - 4$ and $g(x) = -x^2 = 5x$. Find each of the following.

   (a) $f(-3)$ **(b)** $g(-2)$ **(c)** $f(2m)$ **(d)** $g(k - 1)$

   (e) $f(x + h)$

[1.2]   **Graph each of the following.**

   3. $3x + 5y = 15$ **4.** $2x + y = 0$ **5.** $x - 3 = 0$

   6. The line through $(-2, 3)$, $m = -3$

[1.2]   **In Exercises 7–9, find the slope of each line for which the slope is defined.**

   7. Through $(2, -5)$ and $(-1, 7)$ **8.** $3x - 7y = 9$

   9. Through $(9, 5)$ and $(9, 2)$

[1.2]   **In Exercises 10–14, find an equation in the form ax + by = c for each line.**

   10. Through the origin and $(5, -3)$

   11. Undefined slope through $(1, -2)$

   12. Through $(0, 3)$, parallel to $2x - 4y = 1$

   13. Through $(1, -4)$, perpendicular to $3x + y = 1$

   14. Through $(2, 5)$, perpendicular to $x = 5$

[1.2]   15.   Let the supply and demand functions for a certain product be given by the following equations.

Supply: $p = .20q - 5$ Demand: $p = 100 - .15q$

where p represents the price at a supply or demand, respectively, of q units.

(a)   Graph these equations on the same axes.

(b)   Find the equilibrium price.

(c)   Find the equilibrium quantity.

[1.3]   16.   (a)   Find the appropriate linear cost function if eight units cost $450, while forty units cost $770.

(b)   What is the fixed cost?

(c)   What is the marginal cost per item?

[1.4]   17.   The manufacturer of a certain product has determined that his profit in dollars for making x units of this product is given by the equation

$$p = -2x^2 + 120x + 3000.$$

(a)   Find the number of units that will maximize the profit.

(b)   What is the maximum profit for making this product?

Graph each of the following.

[1.4]   18. $y = 2x^2 - 4x - 2$ [1.5]   19. $f(x) = x^3 - 2x^2 - x + 2$

[1.5]   20. $f(x) = \dfrac{3}{2x - 1}$ [1.5]   21. $f(x) = \dfrac{x - 1}{3x + 6}$

[1.5]   22. $f(x) = \dfrac{2x}{(x + 1)(x - 2)}$

[1.5]   23.   Suppose a cost–benefit model is given by the equation

$$y = \frac{20x}{110 - x}$$

where y is the cost in thousands of dollars of removing x percent of a certain pollutant. Find the cost in thousands of dollars of removing each of the following percents of pollution.

(a)   70%   (b)   90%   (c)   100%

Use translations and reflections to graph the following.

[1.6]   24. $y = 4 - x^2$ [1.6]   25. $f(x) = \frac{1}{2}x^3 + 1$

CHAPTER 1   TEST ANSWERS

1. **(a)** (−3, 4), (−2, 10/3), (−1, 8/3),
   (0, 2), (1, 4/3), (2, 2/3), (3, 0);
   range: $\{0, 2/3, 4/3, 2, 8/3, 10/3, 4\}$

   (b) (−3, 1/7), (−2, 1/2), (−1, −1),
   (0, −1/2), (1, −1), (2, 1/2), (3, 1/7);
   range: $\{-1, -1/2, 1/7, 1/2\}$

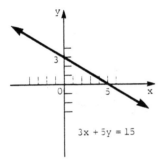

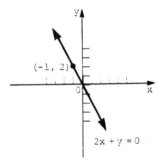

2. **(a)** −13   **(b)** −14   **(c)** 6m − 4   **(d)** $-k^2 + 7k - 6$ **(e)** 3x + 3h − 4

3.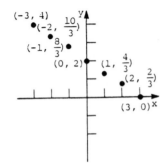

4.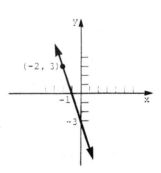

5.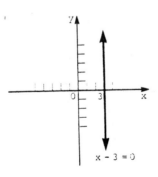

6.

7. −4   **8.** 3/7   **9.** Undefined   **10.** 3x + 5y = 0   **11.** x = 1

12. x − 2y = −6   **13.** x − 3y = 13   **14.** y = 5

15. **(a)**   **(b)** 55   **(c)** 300

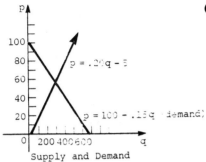

Supply and Demand

16. **(a)** y = 10x + 370   **(b)** $370   **(c)** $10

17. **(a)** 30   **(b)** $4800

18.

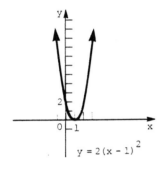

$y = 2(x - 1)^2$

19.

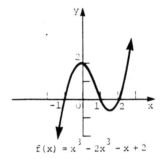

$f(x) = x^3 - 2x^3 - x + 2$

20.

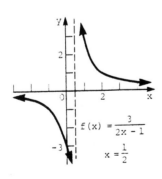

$f(x) = \dfrac{3}{2x - 1}$

$x = \dfrac{1}{2}$

21.

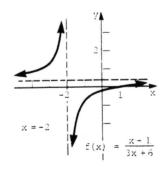

$x = -2$

$f(x) = \dfrac{x - 1}{3x + 6}$

22.

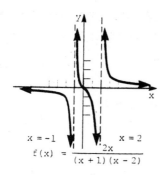

$$f(x) = \frac{2x}{(x + 1)(x - 2)}$$

$x = -1$ $x = 2$

23. **(a)** $35,000

(b) $90,000

(c) $200,000

24.

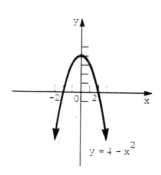

$$y = 4 - x^2$$

25.

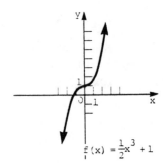

$$f(x) = \frac{1}{2}x^3 + 1$$

CHAPTER 2   THE DERIVATIVE

Section 2.1

1. By reading the graph, as x gets closer to 3 from the left or the right, f(x) gets closer to 3.

$$\lim_{x \to 3} f(x) = 3$$

3. By reading the graph, as x gets closer to -2 from the left, f(x) approaches -1. As x gets closer to -2 from the right, f(x) approaches 1/2. Since these two values of f(x) are not equal,

$$\lim_{x \to -2} f(x)$$

does not exist.

5. By reading the graph, as x approaches 1 from the left or the right, g(x) approaches 1.

$$\lim_{x \to 1} g(x) = 1$$

9. From the table, as x approaches 1 from the left or the right, f(x) approaches 2.

$$\lim_{x \to 1} f(x) = 2$$

11. $k(x) = \dfrac{x^3 - 2x - 4}{x - 2}$, find $\lim_{x \to 2} k(x)$.

x	1.9	1.99	1.999
k(x)	9.41	9.9401	9.9941

x	2.001	2.01	2.1
k(x)	10.006	10.0601	10.61

As x approaches 2 from the left or the right, k(x) approaches 10.

$$\lim_{x \to 2} k(x) = 10$$

13. $h(x) = \dfrac{\sqrt{x} - 2}{x - 1}$, find $\lim_{x \to 1} h(x)$.

x	.9	.99	.999
h(x)	-29.486833	-299.49874	-2999.4999

x	1.001	1.01	1.1
h(x)	3000.49988	300.49876	30.488088

$$\lim_{x \to 1^-} = -\infty$$

$$\lim_{x \to 1^+} = \infty$$

Thus, $\lim_{x \to 1} h(x)$ does not exist.

In Exercises 15–23, $\lim_{x \to 4} f(x) = 16$ and $\lim_{x \to 4} g(x) = 8$.

15. $\lim_{x \to 4} [f(x) - g(x)]$

$$= \lim_{x \to 4} f(x) - \lim_{x \to 4} g(x)$$

$$= 16 - 8 = 8$$

17. $\lim_{x \to 4} \dfrac{f(x)}{g(x)}$

$$= \dfrac{\lim_{x \to 4} f(x)}{\lim_{x \to 4} g(x)}$$

$$= \dfrac{16}{8} = 2$$

19. $\lim\limits_{x \to 4} \sqrt{f(x)}$

$= \lim\limits_{x \to 4} [f(x)]^{1/2}$

$= [\lim\limits_{x \to 4} f(x)]^{1/2}$

$= 16^{1/2} = 4$

21. $\lim\limits_{x \to 4} [g(x)]^3$

$= [\lim\limits_{x \to 4} g(x)]^3$

$= 8^3 = 512$

23. $\lim\limits_{x \to 4} \dfrac{f(x) + g(x)}{2g(x)}$

$= \dfrac{\lim\limits_{x \to 4} [f(x) + g(x)]}{\lim\limits_{x \to 4} 2g(x)}$

$= \dfrac{\lim\limits_{x \to 4} f(x) + \lim\limits_{x \to 4} g(x)}{2 \lim\limits_{x \to 4} g(x)}$

$= \dfrac{16 + 8}{2(8)} = \dfrac{24}{16} = \dfrac{3}{2}$

25. $\lim\limits_{x \to 3} \dfrac{x^2 - 9}{x - 3}$

$= \lim\limits_{x \to 3} \dfrac{(x - 3)(x + 3)}{x - 3}$

$= \lim\limits_{x \to 3} (x + 3)$

$= \lim\limits_{x \to 3} x + \lim\limits_{x \to 3} 3$

$= 3 + 3$

$= 6$

27. $\lim\limits_{x \to -2} \dfrac{x^2 - x - 6}{x + 2}$

$= \lim\limits_{x \to -2} \dfrac{(x - 3)(x + 2)}{x + 2}$

$= \lim\limits_{x \to -2} (x - 3)$

$= \lim\limits_{x \to -2} x + \lim\limits_{x \to -2} (-3)$

$= -2 - 3$

$= -5$

29. $\lim\limits_{x \to 0} \dfrac{x^3 - 4x^2 + 8x}{2x}$

$= \lim\limits_{x \to 0} \dfrac{x(x^2 - 4x + 8)}{2x}$

$= \lim\limits_{x \to 0} \dfrac{x^2 - 4x + 8}{2}$

$= \dfrac{0^2 - 4(0) + 8}{2} = 4$

31. $\lim\limits_{x \to 0} \dfrac{\dfrac{1}{x + 3} - \dfrac{1}{3}}{x}$

$= \lim\limits_{x \to 0} \left(\dfrac{1}{x + 3} - \dfrac{1}{3}\right)\left(\dfrac{1}{x}\right)$

$= \lim\limits_{x \to 0} \left[\dfrac{3}{3(x + 3)} - \dfrac{x + 3}{3(x + 3)}\right]\left(\dfrac{1}{x}\right)$

$= \lim\limits_{x \to 0} \dfrac{3 - x - 3}{3(x + 3)(x)}$

$= \lim\limits_{x \to 0} \dfrac{-x}{3(x + 3)x}$

$= \lim\limits_{x \to 0} \dfrac{-1}{3(x + 3)}$

$= \dfrac{-1}{3(0 + 3)}$

$= -\dfrac{1}{9}$

33. $\lim\limits_{x \to 25} \dfrac{\sqrt{x} - 5}{x - 25}$

$= \lim\limits_{x \to 25} \dfrac{\sqrt{x} - 5}{x - 25} \cdot \dfrac{\sqrt{x} + 5}{\sqrt{x} + 5}$

$= \lim\limits_{x \to 25} \dfrac{x - 25}{(x - 25)(\sqrt{x} + 5)}$

$= \lim\limits_{x \to 25} \dfrac{1}{\sqrt{x} + 5}$

$= \dfrac{1}{\sqrt{25} + 5}$

$= \dfrac{1}{10}$

35. $\lim\limits_{h \to 0} \dfrac{(x + h)^2 - x^2}{h}$

$= \lim\limits_{h \to 0} \dfrac{x^2 + 2hx + h^2 - x^2}{h}$

$= \lim\limits_{h \to 0} \dfrac{2hx + h^2}{h}$

$= \lim\limits_{h \to 0} \dfrac{h(2x + h)}{h}$

$= \lim\limits_{h \to 0} (2x + h)$

$= 2x + 0$

$= 2x$

37. $F(x) = \dfrac{3x}{(x + 2)^3}$

(a)

x	-3	-2.1	-2.01
F(x)	9	6300	6,030,000

x	-1.99	-1.9	-1
F(x)	-5,970,000	-5700	-3

$\lim\limits_{x \to -2^-} F(x) = \infty$

$\lim\limits_{x \to -2^+} F(x) = -\infty$

Thus, $\lim\limits_{x \to -2} F(x)$ does not exist.

(b) The vertical asymptote occurs when $(x + 2)^3 = 0$, or $x = -2$.

(c) If $x = a$ is an asymptote for the graph of $f(x)$, then $\lim\limits_{x \to a} f(x)$ does not exist.

39. By reading the graph, the function is discontinuous at $x = -1$.
$f(-1)$ does not exist, since there is no value shown on the graph for $x = -1$.

$\lim\limits_{x \to -1} f(x) = \dfrac{1}{2}$ because as x approaches -1 from the left or the right, $f(x)$ approaches $\dfrac{1}{2}$.

41. By reading the graph, the function is discontinuous at $x = 1$.
$f(1) = 2$, since the heavy dot above $x = 1$ is at 2.
$\lim\limits_{x \to 1} f(x) = -2$ because as x approaches 1 from the left or the right, $f(x)$ approaches -2.

43. By reading the graph, the function is discontinuous at $x = -5$ and $x = 0$.
$f(-5)$ does not exist since $x = -5$ is a vertical asymptote.
$f(0)$ does not exist since there is no value shown on the graph for $x = 0$.
$\lim\limits_{x \to -5} f(x)$ does not exist since as x approaches -5 from the left, $f(x)$ becomes infinitely large and as x approaches -5 from the right, $f(x)$ becomes infinitely small.
$\lim\limits_{x \to 0} f(x) = 0$ because as x approaches 0 from the left or the right, $f(x)$ approaches 0.

45. $g(x) = \dfrac{1}{x(x - 2)};$

$x = 0, \; x = 2, \; x = 4$
Since g is a rational function, it will not be continuous when $x(x - 2) = 0$. Therefore, g is not

continuous for x = 0 or x = 2, but is continuous at all other points, including x = 4.

47. $k(x) = \dfrac{5 + x}{2 + x}$;

x = 0, x = -2, x = -5
Since k is a rational function, it will not be continuous when 2 + x = 0. Therefore, k is not continuous for x = -2, but is continuous at all other points, including x = 0 and x = -5.

49. $g(x) = \dfrac{x^2 - 4}{x - 2}$;

x = 0, x = 2, x = -2
Since g is a rational function, it will not be continuous when x - 2 = 0. Therefore, g is not continuous for x = 2, but is continuous at all other points, including x = 0 and x = -2.

51. $p(x) = x^2 - 4x + 11$;

x = 0, x = 2, x = -1
Since p is a polynomial function, it is continuous for all real values of x, including x = 0, x = 2, and x = -1.

53. $p(x) = \dfrac{|x + 2|}{x + 2}$; x = -2, x = 0, x = 2

Since p is a rational function, it is not continuous at x = -2, but is continuous at all other points, including x = 0 and x = 2.

55. (a) $\lim\limits_{t \to 12} G(t)$

As t approaches 12, the value of G(t) for the corresponding point on the graph approaches 3. Thus, $\lim\limits_{t \to 12} G(t) = 3$.

(b) $\lim\limits_{t \to 16} G(t)$

$\lim\limits_{t \to 16^+} G(t) = 1.5$

$\lim\limits_{t \to 16^-} G(t) = 2$

Since $\lim\limits_{t \to 16^+} G(t) \neq \lim\limits_{t \to 16^-} G(t)$,

$\lim\limits_{t \to 16} G(t)$ does not exist.

(c) G(16) is the value of function G(t) when t = 16. This value occurs at the heavy dot on the graph.

$$G(16) = 2$$

(d) The tipping point occurs at the break in the graph or when t = 16 months.

57. In dollars,

C(x) = 4x if 0 < x ≤ 150
C(x) = 3x if 150 < x ≤ 400
C(x) = 2.5x if 400 < x.

(a) C(130) = 4(130) = $520

(b) C(150) = 4(150) = $600

(c) C(210) = 3(210) = $630

(d) C(400) = 3(400) = $1200

(e) C(500) = 2.5(500) = $1250

(f) C is discontinuous at x = 150 and x = 400 because those represent points of price change.

59. In dollars,

$C(t) = 30t$ if $0 < t \le 5$

$C(t) = 30(5) = 150$ if $t = 6$ or $t = 7$

$C(t) = 150 + 30(t - 7)$ if $7 < t \le 12$.

The average cost per day is

$$A(t) = \frac{C(t)}{t}.$$

(a) $A(4) = \frac{30(4)}{4} = \$30$

(b) $A(5) = \frac{30(5)}{5} = \$30$

(c) $A(6) = \frac{150}{6} = \$25$

(d) $A(7) = \frac{150}{7} \approx \$21.43$

(e) $A(8) = \frac{150 + 30(8 - 7)}{8}$

$= \frac{180}{8} = \$22.50$

(f) $\lim\limits_{x \to 5^-} A(t) = 30$ because as t

approaches 5 from the left, $A(t)$
approaches 30 (think of the
graph for $t = 1, 2, \ldots, 5$).

(g) $\lim\limits_{x \to 5^+} A(t) = 25$ because as t

approaches 5 from the right,
$A(t)$ approaches 25.

(h) A is discontinuous at $t = 5$,
$t = 6$, $t = 7$, and so on, because
the average cost will differ for
each different rental length.

61. The function is discontinuous at
$t = m$, since the break in the
graph occurs at that point.

Section 2.2

1. $y = x^2 + 2x = f(x)$ between $x = 0$ and

$x = 3$

Average rate of change

$= \frac{f(3) - f(0)}{3 - 0}$

$= \frac{15 - 0}{3}$

$= 5$

3. $y = 2x^3 - 4x^2 + 6x = f(x)$ between

$x = -1$ and $x = 1$

Average rate of change

$= \frac{f(1) - f(-1)}{1 - (-1)}$

$= \frac{4 - (-12)}{2} = \frac{16}{2}$

$= 8$

5. $y = \sqrt{x} = f(x)$ between $x = 1$ and

$x = 4$

Average rate of change

$= \frac{f(4) - f(1)}{4 - 1}$

$= \frac{2 - 1}{3}$

$= \frac{1}{3}$

7. $y = \frac{1}{x - 1} = f(x)$ between $x = -2$ and

$x = 0$

Average rate of change

$= \frac{f(0) - f(-2)}{0 - (-2)}$

$= \frac{-1 - \left(-\frac{1}{3}\right)}{2} = \frac{-\frac{2}{3}}{2}$

$= -\frac{1}{3}$

For Exercises 9 and 11, $s(t) = t^2 + 5t + 2$.

9. $\lim\limits_{h \to 0} \dfrac{s(6 + h) - s(6)}{h}$

$= \lim\limits_{h \to 0} \dfrac{(6+h)^2 + 5(6+h) + 2 - [6^2 + 5(6) + 2]}{h}$

$= \lim\limits_{h \to 0} \dfrac{h^2 + 17h + 68 - 68}{h}$

$= \lim\limits_{h \to 0} \dfrac{h^2 + 17h}{h}$

$= \lim\limits_{h \to 0} \dfrac{h(h + 17)}{h}$

$= \lim\limits_{h \to 0} (h + 17)$

$= 17$

11. $\lim\limits_{h \to 0} \dfrac{s(10 + h) - s(10)}{h}$

$= \lim\limits_{h \to 0} \dfrac{(10+h)^2 + 5(10+h) + 2 - [100 + 50 + 2]}{h}$

$= \lim\limits_{h \to 0} \dfrac{h^2 + 25h + 102 - 102}{h}$

$= \lim\limits_{h \to 0} \dfrac{h(h + 25)}{h}$

$= \lim\limits_{h \to 0} (h + 25)$

$= 25$

13. $s(t) = t^3 + 2t + 9$

$\lim\limits_{h \to 0} \dfrac{s(4 + h) - s(4)}{h}$

$= \lim\limits_{h \to 0} \dfrac{(4+h)^3 + 2(4+h) + 9 - [4^3 + 2(4) + 9]}{h}$

$= \lim\limits_{h \to 0} \dfrac{4^3 + 3(4^2)h + 3(4h^2) + h^3 + 8 + 2h + 9 - (4^3 + 8 + 9)}{h}$

$= \lim\limits_{h \to 0} \dfrac{h^3 + 36h^2 + 48h + 2h}{h}$

$= \lim\limits_{h \to 0} \dfrac{h^3 + 36h^2 + 50h}{h}$

$= \lim\limits_{h \to 0} \dfrac{h(h^2 + 36h + 50)}{h}$

$= \lim\limits_{h \to 0} (h^2 + 36h + 50)$

$= 50$

15. $s(t) = -4t^2 - 6$ at $t = 2$

$\lim\limits_{h \to 0} \dfrac{s(2 + h) - s(2)}{h}$

$= \lim\limits_{h \to 0} \dfrac{-4(2 + h)^2 - 6 - [-4(2)^2 - 6]}{h}$

$= \lim\limits_{h \to 0} \dfrac{-4(4 + 4h + h^2) - 6 + 16 + 6}{h}$

$= \lim\limits_{h \to 0} \dfrac{-16h - 4h^2}{h}$

$= \lim\limits_{h \to 0} \dfrac{h(-16 - 4h)}{h}$

$= \lim\limits_{h \to 0} (-16 - 4h)$

$= -16$

17. $F(x) = x^2 + 2$ at $x = 0$

$\lim\limits_{h \to 0} \dfrac{F(0 + h) - F(0)}{h}$

$= \lim\limits_{h \to 0} \dfrac{(0 + h)^2 + 2 - [0^2 + 2]}{h}$

$= \lim\limits_{h \to 0} \dfrac{h^2}{h}$

$= \lim\limits_{h \to 0} h$

$= 0$

19. The instantaneous rate of change is positive when f is increasing.

21. (a) $s(4) = 10$

     $s(1) = 1$

     Average rate of change

       $= \dfrac{s(4) - s(1)}{4 - 1}$

       $= \dfrac{10 - 1}{4 - 1}$

       $= \dfrac{9}{3} = 3$

   (b) $s(7) = 10$

       $s(4) = 10$

Average rate of change

$$= \frac{s(7) - s(4)}{7 - 4}$$

$$= \frac{10 - 10}{7 - 4}$$

$$= \frac{0}{3} = 0$$

(c) $s(12) = 1$

$s(7) = 10$

Average rate of change

$$= \frac{s(12) - s(7)}{12 - 7}$$

$$= \frac{1 - 10}{12 - 7} = -\frac{9}{5}$$

(d) Sales increase during the first four years, then stay constant until year 7, then decrease.

(e) Many answers are possible; usually this trend occurs with high-tech products or products that mirror trends, such as Walkman radios.

23. (a) $P(1987) = 11$

$P(1985) = 14$

Average rate of change

$$= \frac{P(1987) - P(1985)}{1987 - 1985}$$

$$= \frac{11 - 14}{2}$$

$$= -\frac{3}{2}$$

(b) $P(1988) = 10$

$P(1987) = 11$

Average rate of change

$$= \frac{P(1988) - P(1987)}{1988 - 1987}$$

$$= \frac{10 - 11}{1}$$

$$= -1$$

(c) $P(1989) = 11$

$P(1988) = 10$

Average rate of change

$$= \frac{P(1989) - P(1988)}{1989 - 1988}$$

$$= \frac{11 - 10}{1} = 1$$

(d) $P(1990) = 10.5$

$P(1989) = 11$

Average rate of change

$$= \frac{P(1990) - P(1989)}{1990 - 1989}$$

$$= \frac{10.5 - 11}{1} = -\frac{1}{2}$$

(e) The profit margin increased when the average rate of change was positive, from 1988 to 1989.

(f) The profit margin stabilized after declining from 1985 to 1987.

25.

$$f'(x) = \lim_{h \to 0} (12x + 6h - 4)$$

h	1	.1	.01	.001	.0001
4 + h	5	4.1	4.01	4.001	4.0001
P(4 + h)	31	19.12	18.1102	18.011	18.0011
P(4)	18	18	18	18	18
P(4 + h) − P(4)	13	1.12	.110212	.0109959	.00110817
$\frac{P(4 + h) - P(4)}{h}$	13	11.2	11.0212	10.9959	11.0817

By the table, it appears that

$$\lim_{h \to 0} \frac{P(4 + h) - P(4)}{h} = 11.$$

By the rule for limits, we have

$$\lim_{h \to 0} \frac{P(4 + h) - P(4)}{h}$$

$$= \lim_{h \to 0} \frac{2(4 + h)^2 - 5(4 + h) + 6 - 18}{h}$$

$$= \lim_{h \to 0} \frac{2h^2 + 11h + 18 - 18}{h}$$

$$= \lim_{h \to 0} \frac{2h^2 + 11h}{h}$$

$$= \lim_{h \to 0} 2h + 11$$

$$= 11$$

27. $N(p) = 80 - 5p^2$, $1 \le p \le 4$

(a) Average rate of change of demand is

$$\frac{N(3) - N(2)}{3 - 2}$$

$$= \frac{35 - 60}{1}$$

$$= -25 \text{ boxes per dollar.}$$

(b) Instantaneous rate of change when p is 2 is

$$\lim_{h \to 0} \frac{N(2 + h) - N(2)}{h}$$

$$= \lim_{h \to 0} \frac{80 - 5(2 + h)^2 - [80 - 5(2)^2]}{h}$$

$$= \lim_{h \to 0} \frac{80 - 20 - 20h - 5h^2 - (80 - 20)}{h}$$

$$= \lim_{h \to 0} \frac{-5h^2 - 20h}{h}$$

$$= -20 \text{ boxes per dollar.}$$

(c) Instantaneous rate of change when p is 3 is

$$\lim_{h \to 0} \frac{80 - 5(3 + h)^2 - [80 - 5(3)^2]}{h}$$

$$= \lim_{h \to 0} \frac{80 - 45 - 30h - 5h^2 - 80 + 45}{h}$$

$$= \lim_{h \to 0} \frac{-30h - 5h^2}{h}$$

$$= -30 \text{ boxes per dollar.}$$

(d) As the price increases, the demand decreases; this is an expected change since a higher price usually reduces demand.

29. C(1988) = 36, C(1987) = 25

(a) $\frac{C(1988) - C(1987)}{1988 - 1987}$

$$= \frac{36 - 25}{1}$$

$$= 11$$

The average rate of change was $11 million.

C(1989) = 35, C(1988) = 36

$$\frac{C(1989) - C(1988)}{1989 - 1988}$$

$$= \frac{35 - 36}{1}$$

$$= -1$$

The average rate of change was −$1 million.

(b) $\frac{Cr(1988) - Cr(1987)}{1988 - 1987}$

$$= \frac{1 - 2}{1}$$

$$= -1$$

The average rate of change was −$1 million.

$$\frac{Cr(1989) - Cr(1988)}{1989 - 1988}$$

$$= \frac{10 - 1}{1}$$

$$= 9$$

The average rate of change was $9 million.

(c) Civil penalties were increasing from 1987 to 1988 and decreasing from 1988 to 1989. Criminal penalties decreased slightly from 1987 to 1988, then increased from 1988 to 1989. This indicates that criminal penalties began to replace civil penalties in 1988.

(d) $$\frac{C(1989) - C(1981)}{1989 - 1981}$$

$$= \frac{35 - 6.5}{8}$$

$$= 3.5$$

The average rate of change increased about $3.5 million. The general trend was upward. Penalties for polluting are being more strictly imposed.

31. (a) cocaine

$$\frac{f(1989) - f(1988)}{1989 - 1988}$$

$$= \frac{22.5 - 23}{1}$$

$$= -.5$$

The average rate of change was a decrease of -$.5 billion.

$$\frac{f(1990) - f(1989)}{1990 - 1989}$$

$$= \frac{17.5 - 22.5}{1}$$

$$= -5$$

The average rate of change was a decrease of -$5 billion.

(b) heroin

$$\frac{f(1989) - f(1988)}{1989 - 1988}$$

$$= \frac{15.5 - 16}{1}$$

$$= -\$.5$$

The average rate of change was a decrease of -$.5 billion.

$$\frac{f(1990) - f(1989)}{1990 - 1989}$$

$$= \frac{12.3 - 15.5}{1}$$

$$= -3.2$$

The average rate of change was a decrease of -$3.2 billion.

(c) marijuana

$$\frac{f(1989) - f(1988)}{1989 - 1988}$$

$$= \frac{10 - 11.2}{1}$$

$$= -1.2$$

The average rate of change was a decrease of -$1.2 billion.

$$\frac{f(1990) - f(1989)}{1990 - 1989}$$

$$= \frac{8.8 - 10}{1}$$

$$= -1.2$$

The average rate of change was a decrease of -$1.2 billion.

(d) The greatest decrease in spend-
ing was for cocaine from 1989 –
1990.

33. Let v = the average velocity of the
car.

(a) From 0 sec to 2 sec

$$v = \frac{s(2) - s(0)}{2 - 0}$$

$$= \frac{10 - 0}{2 - 0} = 5 \text{ ft per sec.}$$

(b) From 2 sec to 6 sec

$$v = \frac{s(6) - s(2)}{6 - 2}$$

$$= \frac{14 - 10}{6 - 2} = 1 \text{ ft per sec.}$$

(c) From 6 sec to 10 sec

$$v = \frac{s(10) - s(6)}{10 - 6}$$

$$= \frac{18 - 14}{10 - 6} = 1 \text{ ft per sec.}$$

(d) From 10 sec to 12 sec

$$v = \frac{s(12) - s(10)}{12 - 10}$$

$$= \frac{30 - 18}{12 - 10} = 6 \text{ ft per sec.}$$

(e) From 12 sec to 18 sec

$$v = \frac{s(18) - s(12)}{18 - 12}$$

$$= \frac{36 - 30}{18 - 12} = 1 \text{ ft per sec.}$$

(f) The car is slowing down from 2
to 6 sec and 12 to 18 sec. The
car is speeding up from 0 to 2
sec and 10 to 12 sec. It is
maintaining a constant speed
from 6 to 10 sec.

Section 2.3

1. $f(x) = -4x^2 + 11x; \ x = -2$

Step 1   f(x + h)

$$= -4(x + h)^2 + 11(x + h)$$

$$= -4(x^2 + 2xh + h^2)$$

$$+ 11x + 11h$$

$$= -4x^2 - 8xh - 4h^2 + 11x + 11h$$

Step 2   f(x + h) – f(x)

$$= -4x^2 - 8xh - 4h^2 + 11x + 11h$$

$$- (-4x^2 + 11x)$$

$$= -8xh - 4h^2 + 11h$$

$$= h(-8x - 4h + 11)$$

Step 3 $\dfrac{f(x + h) - f(x)}{h}$

$$= \frac{h(-8x - 4h + 11)}{h}$$

$$= -8x - 4h + 11$$

Step 4 $f'(x) = \lim\limits_{h \to 0} \dfrac{f(x + h) - f(x)}{h}$

$$= \lim\limits_{h \to 0} (-8x - 4h + 11)$$

$$= -8x + 11$$

$f'(-2) = -8(-2) + 11 = 27$ is the
slope of the tangent line at x = –2.

3. $f(x) = \dfrac{-2}{x}; \ x = 4$

$$f(x + h) = \frac{-2}{x + h}$$

$$f(x + h) - f(x) = \frac{-2}{x + h} - \frac{-2}{x}$$

$$= \frac{-2x + 2(x + h)}{x(x + h)}$$

$$= \frac{-2x + 2x + 2h}{x(x + h)}$$

$$= \frac{2h}{x(x + h)}$$

$$\frac{f(x + h) - f(x)}{h} = \frac{2h}{hx(x + h)}$$

$$= \frac{2}{x(x + h)}$$

$$= \frac{2}{x^2 + xh}$$

$$f'(x) = \lim_{h \to 0} \frac{f(x + h) - f(x)}{h}$$

$$= \lim_{h \to 0} \frac{2}{x^2 + xh}$$

$$= \frac{2}{x^2}$$

$f'(4) = \frac{2}{4^2} = \frac{2}{16} = \frac{1}{8}$ is the slope of

the tangent line at x = 4.

5. $f(x) = \sqrt{x}$; x = 16

Steps 1–3 are combined:

$$\frac{f(x + h) - f(x)}{h}$$

$$= \frac{\sqrt{x + h} - \sqrt{x}}{h}$$

$$= \frac{\sqrt{x + h} - \sqrt{x}}{h} \cdot \frac{\sqrt{x + h} + \sqrt{x}}{\sqrt{x + h} + \sqrt{x}}$$

$$= \frac{x + h - x}{h(\sqrt{x + h} + \sqrt{x})}$$

$$= \frac{1}{\sqrt{x + h} + \sqrt{x}}$$

$$f'(x) = \lim_{h \to 0} \frac{f(x + h) - f(x)}{h}$$

$$= \lim_{h \to 0} \frac{1}{\sqrt{x + h} + \sqrt{x}}$$

$$= \frac{1}{2\sqrt{x}}$$

$f'(16) = \frac{1}{2\sqrt{16}} = \frac{1}{8}$ is the slope of

the tangent line at x = 16.

7. $f(x) = x^2 + 2x$; x = 3

$$\frac{f(x + h) - f(x)}{h}$$

$$= \frac{[(x + h)^2 + 2(x + h)] - (x^2 + 2x)}{h}$$

$$= \frac{(x^2 + 2hx + h^2 + 2x + 2h) - (x^2 + 2x)}{h}$$

$$= \frac{2hx + h^2 + 2h}{h} = 2x + h + 2$$

$$f'(x) = \lim_{h \to 0} (2x + h + 2) = 2x + 2$$

$f'(3) = 2(3) + 2 = 8$ is the slope
of the tangent line at x = 3.
Use m = 8 and (3, 15) in the point-
slope form.

$$y - 15 = 8(x - 3)$$
$$y = 8x - 9$$

9. $f(x) = \frac{5}{x}$; x = 2

$$\frac{f(x + h) - f(x)}{h}$$

$$= \frac{\frac{5}{x + h} - \frac{5}{x}}{h}$$

$$= \frac{\frac{5x - 5(x + h)}{(x + h)x}}{h}$$

$$= \frac{5x - 5x - 5h}{h(x + h)(x)}$$

$$= \frac{-5h}{h(x + h)x}$$

$$= \frac{-5}{(x + h)x}$$

$$f'(x) = \lim_{h \to 0} \frac{-5}{(x + h)(x)} = \frac{-5}{x^2}$$

$f'(2) = \frac{-5}{2^2} = -\frac{5}{4}$ is the slope of the

tangent line at x = 2.

Now use $m = -\frac{5}{4}$ and $\left(2, \frac{5}{2}\right)$ in the point-slope form.

$$y - \frac{5}{2} = -\frac{5}{4}(x - 2)$$

$$y - \frac{5}{2} = -\frac{5}{4}x + \frac{10}{4}$$

$$y = -\frac{5}{4}x + 5$$

$$5x + 4y = 20$$

11. $f(x) = 4\sqrt{x};\ x = 9$

$$\frac{f(x + h) - f(x)}{h}$$

$$= \frac{4\sqrt{x + h} - 4\sqrt{x}}{h} \cdot \frac{4\sqrt{x + h} + 4\sqrt{x}}{4\sqrt{x + h} + 4\sqrt{x}}$$

$$= \frac{16(x + h) - 16x}{h(4\sqrt{x + h} + 4\sqrt{x})}$$

$$f'(x) = \lim_{h \to 0} \frac{16(x + h) - 16x}{h(4\sqrt{x + h} + 4\sqrt{x})}$$

$$= \lim_{h \to 0} \frac{16h}{h(4\sqrt{x + h} + 4\sqrt{x})}$$

$$= \lim_{h \to 0} \frac{4}{(\sqrt{x + h} + \sqrt{x})} = \frac{4}{2\sqrt{x}}$$

$$= \frac{2}{\sqrt{x}}$$

$f'(9) = \dfrac{2}{\sqrt{9}} = \dfrac{2}{3}$ is the slope of the tangent line at $x = 9$.

Use $m = \frac{2}{3}$ and $(9, 12)$ in the point-slope form.

$$y - 12 = \frac{2}{3}(x - 9)$$

$$y = \frac{2}{3}x + 6$$

$$3y = 2x + 18$$

For Exercises 13–17, choose any two convenient points on each of the tangent lines.

13. Using the points $(5, 3)$ and $(6, 5)$, we have

$$m = \frac{5 - 3}{6 - 5} = \frac{2}{1}$$

$$= 2.$$

15. Using the points $(-2, 2)$ and $(3, 3)$, we have

$$m = \frac{3 - 2}{3 - (-2)}$$

$$= \frac{1}{5}.$$

17. Using the points $(-3, -3)$ and $(0, -3)$, we have

$$m = \frac{-3 - (-3)}{0 - 3} = \frac{0}{-3}$$

$$= 0$$

19. (a) $f(x) = 5$ is a horizontal line and has slope 0; the derivative is 0.

   (b) $f(x) = x$ has slope 1; the derivative is 1.

   (c) $f(x) = -x$ has slope of -1; the derivative is -1.

   (d) $x = 3$ is vertical and has undefined slope; the derivative does not exist.

   (e) $y = mx + b$ has slope m; the derivative is m.

21. $f(x) = \dfrac{x^2 - 1}{x + 2}$ is not differentiable when $x + 2 = 0$ or $x = -2$ because the function is undefined and a vertical asymptote occurs there.

23. $f(x) = 6x^2 - 4x$

$$\frac{f(x + h) - f(x)}{h}$$

$$= \frac{6(x + h)^2 - 4(x + h) - (6x^2 - 4x)}{h}$$

$$= \frac{12xh + 6h^2 - 4h}{h}$$

$$= 12x + 6h - 4$$

$$f'(x) = \lim_{h \to 0} (12x + 6h - 4)$$

$$= 12x - 4$$

$$f'(2) = 12(2) - 4 = 24 - 4 = 20$$

$$f'(0) = 12(0) - 4 = 0 - 4 = -4$$

$$f'(-3) = 12(-3) - 4 = -36 - 4 = -40$$

25. $f(x) = -9x - 5$

$$\frac{f(x + h) - f(x)}{h}$$

$$= \frac{-9(x + h) - 5 - [-9x - 5]}{h}$$

$$= \frac{-9x - 9h - 5 + 9x + 5}{h} = \frac{-9h}{h} = -9$$

$$f'(x) = \lim_{h \to 0} -9 = -9$$

$$f'(-2) = -9$$

$$f'(0) = -9$$

$$f'(-3) = -9$$

27. $f(x) = \dfrac{6}{x}$

$$\frac{f(x + h) - f(x)}{h} = \frac{\dfrac{6}{x + h} - \dfrac{6}{x}}{h}$$

$$= \frac{6x - 6(x + h)}{hx(x + h)}$$

$$= \frac{-6}{x(x + h)}$$

$$f'(x) = \lim_{h \to 0} \frac{-6}{x(x + h)}$$

$$= \frac{-6}{x^2}$$

$$f'(2) = \frac{-6}{(2)^2} = \frac{-6}{4} = -\frac{3}{2}$$

$$f'(0) = \frac{6}{0} \text{ does not exist.}$$

$$f'(-3) = \frac{-6}{(-3)^2} = \frac{-6}{9} = -\frac{2}{3}$$

29. $f(x) = -3\sqrt{x}$

$$\frac{f(x + h) - f(x)}{h}$$

$$= \frac{-3\sqrt{x + h} + 3\sqrt{x}}{h}$$

$$= \frac{-3\sqrt{x + h} + 3\sqrt{x}}{h} \cdot \frac{-3\sqrt{x + h} - 3\sqrt{x}}{-3\sqrt{x + h} - 3\sqrt{x}}$$

$$= \frac{9(x + h) - 9x}{h(-3\sqrt{x + h} - 3\sqrt{x})}$$

$$= \frac{9}{-3\sqrt{x + h} - 3\sqrt{x}}$$

$$= \frac{3}{-\sqrt{x + h} - \sqrt{x}}$$

$$f'(x) = \lim_{h \to 0} \frac{3}{-\sqrt{x + h} - \sqrt{x}}$$

$$= \frac{3}{-\sqrt{x} - \sqrt{x}} = -\frac{3}{2\sqrt{x}}$$

$$f'(2) = -\frac{3}{2\sqrt{2}}$$

$$f'(0) = \frac{-3}{2\sqrt{0}} \text{ is undefined.}$$

f'(0) does not exist.

$$f'(-3) = \frac{-3}{2\sqrt{-3}} \text{ is not a real number.}$$

f'(-3) does not exist.

31. No derivative exists at x = -6 because the function is not defined at x = -6.

33. This function is continuous for all x. So f(x) has derivatives for all values of x.

35. For x = -5 and x = 0, the function f(x) is not defined. For x = -3 and x = 2 the graph of f(x) has sharp points. For x = 4, the tangent to the graph is vertical. Therefore, no derivative exists for x = -5, x = -3, x = 0, x = 2, or x = 4.

37. (a) The rate of change of f(x) is positive when f(x) is increasing, that is, on (a, 0) and (b, c).

    (b) The rate of change of f(x) is negative when f(x) is decreasing, that is, on (0, b).

    (c) The rate of change is zero when the tangent to the graph is horizontal, that is, at x = 0 and x = b.

39. The zeros of graph (b) correspond to the turning points of graph (a), the points where the derivative is zero. Graph (a) gives the distance while graph (b) gives the velocity.

41. (a) At the first point, the slope of the tangent line is -1. The debt is decreasing at the rate of 1% per month. At the second point, the slope is positive and steeper, about 2. The debt is increasing at the rate of 2% per month. At the third point, the tangent line is horizontal, with zero slope, so the debt is not changing at this point.

    (b) From the graph, during 1990 the debt level decreased from February to March, from April to May, and from August to September. The function is decreasing over these intervals.

    (c) Since the overall direction of the curve is increasing, the average rate of change is also increasing.

43. $R(p) = 20p - \frac{p^2}{500}$

    (a) $R'(p) = 20 - \frac{1}{250}x$

    At p = 1000,

    $$R'(1000) = 20 - \frac{1}{250}(1000)$$
    $$= \$16 \text{ per table.}$$

(b) The marginal revenue for the 1001st table is approximately $R'(1000)$. From (a) this is about $16.

(c) The actual revenue is

$R(1001) - R(1000)$

$$= 20(1001) - \frac{1000^2}{500}$$

$$- \left[20(1000 - \frac{1000^2}{500} \right]$$

$$= 18,015.998 - 18,000$$

$$= \$15.998 \quad \text{or} \quad \$16.$$

(d) The marginal revenue gives a good approximation of the actual revenue from the sale of the 1001st table.

45. $B(t) = 1000 + 50t - 5t^2$

$B'(t) = 50 - 10t$

(a) $B'(3) = 50 - 10(3) = 20$

(b) $B'(5) = 50 - 10(5) = 0$

(c) $B'(6) = 50 - 10(6) = -10$

(d) $B'(t) = 50 - 10t$

The population starts to decline when $B'(t) = 0$.

$$50 - 10t = 0$$
$$50 = 10t$$
$$5 = t$$

So at 5 hr, population begins to decline.

47. (a) From the graph, V_{mp} is just about at the turning point of the curve. Thus, the slope of the tangent line is approximately zero. The power expenditure is not changing.

(b) From the graph, the slope of the tangent line at V_{mr} is approximately .1. The power expended is increasing .1 unit per unit increase in speed.

(c) The slope of the tangent line at V_{opt} is a bit greater than that at V_{mr}, about .12. The power expended increases .12 units for each unit increase in speed.

(d) The power level first decreases to V_{mp}, then increases at greater rates.

49. (a) From Exercise 46, the slope of the tangent at the first point is 1000, the slope of the tangent at the second point is 700, and the slope of the tangent at the third point is 250. Thus,

t	2	10	13
f'(t)	100	700	250

(b) See the answer graph in the textbook.

Section 2.4

1. $y = 10x^3 - 9x^2 + 6x$

$y' = 10(3x^{3-1}) - 9(2x^{2-1}) + 6x^{1-1}$

$= 30x^2 - 18x + 6$

3. $y = x^4 - 5x^3 + \dfrac{y^2}{9} + 5$

$y' = 4x^{4-1} - 5(3x^{3-1}) + \dfrac{1}{9}(2x^{2-1}) + 0$

$= 4x^3 - 15x^2 + \dfrac{2}{9}x$

5. $f(x) = 6x^{1.5} - 4x^{.5}$

$f'(x) = 6(1.5x^{1.5-1}) - 4(.5x^{.5-1})$

$= 9x^{.5} - 2x^{-.5}$ or $9x^{.5} - \dfrac{2}{x^{.5}}$

7. $y = -15x^{3.2} + 2x^{1.9}$

$y' = -15(3.2x^{3.2-1}) + 2(1.9x^{1.9-1})$

$= -48x^{2.2} + 3.8x^{.9}$

9. $y = 8\sqrt{x} + 6x^{3/4}$

$= 8x^{1/2} + 6x^{3/4}$

$y' = 8\left(\dfrac{1}{2}x^{1/2-1}\right) + 6\left(\dfrac{3}{4}x^{3/4-1}\right)$

$= 4x^{-1/2} + \dfrac{9}{2}x^{-1/4}$

or $\dfrac{4}{x^{1/2}} + \dfrac{9}{2x^{1/4}}$

11. $g(x) = 6x^{-5} - x^{-1}$

$g'(x) = 6(-5)x^{-5-1} - (-1)x^{-1-1}$

$= -30x^{-6} + x^{-2}$ or $\dfrac{-30}{x^6} + \dfrac{1}{x^2}$

13. $y = x^{-5} - x^{-2} + 5x^{-1}$

$y' = -5x^{-5-1} - (-2x^{-2-1}) + 5(-1x^{-1-1})$

$= -5x^{-6} + 2x^{-3} - 5x^{-2}$

or $\dfrac{-5}{x^6} + \dfrac{2}{x^3} - \dfrac{5}{x^2}$

15. $f(t) = \dfrac{4}{t} + \dfrac{2}{t^3}$

$= 4t^{-1} + 2t^{-3}$

$f'(t) = 4(-1t^{-1-1}) + 2(-3t^{-3-1})$

$= -4t^{-2} - 6t^{-4}$

or $\dfrac{-4}{t^2} - \dfrac{6}{t^4}$

17. $y = \dfrac{3}{x^6} + \dfrac{1}{x^5} - \dfrac{7}{x^2}$

$= 3x^{-6} + x^{-5} - 7x^{-2}$

$y' = 3(-6x^{-7}) + (-5x^{-6}) - 7(-2x^{-3})$

$= -18x^{-7} - 5x^{-6} + 14x^{-3}$

or $\dfrac{-18}{x^7} - \dfrac{5}{x^6} + \dfrac{14}{x^3}$

19. $h(x) = x^{-1/2} - 14x^{-3/2}$

$h'(x) = -\dfrac{1}{2}x^{-3/2} - 14\left(-\dfrac{3}{2}x^{-5/2}\right)$

$= \dfrac{-x^{-3/2}}{2} + 21x^{-5/2}$

or $\dfrac{-1}{2x^{3/2}} + \dfrac{21}{x^{5/2}}$

21. $y = \dfrac{-2}{\sqrt[3]{x}}$

$= \dfrac{-2}{x^{1/3}} = -2x^{-1/3}$

$y' = -2\left(-\dfrac{1}{3}x^{-4/3}\right)$

$= \dfrac{2x^{-4/3}}{3}$

or $\dfrac{2}{3x^{4/3}}$

23. $y = 8x^{-5} - 9x^{-4}$

$\dfrac{dy}{dx} = 8(-5x^{-6}) - 9(-4x^{-5})$

$= -40x^{-6} + 36x^{-5}$

or $\dfrac{-40}{x^6} + \dfrac{36}{x^5}$

25. $D_x\left[9x^{-1/2} + \dfrac{2}{x^{3/2}}\right]$

$= D_x[9x^{-1/2} + 2x^{-3/2}]$

$= 9\left(-\dfrac{1}{2}x^{-3/2}\right) + 2\left(-\dfrac{3}{2}x^{-5/2}\right)$

$= -\dfrac{9}{2}x^{-3/2} - 3x^{-5/2}$

or $\dfrac{-9}{2x^{3/2}} - \dfrac{3}{x^{5/2}}$

27. $f(x) = \dfrac{x^2}{6} - 4x$

$= \dfrac{1}{6}x^2 - 4x$

$f'(x) = \dfrac{1}{6}(2x) - 4$

$= \dfrac{1}{3}x - 4$

$f'(-2) = \dfrac{1}{3}(-2) - 4$

$= -\dfrac{2}{3} - 4 = -\dfrac{14}{3}$

29. $D_x\left(\dfrac{6}{x}\right) = D_x(6x^{-1})$

$= 6(-1x^{-2})$

$= -6x^{-2}$

$= -\dfrac{6}{x^2}$

Choice (c) equals $D_x\left(\dfrac{6}{x}\right)$.

33. $y = -2x^5 - 7x^3 + 8x^2$

$y' = -2(5x^4) - 7(3x^2) + 8(2x)$

$= -10x^4 - 21x^2 + 16x$

$y'(1) = -10(1)^4 - 21(1)^2 + 16(1)$

$= -10 - 21 + 16$

$= -15$ is the slope of the tangent line at x = 1.

Use (1, -1) to obtain the equation.

$y - (-1) = -15(x - 1)$

$y + 1 = -15x + 15$

$15x + y = 14$

35. $y = -x^{-3} + x^{-2}$

$y' = -3(-x^{-4}) + (-2x^{-3})$

$= 3x^{-4} - 2x^{-3}$

$= \dfrac{3}{x^4} - \dfrac{2}{x^3}$

$y'(1) = \dfrac{3}{(1)^4} - \dfrac{2}{(1)^3} = \dfrac{3}{1} - \dfrac{2}{1}$

$= 3 - 2$

$= 1$ is the slope of the tangent line at x = 1.

37. $f(x) = 6x^2 + 4x - 9$

$f'(x) = 12x + 4$

Let $f'(x) = -2$ to find the point where the slope of the tangent line is -2.

$12x + 4 = -2$

$12x = -6$

$x = -\dfrac{6}{12} = -\dfrac{1}{2}$

Find the y-coordinate.

$f(x) = 6x^2 + 4x - 9$

$$f\left(-\tfrac{1}{2}\right) = 6\left(-\tfrac{1}{2}\right)^2 + 4\left(-\tfrac{1}{2}\right) - 9$$

$$= \tfrac{6}{4} - 2 - 9$$

$$= \tfrac{6}{4} - \tfrac{44}{4}$$

$$= -\tfrac{38}{4} = -\tfrac{19}{2}$$

The slope of the tangent line at $\left(-\tfrac{1}{2}, -\tfrac{19}{2}\right)$ is -2.

41. $C(x) = 3000 - 20x + .03x^2$

$$R(x) = xp = x\left(\frac{5000 - x}{100}\right)$$

$$= x\left(50 - \frac{x}{100}\right)$$

(from Exercise 40)

The profit equation is found as follows.

$P = R - C$

$$= x\left(50 - \frac{x}{100}\right) - [3000 - 20x + .03x^2]$$

$$= 50x - \frac{x^2}{100} - 3000 + 20x - .03x^2$$

$$= 50x + 20x - \frac{x^2}{100} - .03x^2 - 3000$$

$$= 70x - .01x^2 - .03x^2 - 3000$$

Therefore, $P = 70x - .04x^2 - 3000$.
Now, the marginal profit is

$$P'(x) = 70 - .04(2x) - 0$$
$$= 70 - .08x.$$

(a) $P'(500) = 70 - .08(500)$
$$= 70 - 40$$
$$= 30$$

(b) $P'(815) = 70 - .08(815)$
$$= 70 - 65.2$$
$$= 4.8$$

(c) $P'(1000) = 70 - .08(1000)$
$$= 70 - 80$$
$$= -10$$

43. $S(t) = 100 - 100t^{-1}$

$S'(t) = -100(-1t^{-2})$

$$= 100t^{-2}$$

$$= \frac{100}{t^2}$$

(a) $S'(1) = \frac{100}{(1)^2} = \frac{100}{1} = 100$

(b) $S'(10) = \frac{100}{(10)^2} = \frac{100}{100} = 1$

45. $C(x) = 2x;\ R(x) = 6x - \frac{x^2}{1000}$

(a) $C'(x) = 2$

(b) $R'(x) = 6 - \frac{2x}{1000} = 6 - \frac{x}{500}$

(c) $P(x) = R(x) - C(x)$

$$= \left(6x - \frac{x^2}{1000}\right) - 2x$$

$$= 4x - \frac{x^2}{1000}$$

$$P'(x) = 4 - \frac{2x}{1000} = 4 - \frac{x}{500}$$

(d) $P'(x) = 4 - \frac{x}{500} = 0$

$$4 = \frac{x}{500}$$

$$x = 2000$$

(e) Using part (d), we see that marginal profit is 0 when $x = 2000$ units.
The profit for 2000 units produced is

$$P(x) = 4(2000) - \frac{(2000)^2}{1000}$$

$$= 8000 - 4000$$

$$= \$4000.$$

47. $V = \pi r^2 h$

Since h = 80 mm,

$V = \pi r^2 (80) = 80\pi r^2$.

(a) $V'(r) = 80\pi(2r)$

$= 160\pi r$

(b) $V'(4) = 160\pi(4) = 640\pi$ cu mm

(c) $V'(6) = 160\pi(6) = 960\pi$ cu mm

(d) $V'(8) = 160\pi(8) = 1280\pi$ cu mm

(e) As the radius increases, the volume increases.

49. $P(t) = \dfrac{100}{t}$

(a) $P(1) = \dfrac{100}{1} = 100$

(b) $P(100) = \dfrac{100}{100} = 1$

(c) $P(t) = \dfrac{100}{t} = 100t^{-1}$

$P'(t) = 100(-1t^{-2})$

$= -100t^{-2}$

$= \dfrac{-100}{t^2}$

$P'(100) = \dfrac{-100}{(100)^2}$

$= \dfrac{-1}{100}$

$= -.01$

The percent of acid is decreasing at the rate of .01 per day after 100 days.

51. $s(t) = 25t^2 - 9t + 8$

(a) $v(t) = s'(t) = 25(2t) - 9 + 0$

$= 50t - 9$

(b) $v(0) = 50(0) - 9 = -9$

$v(5) = 50(5) - 9 = 241$

$v(10) = 50(10) - 9 = 491$

53. $s(t) = -2t^3 + 4t^2 - 1$

(a) $v(t) = s'(t) = -2(3t^2) + 4(2t) - 0$

$= -6t^2 + 8t$

(b) $v(0) = -6(0)^2 + 8(0) = 0$

$v(5) = -6(5)^2 + 8(5)$

$= -6(25) + 40 = -110$

$v(10) = -6(10)^2 + 8(10)$

$= -6(100) + 80 = -520$

55. $s(t) = -16t^2 + 64t$

(a) $v(t) = s'(t) = -16(2t) + 64$

$= -32t + 64$

$v(2) = -32(2) + 64$

$= -64 + 64$

$= 0$ ft per sec

$v(3) = -32(3) + 64$

$= -96 + 64$

$= -32$ ft per sec

(b) As the ball travels upward, its speed decreases because of the force of gravity until, at maximum height, its speed is 0 ft per sec.

Since v = 0 at maximum height,

$0 = -32t + 64$

$-64 = -32t$

$2 = t$

It takes 2 seconds to reach maximum height.

(c) $s(2) = -16(2)^2 + 64(2)$

$= -16(4) + 128$

$= -64 + 128$

$= 64$

It will go 64 ft high.

Section 2.5

1. $y = (2x - 5)(x + 4)$

$y' = (2x - 5)(1) + (x + 4)(2)$

$= 2x - 5 + 2x + 8$

$= 4x + 3$

3. $y = (3x^2 + 2)(2x - 1)$

$y' = (3x^2 + 2)(2) + (2x - 1)(6x)$

$= 6x^2 + 4 + 12x^2 - 6x$

$= 18x^2 - 6x + 4$

5. $y = (2t^2 - 6t)(t + 2)$

$y' = (2t^2 - 6t)(1) + (4t - 6)(t + 2)$

$= 2t^2 - 6t + 4t^2 + 2t - 12$

$= 6t^2 - 4t - 12$

7. $y = (2x^2 - 4x)(5x^2 + 4)$

$y' = (2x^2 - 4x)(10x)$

$\qquad + (4x - 4)(5x^2 + 4)$

$= 20x^3 - 40x^2 + 20x^3 + 16x$

$\qquad - 20x^2 - 16$

$= 40x^3 - 60x^2 + 16x - 16$

9. $y = (7x - 6)^2 = (7x - 6)(7x - 6)$

$y' = (7x - 6)(7) + (7)(7x - 6)$

$= 49x - 42 + 49x - 42$

$= 98x - 84$

11. $g(t) = (3t^2 + 2)^2$

$\qquad = (3t^2 + 2)(3t^2 + 2)$

$g'(t) = (3t^2 + 2)(6t) + (6t)(3t^2 + 2)$

$\qquad = 18t^3 + 12t + 18t^3 + 12t$

$\qquad = 36t^3 + 24t$

13. $y = (2x - 3)(\sqrt{x} - 1)$

$= (2x - 3)(x^{1/2} - 1)$

$y' = (2x - 3)\left(\frac{1}{2}x^{-1/2}\right) + 2(x^{1/2} - 1)$

$= x^{1/2} - \frac{3}{2}x^{-1/2} + 2x^{1/2} - 2$

$= 3x^{1/2} - \frac{3x^{-1/2}}{2} - 2$

or $3x^{1/2} - \frac{3}{2x^{1/2}} - 2$

15. $g(x) = (-3\sqrt{x} + 6)(4\sqrt{x} - 2)$

$= (-3x^{1/2} + 6)(4x^{1/2} - 2)$

$g'(x) = (-3x^{1/2} + 6)(2x^{-1/2})$

$\qquad + \left(-\frac{3}{2}x^{-1/2}\right)(4x^{1/2} - 2)$

$= -6x^0 + 12x^{-1/2} - 6x^0 + 3x^{-1/2}$

$= -6 + 12x^{-1/2} - 6 + 3x^{-1/2}$

$= -12 + 15x^{-1/2}$

or $-12 + \frac{15}{x^{1/2}}$

17. $f(x) = \frac{6x - 11}{8x + 1}$

$f'(x) = \frac{(8x + 1)(6) - (6x - 11)(8)}{(8x + 1)^2}$

$= \frac{48x + 6 - 48x + 88}{(8x + 1)^2}$

$= \frac{94}{(8x + 1)^2}$

19. $y = \frac{-4}{2x - 11}$

$y' = \frac{(2x - 11)(0) + 4(2)}{(2x - 11)^2}$

$= \frac{8}{(2x - 11)^2}$

21. $y = \dfrac{9 - 7t}{1 - t}$

$y' = \dfrac{(1 - t)(-7) - (9 - 7t)(-1)}{(1 - t)^2}$

$= \dfrac{-7 + 7t + 9 - 7t}{(1 - t)^2}$

$= \dfrac{2}{(1 - t)^2}$

23. $y = \dfrac{x^2 - 4x}{x + 3}$

$y' = \dfrac{(x + 3)(2x - 4) - (x^2 - 4x)(1)}{(x + 3)^2}$

$= \dfrac{2x^2 + 6x - 4x - 12 - x^2 + 4x}{(x + 3)^2}$

$= \dfrac{x^2 + 6x - 12}{(x + 3)^2}$

25. $y = \dfrac{-x^2 + 6x}{4x^2 + 1}$

$y' = \dfrac{(4x^2 + 1)(-2x + 6) - (-x^2 + 6x)(8x)}{(4x^2 + 1)^2}$

$= \dfrac{-8x^3 + 24x^2 - 2x + 6 + 8x^3 - 48x^2}{(4x^2 + 1)^2}$

$= \dfrac{-24x^2 - 2x + 6}{(4x^2 + 1)^2}$

27. $k(x) = \dfrac{x^2 + 7x - 2}{x - 2}$

$k'(x) = \dfrac{(x - 2)(2x + 7) - (x^2 + 7x - 2)(1)}{(x - 2)^2}$

$= \dfrac{2x^2 + 7x - 4x - 14 - x^2 - 7x + 2}{(x - 2)^2}$

$= \dfrac{x^2 - 4x - 12}{(x - 2)^2}$

29. $r(t) = \dfrac{\sqrt{t}}{2t + 3} = \dfrac{t^{1/2}}{2t + 3}$

$r'(t) = \dfrac{(2t + 3)\left(\frac{1}{2}t^{-1/2}\right) - (t^{1/2})(2)}{(2t + 3)^2}$

$= \dfrac{t^{1/2} + \frac{3}{2}t^{-1/2} - 2t^{1/2}}{(2t + 3)^2}$

$= \dfrac{-t^{1/2} + \dfrac{3}{2t^{1/2}}}{(2t + 3)^2}$

$= \dfrac{-\sqrt{t} + \dfrac{3}{2\sqrt{t}}}{(2t + 3)^2}$

or $\dfrac{-2t + 3}{2\sqrt{t}(2t + 3)^2}$

31. The two terms in the numerator are reversed. The correct work follows.

$D_x\left(\dfrac{2x + 5}{x^2 - 1}\right)$

$= \dfrac{(x^2 - 1)(2) - (2x + 5)(2x)}{(x^2 - 1)^2}$

$= \dfrac{2x^2 - 2 - 4x^2 - 10x}{(x^2 - 1)^2}$

$= \dfrac{-2x^2 - 10x - 2}{(x^2 - 1)^2}$

33. $f(x) = x/(x - 2)$, at $(3, 3)$

$m = f'(x) = \dfrac{(x - 2)(1) - x(1)}{(x - 2)^2}$

$= -\dfrac{2}{(x - 2)^2}$

At $(3, 3)$,

$m = -\dfrac{2}{(3 - 2)^2} = -2$

Use the point-slope form.

$y - 3 = -2(x - 3)$

$y = -2x + 9$

35. $C(x) = \dfrac{3x + 2}{x + 4}$

$\overline{C}(x) = \dfrac{C(x)}{x} = \dfrac{3x + 2}{x^2 + 4x}$

(a) $\overline{C}(10) = \dfrac{3(10) + 2}{10^2 + 4(10)} = \dfrac{32}{140}$

$\approx .2286$ hundreds of dollars

or $\$22.86$ per unit

(b) $\bar{C}(20) = \dfrac{3(20) + 2}{(20)^2 + 4(20)} = \dfrac{62}{480}$

$\approx .1292$ hundreds of dollars

or $12.92 per unit

(c) $\bar{C}(x) = \dfrac{3x + 2}{x^2 + 4x}$ per unit

(d) $\bar{C}'(x)$

$= \dfrac{(x^2 + 4x)(3) - (3x + 2)(2x + 4)}{(x^2 + 4x)^2}$

$= \dfrac{3x^2 + 12x - 6x^2 - 12x - 4x - 8}{(x^2 + 4x)^2}$

$= \dfrac{-3x^2 - 4x - 8}{(x^2 + 4x)^2}$

37. $G(x) = \dfrac{1}{200}\left(\dfrac{800 + x^2}{x}\right)$

(a) $G'(x)$

$= \dfrac{1}{200}\left[\dfrac{x(2x) - (800 + x^2)1}{x^2}\right]$

$= \dfrac{1}{200}\left(\dfrac{2x^2 - 800 - x^2}{x^2}\right)$

$= \dfrac{x^2 - 800}{200x^2}$

$G'(20) = \dfrac{(20)^2 - 800}{200(20)^2}$

$= -\dfrac{1}{200} < 0$

Tell the driver to go faster.

(b) $G'(40) = \dfrac{(40)^2 - 800}{200(40)^2}$

$= \dfrac{1}{400} > 0$

Tell the driver to go slower.

39. $s(x) = \dfrac{x}{m + nx}$; m and n constants

(a) $s'(x) = \dfrac{(m + nx)1 - x(n)}{(m + nx)^2}$

$= \dfrac{m + nx - nx}{(m + nx)^2}$

$= \dfrac{m}{(m + nx)^2}$

(b) $x = 50$, $m = 10$, $n = 3$

$s'(50) = \dfrac{m}{(m + 50n)^2}$

$= \dfrac{10}{[10 + 50(3)]^2}$

$= \dfrac{1}{2560}$

$\approx .000391$ mm per ml

41. $f(t) = \dfrac{90t}{99t - 90}$

$f'(t) = \dfrac{(99t - 90)(90) - (90t)(99)}{(99t - 90)^2}$

$= \dfrac{-8100}{(99t - 90)^2}$

(a) $f'(1) = \dfrac{-8100}{(99 - 90)^2}$

$= \dfrac{-8100}{9^2}$

$= \dfrac{-8100}{81} = -100$

$f'(10) = \dfrac{-8100}{[99(10) - 90]^2}$

$= \dfrac{-8100}{(900)^2}$

$= \dfrac{-8100}{810,000}$

$= -\dfrac{1}{100}$ or $-.01$

Section 2.6

For Exercises 1-5, $f(x) = 4x^2 - 2x$ and $g(x) = 8x + 1$.

1. $g(2) = 8(2) + 1 = 17$

    $f[g(2)] = f[17] = 4(17)^2 - 2(17)$
    $$= 1156 - 34 = 1122$$

3. $f(2) = 4(2)^2 - 2(2)$
    $$= 16 - 4 = 12$$

    $g[f(2)] = g[12] = 8(12) + 1$
    $$= 96 + 1 = 97$$

5. $f[g(k)] = 4(8x + 1)^2 - 2(8k + 1)$
    $$= 4(64k^2 + 16k + 1)$$
    $$- 16k - 2$$
    $$= 256k^2 + 48k + 2$$

7. $f(x) = \frac{x}{8} + 12$; $g(x) = 3x - 1$

    $f[g(x)] = \frac{3x - 1}{8} + 12$

    $$= \frac{3x - 1}{8} + \frac{96}{8}$$

    $$= \frac{3x + 95}{8}$$

    $g[f(x)] = 3[\frac{x}{8} + 12] - 1$

    $$= \frac{3x}{8} + 36 - 1$$

    $$= \frac{3x}{8} + 35$$

    $$= \frac{3x}{8} + \frac{280}{8} = \frac{3x + 280}{8}$$

9. $f(x) = \frac{1}{x}$; $g(x) = x^2$

    $f[g(x)] = \frac{1}{x^2}$

    $g[f(x)] = \left(\frac{1}{x}\right)^2$

    $$= \frac{1}{x^2}$$

11. $f(x) = \sqrt{x + 2}$; $g(x) = 8x^2 - 6$

    $f[g(x)] = \sqrt{(8x^2 - 6) + 2}$
    $$= \sqrt{8x^2 - 4}$$

    $g[f(x)] = 8(\sqrt{x + 2})^2 - 6$
    $$= 8x + 16 - 6$$
    $$= 8x + 10$$

13. $f(x) = \sqrt{x + 1}$; $g(x) = \frac{-1}{x}$

    $f[g(x)] = \sqrt{\frac{-1}{x} + 1}$

    $$= \sqrt{\frac{x - 1}{x}}$$

    $g[f(x)] = \frac{-1}{\sqrt{x + 1}}$

17. $y = (5 - x)^{2/5}$

    If $f(x) = x^{2/5}$ and
        $g(x) = 5 - x$,

    then $y = f[g(x)] = (5 - x)^{2/5}$.

19. $y = -\sqrt{13 + 7x}$

    If $f(x) = -\sqrt{x}$ and
        $g(x) = 13 + 7x$,

    then $y = f[g(x)] = -\sqrt{13 + 7x}$.

21. $y = (x^2 + 5x)^{1/3} - 2(x^2 + 5x)^{2/3} + 7$

If $f(x) = x^{1/3} - 2x^{2/3} + 7$ and

$g(x) = x^2 + 5x$,

then $y = f[g(x)] = (x^2 + 5x)^{1/3} - 2(x^2 + 5x)^{2/3} + 7$.

23. $y = (2x^3 + 9x)^5$

Let $f(x) = x^5$ and $g(x) = 2x^3 + 9x$.   Then $(2x^3 + 9x)^5 = f[g(x)]$.

$D_x(2x^3 + 9x)^5 = f'[g(x)] \cdot g'(x)$

$\quad f'(x) = 5x^4$

$f'[g(x)] = 5[g(x)]^4$

$\qquad\quad = 5(2x^3 + 9x)^4$

$g'(x) = 6x^2 + 9$

$D_x(2x^3 + 9x)^5 = 5(2x^3 + 9x)^4(6x^2 + 9)$

25. $f(x) = -8(3x^4 + 2)^3$

Use the generalized power rule with $y = 3x^4 + 2$, $n = 3$ and $u' = 12x^3$.

$f'(x) = -8[3(3x^4 + 2)^{3-1} \cdot 12x^3]$

$\qquad = -8[36x^3(3x^4 + 2)^2]$

$\qquad = -288x^3(3x^4 + 2)^2$

27. $s(t) = 12(2t^4 + 5)^{3/2}$

Use the generalized power rule with $u = 2t^4 + 5$, $n = 3/2$ and $u' = 8t^3$.

$s'(t) = 12[\frac{3}{2}(2t^4 + 5)^{1/2} \cdot 8t^3]$

$\qquad = 12[12t^3(2t^4 + 5)^{1/2}]$

$\qquad = 144t^3(2t^4 + 5)^{1/2}$

29. $f(t) = 8\sqrt{4t^2 + 7}$

$\qquad\quad = 8(4t^2 + 7)^{1/2}$

Use the generalized power rule with $y = 4t^2 + 7$, $n = 1/2$, and $u' = 8t$.

$$f'(t) = 8[\frac{1}{2}(4t^2 + 7)^{-1/2} \cdot 8t]$$

$$= 8[4t(4t^2 + 7)^{-1/2}]$$

$$= 32t(4t^2 + 7)^{-1/2}$$

$$= \frac{32t}{(4t^2 + 7)^{1/2}}$$

$$= \frac{32t}{\sqrt{4t^2 + 7}}$$

31. $r(t) = 4t(2t^5 + 3)^2$

Use the product rule and the power rule.

$$r'(t) = 4t[2(2t^5 + 3) \cdot 10t^4] + (2t^5 + 3)^2 \cdot 4$$

$$= 80t^5(2t^5 + 3) + 4(2t^5 + 3)^2$$

$$= 4(2t^5 + 3)[20t^5 + (2t^5 + 3)]$$

$$= 4(2t^5 + 3)(22t^5 + 3)$$

33. $y = (x^3 + 2)(x^2 - 1)^2$

Use the product rule and the power rule.

$$y' = (x^3 + 2)[2(x^2 - 1) \cdot 2x] + (x^2 - 1)^2(3x^2)$$

$$= (x^3 + 2)[4x(x^2 - 1)] + 3x^2(x^2 - 1)^2$$

$$= (x^2 - 1)[4x(x^3 + 2) + 3x^2(x^2 - 1)]$$

$$= (x^2 - 1)(4x^4 + 8x + 3x^4 - 3x^2)$$

$$= (x^2 - 1)(7x^4 - 3x^2 + 8x)$$

35. $y = (5x^6 + x)^2\sqrt{2x}$

$$= (5x^6 + x)^2(2x)^{1/2}$$

$$y' = (5x^6 + x)^2[\frac{1}{2}(2x)^{-1/2} \cdot 2] + (2x)^{1/2}[2(5x^6 + x)(30x^5 + 1)]$$

$$= (5x^6 + x)^2(2x)^{-1/2} + (2x)^{1/2} \cdot (60x^5 + 2)(5x^6 + x)$$

$$= (5x^6 + x)(2x)^{-1/2}[(5x^6 + x) + (2x)^1(60x^5 + 2)]$$

$$= (5x^6 + x)(2x)^{-1/2}(5x^6 + x + 12x^6 + 4x)$$

$$= \frac{(5x^6 + x)(125x^6 + 5x)}{(2x)^{1/2}}$$

$$= \frac{(5x^6 + x)(125x^6 + 5x)}{\sqrt{2x}}$$

37. $y = \dfrac{1}{(3x^2 - 4)^5} = (3x^2 - 4)^{-5}$

$y' = -5(3x^2 - 4)^{-6} \cdot 6x$

$\quad = -30x(3x^2 - 4)^{-6}$

$\quad = \dfrac{-30x}{(3x^2 - 4)^6}$

39. $p(t) = \dfrac{(2t + 3)^3}{4t^2 - 1}$

$p'(t) = \dfrac{(4t^2 - 1)[3(2t + 3)^2 \cdot 2] - (2t + 3)^3(8t)}{(4t^2 - 1)^2}$

$\quad = \dfrac{6(4t^2 - 1)(2t + 3)^2 - 8t(2t + 3)^3}{(4t^2 - 1)^2}$

$\quad = \dfrac{(2t + 3)^2[6(4t^2 - 1) - 8t(2t + 3)]}{(4t^2 - 1)^2}$

$\quad = \dfrac{(2t + 3)^2[24t^2 - 6 - 16t^2 - 24t]}{(4t^2 - 1)^2}$

$\quad = \dfrac{(2t + 3)^2[8t^2 - 24t - 6]}{(4t^2 - 1)^2}$

$\quad = \dfrac{2(2t + 3)^2(4t^2 - 12t - 3)}{(4t^2 - 1)^2}$

41. $y = \dfrac{x^2 + 4x}{(3x^3 + 2)^4}$

$y' = \dfrac{(3x^3 + 2)^4(2x + 4) - (x^2 + 4x)[4(3x^3 + 2)^3 \cdot 9x^2]}{[(3x^3 + 2)^4]^2}$

$\quad = \dfrac{(3x^3 + 2)^4(2x + 4) - 36x^2(x^2 + 4x)(3x^3 + 2)^3}{(3x^3 + 2)^8}$

$\quad = \dfrac{2(3x^3 + 2)^3[(3x^3 + 2)(x + 2) - 18x^2(x^2 + 4x)]}{(3x^3 + 2)^8}$

$\quad = \dfrac{2(3x^4 + 6x^3 + 2x + 4 - 18x^4 - 72x^3)}{(3x^3 + 2)^5}$

$\quad = \dfrac{-30x^4 - 132x^3 + 4x + 8}{(3x^3 + 2)^5}$

43.   (a) $D_x(f[g(x)])$ at $x = 1$

$\quad = f'[g(1)] \cdot g'(1)$

$\quad = f'(2) \cdot \left(\dfrac{2}{7}\right)$

$\quad = -7\left(\dfrac{2}{7}\right)$

$\quad = -2$

(b) $D_x(f[g(x)])$ at $x = 2$

$= f'[g(2)] \cdot g'(2)$

$= f'(3) \cdot \left(\frac{3}{7}\right)$

$= -8\left(\frac{3}{7}\right)$

$= -\frac{24}{7}$

45. $D(p) = \frac{-p^2}{100} + 500$; $p(c) = 2c - 10$

The demand in terms of the cost is

$D(c) = D[p(c)]$.

$= \frac{-(2c - 10)^2}{100} + 500$

$= \frac{-4(c - 5)^2}{100} + 500$

$= \frac{-c^2 + 10c - 25}{25} + 500$

$= \frac{-c^2 + 10c - 25 + 12{,}500}{25}$

$= \frac{-c^2 + 10c + 12{,}475}{25}$

47. $A = 1500\left(1 + \frac{r}{36{,}500}\right)^{1825}$

dA/dr is the rate of change of A with respect to r.

$\frac{dA}{dr} = 1500(1825)\left(1 + \frac{r}{36{,}500}\right)^{1824}\left(\frac{1}{36{,}500}\right)$

$= 75\left(1 + \frac{r}{36{,}500}\right)^{1824}$

For $r = 6\%$,

$\frac{dA}{dr} = 75\left(1 + \frac{6}{36{,}500}\right)^{1824} = \$101.22.$

For $r = 8\%$,

$\frac{dA}{dr} = 75\left(1 + \frac{8}{36{,}500}\right)^{1824} = \$111.86.$

For $r = 9\%$,

$\frac{dA}{dr} = 75\left(1 + \frac{9}{36{,}500}\right)^{1824} = \$117.59.$

49. $V = \frac{6000}{1 + .3t + .1t^2}$

The rate of change of the value is

$V'(t)$

$= \frac{(1 + .3t + .1t^2)(0) - 6000(.3 + .2t)}{(1 + .3t + .1t^2)^2}$

$= \frac{-6000(.3 + .2t)}{(1 + .3t + .1t^2)^2}.$

(a) 2 years after purchase the rate of decrease in the value is

$= \frac{-6000[.3 + .2(2)]}{[1 + .3(2) + .1(2)^2]^2}$

$= \frac{-6000(.3 + .4)}{(1 + .6 + .4)^2}$

$= \frac{-4200}{4}$

$= -\$1050.$

(b) 4 years after purchase:

$= \frac{-6000[.3 + .2(4)]}{[1 + .3(4) + .1(4)^2]^2}$

$= \frac{-6600}{14.44}$

$= -\$457.06.$

51. The demand function is $p = 300/x^{1/3}$;

$x = 8n$.

The marginal revenue product is

$\frac{dR}{dn} = \left(p + x\frac{dp}{dx}\right)\frac{dx}{dn}$

$= [300x^{-1/3} + 8n(-100x^{-4/3})](8).$

When $n = 8$,

$\frac{dR}{dn} = \left[\frac{300}{2} - 64\left(\frac{100}{256}\right)\right](8)$

$= 125(8)$

$= \$400$ per additional worker.

53. $P(x) = 2x^2 + 1;\ x = f(a) = 3a + 2$

$$P[f(a)] = 2(3a + 2)^2 + 1$$
$$= 2(9a^2 + 12a + 4) + 1$$
$$= 18a^2 + 24a + 9$$

55. $r(t) = 2t;\ A(r) = \pi r^2$

$$A[r(t)] = \pi(2t)^2$$
$$= 4\pi t^2$$

$A = 4\pi t^2$ gives the area of the pollution in terms of the time since the pollutants were first emitted.

57. $C(t) = \frac{1}{2}(2t + 1)^{-1/2}$

$$C'(t) = \frac{1}{2}\left(-\frac{1}{2}\right)(2t + 1)^{-3/2}(2)$$

$$= -\frac{1}{2}(2t + 1)^{-3/2}$$

(a) $C'(0) = -\frac{1}{2}[2(0) + 1]^{-3/2}$

$$= -\frac{1}{2} = -.5$$

(b) $C'(4) = -\frac{1}{2}[2(4) + 1]^{-3/2}$

$$= -\frac{1}{2}(9)^{-3/2}$$

$$= \frac{-1}{2} \cdot \frac{1}{(\sqrt{9})^3}$$

$$= -\frac{1}{54} \approx -.02$$

(c) $C'(7.5) = -\frac{1}{2}[2(7.5) + 1]^{-3/2}$

$$= -\frac{1}{2}(16)^{-3/2}$$

$$= -\frac{1}{2}\left[\frac{1}{(\sqrt{16})^3}\right]$$

$$= -\frac{1}{128} \approx -.008$$

(d) C is always decreasing because

$C' = -\frac{1}{2}(2t + 1)^{-3/2}$ is always

negative for $t \geq 0$.

(The amount of calcium in the bloodstream will continue to decrease over time.)

Chapter 2   Review Exercises

3. As x approaches -3 from the left or the right, $f(x)$ approaches 4.

$$\lim_{x \to -3} f(x) = 4$$

5. Since there is a vertical asymptote at $x = 4$, the limit does not exist.

7. $\lim_{x \to -1} (2x^2 + 3x + 5)$

$$= 2(-1)^2 + 3(-1) + 5$$
$$= 4$$

9. $\lim_{x \to 6} \dfrac{2x + 5}{x - 3}$

$$= \frac{2(6) + 5}{6 + (-3)}$$

$$= \frac{17}{3}$$

11. $\lim_{x \to 4} \dfrac{x^2 - 16}{x - 4}$

$$= \lim_{x \to 4} \frac{(x - 4)(x + 4)}{x - 4}$$

$$= \lim_{x \to 4} (x + 4)$$

$$= 4 + 4$$

$$= 8$$

13. $\lim\limits_{x \to -4} \dfrac{2x^2 + 3x - 20}{x + 4}$

$\quad = \lim\limits_{x \to -4} \dfrac{(2x - 5)(x + 4)}{x - 4}$

$\quad = \lim\limits_{x \to -4} (2x - 5)$

$\quad = 2(-4) - 5$

$\quad = -13$

15. $\lim\limits_{x \to 9} \dfrac{\sqrt{x} - 3}{x - 9}$

$\quad = \lim\limits_{x \to 9} \dfrac{\sqrt{x} - 3}{x - 9} \cdot \dfrac{\sqrt{x} + 3}{\sqrt{x} + 3}$

$\quad = \lim\limits_{x \to 9} \dfrac{x - 9}{(x - 9)(\sqrt{x} + 3)}$

$\quad = \lim\limits_{x \to 9} \dfrac{1}{\sqrt{x} + 3}$

$\quad = \dfrac{1}{(\sqrt{9}) + (3)}$

$\quad = \dfrac{1}{6}$

17. As shown on the graph, f(x) is discontinuous at x_2 and x_4.

19. $f(x) = \dfrac{-5}{3x(2x - 1)};$

    $x = -5,\ 0,\ -1/3,\ 1/2$

    Since f is a rational function, it is discontinuous when $3x(2x - 1) = 0$, for $x = 0$ and $x = 1/2$. It is continuous at $x = -5$ and $x = -1/3$.

21. $f(x) = \dfrac{x - 6}{x + 5};\ x = 6,\ -5,\ 0$

    Since f is a rational function, it is discontinuous when $x + 5 = 0$, or when $x = -5$. It is continuous at $x = 6$ and $x = 0$.

23. $f(x) = x^2 + 3x - 4;\ x = 1,\ -4,\ 0$

    Since f(x) is a polynomial function, it is continuous at all points, including $x = 1$, $x = -4$, and $x = 0$.

25. (a) At $x = 0$, $f(0) = 0$.
    At $x = 4$, $f(4) = 1$.

    Average rate of change

    $\quad = \dfrac{f(0) - f(4)}{0 - 4} = \dfrac{0 - 1}{0 - 4} = \dfrac{1}{4}$ *wrong*

    (b) At $x = 8$, $f(8) = 4$.
    At $x = 2$, $f(2) = 4$.

    Average rate of change

    $\quad = \dfrac{f(8) - f(2)}{8 - 2} = \dfrac{4 - 4}{8 - 2} = 0$

    (c) At $x = 2$, $f(2) = 4$.
    At $x = 4$, $f(4) = 1$.   *wrong*

    Average rate of change

    $\quad = \dfrac{f(2) - f(4)}{2 - 4} = \dfrac{4 - 1}{2 - 4} = -\dfrac{3}{2}$

27. $y = -2x^3 - x^2 + 5 = f(x)$

    Average rate of change

    $\quad = \dfrac{f(6) - f(-2)}{6 - (-2)}$

    $f(6) = -2(6)^3 - (6)^2 + 5 = -463$
    $f(-2) = -2(-2)^3 - (-2)^2 + 5 = 17$

    Average rate of change

    $\quad = \dfrac{-463 - 17}{6 + 2} = \dfrac{-480}{8} = -60$

    $y' = -6x^2 - 2x$

    Instantaneous rate of change at $x = -2$:

    $\quad -6(-2)^2 - 2(-2) = -6(4) + 4 = -20$

29. $y = \dfrac{x + 4}{x - 1} = f(x)$

$f(5) = \dfrac{5 + 4}{5 - 1} = \dfrac{9}{4}$

$f(2) = \dfrac{2 + 4}{2 - 1} = 6$

Average rate of change

$= \dfrac{\frac{9}{4} - 6}{5 - 2} = \dfrac{\frac{-15}{4}}{3} = -\dfrac{5}{4}$

$y' = \dfrac{(x - 1)(1) - (x + 4)(1)}{(x - 1)^2}$

$= \dfrac{x - 1 - x - 4}{(x - 1)^2} = \dfrac{-5}{(x - 1)^2}$

Instantaneous rate of change at

x = 2:

$\dfrac{-5}{(2 - 1)^2} = \dfrac{-5}{1} = -5$

31. $y = 5x^2 + 6x$

$y' = \lim_{h \to 0} \dfrac{f(x + h) - f(x)}{h}$

$= \lim_{h \to 0} \dfrac{[5(x + h)^2 + 6(x + h)] - [5x^2 + 6x]}{h}$

$= \lim_{h \to 0} \dfrac{5(x^2 + 2xh + h^2) + 6x + 6h - 5x^2 - 6x}{h}$

$= \lim_{h \to 0} \dfrac{5x^2 + 10xh + 5h^2 + 6x + 6h - 5x^2 - 6x}{h}$

$= \lim_{h \to 0} \dfrac{10xh + 5h^2 + 6h}{h}$

$= \lim_{h \to 0} \dfrac{h(10x + 5h + 6)}{h}$

$= \lim_{h \to 0} (10x + 5h + 6)$

$= 10x + 6$

33. $y = x^2 - 6x$, tangent at x = 2

$y' = 2x - 6$

Slope $= y'(2) = 2(2) - 6 = -2$

Use (2, −8) and point-slope form.

$y - (-8) = -2(x - 2)$

$y + 8 = -2x + 4$

$y + 2x = -4$

35. $y = \dfrac{3}{x - 1}$, tangent at x = −1

$y = \dfrac{3}{x - 1} = 3(x - 1)^{-1}$

$y' = 3(-1)(x - 1)^{-2}(1)$

$= -3(x - 1)^{-2}$

Slope $= y'(-1) = -3(-1 - 1)^{-2} = -\dfrac{3}{4}$

Use (−1, −3/2) and point-slope form.

$y - \left(\dfrac{-3}{2}\right) = -\dfrac{3}{4}[x - 1(-1)]$

$y + \dfrac{3}{2} = -\dfrac{3}{4}(x + 1)$

$y = -\dfrac{3}{4}x - \dfrac{9}{4}$

$3x + 4y = -9$

37. $y = \dfrac{3}{x^2 - 1}$

$= 3(x^2 - 1)^{-1}$

$y' = 3[-1(x^2 - 1)^{-2}2x]$

$= -6x(x^2 - 1)^{-2}$

$= \dfrac{-6x}{(x^2 - 1)^2}$

slope $= y'(2) = \dfrac{-6(2)}{(4 - 1)^2}$

$= \dfrac{-12}{9}$

$= -\dfrac{4}{3}$

Use (2, 1) and point-slope form.

$y - 1 = -\dfrac{4}{3}(x - 2)$

$3y - 3 = -4x + 8$

$4x + 3y = 11$

39. $y = -\sqrt{8x + 1}$

$\qquad = -(8x + 1)^{1/2}$

$y' = -\left[\frac{1}{2}(8x + 1)^{-1/2} \cdot 8\right]$

$\qquad = -4(8x + 1)^{-1/2}$

$\qquad = \dfrac{-4}{(8x + 1)^{1/2}}$

slope $= y'(3) = \dfrac{-4}{(8(3) + 1)^{1/2}}$

$\qquad\qquad = -\dfrac{4}{5}$

Use $(3, -5)$ and point-slope form.

$y - (-5) = -\dfrac{4}{5}(x - 3)$

$5y + 25 = -4x + 12$

$4x + 5y = -13$

43. $y = x^3 - 4x^2$

$y' = 3x^2 - 4(2x) = 3x^2 - 8x$

45. $y = -3x^{-2}$

$y' = (-3)(-2)x^{-3}$

$\qquad = 6x^{-3}$ or $\dfrac{6}{x^3}$

47. $f(x) = 6x^{-1} - 2\sqrt{x} = 6x^{-1} - 2(x)^{1/2}$

$f'(x) = 6(-x^{-2}) - 2\left(\frac{1}{2}x^{-1/2}\right)$

$\qquad = -6x^{-2} - x^{-1/2}$

$\qquad$ or $\dfrac{-6}{x^2} - \dfrac{1}{x^{1/2}}$

49. $y = (-5t + 4)(t^3 - 2t^2)$

$y' = (-5t + 4)(3t^2 - 4t) + (t^3 - 2t^2)(-5)$

$\qquad = -15t^3 + 20t^2 + 12t^2 - 16t - 5t^3 + 10t^2$

$\qquad = -20t^3 + 42t^2 - 16t$

51. $p(t) = 8t^{3/4}(7t - 2)$

$p'(t) = 8t^{3/4}(7) + (7t - 2)\left[8\left(\frac{3}{4}\right)t^{-1/4}\right]$

$\qquad = 56t^{3/4} + (7t - 2)(6t^{-1/4})$

$\qquad = 56t^{3/4} + 42t^{3/4} - 12t^{-1/4}$

$\qquad = 98t^{3/4} - 12t^{-1/4}$

$\qquad$ or $98t^{3/4} - \dfrac{12}{t^{1/4}}$

53. $y = 15x^{-3/5}\left(6 - \dfrac{x}{3}\right)$

$y' = 15x^{-3/5}\left(-\dfrac{1}{3}\right) + \left(6 - \dfrac{x}{3}\right)\left[15\left(-\dfrac{3}{5}x^{-8/5}\right)\right]$

$\qquad = -5x^{-3/5} - 9x^{-8/5}\left(6 - \dfrac{x}{3}\right)$

$\qquad = -5x^{-3/5} - 54x^{-8/5} + 3x^{-3/5}$

$\qquad = -2x^{-3/5} - 54x^{-8/5}$

$\qquad$ or $-\dfrac{2}{x^{3/5}} - \dfrac{54}{x^{8/5}}$

55. $r(x) = \dfrac{-8}{2x + 1} = -8(2x + 1)^{-1}$

$r'(x) = -8[(-1)(2x + 1)^{-2}(2)]$

$\qquad = 16(2x + 1)^{-2}$

$\qquad = \dfrac{16}{(2x + 1)^2}$

57. $y = \dfrac{2x^3 - 5x^2}{x + 2}$

$y' = \dfrac{(x + 2)(6x^2 - 10x) - (2x^3 - 5x^2)(1)}{(x + 2)^2}$

$\qquad = \dfrac{6x^3 + 12x^2 - 10x^2 - 20x - 2x^3 + 5x^2}{(x + 2)^2}$

$\qquad = \dfrac{4x^3 + 7x^2 - 20x}{(x + 2)^2}$

59. $k(x) = (5x - 1)^6$

$k'(x) = 6(5x - 1)^5(5)$

$\qquad = 30(5x - 1)^5$

61. $y = -3\sqrt{8t - 1} = -3(8t - 1)^{1/2}$

$\quad y' = -3\left[\frac{1}{2}(8t - 1)^{-1/2}(8)\right]$

$\quad\quad = -12(8t - 1)^{-1/2}$

$\quad\quad$ or $\dfrac{-12}{(8t - 1)^{1/2}}$

63. $y = 4x^2(3x - 2)^5$

$\quad y' = (4x^2)[5(3x - 2)^4(3)] + (3x - 2)^5(8x)$

$\quad\quad = 60x^2(3x - 2)^4 + 8x(3x - 2)^5$

$\quad\quad = 4x(3x - 2)^4[15x + 2(3x - 2)]$

$\quad\quad = 4x(3x - 2)^4(15x + 6x - 4)$

$\quad\quad = 4x(3x - 2)^4(21x - 4)$

65. $s(t) = \dfrac{t^3 - 2t}{(4t - 3)^4}$

$s'(t)$

$= \dfrac{(4t - 3)^4(3t^2 - 2) - (t^3 - 2t)4(4t - 3)^3(4)}{[(4t - 3)^4]^2}$

$= \dfrac{(4t - 3)^4(3t^2 - 2) - 16(t^3 - 2t)(4t - 3)^3}{(4t - 3)^8}$

$= \dfrac{(4t - 3)^3[(4t - 3)(3t^2 - 2) - 16(t^3 - 2t)]}{(4t - 3)^8}$

$= \dfrac{(4t - 3)^3(12t^3 - 9t^2 - 8t + 6 - 16t^3 + 32t)}{(4t - 3)^8}$

$= \dfrac{-4t^3 - 9t^2 + 24t + 6}{(4t - 3)^5}$

67. $D_x\left[\dfrac{2x + \sqrt{x}}{1 - x}\right]$

$= D_x\left[\dfrac{2x + x^{1/2}}{1 - x}\right]$

$= \dfrac{(1 - x)\left(2 + \frac{1}{2}x^{-1/2}\right) - (2x + x^{1/2})(-1)}{(1 - x)^2}$

$= \dfrac{2 + \frac{1}{2}x^{-1/2} - 2x - \frac{1}{2}x^{1/2} + 2x + x^{1/2}}{(1 - x)^2}$

$= \dfrac{2 + \frac{1}{2}x^{-1/2} + \frac{1}{2}x^{1/2}}{(1 - x)^2}$

$= \dfrac{\frac{1}{2}x^{-1/2}[4x^{1/2} + x + 1]}{(1 - x)^2}$

$= \dfrac{4x^{1/2} + x + 1}{2x^{1/2}(1 - x)^2}$

69. $y = \dfrac{\sqrt{x - 1}}{x} = \dfrac{(x - 1)^{1/2}}{x}$

$\dfrac{dy}{dx} = \dfrac{x\left[\frac{1}{2}(x - 1)^{-1/2}\right] - (x - 1)^{1/2}(1)}{x^2}$

$= \dfrac{x(x - 1)^{-1/2} - 2(x - 1)^{1/2}}{2x^2}$

$= \dfrac{(x - 1)^{-1/2}[x - 2(x - 1)]}{2x^2}$

$= \dfrac{(x - 1)^{-1/2}(x - 2x + 2)}{2x^2}$

$= \dfrac{2 - x}{2x^2(x - 1)^{1/2}}$

71. $f'(-2)$ if $f(t) = \dfrac{2 - 3t}{\sqrt{2 + t}}$

When $t = -2$, $f(t)$ is not defined. Therefore, $f'(-2)$ does not exist.

73. (d) $F(x) = \dfrac{x^2}{1500}$ can be differentiated using the power rule.

75. $C(x) = \sqrt{3x + 2}$

$\overline{C}(x) = \dfrac{C(x)}{x} = \dfrac{\sqrt{3x + 2}}{x} = \dfrac{(3x + 2)^{1/2}}{x}$

$\overline{C}'(x) = \dfrac{x\left[\frac{1}{2}(3x + 2)^{-1/2}(3)\right] - (3x + 2)^{1/2}(1)}{x^2}$

$= \dfrac{\frac{3}{2}x(3x + 2)^{-1/2} - (3x + 2)^{1/2}}{x^2}$

$= \dfrac{3x(3x + 2)^{-1/2} - 2(3x + 2)^{1/2}}{2x^2}$

$= \dfrac{(3x + 2)^{-1/2}[3x - 2(3x + 2)]}{2x^2}$

$= \dfrac{3x - 6x - 4}{2x^2(3x + 2)^{1/2}} = \dfrac{-3x - 4}{2x^2(3x + 2)^{1/2}}$

77. $C(x) = (4x + 3)^4$

$\overline{C}(x) = \dfrac{C(x)}{x} = \dfrac{(4x + 3)^4}{x}$

$\overline{C}'(x) = \dfrac{x[4(4x+3)^3(4)] - (4x+3)^4(1)}{x^2}$

$= \dfrac{16x(4x+3)^3 - (4x+3)^4}{x^2}$

$= \dfrac{(4x+3)^3[16x - (4x+3)]}{x^2}$

$= \dfrac{(4x+3)^3(12x - 3)}{x^2}$

79. $P(x) = \dfrac{x^2}{x - 1}$

$P'(x) = \dfrac{(x-1)2x - (x^2)(1)}{(x-1)^2}$

$= \dfrac{2x^2 - 2x - x^2}{(x-1)^2}$

$= \dfrac{x^2 - 2x}{(x-1)^2}$

(a) $P'(4) = \dfrac{(4)^2 - 2(4)}{(4-1)^2}$

$= \dfrac{16 - 8}{9}$

$= \dfrac{8}{9}$

In dollars, this is $\dfrac{8}{9}(100) = \$88.89$, which represents the approximate increase in profit from selling the fifth unit.

(b) $P'(12) = \dfrac{(12)^2 - 2(12)}{(12-1)^2}$

$= \dfrac{144 - 24}{121}$

$= \dfrac{120}{121}$

In dollars, this is $\dfrac{120}{121}(100) =$ $99.17, which represents the approximate increase in profit from selling the thirteenth unit.

(c) $P'(20) = \dfrac{(20)^2 - 2(20)}{(20-1)^2}$

$= \dfrac{400 - 40}{361}$

$= \dfrac{360}{361}$

In dollars, this is $\dfrac{360}{361}(100) =$ $99.72, which represents the approximate increase in profit from selling the twenty-first unit.

(d) As the number of units sold increases, the marginal profit increases.

81. $R(x) = 5000 + 16x - 3x^2$

(a) $R'(x) = 16 - 6x$

(b) Since x is in hundreds of dollars, $1000 corresponds to $x = 10$.

$R'(10) = 16 - 6(10)$

$= 16 - 60 = -44$

So an increase of $100 on advertising when advertising expenditures are $1000 will result in the revenue decreasing by $44.

83. $P(x) = 15x + 25x^2$

(a) $P(6) = 15(6) + 25(6)^2$

$= 90 + 900 = 990$

$P(7) = 15(7) + 25(7)^2$

$= 105 + 1225 = 1330$

Average rate of change

$= \dfrac{P(7) - P(6)}{7 - 6}$

$= \dfrac{1330 - 990}{1}$

$= 340$ cents or $3.40

(b)   P(6) = 990

P(6.5) = 15(6.5) + 25(6.5)²

= 97.5 + 1056.25

= 1153.75

Average rate of change

$$= \frac{P(6.5) - P(6)}{6.5 - 6}$$

$$= \frac{1153.75 - 990}{.5}$$

= 327.5 cents or \$3.28

(c)   P(6) = 990

P(6.1) = 15(6.1) + 25(6.1)²

= 91.5 + 930.25

= 1021.75

Average rate of change

$$= \frac{P(6.1) - P(6)}{6.1 - 6}$$

$$= \frac{1021.75 - 990}{.1}$$

= 317.5 cents or \$3.18

(d) P'(x) = 15 + 50x

P'(6) = 15 + 50(6)

= 15 + 300

= 315 cents or \$3.15

(e) P'(20) = 15 + 50(20)

= 1015 cents or \$10.15

(f) P'(30) = 15 + 50(30)

= 1515 cents or \$15.15

(g) The domain of x is [0, ∞) since pounds cannot be measured with negative numbers.

(h) Since P'(x) = 15 + 50x gives the marginal profit, and x ≥ 0, P'(x) can never be negative.

(i) $\overline{P}(x) = \dfrac{P(x)}{x}$

$$= \frac{15x + 25x^2}{x}$$

= 15 + 25x

(j) $\overline{P}'(x) = 25$

(k) The marginal average profit cannot change since $\overline{P}'(x)$ is constant. The profit per pound never changes.

For Exercises 85 and 87, use a computer or graphing calculator. Solutions will vary according to the program that is used. Answers are given.

85. (a) The graph shows that there are no values of x where the derivative is positive.

(b) The graph shows that there are no values of x where the derivative is zero.

(c) The graph shows that the derivative is negative for all x-values, that is, for x in the interval (−∞, ∞).

(d) Since the derivative is always negative, the graph of g(x) is always decreasing.

87. (a) The graph shows that the derivative is positive where x-values are in the interval (−1, 1).

(b) G'(x) = 0 where x = −1.

(c) The graph shows that the derivative is negative for (−∞, −1) and (1, ∞).

(d) The derivative is 0 when $G(x)$
is at a low point. It is
positive where $G(x)$ is increasing
and negative where $G(x)$ is
decreasing.

CHAPTER 2   TEST

[2.1]   1.   Find each of the following limits, if it exists.

(a) $\lim_{x \to 1} f(x)$ (b) $\lim_{x \to -1} f(x)$

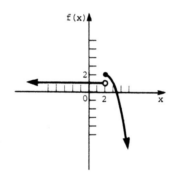   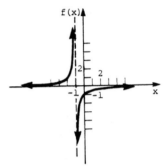

(c) $\lim_{x \to 2} f(x)$ (d) $\lim_{x \to -2} f(x)$

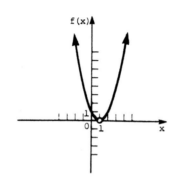   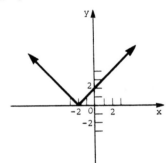

[2.1]   **Find each of the following limits, if it exists.**

2. $\lim_{x \to 2} \left(\dfrac{1}{x} + 1\right)(3x - 2)$ 3. $\lim_{x \to 0} \dfrac{3x + 5}{4x}$

4. $\lim_{x \to 2} \dfrac{x^2 - 5x + 6}{x - 2}$ 5. $\lim_{x \to 1} \dfrac{x - 1}{\sqrt{x} - 1}$

[2.2]   6.   Use the graph to find the average rate of change of f on the given intervals.

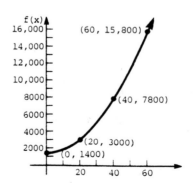

(a) From x = 0 to x = 20   (b) From x = 20 to x = 60

(c) From x = 0 to x = 40   (d) From x = 0 to x = 60

[2.2]   7. Find the average rate of change of $f(x) = x^3 - 5x$ from x = 1 to x = 5.

[2.3]   **Use the definition of the derivative to find the derivative of each function.**

   8. $y = x^3 - 5x^2$ 9. $y = \dfrac{3}{x}$

[2.3]   10. Suppose the total profit in thousands of dollars from selling x units is given by

   $$P(x) = 2x^2 - 6x + 9.$$

   (a) Find the average rate of change of profit as x increases from 2 to 4.

   (b) Find the marginal profit when 10 units are sold.

   (c) Find the average rate of change of profit when sales are increased from 100 to 200 units.

[2.4]   **Find the derivative of each function.**

   11. $y = 2x^4 - 3x^3 + 4x^2 - x + 1$ 12. $f(x) = x^{3/4} - x^{-2/3} + x^{-1}$

   13. $f(x) = -2x^{-2} - 3\sqrt{x}$

[2.4]   14. Find the slope of the tangent line to $y = -2x + \dfrac{1}{x} + \sqrt{x}$ at x = 1.

   Find the equation of the tangent line.

[2.5]   **Find the derivative of each function.**

   15. $f(x) = (2x^2 - 5x)(3x^2 + 1)$ 16. $y = \dfrac{x^2 + x}{x^3 - 1}$

[2.5]   17. Management has determined that the cost in thousands of dollars for producing x units is given by the equation

   $$C(x) = \frac{3x^2}{x^2 + 1} = 200.$$

   Find and interpret the marginal cost when x = 3.

[2.4]   18. At the beginning of an experiment, a culture is determined to have 2×10^4 bacteria. Thereafter, the number of bacteria observed at time t (in hours) is given by the equation

$$B(t) = 10^4(2 - 3\sqrt{t} + 2t + t^2).$$

How fast is the population growing at the end of 4 hours?

[2.6]   19. Find each of the following.

(a) $\dfrac{dy}{dt}$ if $y = \dfrac{\sqrt{x - 2}}{x + 1}$

(b) $D_x \left[\sqrt{(3x^2 - 1)^3}\right]$

(c) $f'(2)$ if $f(x) = \dfrac{2t - 1}{\sqrt{t + 2}}$

[2.6]   Find the derivative of each function.

20. $y = -2x\sqrt{3x - 1}$

21. $y = (5x^3 - 2x)^5$

[2.1]   22. Is $f(x) = \dfrac{x - 2}{x(3 - x)(x + 4)}$ continuous at the given values of x?

(a) $x = 2$ (b) $x = 0$ (c) $x = 5$ (d) $x = 3$

[2.1]   23. Use the graph to answer the following questions.

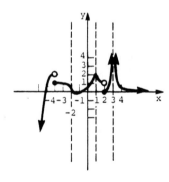

(a) On which of the following intervals is the graph continuous?

$(-5, -2)$, $(-3, 2)$, $(1, 4)$

(b) Where is the function discontinuous?

[2.4]   24. Explain the difference between the average cost function and the marginal cost function.

[2.1]   25. Under what circumstances could $\lim\limits_{x \to a} f(x)$ fail to exist?

CHAPTER 2   TEST ANSWERS

1. (a) 0   (b) Does not exist   (c) Does not exist   (d) 0   **2.** 6

3. Does not exist   **4.** −1   **5.** 2   **6.** (a) 80   (b) 320   (c) 160

(d) 240   **7.** 26   **8.** $y' = 3x^2 - 10x$ **9.** $y' = -3/x^2$

10. (a) $6000 per unit   (b) $34,000 per unit   (c) $594,000 per unit

11. $y' = 8x^3 - 9x^2 + 8x - 1$ **12.** $f'(x) = \dfrac{3}{4x^{1/4}} + \dfrac{2}{3x^{5/3}} - \dfrac{1}{x^2}$ **13.** $f'(x) = \dfrac{4}{x^3} - \dfrac{3}{2\sqrt{x}}$

14. $-\dfrac{5}{2}$; $5x + 2y = 5$ **15.** $f'(x) = 24x^3 - 45x^2 + 4x - 5$ **16.** $y' = \dfrac{-x^4 - 2x^3 - 2x - 1}{(x^3 - 1)^2}$

17. $C'(3) = .18$; after three units have been produced, the cost to produce one more unit will be approximately $.18(1000)$ or $180.

18. 9.25×10^4 bacteria per hour   **19.** (a) $\dfrac{5 - x}{2\sqrt{x} - 2(x + 1)^2}$ (b) $9x\sqrt{3x^2 - 1}$

(c) $\dfrac{13}{24}$ **20.** $y' = \dfrac{2 - 9x}{\sqrt{3x - 1}}$ **21.** $y' = 5(5x^3 - 2x)(15x^2 - 2)$

22. (a) Yes   (b) No   (c) Yes   (d) No   **23.** (a) $(-3, 2)$ (b) At $x = -4$, $x = 2$, and $x = 3$

24. The average cost function gives the cost per item when n items are produced. The marginal cost function gives the cost of the (n + 1)th item when n items are produced.

25. The limit may not exist if $x = a$ is a vertical asymptote or if the limit from the left does not equal the limit from the right.

CHAPTER 3 CURVE SKETCHING

Section 3.1

1. By reading the graph, f(x) is

   (a) increasing on (1, ∞) and

   (b) decreasing on (-∞, 1).

3. By reading the graph, g(x) is

   (a) increasing on (-∞, -2) and

   (b) decreasing on (-2, ∞).

5. By reading the graph, h(x) is

   (a) increasing on (-∞, -4) and
   (-2, ∞) and

   (b) decreasing on (-4, -2).

7. By reading the graph, f(x) is

   (a) increasing on (-7, -4) and
   (-2, ∞) and

   (b) decreasing on (-∞, -7) and
   (-4, -2).

9. $f(x) = x^2 + 12x - 6$

   $f'(x) = 2x + 12$

   $f'(x)$ is zero when

   $$2x(x + 6) = 0$$
   $$x = -6.$$

   Since f(x) is polynomial function, there are no values where f fails to exist. -6 is the only critical number.

   -6 determines two intervals on a number line.

Testing a point in each interval (any point will do other than the one where x = -6), we find the following.

   (a) $f'(0) = 12 > 0$, so f(x) is increasing on (-6, ∞).

   (b) $f'(-8) = -4 < 0$, so f(x) is decreasing on (-∞, -6).

11. $y = 2 + 3.6x - 1.2x^2$

    $y' = 3.6 - 2.4x$

    y' is zero when

    $$3.6 - 2.4x = 0$$
    $$x = \frac{3}{2}.$$

Test a point in each interval.

   (a) When x = 1, $y' = 1.2 > 0$, so y is increasing on (-∞, 3/2).

   (b) When x = 2, $y' = -1.2 < 0$, so y is decreasing on (3/2, ∞).

13. $f(x) = \frac{2}{3}x^3 - x^2 - 24x - 4$

    $f'(x) = 2x^2 - 2x - 24$

    $\qquad = 2(x^2 - x - 12)$

    $\qquad = 2(x + 3)(x - 4)$

    $f'(x)$ is zero when x = -3 or x = 4.

Test a point in each interval.

$f'(-4) = 16 > 0$

$f'(0) = -24 < 0$

$f'(5) = 16 > 0$

(a) $f'(x)$ is increasing on $(-\infty, -3)$ and $(4, \infty)$.

(b) $f(x)$ is decreasing on $(-3, 4)$.

15. $f(x) = 4x^3 - 15x^2 - 72x + 5$

$f'(x) = 12x^2 - 30x - 72$

$\qquad = 6(2x^2 - 5x - 12)$

$\qquad = 6(2x + 3)(x - 4)$

$f'(x)$ is zero when $x = -\dfrac{3}{2}$ or $x = 4$.

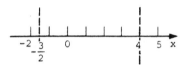

$f'(-2) = 36 > 0$

$f'(0) = -72 < 0$

$f'(5) = 78 > 0$

(a) $f(x)$ is increasing on $(-\infty, -3/2)$ and $(4, \infty)$.

(b) $f(x)$ is decreasing on $(-3/2, 4)$.

17. $y = -3x + 6$

$y' = -3 < 0$

(a) Since y' is always negative, the function is increasing on no interval.

(b) y' is always negative so the function is decreasing everywhere, or on the interval $(-\infty, \infty)$.

19. $f(x) = \dfrac{x + 2}{x + 1}$

$f'(x) = \dfrac{(x + 1)(1) - (x + 2)1}{(x + 1)^2}$

$\qquad = \dfrac{-1}{(x + 1)^2}$

Since $f(x)$ is a rational function, it fails to exist when $x + 1 = 0$, that is, when $x = -1$. Since the derivative is never zero, there are no other critical points.

$f'(-2) = -1 < 0$

$f'(0) = -1 < 0$

(a) $f(x)$ is increasing on no interval.

(b) $f(x)$ is decreasing everywhere that it is defined, on $(-\infty, -1)$ and on $(-1, \infty)$.

21. $y = |x + 4|$

$y = |x + 4|$ is equivalent to

$y = \begin{cases} x + 4 & \text{if } x \geq -4 \\ -x - 4 & \text{if } x < -4. \end{cases}$

Thus, we can think of y as a polynomial function over the entire real number line. (The student should test a few values of x to be sure that these functions are equivalent.) Now we can take the derivative:

$y' = \begin{cases} 1 & \text{if } x \geq -4 \\ -1 & \text{if } x < -4. \end{cases}$

(a) y is increasing on $(-4, \infty)$.

(b) y is decreasing on $(-\infty, -4)$.

23. $f(x) = -\sqrt{x-1} = -(x-1)^{1/2}$

Note that $f(x)$ is defined only for $x \geq 1$ and exists for all such values.

$f'(x) = -\frac{1}{2}(x-1)^{-1/2}$

$= \frac{-1}{2\sqrt{x-1}} < 0$ for all x.

Since $f'(x)$ is never zero, there are no critical points.

(a) $f(x)$ is increasing on no interval.

(b) $f(x)$ is decreasing on $(1, \infty)$.

25. $y = \sqrt{x^2 + 1} = (x^2 + 1)^{1/2}$

$y' = \frac{1}{2}(x^2 + 1)^{-1/2}(2x)$

$= x(x^2 + 1)^{-1/2}$

$= \frac{x}{\sqrt{x^2 + 1}}$

$y' = 0$ when $x = 0$.

Since y does not fail to exist for any x, and since $y' = 0$ when $x = 0$, $x = 0$ is the only critical point.

$y'(1) = \frac{1}{\sqrt{2}} > 0$

$y'(-1) = \frac{-1}{\sqrt{2}} < 0$

(a) y is increasing on $(0, \infty)$.

(b) y is decreasing on $(-\infty, 0)$.

27. $f(x) = x^{2/3}$

$f'(x) = \frac{2}{3}x^{-1/3} = \frac{2}{3x^{1/3}}$

$f'(x)$ is never zero, but fails to exist when $x = 0$.

$f'(-1) = -\frac{2}{3} < 0$

$f'(1) = \frac{2}{3} > 0$

(a) $f(x)$ is increasing on $(0, \infty)$.

(b) $f(x)$ is decreasing on $(-\infty, 0)$.

29. $f(x) = ax^2 + bx + c$

$f'(x) = 2ax + b$

Let $f'(x) = 0$ to find the critical number.

$2ax + b = 0$

$2ax = -b$

$x = \frac{-b}{2a}$

Choose a value in the interval $(-\infty, -b/2a)$. Since $a > 0$,

$\frac{-b}{2a} - \frac{1}{2a} = \frac{-b-1}{2a} < \frac{-b}{2a}$.

$f'\left(\frac{-b-1}{2a}\right) = 2a\left(\frac{-b-1}{2a}\right) + b$

$= -1 < 0$

Choose a value in the interval $(-b/2a, \infty)$. Since $a > 0$,

$\frac{-b}{2a} + \frac{1}{2a} = \frac{-b+1}{2a} > \frac{-b}{2a}$.

$f'\left(\frac{-b+1}{2a}\right) = 1 > 0$.

f(x) is increasing on (-b/2a, ∞) and decreasing on (-∞, -b/2a).

This tells us that the curve opens upward and x = -b/2a is the x-coordinate of the vertex.

$$f\left(\frac{-b}{2a}\right) = a\left(\frac{-b}{2a}\right)^2 + b\left(\frac{-b}{2a}\right) + c$$

$$= \frac{ab^2}{4a^2} - \frac{b^2}{2a} + c$$

$$= \frac{b^2}{4a} - \frac{2b^2}{4a} + \frac{4ac}{4a}$$

$$= \frac{4ac - b^2}{4a}$$

The vertex is $\left(-\dfrac{b}{2a}, \dfrac{4ac - b^2}{4a}\right)$.

31. $C(x) = x^3 - 2x^2 + 8x + 50$

$C'(x) = 3x^2 - 4x + 8$

Since C(x) is a polynomial function, the only critical points are found by solving C'(x) = 0.

$3x^2 - 4x + 8 = 0$

$$x = \frac{4 \pm \sqrt{16 - 96}}{6} = \frac{4 \pm \sqrt{-80}}{6}$$

Since $\sqrt{-80}$ is not a real number, there are no critical points. The function either always increases or always decreases.

Test any point, say x = 0.

$$C'(0) = 8 > 0$$

(a) C(x) is decreasing nowhere.

(b) C(x) is increasing everywhere.

33. $C(x) = 4.8x - .0004x^2, \ 0 \le x \le 2250$

$R(x) = 8.4x - .002x^2, \ 0 \le x \le 2250$

$P(x) = R(x) - C(x)$

$\quad = 8.4x - .002x^2 - (4.8x - .0004x^2)$

$\quad = 3.6x - .0016x^2$

$P'(x) = 3.6 - .0032x$

$P'(x) = 0$ when

$$x = \frac{3.6}{.0032}$$

$$= 1125.$$

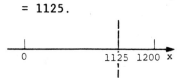

$P'(0) = 3.6 > 0$

$P'(1200) = -.24 < 0$

P is increasing on [0, 1125).

35. $P(t) = 2 + 50t - \dfrac{5}{2}t^2$

$P'(t) = 50 - 5t$

$\quad = 5(10 - t)$

P'(t) is zero when t = 10.

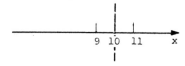

$P'(9) = 50 - 5(9) = 5 > 0$

$P'(11) = 50 - 5(11) = -5 < 0$

P(t) is increasing on (-∞, 10) and decreasing on (10, ∞), so the number of people infected starts to decline after 10 days.

37. $K(x) = \dfrac{4x}{3x^2 + 27}$ for x ≥ 0

$$K'(x) = \frac{(3x^2 + 27)4 - 4x(6x)}{(3x^2 + 27)^2}$$

$$= \frac{108 - 12x^2}{(3x^2 + 27)^2}$$

K'(x) is zero when

$$108 - 12x^2 = 0$$

$$12(9 - x^2) = 0$$

$$(3 - x)(3 + x) = 0$$

$$x = 3 \ \text{ or } \ x = -3.$$

However, $x \geq 0$, so $x = 3$ is the only critical point.

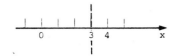

$K'(0) = \dfrac{108}{27^2} > 0$

$K'(4) = \dfrac{-84}{(48 + 27)^2} < 0$

(a) $K(x)$ is increasing on $(0, 3)$.
(Note: x must be at least 0.)

(b) $K(x)$ is decreasing on $(3, \infty)$.

39. As shown on the graph,

(a) horsepower increases with engine speed on $(1000, 6100)$

(b) horsepower decreases with engine speed on $(6100, 6500)$

(c) torque increases with engine speed on $(1000, 3000)$ and $(3600, 4200)$

(d) torque decreases with engine speed on $(3000, 3600)$ and $(4200, 6500)$.

Section 3.2

1. As shown on the graph, the relative minimum of -4 occurs when $x = 1$.

3. As shown on the graph, the relative maximum of 3 occurs when $x = -2$.

5. As shown on the graph, the relative maximum of 3 occurs when $x = -4$ and the relative minimum of 1 occurs when $x = -2$.

7. As shown on the graph, the relative maximum of 3 occurs when $x = -4$; the relative minimum of -2 occurs when $x = -7$ and $x = -2$.

9. $f(x) = x^2 + 12x - 8$
    $f'(x) = 2x + 12$
    $\qquad = 2(x + 6)$

    $f'(x)$ is zero when $x = -6$.

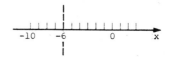

Test $f'(x)$ at -10 and 0.

$f'(-10) = -8 < 0$
$\quad f'(0) = 12 > 0$

Thus, we see that $f(x)$ is decreasing on $(-\infty, -6)$ and increasing on $(-6, \infty)$ so $f(-6)$ is a relative minimum.

$$f(-6) = (-6)^2 + 12(-6) - 8$$
$$= 36 - 72 - 8$$
$$= -44$$

Relative minimum of -44 at -6

11. $f(x) = 4 - 3x - .5x^2$
     $f'(x) = -3 - x$

     $f'(x)$ is zero when $x = -3$.

$f'(-4) = 1 > 0$

$f'(0) = -3 < 0$

$f(x)$ is increasing on $(-\infty, -3)$ and decreasing on $(-3, \infty)$. $f(-3)$ is a relative maximum.

$$f(-3) = 4 - 3(-3) - .5(-3)^2$$
$$= 4 + 9 - .5(9)$$
$$= 8.5$$

Relative maximum of 8.5 at -3

13. $f(x) = x^3 + 6x^2 + 9x - 8$

$f'(x) = 3x^2 + 12x + 9$
$\qquad = 3(x^2 + 4x + 3)$
$\qquad = 3(x + 3)(x + 1)$

$f'(x)$ is zero when $x = -1$ and $x = -3$.

$f'(-4) = 9 > 0$

$f'(-2) = -3 < 0$

$f'(0) = 9 > 0$

Thus, $f(x)$ is increasing on $(-\infty, -3)$, decreasing on $(-3, -1)$ and increasing on $(-1, \infty)$.
$f(x)$ has a relative maximum at -3 and a relative minimum at -1.

$$f(-3) = -8$$
$$f(-1) = -12$$

Relative maximum of -8 at -3;
relative minimum of -12 at -1

15. $f(x) = -\dfrac{4}{3}x^3 - \dfrac{21}{2}x^2 - 5x + 8$

$f'(x) = -4x^2 - 21x - 5$
$\qquad = (-4x - 1)(x + 5)$

$f'(x)$ is zero when $x = -5$, or $x = -\dfrac{1}{4}$.

$f'(-6) = -23 < 0$

$f'(-4) = 15 > 0$

$f'(0) = -5 < 0$

$f(x)$ is decreasing on $(-\infty, -5)$, increasing on $(-5, -1/4)$, and decreasing on $(-1/4, \infty)$. $f(x)$ has a relative minimum at -5 and a relative maximum at $-1/4$.

$$f(-5) = -\frac{377}{6}$$
$$f\left(-\frac{1}{4}\right) = \frac{827}{96}$$

Relative maximum of 827/96 at $-1/4$;
relative minimum of $-377/6$ at -5

17. $f(x) = 2x^3 - 21x^2 + 60x + 5$

$f'(x) = 6x^2 - 42x + 60$
$\qquad = 6(x^2 - 7x + 10)$
$\qquad = 6(x - 5)(x - 2)$

$f'(x)$ is zero when $x = 5$ or $x = 2$.

$f'(0) = 60 > 0$

$f'(3) = -12 < 0$

$f'(6) = 24 > 0$

$f(x)$ is increasing on $(-\infty, 2)$ and $(5, \infty)$; $f(x)$ is decreasing on $(2, 5)$.

f(2) = 57

f(5) = 30

Relative maximum of 57 at 2;
relative minimum of 30 at 5

19. $f(x) = x^4 - 18x^2 - 4$

$f'(x) = 4x^3 - 36x$

$= 4x(x^2 - 9)$

$= 4x(x + 3)(x - 3)$

$f'(x)$ is zero when $x = 0$ or $x = -3$
or $x = 3$.

$f'(-4) = 4(-4)^3 - 36(-4) = -112 < 0$

$f'(-1) = -4 + 36 = 32 > 0$

$f'(1) = 4 - 36 = -32 < 0$

$f'(4) = 4(4)^3 - 36(4) = 112 > 0$

$f(x)$ is decreasing on $(-\infty, -3)$ and
$(0, 4)$; $f(x)$ is increasing on
$(-3, 0)$ and $(3, \infty)$.

$f(-3) = -85$

$f(0) = -4$

$f(3) = -85$

Relative maximum of -4 at 0;
relative minimum of -85 at 3
and -3

21. $f(x) = -(8 - 5x)^{2/3}$

$f'(x) = -\frac{2}{3}(8 - 5x)^{-1/3}(-5)$

$= \frac{10}{3(8 - 5x)^{1/3}}$

$f'(x)$ is never zero but fails to
exist if $8 - 5x = 0$. 8/5 is a
critical number.

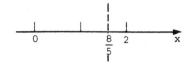

$f'(0) = \frac{10}{3(8)^{1/3}} = \frac{5}{3} > 0$

$f'(2) = \frac{10}{3(8 - 10)^{1/3}} \approx -2.6 < 0$

$f(x)$ is increasing on $(-\infty, 8/5)$ and
decreasing on $(8/5, \infty)$.

$f\left(\frac{8}{5}\right) = -\left[8 - 5\left(\frac{8}{5}\right)\right]^{2/3} = 0$

Relative maximum of 0 at 8/5

23. $f(x) = 2x + 3x^{2/3}$

$f'(x) = 2 + 2x^{-1/3}$

$= 2 + \frac{2}{\sqrt[3]{x}}$

Critical numbers:

$2 + \frac{2}{\sqrt[3]{x}} = 0$ $\sqrt[3]{x} = 0$

$\frac{2}{\sqrt[3]{x}} = -2$ $x = 0$

$\frac{1}{\sqrt[3]{x}} = -1$

$x = (-1)^3$

$x = -1$

f'(-2) = 2 + \frac{2}{\sqrt[3]{-2}} \approx .41 > 0

$f'\left(-\frac{1}{2}\right) = 2 + \frac{2}{\sqrt[3]{-\frac{1}{2}}}$

$= 2 + \frac{2\sqrt[3]{2}}{-1} \approx -.52 < 0$

$$f'(1) = 2 + \frac{2}{\sqrt[3]{1}} = 4 > 0$$

$f(x)$ is increasing on $(-\infty, -1)$ and $(0, \infty)$.

$f(x)$ is decreasing on $(-1, 0)$.

$$f(-1) = 2(-1) + 3(-1)^{2/3} = 1$$
$$f(0) = 0$$

Relative maximum of 1 at -1; relative minimum of 0 at 0

25. $f(x) = x - \dfrac{1}{x}$

$f'(x) = 1 + \dfrac{1}{x^2}$ is never zero, but fails to exist at $x = 0$.

Since $f(x)$ also fails to exist at $x = 0$, there are no critical numbers and no relative extrema.

27. $f(x) = \dfrac{x^2}{x^2 + 1}$

$$f'(x) = \frac{(x^2 + 1)2x - x^2(2x)}{(x^2 + 1)^2}$$

$$= \frac{2x}{(x^2 + 1)^2}$$

$f'(x)$ is zero when $x = 0$.

$$f'(-1) = -\frac{1}{2} < 0$$

$$f'(1) = \frac{1}{2} > 0$$

$f(x)$ is decreasing on $(-\infty, 0)$ and increasing on $(0, \infty)$.

$$f(0) = 0$$

Relative minimum of 0 at 0

29. $f(x) = \dfrac{x^2 - 2x + 1}{x - 3}$

$$f'(x) = \frac{(x - 3)(2x - 2) - (x^2 - 2x + 1)(1)}{(x - 3)^2}$$

$$= \frac{x^2 - 6x + 5}{(x - 3)^2}$$

Find critical numbers:

$$x^2 - 6x + 5 = 0$$
$$(x - 5)(x - 1) = 0$$
$$x = 5 \quad \text{or} \quad x = 1$$

Note that $f(x)$ and $f'(x)$ do not exist at $x = 3$.

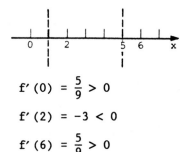

$$f'(0) = \frac{5}{9} > 0$$

$$f'(2) = -3 < 0$$

$$f'(6) = \frac{5}{9} > 0$$

$f(x)$ is increasing on $(-\infty, 1)$ and $(5, \infty)$.

$f(x)$ is decreasing on $(1, 5)$

$$f(1) = 0$$
$$f(5) = 8$$

Relative maximum of 0 at 1; relative minimum of 8 at 5

31. $y = -2x^2 + 8x - 1$

$y' = -4x + 8$

$\quad = 4(2 - x)$

The vertex occurs when $y' = 0$ or $x = 2$.

$$-2(2)^2 + 8(2) - 1 = 7$$

The vertex is at $(2, 7)$.

33. $y = 2x^2 - 5x + 2$

$y' = 4x - 5 = 0$ when

$$x = \frac{5}{4}$$

$$2\left(\frac{5}{4}\right)^2 - 5\left(\frac{5}{4}\right) + 2 = -\frac{9}{8}$$

The vertex is at $\left(\frac{5}{4}, -\frac{9}{8}\right)$.

35. $C(x) = 75 + 10x$; $p = 70 - 2x$

$P(x) = R(x) - C(x)$

$\qquad = px - C(x)$

$\qquad = (70 - 2x)x - (75 + 10x)$

$\qquad = 60x - 2x^2 - 75$

$P'(x) = 60 - 4x$

$P'(x) = 0$ when $x = 15$.

(a) $p = 70 - 2x$

$\qquad = 70 - 2(15)$

$\qquad = 40$, the price per unit that produces maximum profit.

(b) As shown above, the maximum profit occurs when 15 units are sold.

(c) $P(x) = 60x - 2x^2 - 75$

$P(15) = 60(15) - 2(15)^2 - 75$

$\qquad = 375$

The maximum profit is 375.

37. $C(x) = -13x^2 + 300x$;

$p = -x^2 - x + 360$

$P(x) = R(x) - C(x)$

$\qquad = px - C(x)$

$\qquad = (-x^2 - x + 360)x$

$\qquad\quad - (-13x^2 + 300x)$

$\qquad = -x^3 + 12x^2 + 60x$

$P'(x) = -3x^2 + 24x + 60$

$\qquad\quad = -3(x^2 - 8x - 20)$

$\qquad\quad = -3(x + 2)(x - 10)$

$P'(x) = 0$ when $x = -2$ or $x = 10$.

Disregard the negative value.

(a) $p = -x^2 - x + 360$

$\qquad = -(10)^2 - 10 + 360$

$\qquad = 250$, the price per unit that produces maximum profit.

(b) As shown above, the maximum profit occurs when 10 units are sold.

(c) $P(x) = -x^3 + 12x^3 + 60x$

$P(10) = -1000 + 1200 + 600$

$\qquad\quad = 800$

The maximum profit is 800.

39. $p = D(q) = \frac{1}{3}q^2 - \frac{25}{2}q + 100,$

$0 \leq q \leq 16$

$R = pq$

$\quad = \left(\frac{1}{3}q^2 - \frac{25}{2}q + 100\right)q$

$\quad = \frac{1}{3}q^3 - \frac{25}{2}q^2 + 100q$

$R' = q^2 - 25q + 100$

Set $R' = 0$.

$0 = q^2 - 25q + 100$

$\quad = (q - 20)(q - 5)$

$q = 20$ or $q = 5$

Since $0 \leq q \leq 16$, test only $q = 5$ for maximum.

$R'(4) = 16 - 100 + 100$

$\qquad\quad = 16 > 0$

$R'(6) = 36 - 150 + 100$

$\qquad\quad = -14 < 0$

So q = 5 maximizes R.

$$p = \frac{1}{3}q^2 - \frac{25}{2}q + 100$$

$$= \frac{1}{3}(25) - \frac{1}{2}(25) + 100$$

$$= \frac{275}{6}$$

41. $a(t) = .008t^3 - .27t^2 + 2.02t + 7$

t = the number of hours after 12 noon

$a'(t) = 3(.008t^2) - 2(.27t) + 2.02$

Set a' = 0 and use the quadratic formula to solve for t.

$.024t^2 - .54t + 2.02 = 0$

$.012t^2 - .27t + 1.01 = 0$

$$t = \frac{.27 \pm \sqrt{(-.27)^2 - 4(.012)(1.01)}}{2(.012)}$$

$t \approx 17.76$ or $t \approx 4.74$

Test for maximum or minimum:

$a'(17) = .012(17)^2 - .27(17) + 1.01$

$\qquad = 3.468 - 4.59 + 1.01 < 0$

$a'(18) = .012(18)^2 - .27(18) + 1.01$

$\qquad = 3.888 - 4.59 + 1.01 > 0$

Minimum activity occurs when t = 17.76.

$a'(4) = .012(4)^2 - .27(4) + 1.01$

$\qquad = .192 - 1.08 + 1.01 > 0$

$a'(5) = .012(5)^2 - .27(5) + 1.01 < 0$

$\qquad = .3 - 1.35 + 1.01 < 0$

Maximum activity occurs when t = 4.74.

When t = 4.74, the time is 4.74 hours past noon.

Since .74 hours = .74(60)

$\qquad\qquad\qquad = 44$ minutes,

the time is 4:44 P.M.

When t = 17.76, the time is 17.76 hours past noon or

17.76 - 12 = 5.76 hours past midnight.

Since .76 hours = .76(60)

$\qquad\qquad\qquad = 46$ minutes,

the time is 5:46 A.M.

Activity level is highest at 4:44 P.M. and lowest at 5:46 A.M.

43. $R(t) = \dfrac{20t}{t^2 + 100}$

$R'(t) = \dfrac{20(t^2 + 100) - 20t(2t)}{(t^2 + 100)^2}$

$\qquad = \dfrac{2000 - 20t^2}{(t^2 + 100)^2}$

$R'(t) = 0$ when $2000 - 20t^2 = 0$.

$$-20t^2 = -2000$$

$$t^2 = 100$$

$$t = \pm 10$$

Disregard the negative value. Verify that t = 10 gives a maximum rating:

$$R'(9) = .0116 > 0$$

$$R'(11) = -.0086 < 0$$

The film should be 10 minutes long.

Exercises 45 and 47 should be solved by computer or graphing calculator methods. The solutions will vary according to the method or computer program that is used.

45. $f(x) = x^5 - x^4 + 4x^3 - 30x^2 + 5x + 6$

The answer is that there is a relative maximum of 6.2 at .085 and a relative minimum of −57.7 at 2.2.

47. $f(x) = .001x^6 - .02x^5 - x^4 - 4x^3 + 12x^2 - 9x - 17$

The answer is that there are relative maxima of 280 at -5.1 and of -18.96 at $.89$ and a relative minimum of -19.08 at $.56$.

Section 3.3

1. As shown on the graph, the absolute maximum occurs at x_3; there is no absolute minimum. (There is no functional value that is less than all others.)

3. As shown on the graph, there are no absolute extrema.

5. As shown on the graph, the absolute minimum occurs at x_1; there is no absolute maximum.

7. As shown on the graph, the absolute maximum occurs at x_1; the absolute minimum occurs at x_2.

9. $f(x) = x^2 + 6x + 2$; $[-4, 0]$

Find critical number(s):

$$f'(x) = 2x + 6 = 0$$
$$2(x + 3) = 0$$
$$x = -3$$

Evaluate the function at -3 and at the endpoints.

x	f(x)	
-4	-6	
-3	-7	Absolute minimum
0	0	Absolute maximum

11. $f(x) = 5 - 8x - 4x^2$; $[-5, 1]$

Find critical number(s).

$$f'(x) = -8 + 8x = 0$$
$$8(x - 1) = 0$$
$$x = 1$$

Here, an endpoint is also a critical number.

x	f(x)	
-5	-55	Absolute minimum
1	-7	Absolute maximum

13. $f(x) = x^3 - 3x^2 - 24x + 5$; $[-3, 6]$

Find critical numbers:

$$f'(x) = 3x^2 - 6x - 24 = 0$$
$$3(x^2 - 2x - 8) = 0$$
$$3(x + 2)(x - 4) = 0$$
$$x = -2 \text{ or } 4$$

x	f(x)	
-3	23	
-2	32	Absolute maximum
4	-75	Absolute minimum
6	-31	

15. $f(x) = \frac{1}{3}x^3 - \frac{1}{2}x^2 - 6x + 3$; $[-4, 4]$

Find critical numbers:

$f'(x) = x^2 - x - 6 = 0$

$(x + 2)(x - 3) = 0$

$x = -2$ or 3

x	f(x)	
-4	$-\frac{7}{3} \approx -2.3$	
-2	$\frac{31}{3} \approx 10.3$	Absolute maximum
3	$-\frac{21}{2} \approx -10.5$	Absolute minimum
4	$-\frac{23}{3} \approx -7.7$	

17. $f(x) = x^4 - 32x^2 - 7$; $[-5, 6]$

$f'(x) = 4x^3 - 64x = 0$

$4x(x^2 - 16) = 0$

$4x(x - 4)(x + 4) = 0$

$x = 0, 4,$ or -4

x	f(x)	
-5	-182	
-4	-263	Absolute minimum
0	-7	
4	-263	Absolute minimum
6	137	Absolute maximum

19. $f(x) = \frac{1}{1 + x}$; $[0, 2]$

$f'(x) = \frac{-1}{(1 + x)^2}$

$f'(x)$ is never zero. Although $f'(x)$ does not exist if $x = -1$, -1 is not in the given interval.

x	f(x)	
0	1	Absolute maximum
2	$\frac{1}{3}$	Absolute minimum

21. $f(x) = \frac{8 + x}{8 - x}$; $[4, 6]$

$f'(x) = \frac{(8 - x)(1) - (8 + x)(-1)}{(8 - x)^2}$

$= \frac{16}{(8 - x)^2}$

$f'(x)$ is never zero. Although $f'(x)$ fails to exist if $x = 8$, 8 is not in the given interval.

x	f(x)	
4	3	Absolute minimum
6	7	Absolute maximum

23. $f(x) = \frac{x}{x^2 + 2}$; $[0, 4]$

$f'(x) = \frac{(x^2 + 2)1 - x(2x)}{(x^2 + 2)^2}$

$= \frac{-x^2 + 2}{(x^2 + 2)^2} = 0$

$-x^2 + 2 = 0$

$x^2 = 2$

$x = \sqrt{2}$ or $x = -\sqrt{2}$ but $-\sqrt{2}$ is not in $[0, 4]$.

$f'(x)$ is defined for all x.

x	f(x)	
0	0	Absolute minimum
$\sqrt{2}$	$\frac{\sqrt{2}}{4} \approx .35$	Absolute maximum
4	$\frac{2}{9} \approx .22$	

25. $f(x) = (x^2 + 18)^{2/3}$; $[-3, 3]$

$f'(x) = \frac{2}{3}(x^2 + 18)^{-1/3}(2x)$

$= \frac{4x}{3(x^2 + 18)^{1/3}}$

This derivative always exists, and is 0 when

$$\frac{4x}{3(x^2 + 18)^{1/3}} = 0,$$

$$4x = 0$$

$$x = 0.$$

x	f(x)	
0	$18^{2/3} \approx 6.87$	Absolute minimum
-3	9	Absolute maximum
3	9	Absolute maximum

27. $f(x) = (x + 1)(x + 2)^2$; $[-4, 0]$

$f'(x) = (x + 1)(2)(x + 2)(1)$
$\qquad + (x + 2)^2(1)$
$\qquad = 2(x + 1)(x + 2) + (x + 2)^2$
$\qquad = [2(x + 1) + x + 2](x + 2) = 0$
$\qquad\qquad (3x + 4)(x + 2) = 0$

$$x = -\frac{4}{3} \quad \text{or} \quad -2$$

x	f(x)	
-4	-12	Absolute minimum
-2	0	
$-\frac{4}{3}$	$-\frac{4}{9}$	
0	4	Absolute maximum

29. $f(x) = \dfrac{1}{\sqrt{x^2 + 1}}$; $[-1, 1]$

$f'(x)$

$= \dfrac{(x^2 + 1)^{1/2}(0) - 1\left(\frac{1}{2}\right)(x^2 + 1)^{-1/2}(2x)}{x^2 + 1}$

$= \dfrac{-x(x^2 + 1)^{-1/2}}{x^2 + 1} = 0$

$-x(x^2 + 1)^{-3/2} = 0$

$\dfrac{-x}{(\sqrt{x^2 + 1})^3} = 0$

$x = 0$

$f'(x)$ is defined for all x.

x	f(x)	
-1	$1/\sqrt{2} \approx .71$	Absolute minimum
0	1	Absolute maximum
1	$1/\sqrt{2} \approx .71$	Absolute minimum

31. $C(x) = x^2 + 200x + 100$

Since x is given in hundreds, the domain is $[1, 12]$.

Average cost per unit is

$$\overline{C}(x) = \frac{C(x)}{x} = \frac{x^2 + 200x + 100}{x}$$

$$= x + 200 + \frac{100}{x}.$$

$$\overline{C}'(x) = 1 - \frac{100}{x^2} = 0 \text{ when}$$

$$1 = \frac{100}{x^2}$$

$$x^2 = 100$$

$$x = 10.$$

Test for minimum.

$$\overline{C}'(5) = 1 - \frac{100}{25} = -3 < 0$$

$$\overline{C}'(15) = 1 - \frac{100}{225} = 1 - \frac{4}{9} = \frac{5}{9} > 0$$

The average cost per unit is as small as possible when x = 10, or 1000 manuals are produced. If the student produces 1000 manuals, his cost will be

$$C(10) = 10^2 + 200(10) + 100$$
$$= \$2200.$$

Thus, in order to make a profit, for each copy the student will have to charge at least

$$\frac{2200}{1000} = 2.2,$$

or more than $2.20.

33. $C(x) = 81x^2 + 17x + 324$

(a) $1 \le x \le 10$

$$\overline{C}(x) = \frac{C(x)}{x} = \frac{81x^2 + 17x + 324}{x}$$

$$= 81x + 17 + \frac{324}{x}$$

$$\overline{C}'(x) = 81 - \frac{324}{x^2} = 0 \text{ when}$$

$$81 = \frac{324}{x^2}$$

$$x^2 = 4$$

$$x = 2.$$

Test for minimum.

$\overline{C}'(1) = -243 < 0$

$\overline{C}'(3) = 45 > 0$

$\overline{C}(x)$ is a minimum when $x = 2$.

$$\overline{C}(2) = 81(2) + 17 + \frac{324}{2} = 341$$

The minimum for $1 \le x \le 10$ is 341.

(b) $10 \le x \le 20$

There are no critical values in this interval. Test the end-points.

$\overline{C}(10) = 859.4$

$\overline{C}(20) = 1620 + 17 + 16.2$

$\qquad = 1653.2$

The minimum for $10 \le x \le 20$ is 859.4.

35. $f(x) = \frac{x^2 + 36}{2x}, \ 1 \le x \le 12$

$$f'(x) = \frac{2x(2x) - (x^2 + 36)(2)}{(2x)^2}$$

$$= \frac{4x^2 - 2x^2 - 72}{4x^2}$$

$$= \frac{2x^2 - 72}{4x^2}$$

$$= \frac{2(x^2 - 36)}{4x^2}$$

$$= \frac{(x + 6)(x - 6)}{2x^2}$$

$f'(x) = 0$ when $x = 6$ and when $x = -6$. Only 6 is in the interval $1 \le x \le -12$.

Test for relative maximum or minimum.

$$f'(5) = \frac{(11)(-1)}{50} < 0$$

$$f'(7) = \frac{(13)(1)}{98} > 0$$

So the minimum is at $x = 6$, or at 6 months. Since $f(6) = 6$, the minimum percent is 6%.

37. $M(x) = -\frac{1}{45}x^2 + 2x - 20, \ 30 \le x \le 65$

$$M'(x) = -\frac{1}{45}(2x) + 2$$

$$= -\frac{2x}{45} + 2$$

When $M'(x) = 0$,

$$-\frac{2x}{45} + 2 = 0$$

$$2 = \frac{2x}{45}$$

$$45 = x.$$

x	M(x)
30	20
45	25
65	$145/9 \approx 16.1$

The absolute maximum miles per gallon is 25 and the absolute minimum miles per gallon is about 16.1.

39. Total area $A(x) = \pi\left(\dfrac{x}{2\pi}\right)^2 + \left(\dfrac{12 - x}{4}\right)^2$

$$= \frac{x^2}{4\pi} + \frac{(12 - x)^2}{16}$$

$A'(x) = \dfrac{x}{2\pi} - \dfrac{12 - x}{8} = 0$

$$\frac{4x - \pi(12 - x)}{8\pi} = 0$$

$$x = \frac{12\pi}{4 + \pi} \approx 5.28$$

x	Area
0	9
5.28	5.04
12	11.46

The total area is minimized when the piece used to form the circle is $\dfrac{12\pi}{4 + \pi}$ feet, or about 5.28 feet.

Exercises 41 and 43 should be solved by computer or graphing calculator methods. The solutions will vary according to the method or computer program that is used.

41. $f(x) = \dfrac{x^3 + 2x + 5}{x^4 + 3x^3 + 10}; \ [-3, 0]$

The answer is the absolute maximum is at 0 and the absolute minimum is at -2.4

43. $f(x) = x^{4/5}$ at $x^2 - 4\sqrt{x}; \ [0, 4]$

The answer is the absolute maximum is at 0 and the absolute minimum is at .74.

Section 3.4

1. $f(x) = 3x^3 - 4x + 5$

$f'(x) = 9x^2 - 4$

$f''(x) = 18x$

$f''(0) = 18(0) = 0$

$f''(2) = 18(2) = 36$

$f''(-3) = 18(-3) = -54$

3. $f(x) = 3x^4 - 5x^3 + 2x^2$

$f'(x) = 12x^3 - 15x^2 + 4x$

$f''(x) = 36x^2 - 30x + 4$

$f''(0) = 36(0)^2 - 30(0) + 4 = 4$

$f''(2) = 36(2)^2 - 30(2) + 4 = 88$

$f''(-3) = 36(-3)^2 - 30(-3) + 4 = 418$

5. $f(x) = 3x^2 - 4x + 8$

$f'(x) = 6x - 4$

$f''(x) = 6$

$f''(0) = 6$

$f''(2) = 6$

$f''(-3) = 6$

7. $f(x) = (x + 4)^3$

$f'(x) = 3(x + 4)^2(1)$

$\quad = 3(x + 4)^2$

$f''(x) = 6(x + 4)(1)$

$\quad = 6(x + 4)$

$$f''(0) = 6(0 + 4) = 24$$
$$f''(2) = 6(2 + 4) = 36$$
$$f''(-3) = 6(-3 + 4) = 6$$

9. $f(x) = \dfrac{2x + 1}{x - 2}$

$$f'(x) = \dfrac{(x - 2)(2) - (2x + 1)(1)}{(x - 2)^2}$$

$$= \dfrac{-5}{(x - 2)^2}$$

$$f''(x) = \dfrac{(x - 2)^2(0) - (-5)2(x - 2)(1)}{(x - 2)^4}$$

$$= \dfrac{10}{(x - 2)^3}$$

$$f''(0) = \dfrac{10}{(0 - 2)^3} = -\dfrac{5}{4}$$

$f''(2)$ does not exist.

$$f''(-3) = \dfrac{10}{(-3 - 2)^3} = -\dfrac{2}{25}$$

11. $f(x) = \dfrac{x^2}{1 + x}$

$$f'(x) = \dfrac{(1 + x)(2x) - x^2(1)}{(1 + x)^2}$$

$$= \dfrac{2x + x^2}{(1 + x)^2}$$

$$f''(x) = \dfrac{(1+x)^2(2+2x) - (2x+x^2)(2)(1+x)}{(1 + x)^4}$$

$$= \dfrac{(1 + x)(2 + 2x) - (2x + x^2)(2)}{(1 + x)^3}$$

$$= \dfrac{2}{(1 + x)^3}$$

$$f''(0) = 2$$

$$f''(2) = \dfrac{2}{27}$$

$$f''(-3) = -\dfrac{1}{4}$$

13. $f(x) = \sqrt{x + 4} = (x + 4)^{1/2}$

$$f'(x) = \dfrac{1}{2}(x + 4)^{-1/2}$$

$$f''(x) = \dfrac{1}{2}\left(-\dfrac{1}{2}\right)(x + 4)^{-3/2}$$

$$= -\dfrac{1}{4}(x + 4)^{-3/2}$$

$$= \dfrac{-1}{4(x + 4)^{3/2}}$$

$$f''(0) = \dfrac{-1}{4(0 + 4)^{3/2}} = \dfrac{-1}{4(4^{3/2})}$$

$$= \dfrac{-1}{4(8)}$$

$$= -\dfrac{1}{32}$$

$$f''(2) = \dfrac{-1}{4(2 + 4)^{3/2}}$$

$$= \dfrac{-1}{4(6^{3/2})} \approx -.0170$$

$$f''(-3) = \dfrac{-1}{4(-3 + 4)^{3/2}} = \dfrac{-1}{4(1^{3/2})}$$

$$= -\dfrac{1}{4}$$

15. $f(x) = 5x^{3/5}$

$$f'(x) = 3x^{-2/5}$$

$$f''(x) = -\dfrac{6}{5}x^{-7/5} \quad \text{or} \quad \dfrac{-6}{5x^{7/5}}$$

$f''(0)$ does not exist.

$$f''(2) = -\dfrac{6}{5}(2^{-7/5}) = \dfrac{-6}{5(2^{7/5})}$$

$$\approx -.4547$$

$$f''(-3) = -\dfrac{6}{5}(-3)^{-7/5} = \dfrac{-6}{5(-3)^{7/5}}$$

$$\approx .2578$$

17. $f(x) = -x^4 + 2x^2 + 8$

$f'(x) = -4x^3 + 4x$

$f''(x) = -12x^2 + 4$

$f'''(x) = -24x$

$f^{(4)}(x) = -24$

19. $f(x) = 4x^5 + 6x^4 - x^2 + 2$

$f'(x) = 20x^4 + 24x^3 - 2x$

$f''(x) = 80x^3 + 72x^2 - 2$

$f'''(x) = 240x^2 + 144x$

$f^{(4)}(x) = 480x + 144$

21. $f(x) = \dfrac{x - 1}{x + 2}$

$f'(x) = \dfrac{(x + 2) - (x - 1)}{(x + 2)^2}$

$= \dfrac{3}{(x + 2)^2}$

$f''(x) = \dfrac{-3(2)(x + 2)}{(x + 2)^4}$

$= \dfrac{-6}{(x + 2)^3}$

$f'''(x) = \dfrac{(-6)(-3)(x + 2)^2}{(x + 2)^6}$

$= \dfrac{18}{(x + 2)^4}$ or $18(x + 2)^{-4}$

$f^{(4)}(x) = \dfrac{-18(4)(x + 2)^3}{(x + 2)^8}$

$= \dfrac{-72}{(x + 2)^5}$

or $-72(x + 2)^{-5}$

23. $f(x) = \dfrac{3x}{x - 2}$

$f'(x) = \dfrac{(x - 2)(3) - 3x(1)}{(x - 2)^2}$

$= \dfrac{-6}{(x - 2)^2}$

$f''(x) = \dfrac{-6(-2)(x - 2)}{(x - 2)^4}$

$= \dfrac{12}{(x - 2)^3}$

$f'''(x) = \dfrac{-12(3)(x - 2)^2}{(x - 2)^6}$

$= -36(x - 2)^{-4}$

or $\dfrac{-36}{(x - 2)^4}$

$f^{(4)}(x) = \dfrac{-36(-4)(x - 2)^3}{(x - 2)^8}$

$= 144(x - 2)^{-5}$

or $\dfrac{144}{(x - 2)^5}$

25. Concave upward on $(2, \infty)$

Concave downward on $(-\infty, 2)$

Point of inflection at $(2, 3)$

27. Concave upward on $(-\infty, -1)$ and $(8, \infty)$.

Concave downward on $(-1, 8)$

Points of inflection at $(-1, 7)$ and $(8, 6)$

29. Concave upward on $(2, \infty)$

Concave downward on $(-\infty, 2)$

No points of inflection

31. $f(x) = x^2 + 10x - 9$

$f'(x) = 2x + 10$

$f''(x) = 2 > 0$ for all x.

Always concave upward

No points of inflection

33. $f(x) = x^3 + 3x^2 - 45x - 3$

$f'(x) = 3x^2 + 6x - 45$

$f''(x) = 6x + 6$

$f''(x) = 6x + 6 > 0$ when

$6(x + 1) > 0$

$x > -1.$

Concave upward on $(-1, \infty)$

$f''(x) = 6x + 6 < 0$ when

$$6(x + 1) < 0$$

$$x < -1.$$

Concave downward on $(-\infty, -1)$

$f''(x) = 6x + 6 = 0$

$$6(x + 1) = 0$$

$$x = -1$$

$f(1) = 44$

Point of inflection at $(-1, 44)$

35. $f(x) = -2x^3 + 9x^2 + 168x - 3$

$f'(x) = -6x^2 + 18x + 168$

$f''(x) = -12x + 18$

$f''(x) = -12x + 18 > 0$ when

$$-6(2x - 3) > 0$$

$$2x - 3 < 0$$

$$x < \frac{3}{2}.$$

Concave upward on $(-\infty, 3/2)$

$f''(x) = -12x + 18 < 0$ when

$$-6(2x - 3) < 0$$

$$2x - 3 > 0$$

$$x > \frac{3}{2}.$$

Concave downward on $(3/2, \infty)$

$f''(x) = -12x + 18 = 0$ when

$$-6(2x + 3) = 0$$

$$2x + 3 = 0$$

$$x = \frac{3}{2}.$$

$$f\left(\frac{3}{2}\right) = \frac{525}{2}$$

Point of inflection at $(3/2, 525/2)$

37. $f(x) = \dfrac{3}{x - 5}$

$f'(x) = \dfrac{-3}{(x - 5)^2}$

$f''(x) = \dfrac{-3(-2)(x - 5)}{(x - 5)^4}$

$$= \dfrac{6}{(x - 5)^3}$$

$f''(x) = \dfrac{6}{(x - 5)^3} > 0$ when

$$(x - 5)^3 > 0$$

$$x - 5 > 0$$

$$x > 5.$$

Concave upward on $(5, \infty)$

$f''(x) = \dfrac{6}{(x - 5)^3} < 0$ when

$$(x - 5)^3 < 0$$

$$x - 5 < 0$$

$$x < 5.$$

Concave downward on $(-\infty, 5)$

$f''(x) \neq 0$ for any value of x; it does not exist when $x = 5$. There is a change of concavity there, but no point of inflection since $f(5)$ does not exist.

39. $f(x) = x(x + 5)^2$

$f'(x) = x(2)(x + 5) + (x + 5)^2$

$$= (x + 5)(2x + x + 5)$$

$$= (x + 5)(3x + 5)$$

$f''(x) = (x + 5)(3) + (3x + 5)$

$$= 3x + 15 + 3x + 5$$

$$= 6x + 20$$

$f''(x) = 6x + 20 > 0$ when

$$2(3x + 10) > 0$$

$$3x > -10$$

$$x > -\frac{10}{3}.$$

Concave upward on $(-10/3, \infty)$

$f''(x) = 6x + 20 < 0$ when

$$2(3x + 10) < 0$$

$$3x < -10$$

$$x < -\frac{10}{3}.$$

Concave downward on $(-\infty, -10/3)$

$$f\left(-\frac{10}{3}\right) = -\frac{10}{3}\left(-\frac{10}{3} + 5\right)^2$$

$$= \frac{-10}{3}\left(\frac{-10 + 15}{3}\right)^2$$

$$= -\frac{10}{3} \cdot \frac{25}{9}$$

$$= -\frac{250}{27}$$

Point of inflection at

$$(-10/3, -250/27)$$

41. $f(x) = -x^2 - 10x - 25$

$f'(x) = -2x - 10$

$$= -2(x + 5) = 0$$

Critical number:

$$x = -5$$

$f''(x) = -2 < 0$ for all x.

The curve is concave downward, which means a relative maximum occurs at $x = -5$.

43. $f(x) = 3x^3 - 3x^2 + 1$

$f'(x) = 9x^2 - 6x$

$$= 3x(3x - 2) = 0$$

Critical numbers: $x = 0$ or $x = \frac{2}{3}$

$f''(x) = 18x - 6$

$f''(0) = -6 < 0$, which means that a relative maximum occurs at $x = 0$.

$f''\left(\frac{2}{3}\right) = 6 > 0$, which means that a relative minimum occurs at $x = 2/3$.

45. $f(x) = (x + 3)^4$

$f'(x) = 4(x + 3)^3$

$$= 4(x + 3)^3 = 0$$

Critical number: $x = -3$

$f''(x) = 12(x + 3)^2$

$f''(-3) = 12(-3 + 3)^2 = 0$

The second derivative test fails.

Use the first derivative test.

$f'(-4) = 4(-4 + 3)^3$

$$= 4(-1)^3 = -4 < 0$$

This indicates that f is decreasing on $(-\infty, -3)$.

$f'(0) = 4(0 + 3)^3 > 0$

This indicates that f is increasing on $(-3, \infty)$.

A relative minimum occurs at -3.

47.   (a) The left side of the graph changes from concave upward to concave downward at the point of inflection between car phones and CD players. The rate of growth of sales begins to decline at the point of inflection.

(b) Food processors are closest to the right-hand point of inflection. This inflection point indicates that the rate of decline of sales is beginning to slow.

49. $R(x) = \frac{4}{27}(-x^3 + 66x^2 + 1050x - 400)$;

$0 \le x \le 25$

$R'(x) = \frac{4}{27}(-3x^2 + 132x + 1050)$

$R''(x) = \frac{4}{27}(-6x + 132)$

A point of diminishing returns occurs at a point of inflection, or where $R''(x) = 0$.

$$\frac{4}{27}(-6x + 132) = 0$$

$$-6x + 132 = 0$$

$$6x = 132$$

$$x = 22$$

Test $R''(x)$ to determine whether concavity changes at $x = 22$.

$R''(20) = \frac{4}{27}(-6 \cdot 20 + 132) = \frac{16}{9} > 0$

$R''(24) = \frac{4}{27}(-6 \cdot 24 + 132) = -\frac{16}{9} < 0$

$R(x)$ is concave up on $(0, 22)$ and concave down on $(22, 25)$.

$R(22)$

$= \frac{4}{27}[-22^3 + 66(22)^2 + 1050(22) - 400]$

≈ 6517.9

The point of diminishing returns is $(22, 6517.9)$.

51. Let $D(q)$ represent the demand function.
The revenue function, $R(q)$, is
$R(q) = qD(q)$.

The marginal revenue is given by

$R'(q) = qD'(q) + D(q)(1)$

$\qquad = qD'(q) + D(q)$.

$R''(q) = qD''(q) + D'(q)(1) + D'(q)$

$\qquad = qD''(q) + 2D'(q)$

gives the rate of decline of marginal revenue.

$D'(q)$ gives the rate of decline of price.

If marginal revenue declines more quickly than price,

$$qD''(q) + 2D'(q) - D'(q) < 0$$

or $\quad qD''(q) + D'(q) < 0$.

53. (a) $R(t) = t^2(t - 18) + 96t + 1000$;

$0 < t < 8$

$\qquad = t^3 - 18t^2 + 96t + 1000$

$R'(t) = 3t^2 - 36t + 96$

Set $R'(t) = 0$.

$$3t^2 - 36t + 96 = 0$$

$$t^2 - 12t + 32 = 0$$

$$(t - 8)(t - 4) = 0$$

$$t = 8 \quad \text{or} \quad t = 4$$

8 is not in the domain of $R(t)$.

$R'' = 6t - 36$

$R''(4) = -12 < 0$ implies that $R(t)$ is maximized at $t = 4$ hr.

(b) $R(4) = 16(-14) + 96(4) + 1000$

$\qquad = -224 + 384 + 1000$

$\qquad = 1160$

The maximum population is 1160 million.

55. (a) $K(x) = \dfrac{4x}{3x^2 + 27}$; $x > 0$

$K'(x) = \dfrac{(3x^2 + 27)(4) - 4x(6x)}{(3x^2 + 27)^2}$

$= \dfrac{-12x^2 + 108}{(3x^2 + 27)^2}$

Set $K'(x) = 0$.

$-12(x^2 - 9) = 0$

$x = -3$ or $x = 3$

-3 is not in the domain of $K(x)$.

$K''(x) = \dfrac{(3x^2 + 27)^2(-24x) - (-12x^2 + 108)(2)(3x^2 + 27)(6x)}{(3x^2 + 27)^4}$

$= \dfrac{(3x^2 + 27)^2(-24x) - (-2x(3x^2 + 27)(-12x^2 + 108)}{(3x^2 + 27)^4}$

$K''(3) = -\dfrac{72}{54} < 0$ implies that $K(x)$ is maximized at $x = 3$ hours.

(b) $K(3) = \dfrac{4(3)}{3(3)^2 + 27}$

$= \dfrac{2}{9}$

The maximum concentration is 2/9%.

57. $s(t) = 8t^2 + 4t$

$v(t) = s'(t) = 16t + 4$

$a(t) = v'(t) = s''(t) = 16$

$v(0) = 4$ cm/sec

$v(4) = 68$ cm/sec

$a(0) = 16$ cm/sec^2

$a(4) = 16$ cm/sec^2

59. $s(t) = -5t^3 - 8t^2 + 6t - 3$

$v(t) = s'(t) = -15t^2 - 16t + 6$

$a(t) = v'(t) = s''(t) = -30t - 16$

$v(0) = 6$ cm/sec

$v(4) = -298$ cm/sec

$a(0) = -16$ cm/sec^2

$a(4) = -136$ cm/sec^2

61. $s(t) = \dfrac{-2}{3t + 4} = -2(3t + 4)^{-1}$

$v(t) = s'(t) = 2(3t + 4)^{-2}(3)$

$= 6(3t + 4)^{-2}$

$= \dfrac{6}{(3t + 4)^2}$

$$a(t) = v'(t) = s''(t)$$
$$= -12(3t + 4)^{-3}(3)$$
$$= -36(3t + 4)^{-3}$$
$$= \frac{-36}{(3t + 4)^3}$$

$$v(0) = \frac{6}{(0 + 4)^2} = \frac{3}{8} \text{ cm/sec}$$

$$v(4) = \frac{6}{(12 + 4)^2} = \frac{3}{128} \text{ cm/sec}$$

$$a(0) = \frac{-36}{(0 + 4)^3} = -\frac{9}{16} \text{ cm/sec}^2$$

$$a(4) = \frac{-36}{(12 + 4)^3} = -\frac{9}{1024} \text{ cm/sec}^2$$

63. $s(t) = -16t^2$

   $v(t) = s'(t) = -32t$

   (a) $v(3) = -32(3) = -96$ ft/sec

   (b) $v(5) = -32(5) = -160$ ft/sec

   (c) $v(8) = -32(8) = -256$ ft/sec

   (d) $a(t) = v'(t) = s''(t)$
   $$= -32 \text{ ft/sec}^2$$

Exercises 65 and 67 should be solved by computer or graphing calculator methods. The solutions will vary according to the method or computer program that is used.

65. $f(x) = .25x^4 - 2x^3 + 3.5x^2 + 4x - 1;$
   $(-5, 5)$ in steps of .5

   The answers are

   (a) increasing on $(0, 2)$ and $(4, 5)$; decreasing on $(-5, -.5)$ and $(2.5, 3.5)$

   (b) minima between $-.5$ and 0, and between 3.5 and 4; a maximum between 2 and 2.5

(c) concave upward on $(-5, .5)$ and $(3.5, 5)$; concave downward on $(.5, 3.5)$

(d) inflection points between .5 and 1, and between 3 and 3.5.

67. $f(x) = 3.1x^4 - 4.3x^3 + 5.82;$ $(-1, 2)$ in steps of .2

   The answers are

   (a) decreasing on $(-1, 1)$; increasing on $(1.2, 2)$

   (b) minimum between 1 and 1.2

   (c) concave upward on $(-1, 0)$ and $(.8, 2)$; concave downward on $(0, .8)$

   (d) inflection points between .6 and .8, and at 0.

Section 3.5

1. As x increases without bound $(x \to \infty)$, the graph approaches the line $y = 3$. Thus $\lim\limits_{x \to \infty} f(x)$ exists and has the value 3.

3. $\lim\limits_{x \to \infty} \dfrac{3x}{5x - 1}$

   $= \lim\limits_{x \to \infty} \dfrac{\dfrac{3x}{x}}{\dfrac{5x - 1}{x}}$

   $= \lim\limits_{x \to \infty} \dfrac{3}{5 - \dfrac{1}{x}}$

$$= \frac{\lim\limits_{x \to \infty} 3}{\lim\limits_{x \to \infty} 5 + \lim\limits_{x \to \infty} \left(-\frac{1}{x}\right)}$$

$$= \frac{3}{5 + 0} = \frac{3}{5}$$

5. $\lim\limits_{x \to \infty} \dfrac{2x + 3}{4x - 7}$

$$= \lim\limits_{x \to \infty} \frac{2 + \dfrac{3}{x}}{4 - \dfrac{7}{x}}$$

$$= \frac{\lim\limits_{x \to \infty} 2 + \lim\limits_{x \to \infty} \dfrac{3}{x}}{\lim\limits_{x \to \infty} 4 + \lim\limits_{x \to \infty} \left(-\dfrac{7}{x}\right)}$$

$$= \frac{2 + 0}{4 + 0}$$

7. $\lim\limits_{x \to \infty} \dfrac{x^2 + 2x}{2x^2 - 2x + 1}$

$$= \lim\limits_{x \to \infty} \frac{1 + \dfrac{2}{x}}{2 - \dfrac{2}{x} + \dfrac{1}{x^2}}$$

$$= \frac{\lim\limits_{x \to \infty} \left(1 + \dfrac{2}{x}\right)}{\lim\limits_{x \to \infty} \left(2 - \dfrac{2}{x} + \dfrac{1}{x^2}\right)}$$

$$= \frac{\lim\limits_{x \to \infty} 1 + \lim\limits_{x \to \infty} \dfrac{2}{x}}{\lim\limits_{x \to \infty} 2 + \lim\limits_{x \to \infty} \left(-\dfrac{2}{x}\right) + \lim\limits_{x \to \infty} \dfrac{1}{x^2}}$$

$$= \frac{1 + 0}{2 + 0 + 0}$$

$$= \frac{1}{2}$$

9. $\lim\limits_{x \to \infty} \dfrac{3x^3 + 2x - 1}{2x^4 - 3x^3 - 2}$

$$= \lim\limits_{x \to \infty} \frac{\dfrac{3}{x} + \dfrac{2}{x^3} - \dfrac{1}{x^4}}{2 - \dfrac{3}{x} - \dfrac{2}{x^4}}$$

$$= \frac{\lim\limits_{x \to \infty} \dfrac{3}{x} + \lim\limits_{x \to \infty} \dfrac{2}{x^3} + \lim\limits_{x \to \infty} \left(\dfrac{-1}{x^4}\right)}{\lim\limits_{x \to \infty} 2 + \lim\limits_{x \to \infty} \left(\dfrac{-3}{x}\right) + \lim\limits_{x \to \infty} \left(\dfrac{-2}{x^4}\right)}$$

$$= \frac{0 + 0 + 0}{2 + 0 + 0}$$

$$= 0$$

11. $\lim\limits_{x \to \infty} \left(\sqrt{x^2 + 4} - x\right)$

$$= \lim\limits_{x \to \infty} \left(\sqrt{x^2 + 4} - x\right) \cdot \frac{\left(\sqrt{x^3 + 4} + x\right)}{\left(\sqrt{x^2 + 4} + x\right)}$$

$$= \lim\limits_{x \to \infty} \frac{(x^2 + 4) - x^2}{\sqrt{x^2 + 4} + x}$$

$$= \lim\limits_{x \to \infty} \frac{4}{\sqrt{x^2 + 4} + x}$$

$$= \lim\limits_{x \to \infty} \frac{\dfrac{4}{x}}{\sqrt{\dfrac{x^2 + 4}{x^2}} + \dfrac{x}{x}}$$

$$= \lim\limits_{x \to \infty} \frac{\dfrac{4}{x}}{\sqrt{1 + \dfrac{4}{x^2}} + 1}$$

$$= \frac{\lim\limits_{x \to \infty} \dfrac{4}{x}}{\lim\limits_{x \to \infty} \left(\sqrt{1 + \dfrac{4}{x^2}} + 1\right)}$$

$$= \frac{0}{\sqrt{\lim\limits_{x \to \infty} \left(1 + \dfrac{4}{x^2}\right)} + \lim\limits_{x \to \infty} 1}$$

$$= \frac{0}{1 + 1} = \frac{0}{2}$$

$$= 0$$

For Exercises 13-33, see the answer graphs in the back of the textbook.

13. $f(x) = -2x^3 - 9x^2 + 108x - 10$

$f'(x) = -6x^2 - 18x + 108$

$= -6(x^2 + 3x - 18) = 0$

$(x + 6)(x - 3) = 0$

Critical numbers: $x = -6$ and $x = 3$

Critical points: $(-6, -550)$ and $(3, 179)$

$f''(x) = -12x - 18$

$f''(-6) = 54 > 0$

$f''(3) = -54 < 0$

Relative maximum at 3, relative minimum at -6

Increasing on $(-6, 3)$

Decreasing on $(-\infty, -6)$ and $(3, \infty)$

$f''(x) = -12x - 18 = 0$

$-6(2x + 3) = 0$

$x = -\dfrac{3}{2}$

Point of inflection at $(-1.5, -185.5)$

Concave up on $(-\infty, -1.5)$

Concave down on $(-1.5, \infty)$

15. $f(x) = 2x^3 + \dfrac{7}{2}x^2 - 5x + 3$

$f'(x) = 6x^2 + 7x - 5 = 0$

$(2x - 1)(3x + 5) = 0$

Critical numbers: $x = \dfrac{1}{2}$ and $x = -\dfrac{5}{3}$

Critical points: $1/2, 1.625)$ and $(-5/3, 11.80)$

$f''(x) = 12x + 7$

$f''\left(\dfrac{1}{2}\right) = 13 > 0$

$f''\left(-\dfrac{5}{3}\right) = -13 < 0$

Relative maximum at $-5/3$, relative minimum at $1/2$

Increasing on $(-\infty, -5/3)$ and $(1/2, \infty)$

Decreasing on $(-5/3, 1/2)$

$f''(x) = 12x + 7 = 0$

$x = -\dfrac{7}{12}$

Point of inflection at $(-7/12, 6.71)$

Concave up on $(-7/12, \infty)$

Concave down on $(-\infty, -7/12)$

17. $f(x) = x^4 - 18x^2 + 5$

$f'(x) = 4x^3 - 36x = 0$

$4x(x^2 - 9) = 0$

$4(x)(x - 3)(x + 3) = 0$

Critical numbers $x = 0$, $x = 3$, and $x = -3$

Critical points: $(0, 5)$, $(3, -76)$ and $(-3, -76)$

$f''(x) = 12x^2 - 36$

$f''(-3) = 72 > 0$

$f''(0) = -36 < 0$

$f''(3) = 72 > 0$

Relative maximum at 0, relative minimum at -3 and 3

Increasing on $(-3, 0)$ and $(3, \infty)$

Decreasing on $(-\infty, -3)$ and $(0, 3)$

$f''(x) = 12x^2 - 36 = 0$

$12(x^2 - 3) = 0$

$x = \pm\sqrt{3}$

Points of inflection at $(\sqrt{3}, -40)$, $(-\sqrt{3}, -40)$

Concave up on $(-\infty, -\sqrt{3})$ and $(\sqrt{3}, \infty)$

Concave down on $(-\sqrt{3}, \sqrt{3})$

x-intercepts: $0 = x^4 - 18x^2 + 5$

$\qquad$ Let $y = x^2$.

$\qquad 0 = y^2 - 18y + 5$

Use the quadratic formula

$y = 9 \pm 2\sqrt{19} \approx 17.72$ or $.28$

$x = \pm 4.2$ or $\pm .53$

y-intercept: $y = 0^4 - 18(0)^2 + 5 = 5$

19. $\quad f(x) = x^4 - 2x^3$

$\quad f'(x) = 4x^3 - 6x^2 = 0$

$\qquad 2x^2(2x - 3) = 0$

Critical numbers: $x = 0$ and $x = 3/2$

Critical point: $(0, 0)$ and

$(3/2, -27/16)$

$f''(x) = 12x^2 - 12x$

$f''(0) = 0$

$f''\left(\dfrac{3}{2}\right) = 9 > 0$

Relative minimum at $3/2$

No relative extremum at 0

Increasing on $(3/2, \infty)$

Decreasing on $(-\infty, 3/2)$

$f''(x) = 12x^2 - 12x = 0$

$\qquad 12x(x - 1) = 0$

$\qquad x = 0, \ x = 1$

Points of inflection at $(0, 0)$ and

$(1, -1)$

Concave up on $(-\infty, 0)$ and $(1, \infty)$

Concave down on $(0, 1)$

x-intercepts: $0 = x^4 - 2x^3$

$\qquad 0 = x^3(x - 2)$

$\qquad x = 0, \ x = 2$

y-intercept: $\ y = 0^4 - 2(0)^3 = 0$

21. $\quad f(x) = x + \dfrac{2}{x} = x + 2x^{-1}$

Vertical asymptote at $x = 0$

$f'(x) = 1 - 2x^{-2} = 1 - \dfrac{2}{x^2}$

$\qquad 1 - \dfrac{2}{x^2} = 0$

$\qquad \dfrac{x^2 - 2}{x^2} = 0$

$\qquad x = \pm\sqrt{2}$

Critical numbers: $x = \sqrt{2}$ and $x = -\sqrt{2}$

Critical points: $(\sqrt{2}, 2\sqrt{2})$ and

$(-\sqrt{2}, -2\sqrt{2})$

$f''(x) = 4x^{-3} = \dfrac{4}{x^3}$

$f''(\sqrt{2}) = \dfrac{2}{\sqrt{2}} > 0$

$f''(-\sqrt{2}) = -\dfrac{2}{\sqrt{2}} < 0$

Relative maximum at $-\sqrt{2}$

Relative minimum at $\sqrt{2}$

Increasing on $(-\infty, -\sqrt{2})$ and $(\sqrt{2}, \infty)$

Decreasing on $(-\sqrt{2}, 0)$ and $(0, \sqrt{2})$

(Recall that $f(x)$ does not exist at

$x = 0$.)

$f''(x) = \dfrac{4}{x^3}$ is never zero.

There are no points of inflection.

Concave up on $(0, \infty)$

Concave down on $(-\infty, 0)$

$f(x)$ is never zero, so there are no

x-intercepts.

$f(x)$ does not exist for $x = 0$ so

there is no y-intercept.

$y = x$ is an oblique asymptote.

23. $f(x) = \dfrac{x^2 + 25}{x}$

Vertical asymptote at x = 0

$f'(x) = \dfrac{x(2x) - (x^2 + 25)}{x^2}$

$\qquad = \dfrac{x^2 - 25}{x^2}$

$f'(x) = \dfrac{(x - 5)(x + 5)}{x^2} = 0$

Critical numbers: x = 5 and x = -5
Critical points: (5, 10) and
(-5, -10)

$f''(x) = \dfrac{x^2(2x) - 2x(x^2 - 25)}{x^4}$

$\qquad = \dfrac{50}{x^3}$

$f''(-5) = -\dfrac{2}{5} < 0$

$f''(5) = \dfrac{2}{5} > 0$

Relative maximum at -5,
relative minimum at 5
Increasing on (-∞, -5) and (5, ∞)
Decreasing on (-5, 0) and (0, 5)
(Recall that f(x) does not exist at
x = 0.)

$f''(x) = \dfrac{50}{x^3}$ is never zero.

There are no points of inflection.
Concave up on (0, ∞)
Concave down on (-∞, 0)
f(x) is never zero, so there are no
x-intercepts.
f(x) doe not exist at x = 0 so there
is no y-intercept.
y = x is an oblique asymptote.

25. $f(x) = \dfrac{x - 1}{x + 1}$

Vertical asymptote at x = -1
Horizontal asymptote at y = 1

$f'(x) = \dfrac{(x + 1) - (x - 1)}{(x + 1)^2}$

$\qquad = \dfrac{2}{(x + 1)^2}$

f'(x) is never zero.
f'(x) fails to exist for x = -1.

$f''(x) = \dfrac{(x + 1)^2(0) - 2(2)(x + 1)}{(x + 1)^4}$

$\qquad = \dfrac{x - 3}{(x + 1)^3}$

f''(x) fails to exist for x = -1.
No critical values, no maximum or
minimum
No points of inflection

$f''(-2) = 5 > 0$

$f''(0) = -8 < 0$

Concave up on (-∞, -1)
Concave down on (-1, ∞)

x-intercept: $\dfrac{x - 1}{x + 1} = 0$

$\qquad\qquad\qquad x = 1$

y-intercept: $y = \dfrac{0 - 1}{0 + 1} = -1$

27. $f(x) = \dfrac{x}{x^2 + 1}$

Horizontal asymptote at y = 0

$f'(x) = \dfrac{(x^2 + 1)(1) - x(2x)}{(x^2 + 1)^2}$

$\qquad = \dfrac{1 - x^2}{(x^2 + 1)^2} = 0$

$\qquad\qquad 1 - x^2 = 0$

Critical numbers: x = 1 and x = -1

Critical points: (1, 1/2) and
(-1, -1/2)

f''(x)

$$= \frac{(x^2 + 1)^2(-2x) - (1 - x^2)(2)(x^2 + 1)(2x)}{(x^2 + 1)^3}$$

$$= \frac{-2x^3 - 2x - 4x + 4x^3}{(x^2 + 1)^3}$$

$$= \frac{2x^3 - 6x}{(x^2 + 1)^3}$$

$$f''(1) = -\frac{1}{2} < 0$$

$$f''(-1) = \frac{1}{2} > 0$$

Relative maximum at 1

Relative minimum at -1

Increasing on (-1, 1)

Decreasing on (-∞, -1) and (1, ∞)

$$f''(x) = \frac{2x^3 - 6x}{(x^2 + 1)^3} = 0$$

$$2x^3 - 6x = 0$$

$$2x(x^2 - 3) = 0$$

$$x = 0, \ x = \pm\sqrt{3}$$

Points of inflection at (0, 0),

$$\left(\sqrt{3}, \frac{\sqrt{3}}{4}\right) \text{ and } \left(-\sqrt{3}, -\frac{\sqrt{3}}{4}\right)$$

Concave up on (-√3, 0) and (√3, ∞)

Concave down on (-∞, -√3) and (0, √3)

x-intercept: $0 = \frac{x}{x^2 + 1}$

$$0 = x$$

y-intercept: $y = \frac{0}{0^2 + 1} = 0$

29. $f(x) = \frac{1}{x^2 - 4}$

Vertical asymptotes at x = 2 and
x = -2

Horizontal asymptote at y = 0

$$f'(x) = \frac{(x^2 - 4)(0) - 1(2x)}{(x^2 - 4)^2}$$

$$= \frac{-2x}{(x^2 - 4)^2} = 0$$

Critical number: x = 0

Critical point: $\left(0, -\frac{1}{4}\right)$

$$f''(x) = \frac{(x^2 - 4)^2(-2) - (-2x)(2)(x^2 - 4)(2x)}{(x^2 - 4)^4}$$

$$= \frac{6x^2 + 8}{(x^2 - 4)^3}$$

$$f''(0) = -\frac{1}{8} < 0$$

Relative maximum at 0

Increasing on (-∞, -2) and (-2, 0)

Decreasing on (0, 2) and (2, ∞)

(Recall that f(x) does not exist at
x = 2 and x = -2.)

$f''(x) = \frac{6x^2 + 8}{(x^2 - 4)^3}$ can never be zero.

There are no points of inflection.

$$f''(-3) = \frac{62}{125} > 0$$

$$f''(3) = \frac{62}{125} > 0$$

Concave up on (-∞, -2) and (2, ∞)

Concave down on (-2, 2)

f(x) is never zero so there is no
x-intercept.

y-intercept: $y = \frac{1}{0^2 - 4} = -\frac{1}{4}$

Graphs for Exercises 31 and 33 may vary. Examples are given in the answers at the back of the textbook.

31. (a) indicates that there can be no asymptotes, sharp "corners", holes, or jumps. The graph must be one smooth curve.

    (b) and (c) indicate relative maxima at -3 and 4 and a relative minimum at 1.

    (d) and (e) are consistant with (g).

    (f) indicates turning points at the critical numbers -3 and 4.

33. (a) indicates that the curve may not contain breaks.

    (b) and (c) indicate relative minima at -6 and 3 and a relative maximum at 1.

    (d) and (e), when combined with (b) and (c) show that concavity does not change between relative extrema.

    (f) gives the y-intercept.

35. $\lim\limits_{x \to \infty} \bar{c}(x)$

    $= \lim\limits_{x \to \infty} \dfrac{15,000 + 64}{x}$

    $= \lim\limits_{x \to \infty} \dfrac{\dfrac{15,000}{x} + 6}{1}$

    $= \dfrac{\lim\limits_{x \to \infty} \dfrac{15,000}{x} + \lim\limits_{x \to \infty} 6}{\lim\limits_{x \to \infty} 1}$

    $= \dfrac{0 + 6}{1}$

    $= 6$

The average cost approaches 6 as the number of tapes produced becomes very large.

37. $A(h) = \dfrac{.17h}{h^2 + 2}$

    $\lim\limits_{h \to \infty} A(h) = \lim\limits_{h \to \infty} \dfrac{.17h}{\left(h + \dfrac{2}{h}\right)h}$

    $= \lim\limits_{h \to \infty} \dfrac{.17}{h + \dfrac{2}{h}}$

    $= 0$

The concentration of the drug in the bloodstream approaches 0 as the number of hours after injection increases, which means that the drug in the bloodstream eventually dissipates.

Exercises 39-43 should be solved by computer or graphing calculator methods. The solutions will vary according to the method or computer program that is used.

39. $\lim\limits_{x \to \infty} \dfrac{\sqrt{9x^2 + 5}}{2x}$

    (a) The answer is 1.5.

41. $\lim\limits_{x \to -\infty} \dfrac{\sqrt{36x^2 + 2x + 7}}{3x}$

    (a) The answer is -2.

43. $\lim\limits_{x \to \infty} \dfrac{(1 + 5x^{1/3} + 2x^{5/3})^3}{x^5}$

    (a) The answer is 8.

Chapter 3 Review Exercises

5. $f(x) = x^2 - 5x + 3$

$f'(x) = 2x - 5 = 0$

Critical number: $x = \dfrac{5}{2}$

$f''(x) = 2 > 0$ for all x.

$f(x)$ is a minimum at $x = 5/2$.

$f(x)$ is increasing on $(5/2, \infty)$

and decreasing on $(-\infty, 5/2)$.

7. $f(x) = -x^3 - 5x^2 + 8x - 6$

$f'(x) = -3x^2 - 10x + 8$

$\qquad 3x^2 + 10x - 8 = 0$

$\qquad (3x - 2)(x + 4) = 0$

Critical numbers: $x = \dfrac{2}{3}$ or $x = -4$

$f''(x) = -6x - 10$

$f''\!\left(\dfrac{2}{3}\right) = -14 < 0$

$f(x)$ is a maximum at $x = 2/3$.

$f''(-4) = 14 > 0$

$f(x)$ is a minimum at $x = -4$.

$f(x)$ is increasing on $(-4, 2/3)$;

decreasing on $(-\infty, -4)$ and on

$(2/3, \infty)$.

9. $f(x) = \dfrac{6}{x - 4}$

$f'(x) = \dfrac{-6}{(x - 4)^2} < 0$ for all x, but

not defined for $x = 4$.

$f(x)$ is never increasing, it is

decreasing on $(-\infty, 4)$ and $(4, \infty)$.

11. $f(x) = -x^2 + 4x - 8$

$f'(x) = -2x + 4 = 0$

Critical number: $x = 2$

$f''(x) = -2 < 0$ for all x, so $f(2)$ is

a relative maximum.

$\qquad f(2) = -4$

Relative maximum of -4 at 2

13. $f(x) = 2x^2 - 8x + 1$

$f'(x) = 4x - 8 = 0$

Critical number: $x = 2$

$f''(x) = 4 > 0$ for all x, $f(2)$ is

a relative minimum.

$\qquad f(2) = -7$

Relative minimum of -7 at 2

15. $f(x) = 2x^3 + 3x^2 - 36x + 20$

$f'(x) = 6x^2 + 6x - 36 = 0$

$\qquad 6(x^2 + x - 6) = 0$

$\qquad (x + 3)(x - 2) = 0$

Critical numbers: $x = -3$, $x = 2$

$f''(x) = 12x + 6$

$f''(-3) = -30 < 0$, so a maximum occurs

$\qquad$ at $x = -3$.

$f''(2) = 30 > 0$, so a minimum occurs

$\qquad$ at $x = 2$.

$f(-3) = 101$

$f(2) = -24$

Relative maximum of 101 at -3

Relative minimum of -24 at 2

17. $f(x) = 3x^4 - 5x^2 - 11x$

$f'(x) = 12x^3 - 10x - 11$

$f''(x) = 36x^2 - 10$

$f''(1) = 36(1)^2 - 10 = 26$

$f''(-3) = 36(-3)^2 - 10 = 314$

19. $f(x) = \dfrac{5x - 1}{2x + 3}$

$f'(x) = \dfrac{(2x + 3)5 - (5x - 1)2}{(2x + 3)^2}$

$= \dfrac{17}{(2x + 3)^2}$

$f''(x) = \dfrac{(2x + 3)^2(0) - 17(2)(2x + 3)(2)}{(2x + 3)^4}$

$= \dfrac{-68}{(2x + 3)^3}$ or $-68(2x + 3)^{-3}$

$f''(1) = \dfrac{-68}{[2(1) + 3]^3} = -\dfrac{68}{125}$

$f''(-3) = \dfrac{-68}{[2(-3) + 3]^3}$

$= \dfrac{-68}{-27} = \dfrac{68}{27}$

21. $f(t) = \sqrt{t^2 + 1} = (t^2 + 1)^{1/2}$

$f'(t) = \dfrac{1}{2}(t^2 + 1)^{-1/2}(2t)$

$= t(t^2 + 1)^{-1/2}$

$f''(t) = (t^2 + 1)^{-1/2}(1)$

$+ t[(-\dfrac{1}{2})(t^2 + 1)^{-3/2}(2t)]$

$= (t^2 + 1)^{-1/2} - t^2(t^2 + 1)^{-3/2}$

$= \dfrac{1}{(t^2 + 1)^{1/2}} - \dfrac{t^2}{(t^2 + 1)^{3/2}}$

$= \dfrac{t^2 + 1 - t^2}{(t^2 + 1)^{3/2}}$

$= (t^2 + 1)^{-3/2}$

or $\dfrac{1}{(t^2 + 1)^{3/2}}$

$f''(1) = \dfrac{1}{(1 + 1)^{3/2}}$

$= \dfrac{1}{2^{3/2}} \approx .354$

$f''(-3) = \dfrac{1}{(9 + 1)^{3/2}}$

$= \dfrac{1}{10^{3/2}} \approx .032$

23. $f(x)\ -x^2 + 5x + 1;\ [1, 4]$

$f'(x) = -2x + 5 = 0$ when

$x = \dfrac{5}{2}$

$f(1) = 5$

$f(\dfrac{5}{2}) = \dfrac{29}{4}$

$f(4) = 5$

Absolute maximum of 29/4 at 5/2;

absolute minimum of 5 at 1 and 4

25. $f(x) = x^3 + 2x^2 - 15x + 3;\ [-4, 2]$

$f'(x) = 3x^2 + 4x - 15 = 0$ when

$(3x - 5)(x + 3) = 0$

$x = \dfrac{5}{3}$ or $x = -3$

$f(-4) = 31$

$f(-3) = 39$

$f(\dfrac{5}{3}) = -\dfrac{319}{27}$

$f(2) = -11$

Absolute maximum of 39 at -3;

absolute minimum of $-319/27$ at 5/3

27. $\lim\limits_{x \to \infty} g(x)$ does not exist because $g(x)$

increases without bound as x gets

smaller and smaller.

29. $\lim\limits_{x \to \infty} \dfrac{x^2 + 5}{5x^2 - 1}$

$= \lim\limits_{x \to \infty} \dfrac{\dfrac{x^2}{x^2} - \dfrac{5}{x^2}}{\dfrac{5x^2}{x^2} - \dfrac{1}{x^2}}$

$= \dfrac{\lim\limits_{x \to \infty} 1 - \lim\limits_{x \to \infty} \dfrac{5}{x^2}}{\lim\limits_{x \to \infty} 5 - \lim\limits_{x \to \infty} \dfrac{1}{x^2}}$

$= \dfrac{1 - 0}{5 - 0} = \dfrac{1}{5}$

31. $\lim\limits_{x \to -\infty} \left(\dfrac{3}{4} + \dfrac{2}{x} - \dfrac{5}{x^2} \right)$

$= \lim\limits_{x \to -\infty} \dfrac{3}{4} + \lim\limits_{x \to -\infty} \dfrac{2}{x} + \lim\limits_{x \to -\infty} \dfrac{-5}{x^2}$

$= \dfrac{3}{4} + 0 + 0$

$= \dfrac{3}{4}$

For Exercises 33–45, see the answer graphs in the back of the textbook.

33. $f(x) = -2x^3 - \dfrac{1}{2}x^2 + x - 3$

$f'(x) = -6x^2 - x + 1 = 0$

$(3x - 1)(2x + 1) = 0$

Critical numbers: $x = 1/3$ and $x = -1/2$

Critical points: $(1/3, -2.80)$ and $(-1/2, -3.375)$

$f''(x) = -12x - 1$

$f''\left(\dfrac{1}{3}\right) = -5 < 0$

$f''\left(-\dfrac{1}{2}\right) = 5 > 0$

Relative maximum at $1/3$

Relative minimum at $-1/2$

Increasing on $(-1/2, 1/3)$

Decreasing on $(-\infty, -1/2)$ and $(1/3, \infty)$

$f''(x) = -12x - 1 = 0$

$x = -\dfrac{1}{12}$

Point of inflection at $(-1/12, -3.09)$

Concave up on $(-\infty, -1/12)$

Concave down on $(-1/12, \infty)$

y–intercept:

$y = -2(0)^3 - \dfrac{1}{2}(0)^2 + (0) - 3 = -3$

35. $f(x) = x^4 - \dfrac{4}{3}x^3 - 4x^2 + 1$

$f'(x) = 4x^3 - 4x^2 - 8x = 0$

$4x(x^2 - x - 2) = 0$

$4x(x - 2)(x + 1) = 0$

Critical numbers: $x = 0$, $x = 2$, or $x = -1$

Critical points: $(0, 1)$, $(2, -29/3)$ and $(-1, -2/3)$

$f''(x) = 12x^2 - 8x - 8$

$= 4(3x^2 - 2x - 2)$

$f''(-1) = 12 > 0$

$f''(0) = -8 < 0$

$f''(2) = 8 > 0$

Relative maximum at 0

Relative minima at -1 and 2

Increasing on $(-1, 0)$ and $(2, \infty)$

Decreasing on $(-\infty, -1)$ and $(0, 2)$

$f''(x) = 4(3x^2 - 2x - 2) = 0$

$x = \dfrac{2 \pm \sqrt{4 - (-24)}}{6}$

$= \dfrac{1 \pm \sqrt{7}}{3}$

Points of inflection at

$(\frac{1 + \sqrt{7}}{3}, -3.47)$ and $(\frac{1 - \sqrt{7}}{3}, .11)$

Concave up on $(-\infty, \frac{1 - \sqrt{7}}{3})$ and

$(\frac{1 + \sqrt{7}}{3}, \infty)$

Concave down on $(\frac{1 - \sqrt{7}}{3}, \frac{1 + \sqrt{7}}{3})$

y-intercept:

$y = (0)^4 - \frac{4}{3}(0)^3 - 4(0)^2 + 1 = 1$

37. $f(x) = \frac{x - 1}{2x + 1}$

Vertical asymptote at $x = -\frac{1}{2}$

Horizontal asymptote at $y = \frac{1}{2}$

$f'(x) = \frac{(2x + 1) - (x - 1)(2)}{(2x + 1)^2}$

$= \frac{3}{(2x + 1)^2}$

f' is never zero.

$f(x)$ has no extrema.

$f''(x) = \frac{-12}{(2x + 1)^3}$

$f''(0) = -12 < 0$

$f''(-1) = 12 > 0$

Concave up on $(-\infty, -\frac{1}{2})$

Concave down on $(-\frac{1}{2}, \infty)$

x-intercept: $\frac{x - 1}{2x + 1} = 0$

$x = 1$

y-intercept: $y = \frac{0 - 1}{2(0) + 1} = -1$

39. $f(x) = -4x^3 - x^2 + 4x + 5$

$f'(x) = -12x^2 - 2x + 4$

$= -2(6x^2 + x - 2) = 0$

$(3x + 2)(2x - 1) = 0$

Critical numbers: $x = -2/3$ and $x = 1/2$

Critical points: $(-2/3, 3.07)$ and $(1/2, 6.25)$

$f''(x) = -24x - 2$

$= -2(12x + 1)$

$f''(-\frac{2}{3}) = 14 > 0$

$f''(\frac{1}{2}) = -14 < 0$

Relative maximum at $1/2$

Relative minimum at $-2/3$

Increasing on $(-2/3, 1/2)$

Decreasing on $(-\infty, -2/3)$ and $(1/2, \infty)$

$f''(x) = -2(12x + 1) = 0$

$x = -\frac{1}{12}$

Point of inflection at $(-\frac{1}{12}, 4.66)$

Concave up on $(-\infty, -\frac{1}{12})$

Concave down on $(-\frac{1}{12}, \infty)$

y-intercept:

$y = -4(0)^3 - (0)^2 + 4(0) + 5 = 5$

41. $f(x) = x^4 + 2x^2$

$f'(x) = 4x^3 + 4x$

$= 4x(x^2 + 1) = 0$

Critical number: $x = 0$

Critical point: $(0, 0)$

$f''(x) = 12x^2 + 4$

$f''(0) = 4 > 0$

Relative minimum at 0

Increasing on $(0, \infty)$

Decreasing on $(-\infty, 0)$

$f''(x) = 12x^2 + 4$

$\qquad 4(3x^2 + 1) \neq 0$ for any x

No points of inflection

$f''(-1) = 16 > 0$

$f''(1) = 16 > 0$

Concave up on $(-\infty, \infty)$

x-intercept and y-intercept at $(0, 0)$

43. $f(x) = \dfrac{x^2 + 4}{x}$

Vertical asymptote at $x = 0$

Oblique asymptote $y = x$

$f'(x) = \dfrac{x(2x) - (x^2 + 4)}{x^2}$

$\qquad = \dfrac{x^2 - 4}{x^2} = 0$

Critical numbers: $x = -2$ and $x = 2$

Critical points: $(-2, -4)$ and $(2, 4)$

$f''(x) = \dfrac{8}{x^3}$

$f''(-2) = -1 < 0$

$f''(2) = 1 > 0$

Relative maximum at -2

Relative minimum at 2

Increasing on $(-\infty, -2)$ and $(2, \infty)$

Decreasing on $(-2, 0)$ and $(0, 2)$

$f''(x) = \dfrac{8}{x^3} > 0$ for all x.

No inflection points

Concave up on $(0, \infty)$

Concave down on $(-\infty, 0)$

No x- or y-intercepts

45. $f(x) = \dfrac{2x}{3 - x}$

Vertical asymptote at $x = 3$

Horizontal asymptote at $y = -2$

$f'(x) = \dfrac{(3 - x)2 - (2x)(-1)}{(3 - x)^2}$

$\qquad = \dfrac{6}{(3 - x)^2}$

$f'(x)$ is never zero.

No critical values, no relative extrema

$f'(0) = \dfrac{2}{3} > 0$

$f'(4) = 6 > 0$

Increasing on $(-\infty, 3)$ and $(3, \infty)$

$f''(x) = \dfrac{12}{(3 - x)^3}$

$f''(x)$ is never zero; no inflection points.

$f''(0) = \dfrac{12}{27} > 0$

$f''(4) = -12 < 0$

Concave up on $(-\infty, 3)$

Concave down on $(3, \infty)$

x-intercept and y-intercept at $(0, 0)$

CHAPTER 3   TEST

[3.1]   Find the largest open intervals where each of the following functions is
(a) increasing or (b) decreasing.

1.

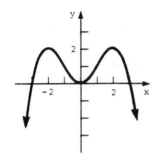

2. $f(x) = 3x^3 - 3x^2 - 3x + 5$

3. $f(x) = \dfrac{x - 2}{x + 3}$

[3.1]   **4.**   For a certain product, the demand equation is p = 10 − .05x. The
cost function for this product is C = 5x − .75. Over what inter-
val(s) is the profit increasing?

[3.1]   **5.**   What are critical numbers? Why are they significant?

[3.2]   Find the locations and values of all relative maxima and minima.

6. $f(x) = 4x^3 - \dfrac{9}{2}x^2 - 3x - 1$ **7.** $f(x) = 4x^{2/3}$

[3.2]   **8.**   A manufacturer estimates that the cost in dollars per unit for a
production run of x thousand units is given by C = $3x^2$ − 60x + 320.
How many thousand units should be produced during each run to
minimize the cost per unit, and what is the minimum cost per unit?

[3.2]   **9.**   Use the derivative to find the vertex of the parabola y = −$2x^2$ + 6x + 9.

[3.3]   **10.**   What conditions must a function, f(x), meet for the Extreme Value
Theorem to hold?

[3.3] **Find the locations of any absolute extrema for the functions with graphs as follows.**

11.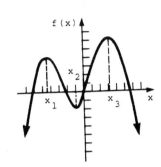

12.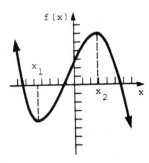

[3.3] **Find the locations of all absolute extrema for the functions defined as follow, with the specified domains.**

13. $f(x) = x^3 - 12x$; $[0, 4]$

14. $f(x) = \dfrac{x^2 + 4}{x}$; $[-3, -1]$

[3.3] **15.** Why is it important to check the endpoints of the domain when checking for absolute extrema?

[3.4] **Find the second derivative of each function, then find $f''(-3)$.**

16. $f(x) = -5x^3 + 3x^2 - x + 1$

17. $f(t) = \sqrt{t^2 - 5}$

[3.4] **18.** Find the largest open intervals where $f(x) = x^3 - 3x^2 - 9x - 1$ is concave upward or concave downward. Find the location of any points of inflection. Graph the function.

[3.4] **19.** The function $s(t) = t^3 - 9t^2 + 15t + 25$ gives the displacement in centimeters at time t (in seconds) of a particle moving along a line. Find the velocity and acceleration functions. Then find the velocity and acceleration at $t = 0$ and $t = 2$.

[3.4] **20.** What is the difference between velocity and acceleration?

[3.5] **Evaluate each limit that exists.**

21. $\lim\limits_{x \to \infty} \dfrac{2x}{7x + 1}$

22. $\lim\limits_{x \to -\infty} \dfrac{3x^2 + 2x}{4x^3 - x}$

[3.5]   23. Graph $f(x) = \frac{2}{3}x^3 - \frac{5}{2}x^2 - 3x + 1$. Give critical points, regions where the function is increasing or decreasing, points of inflection, regions where the function is concave up or concave down.

[3.5]   24. Sketch a graph of a single function that has all the properties listed.

(a) continuous for all real numbers

(b) increasing on $(-3, 2)$ and $(4, \infty)$

(c) decreasing on $(-\infty, -3)$ and $(2, 4)$

(d) concave upward on $(-\infty, 0)$

(e) concave downward on $(0, 4)$ and $(4, \infty)$

(f) differentiable everywhere except $x = 4$

(g) $f'(-3) = f'(2) = 0$

(h) inflection point at $(0, 0)$

[3.5]   25. The cost for manufacturing a certain computer component is

$$C(x) = 12{,}000 + 11x$$

where x is the number of components manufactured. The average cost per component, denoted by $\overline{C}(x)$, is found by dividing $C(x)$ by x. Find and interpret $\lim\limits_{x \to \infty} \overline{C}(x)$.

CHAPTER 3 TEST ANSWERS

1. (a) $(-\infty, -2)$ and $(0, 2)$ (b) $(-2, 0)$ and $(2, \infty)$ 2. (a) $(-\infty, -1/3)$ and

    $(1, \infty)$ (b) $(-1/3, 1)$ 3. (a) $(-\infty, -3)$ and $(-3, \infty)$ (b) never decreasing

4. $(0, 50)$

5. Critical numbers are x–values for which $f'(x) = 0$ or $f'(x)$ does not exist. They
    tell where the derivative may change signs, and are used to tell where a
    function is increasing or decreasing.

6. Relative maximum of $-19/32$ at $1/4$; relative minimum of $-9/2$ at 1

7. Relative minimum of 0 at 0   8. 10 thousand units; $20 per unit

9. $(3/2, 27/2)$ 10. $f(x)$ must be continuous on a closed interval.

11. Absolute maximum at x_3 12. No absolute extrema

13. Absolute maximum of 16 at 4; absolute minimum of -16 at 2

14. Absolute maximum of -4 at -2; absolute minimum of -5 at -1

15. The smallest or largest value in a closed interval may occur at the endpoints.

16. $f''(x) = -30x + 6$; $f''(-3) = 96$ 17. $f''(t) = \dfrac{-5}{(t^2 - 5)^{3/2}}$; $f''(-3) = -\dfrac{5}{8}$

18. Concave upward on $(1, \infty)$; concave   19. $v(t) = 3t^2 - 18t + 15$;
     downward on $(-\infty, 1)$; point of in- $a(t) = 6t - 18$; $v(0) = 15$ cm/sec;
     flection $(1, -12)$ $a(0) = -18$ cm/sec^2; $v(2) = -9$
                                                           cm/sec; $a(2) = -6$ cm/sec^2

20. Velocity gives the rate of change
     of position relative to time.
     Accelation gives the rate of
     change of velocity.

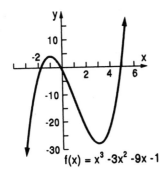
$f(x) = x^3 - 3x^2 - 9x - 1$

21. $2/7$ 22. 0

23. Critical points: $\left(-\frac{1}{2}, \frac{43}{24}\right)$ and $\left(3, -\frac{25}{2}\right)$; increasing on $\left(-\infty, -\frac{1}{2}\right)$ and $(3, \infty)$;

decreasing on $\left(-\frac{1}{2}, 3\right)$; point of inflection at $\left(\frac{5}{4}, -\frac{257}{48}\right)$; concave up on $\left(\frac{5}{4}, \infty\right)$;

concave down on $\left(-\infty, \frac{5}{4}\right)$

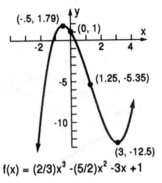

$f(x) = (2/3)x^3 - (5/2)x^2 - 3x + 1$

24.

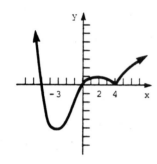

25. 11; the average cost approaches 11 as the number of components manufactured becomes very large.

CHAPTER 4   APPLICATIONS OF THE DERIVATIVE

Section 4.1

1. $x + y = 100$, $P = xy$

   (a) $y = 100 - x$

   (b) $P = xy = x(100 - x)$

   $\qquad = 100x - x^2$

   (c) Since $y = 100 - x$ and x and y are nonnegative numbers, $x \geq 0$ and $100 - x \geq 0$ or $x \leq 100$. The domain of P is $[0, 100]$

   (d) $P' = 100 - 2x$

   $\qquad 100 - 2x = 0$

   $\qquad 2(50 - x) = 0$

   $\qquad\qquad x = 50$

   (e)

x	P
0	0
50	2500
100	0

   (f) From the chart, the maximum value of P is 2500; this occurs when $x = 50$ and $y = 50$.

3. $x + y = 200$

   $P = x^2 + y^2$

   (a) $y = 200 - x$

   (b) $P = x^2 + (200 - x)^2$

   $\qquad = x^2 + 40,000 - 400x + x^2$

   $\qquad = 2x^2 - 400x + 40,000$

   (c) Since $y = 200 - x$ and x and y are nonnegative numbers, the domain of P is $[0, 200]$.

   (d) $P' = 4x - 400$

   $\qquad 4x - 400 = 0$

   $\qquad 4(x - 100) = 0$

   $\qquad\qquad x = 100$

   (e)

x	P
0	40,000
100	20,000
200	40,000

   (f) The minimum value of P is 20,000; this occurs when $x = 100$ and $y = 100$.

5. $x + y = 150$

   Maximize $x^2 y$.

   (a) $y = 150 - x$

   (b) Let $P = x^2 y = x^2(150 - x)$

   $\qquad\qquad = 150x^2 - x^3$

   (c) Since $y = 150 - x$ and x and y are nonnegative, the domain of P is $[0, 150]$.

   (d) $P' = 300x - 3x^2$

   $\qquad 300x - 3x^2 = 0$

   $\qquad 3x(100 - x) = 0$

   $\qquad\qquad x = 0 \quad$ or $\quad x = 100$

   (e)

x	P
0	0
100	500,000
150	0

   (f) The maximum value of $x^2 y$ occurs when $x = 100$ and $y = 50$. The maximum value is 500,000.

7. $x - y = 10$

    Minimize xy.

    (a) $y = x - 10$

    (b) Let $P = xy = x(x - 10)$
        $$= x^2 - 10x$$

    (c) Since $y = x - 10$ and x and y are nonnegative numbers, the domain of P is $[10, \infty)$.

    (d) $P' = 2x - 10$
        $$2x - 10 = 0$$
        $$2(x - 5) = 0$$
        $$x = 5$$

        Note that 5 is not in the domain.

    (e)

x	P
10	0

    (f) $P'(10) = 10 > 0$ implies that P(10) is a minimum. The minimum value of xy occurs when x = 10 and y = 0. The minimum value is 0.

9. Let x = the width and y = the length.

    (a) The perimeter is
        $$P = 2x + y$$
        $$= 1200.$$
        So, $y = 1200 - 2x$.

    (b) Area $= xy = x(1200 - 2x)$
        $$A(x) = 1200x - 2x^2$$

    (c) $A' = 1200 - 4x$
        $$1200 - 4x = 0$$
        $$1200 = 4x$$
        $$300 = x$$

$A'' = -4$, which implies that x = 300 m leads to the maximum area.

(d) If x = 300,
    $$y = 1200 - 2(300) = 600.$$

    The maximum area is
    $$(300)(600) = 180,000 \text{ m}^2.$$

11. Let x = the width of the rectangle and y = the total length of the rectangle.

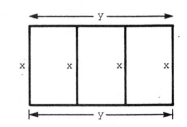

An equation for the fencing is
    $$3600 = 4x + 2y$$
    $$2y = 3600 - 4x$$
    $$y = 1800 - 2x.$$

Area $= xy = x(1800 - 2x)$
    $$A(x) = 1800x - 2x^2$$

$A' = 1800 - 4x$
$1800 - 4x = 0$
    $$1800 = 4x$$
    $$450 = x$$

$A'' = -4$, which implies that x = 450 is the location of a maximum.
If x = 450, y = 1800 - 2(450) = 900.
The maximum area is

    $$(450)(900) = 405,000 \text{ m}^2.$$

13. Let $8 - x$ = the distance the hunter will travel on the river.

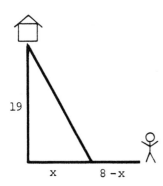

Then $\sqrt{19^2 + x^2}$ = the distance he will travel on land.

Since the rate on the river is 5 mph, the rate on land is 2 mph, and $t = d/r$,

$\dfrac{8 - x}{5}$ = the time on the river

$\dfrac{\sqrt{361 + x^2}}{2}$ = the time on land.

The total time is

$$T(x) = \frac{8 - x}{5} + \frac{\sqrt{361 + x^2}}{2}$$

$$= \frac{8}{5} - \frac{1}{5}x + \frac{1}{2}(361 + x^2)^{1/2}.$$

$$T'(x) = -\frac{1}{5} + \frac{1}{4} \cdot 2x(361 + x^2)^{-1/2}$$

$$-\frac{1}{5} + \frac{x}{2(361 + x^2)^{1/2}} = 0$$

$$\frac{1}{5} = \frac{x}{2(361 + x^2)^{1/2}}$$

$$2(361 + x^2)^{1/2} = 5x$$

$$4(361 + x^2) = 25x^2$$

$$1444 + 4x^2 = 25x^2$$

$$1444 = 21x^2$$

$$\frac{38}{\sqrt{21}} = x$$

$$8.29 \approx x$$

8.29 is not possible, since the cabin is only 8 miles west. Check the endpoints.

y	T(x)
0	17.5
8	10.3

$T(x)$ is minimized when $x = 8$.

The distance along the river is given by $8 - x$, so the hunter should travel $8 - 8 = 0$ miles along the river. He should complete the entire trip on land.

15. $p(x) = 100 - \dfrac{x}{10}$

(a) Revenue from sale of x thousand candy bars:

$$R(x) = 1000xp$$

$$= 1000x\left(100 - \frac{x}{10}\right)$$

$$= 100{,}000x - 100x^2$$

(b) $R'(x) = 100{,}000 - 200x$

$$100{,}000 - 200x = 0$$

$$100{,}000 = 200x$$

$$500 = x$$

The maximum revenue occurs when 500 thousand bars are sold.

(c) $R(500) = 100{,}000(500)$

$$- 100(500)^2$$

$$= 25{,}000{,}000$$

The maximum revenue is 25,000,000 cents.

17. $G(x) = \frac{1}{32}\left(\frac{64}{x} + \frac{x}{50}\right)$

The cost of fuel on the 400-mile
trip is $1.60 per gallon.
The total cost of fuel is

$C(x)$ = (Cost per gallon)

$\qquad \times G(x) \times$ (miles traveled)

$\qquad = 1.60\left(\frac{1}{32}\right)\left(\frac{64}{x} + \frac{x}{50}\right)$ (400)

$\qquad = 20\left(\frac{64}{x} + \frac{x}{50}\right)$

$\qquad = \frac{1280}{x} + \frac{2}{5}x$ dollars.

(a) $C'(x) = \frac{-1280}{x^2} + \frac{2}{5} = 0$

$\qquad x = \sqrt{3200}$

$\qquad \approx 56.6$ mi/hr

$\qquad C''(x) = \frac{-1280(-2)}{x^3} > 0$ for all

$x > 0$ which implies $C(x)$ is
minimized when $x \approx 56.6$ mph.

(b) $C(10\sqrt{32}) = \frac{1280}{10\sqrt{32}} + \frac{2}{5}(10\sqrt{32})$

$\qquad\qquad \approx 22.61 + 22.63$

$\qquad\qquad = 45.24$

The minimum total cost is
$45.24.

19. Let x = the length at $3 per foot
$\qquad y$ = the width at $6 per foot

$\qquad xy = 20,000$

$\qquad y = \frac{20,000}{x}$

Perimeter $= 2x + 2y = 2x + \frac{40,000}{x}$

Cost $= 2x(\$3) + \frac{40,000}{x}(\$6)$

$\qquad = 6x + \frac{240,000}{x}$

Minimize cost.

$C' = 6 - \frac{240,000}{x^2}$

$6 - \frac{240,000}{x^2} = 0$

$\qquad 6 = \frac{240,000}{x^2}$

$\qquad 6x^2 = 240,000$

$\qquad x^2 = 40,000$

$\qquad x = 200$

$\qquad y = \frac{20,000}{200} = 100$

400 ft at $3 per foot will cost
$1200. 200 ft at $6 per foot will
cost $1200. The entire cost will
be $2400.

21. Profit is 5 dollars per seat for
$60 \le x \le 80$ seats.
Profit (in dollars) is

$\qquad 5 - .05(x - 80)$

per seat for $x > 80$ seats.
We expect that the number of seats
which makes the total profit a maximum will be greater than 80 because
after 80 the profit is still increasing, though at a slower rate.
(Thus we know the function is concave down and its one extrema will
be a maximum.)

(a) The total profit for x seats is

$\qquad P(x) = [5 - .05(x - 80)]x$

$\qquad\qquad = (5 - .05x + 4)x$

$\qquad\qquad = (9 - .05x)x$

$\qquad\qquad = 9x - .05x^2.$

$P'(x) = 9 - .10x$

$9 - .10x = 0$

$9 = .10x$

$x = 90$

90 seats will produce maximum profit.

(b) $P(90) = 9(90) - .05(90)^2$

$= 810 - .05(8100)^2$

$= 405$

The maximum profit is $405.

23. Let x = a side of the base

h = the height of the box.

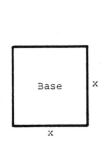 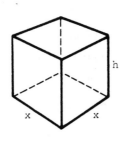

An equation for the volume of the box is

$$V = x^2 h.$$

$$32 = x^2 h$$

$$h = \frac{32}{x^2}.$$

The box is open at the top so the area of the surface material m(x) in square inches is the area of the base plus the area of the four sides.

$m(x) = x^2 + 4xh$

$= x^2 + 4x\left(\frac{32}{x^2}\right)$

$= x^2 + \frac{128}{x}$

$m'(x) = 2x - \frac{128}{x^2}$

$\frac{2x^3 - 128}{x^2} = 0$

$2x^3 - 128 = 0$

$2(x^3 - 64) = 0$

$x = 4$

$m'(x) = 2 + \frac{256}{x^3} > 0$ since $x > 0$.

So, x = 4 minimizes the surface material.

If $x = 4$, $h = \frac{32}{x^2} = \frac{32}{16} = 2$.

The dimensions that will minimize the surface material are 4 in by 4 in by 2 in.

25. Let x = the width.

Then 2x = the length

h = the height.

An equation for volume is

$$V = (2x)(x)h = 2x^2 h$$

$$36 = 2x^2 h.$$

So, $h = \frac{18}{x^2}$.

The surface area S(x) is the sum of the areas of the base and the four sides.

$S(x) = (2x)(x) + 2xh + 2(2x)h$

$\quad = 2x^2 + 6xh$

$\quad = 2x^2 + 6x\left(\frac{18}{x^2}\right)$

$\quad = 2x^2 + \frac{108}{x}$

$S'(x) = 4x - \frac{108}{x^2}$

$\frac{4x^3 - 108}{x^2} = 0$

$4(x^3 - 27) = 0$

$\quad\quad x = 3$

$S''(x) = 4 + \frac{108(2)}{x^3}$

$\quad\quad = 4 + \frac{216}{x^3} > 0$ and $x > 0$.

So $x = 3$ minimizes the surface material.

If $x = 3$, $h = \frac{18}{x^2} = \frac{18}{9} = 2$.

The dimensions are 3 ft by 6 ft by 2 ft.

27. 120 centimeters of ribbon are available; it will cover 4 heights and 8 radii.

$\quad 4h + 8r = 120$

$\quad\quad h + 2r = 30$

$\quad\quad\quad\quad h = 30 - 2r$

$V = \pi r^2 h$

$V = \pi r^2(30 - 2r)$

$\quad = 30\pi r^2 - 2\pi r^3$

Maximize volume.

$V' = 60\pi r - 6\pi r^2$

$60\pi r - 6\pi r^2 = 0$

$6\pi r(10 - r) = 0$

$r = 0$ or $r = 10$

r	V
0	0
10	1000π

$V'' = 6\pi - 12\pi r < 0$ for $r = 10$, which implies that $r = 10$ gives maximum volume.

When $r = 10$, $h = 30 - 2(10) = 10$. The volume is maximized when the radius and height are both 10 cm.

29. $V = \pi r^2 h$

$16 = \pi r^2 h$,

so

$h = \frac{16}{\pi r^2}$.

The total cost is the sum of the cost of the top and bottom and the cost of the sides, or

$C = \$2(2)(\pi r^2) + \$1(2\pi rh)$

$\quad = 4(\pi r^2) + 1(2\pi r)\left(\frac{16}{\pi r^2}\right)$

$\quad = 4\pi r^2 + \frac{32}{r}$.

Minimize cost.

$C' = 8\pi r - \frac{32}{r^2}$

$8\pi r - \frac{32}{r^2} = 0$

$\quad\quad 8\pi r^3 = 32$

$\quad\quad\quad \pi r^3 = 4$

$\quad\quad\quad\quad r = \sqrt[3]{\frac{4}{\pi}}$

$\quad\quad\quad\quad\quad \approx 1.08$

$h = \frac{16}{\pi(1.08)^2} \approx 4.34$

The radius should be 1.08 ft and the height should be 4.34 ft. If these rounded values for the height and radius are used, the cost is

$2(2)(\pi r^2)$ + $1(2\pi rh)$

$= 4\pi(1.08)^2 + 2\pi(1.08)(4.34)$

$= \$44.11$.

31. Let x = the width of printed material

and y = the length of printed material.

Then, the area of the printed material is

$$xy = 36$$

$$y = \frac{36}{x}.$$

Also, x + 2 = the width of a page
and y + 3 = the length of a page.

The area of a page is

A = (x + 2)(y + 3)

$= xy + 2y + 3x + 6$

$= 36 + 2\left(\frac{36}{x}\right) + 3x + 6$

$= 42 + \frac{72}{x} + 3x.$

$A' = -\frac{72}{x^2} + 3 = 0$

$x^2 = 24$

$x = \sqrt{24}$

$= 2\sqrt{6}$

$A'' = \frac{216}{x^3} > 0$ when $x = 2\sqrt{6}$, which

implies that A is minimized when

$x = 2\sqrt{6}$.

$y = \frac{36}{x} = \frac{36}{2\sqrt{6}} = \frac{18}{\sqrt{6}} = \frac{18\sqrt{6}}{6} = 3\sqrt{6}$

The width of a page is

x + 2 = $2\sqrt{6}$ + 2

$\approx$ 6.9 in.

The length of a page is

y + 3 = $3\sqrt{6}$ + 3

$\approx$ 10.3 in.

33. Distance on shore: 7 − x miles

Cost on shore: $400 per mile

Distance underwater: $\sqrt{x^2 + 36}$

Cost underwater: $500 per mile

Find the distance from A, that is, (7 − x), to minimize cost, C(x).

C(x) = (7 − x)(400) + ($\sqrt{x^2 + 36}$)(500)

$= 2800 - 400x + 500(x^2 + 36)^{1/2}$

$C'(x) = -400 + 500\left(\frac{1}{2}\right)(x^2 + 36)^{-1/2}(2x)$

$= -400 + \frac{500x}{\sqrt{x^2 + 36}}$

If C'(x) = 0,

$\frac{500x}{\sqrt{x^2 + 36}} = 400$

$\frac{5x}{4} = \sqrt{x^2 + 36}$

$\frac{25}{16}x^2 = x^2 + 36$

$\frac{9}{16}x^2 = 36$

$x^2 = \frac{36 \cdot 16}{9}$

$x = \frac{6 \cdot 4}{3} = 8.$

(Discard the negative solution.)
x = 8 is impossible since Point A is only 7 miles from point C.

Check the endpoints.

x	C(x)
0	5800
7	4610

The cost is minimized when x = 7.
7 − x = 7 − 7 = 0, so the company
should angle the cable at Point A.

35. $H(S) = f(S) - S$

$f(S) = -.1S^2 + 11S$

$H(S) = -.1S^2 + 11S - S$

$\quad\quad = -.1S^2 + 10S$

$H'(S) = -.2S + 10$

If $H'(S) = -.2S + 10 = 0$,

$\quad\quad .2S = 10$

$\quad\quad\quad S = 50.$

The number of creatures needed to
sustain the population is $S_0 = 50$
thousand.

$H''(S) = -.2 < 0$,

so $H(S)$ is a maximum at $S_0 = 50$.

$H(S_0) = -.1(50)^2 + 10(50)$

$\quad\quad = -250 + 500$

$\quad\quad = 250$

The maximum sustainable harvest is
250 thousand.

37. $H(S) = f(S) - S$

$f(S) = 15\sqrt{S}$

$H(S) = 15S^{1/2} - S$

$H'(S) = \frac{1}{2} \cdot 15S^{-1/2} - 1$

If $H'(S) = 0$,

$\quad 7.5 = \sqrt{S}$

$\quad\quad S = 56.25.$

The number of creatures needed to
sustain the population is $S_0 = 56.25$
thousand.

$H''(S) = -\frac{1}{4} \cdot 15S^{-3/2} < 0$,

so H is maximum at $S_0 = 56.25$.

$H(S_0) = 15\sqrt{56.25} - 56.25$

$\quad\quad = 56.25$

The maximum sustainable harvest is
56.25 thousand.

39. $H(S) = f(S) - S$

$f(S) = \frac{25S}{S + 2}$

$H'(S) = \frac{(S + 2)(25) - 25S}{(S + 2)^2} - 1$

$\quad\quad = \frac{25S + 50 - 25S - (S + 2)^2}{(S + 2)^2}$

$\quad\quad = \frac{50 - (S^2 + 4S + 4)}{(S + 2)^2}$

$\quad\quad = \frac{-S^2 - 4S + 46}{(S + 2)^2}$

$H'(S) = 0$

$S^2 + 4S - 46 = 0$

$S = \frac{-4 \pm \sqrt{16 + 184}}{2}$

$\quad = 5.071$

(Discard the negative solution.)
The number of creatures needed to
sustain the population is $S_0 = 5.071$
thousand.

$H'' = \frac{(S+2)^2(-2S-4)-(-S^2-4S+46)(2S+4)}{(S + 2)^4} < 0$,

so H is a maximum at $S_0 = 5.017$.

$H(S_0) = \dfrac{25(5.017)}{7.017} - 5.017$

$= 12.86$

The maximum sustainable harvest is 12.86 thousand.

41. Let x = distance from P to A.

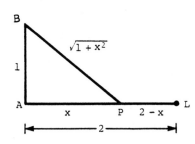

Energy used over land: 1 unit per mile

Energy used over water: 4/3 units per mile

Distance over land: (2 − x) mi

Distance over water: $\sqrt{1 + x^2}$ mi

Find the location of P to minimize energy used.

$E(x) = 1(2 - x) + \dfrac{4}{3}\sqrt{1 + x^2}$

$E'(x) = -1 + \dfrac{4}{3}\left(\dfrac{1}{2}\right)(1 + x^2)^{-1/2}(2x)$

If $E'(x) = 0$,

$\dfrac{4}{3}x(1 + x^2)^{-1/2} = 1$

$\dfrac{4x}{3(1 + x^2)^{1/2}} = 1$

$\dfrac{4}{3}x = (1 + x^2)^{1/2}$

$\dfrac{16}{9}x^2 = 1 + x^2$

$\dfrac{7}{9}x^2 = 1$

$x^2 = \dfrac{9}{7}$

$x = \dfrac{3}{\sqrt{7}}$

$= \dfrac{3\sqrt{7}}{7}.$

Point P is $3\sqrt{7}/7 \approx 1.134$ mi from Point A.

43. (a) Distance traveled by salmon = d

Net speed, or rate of salmon = $v - v_0$

Time for journey $= \dfrac{\text{distance}}{\text{rate}}$

$= \dfrac{d}{v - v_0} = t$

(b) Let k_1 = constant of proportionality in expression for energy expended per hour, or

$k_1 v^\alpha.$

Then,

$$\begin{pmatrix}\text{total energy}\\ \text{expended over}\\ \text{journey}\end{pmatrix} = \begin{pmatrix}\text{energy}\\ \text{expended}\\ \text{per hour}\end{pmatrix} \times \begin{pmatrix}\text{time}\\ \text{for}\\ \text{journey}\end{pmatrix}$$

$T = (k_1 v^\alpha)\left(\dfrac{d}{v - v_0}\right)$

$= k_1 d\left(\dfrac{v^\alpha}{v - v_0}\right).$

Since d is constant for each journey, let $k = k_1 d$.

Then,

$T = k\left(\dfrac{v^\alpha}{v - v_0}\right).$

(c)

$$T' = \frac{k\alpha v^{\alpha-1}(v - v_0) - (1)(kv^\alpha)}{(v - v_0)^2} = 0$$

$$k\alpha v^{\alpha-1}(v - v_0) - (kv^\alpha) = 0$$

$$\alpha kv^\alpha v^{-1}(v - v_0) - kv^\alpha = 0$$

$$kv^\alpha[\alpha v^{-1}(v - v_0) - 1] = 0$$

$$kv^\alpha = 0 \quad \text{or} \quad \alpha v^{-1}(v - v_0) - 1 = 0$$

$$v = 0 \quad \text{or} \quad \frac{\alpha(v - v_0)}{v} = 1$$
$$\text{(impossible)}$$

$$\alpha(v - v_0) = v$$

$$\alpha v - \alpha v_0 = v$$

$$\alpha v - v = \alpha v_0$$

$$\alpha(\alpha - 1) = \alpha v_0$$

$$v = \frac{\alpha v_0}{\alpha - 1} \quad \text{(minimize time)}$$

(d)

$$t = \frac{d}{v - v_0}$$

$$40 = \frac{20}{v - 2}$$

$$40(v - 2) = 20$$

$$v = 2.5 \text{ mph}$$

Now, using $v = \frac{\alpha v_0}{\alpha - 1}$, we have

$$2.5 = \frac{\alpha(2)}{\alpha - 1}$$

$$2.5(\alpha - 1) = 2\alpha$$

$$2.5\alpha - 2.5 = 2\alpha$$

$$.5\alpha = 2.5$$

$$\alpha = 5.$$

Assume that the current of the river stays constant and that salmon swim at constant velocity.

Exercises 45–51 should be solved by computer methods. The solutions will vary according to the computer programs that are used.

45. The answer is the radius is 5.206 cm, the height is 11.75 cm.

47. The answer is the radius is 5.242 cm, the height is 11.58 cm

49. $C(x) = .01x^3 + .05x^2 + .2x + 28$

    The answer is $A(x) = .01x^2 + .05x + .2 + 28/x$, $x = 10.41$.

51. $C(x) = 30 + 42x^{1/2} + .2x^{3/2} + .03x^{5/2}$

    The answer is $A(x) = 30/x + 42x^{-1/2} + .2x^{1/2} + .03x^{3/2}$, $x = 23.49$.

Section 4.2

1. $x = \sqrt{\dfrac{kM}{2f}}$ *Equation 3*

    $$T(x) = \left(f + \frac{gM}{x}\right)x + \frac{kM}{2x}$$

    $$= f \cdot x + gM + \frac{kM}{2x}$$

    $$T'(x) = f - \frac{kM}{2}x^{-2}$$

    When $T'(x) = 0$,

    $$f = \frac{kM}{2}x^{-2}$$

    $$x^2 = \frac{kM}{2f}$$

    $$x = \sqrt{\frac{kM}{2f}}.$$

$$T'' = -\frac{kM}{2}(-2x^{-3})$$

$$= \frac{kM}{x^3}$$

Since $x > 0$, $T'' > 0$ and therefore

$x = \sqrt{\frac{kM}{2f}}$ is minimum.

3. Use equation (3) with

M = 100,000 units produced annually

k = \$1 cost to store one unit for one year

f = \$500 cost to set up, fixed cost

x = the number of batches per year.

$$x = \sqrt{\frac{kM}{2f}} = \sqrt{\frac{(1)(100,000)}{2(500)}}$$

$$= \sqrt{100} = 10$$

5. From Exercise 3, the number of batches is 10 and the number of units is 100,000.

The number of units per batch is $\frac{100,000}{10} = 10,000$.

7. Use equation (3) with

f = the fixed cost for one order

 = \$60

M = the total number of units in one year

 = 100,000

k = the cost of storing one unit for one year

 = \$.50

x = the number of units annually

$$x = \sqrt{\frac{kM}{2f}} = \sqrt{\frac{(.50)(100,000)}{2(60)}}$$

≈ 20 (The number of units must be an integer.)

The optimum number of copies per order is

$$\frac{M}{x} = \frac{100,000}{20} = 5000.$$

9. In the inventory problem, the manufacturing cost is

$$f + \frac{gM}{x}.$$

The total manufacturing cost is

$$\left(f + \frac{gM}{x}\right)x.$$

Total storage cost is $\left(\frac{M}{x}\right)k$, where $\frac{M}{x}$ is the size of each batch and k is the cost of storing one unit. The total cost is

$$T(x) = \left(f + \frac{gM}{x}\right)x + \frac{kM}{x}$$

$$= fx + gM + \frac{kM}{x}$$

$$T'(x) = f - kMx^{-2}$$

$$f - kMx^{-2} = 0$$

$$f = \frac{kM}{x^2}$$

$$x^2f = kM$$

$$x^2 = \frac{kM}{f}$$

$$x = \sqrt{\frac{kM}{f}}$$

$T''(x) = kMx^{-3} > 0$, so $x = \sqrt{\frac{kM}{f}}$

minimizes the total cost.

11. In the inventory problem, total manufacturing cost is $\left(f + \frac{gM}{x}\right)x$.

In this case, storage cost is $\left(\frac{M}{2x}\right)k_1 + \left(\frac{M}{x}\right)k_2$.

The total cost is

$$T(x) = \left(f + \frac{gM}{x}\right)x + \left(\frac{M}{2x}\right)k_1 + \left(\frac{M}{x}\right)k_2$$

$$= fx + gM + \frac{Mk_1}{2x} + \frac{Mk_2}{x}$$

$$= fx + gM + \frac{Mk_1 + 2Mk_2}{2x}$$

$$= fx + gM + \frac{(k_1 + 2k_2)M}{2x}$$

$$T'(x) = f - \frac{(k_1 + 2k_2)M}{2x^2}$$

$$f - \frac{(k_1 + 2k_2)M}{2x^2} = 0$$

$$2fx^2 = (k_1 + 2k_2)M$$

$$x^2 = \frac{(k_1 + 2k_2)M}{2f}$$

$$x = \sqrt{\frac{(k_1 + 2k_2)M}{2f}}$$

$$T''(x) = \frac{(k_1 + k_2)M}{2x^3} > 0, \text{ so}$$

$$x = \sqrt{\frac{(k_1 + 2k_2)M}{2f}} \text{ minimizes the}$$

total cost.

13. $q = 25,000 - 50p$

(a) $\dfrac{dq}{dp} = -50$

$$E = -\frac{p}{q} \cdot \frac{dq}{dp}$$

$$= -\frac{p}{25,000 - 50p}(-50)$$

$$= \frac{p}{500 - p}$$

(b) $R = pq$

$$\frac{dR}{dp} = q(1 - E)$$

When R is maximum, $q(1 - E) = 0$.
Since $q = 0$ means no revenue,
set $1 - E = 0$.

$$E = 1$$

From (a),

$$\frac{p}{500 - p} = 1$$

$$p = 500 - p$$

$$p = 250.$$

$$q = 25,000 - 50p$$

$$= 25,000 - 50(250)$$

$$= 12,500$$

Total revenue is maximized if
$q = 12,500$.

15. $q = \dfrac{3000}{p}$

$$\frac{dq}{dp} = -\frac{3000}{p^2}$$

(a) $E = -\dfrac{p}{q} \cdot \dfrac{dq}{dp}$

$$= -\frac{p}{\frac{3000}{p}}\left(-\frac{3000}{p^2}\right)$$

$$= \left(-\frac{p^2}{3000}\right)\left(-\frac{3000}{p^2}\right)$$

$$= 1$$

(b) Since E is a constant, the demand always has unit elasticity.

$$\frac{dR}{dp} = q(1 - E)$$

$$= q(1 - 1)$$

$$= 0$$

So no value of q maximizes total revenue.

17. $q = 300 - 2p$

$\dfrac{dq}{dp} = -2$

$E = -\dfrac{p}{q} \cdot \dfrac{dq}{dp}$

$E = -\dfrac{p(-2)}{300 - 2p}$

$\quad = \dfrac{2p}{300 - 2p}$

(a) When $p = \$100$,

$\quad E = \dfrac{200}{300 - 200}$

$\qquad = 2 > 1.$

$E = 2$; elastic; a percentage in-crease in price will result in a greater percentage decrease in demand.

(b) When $p = \$50$,

$\quad E = \dfrac{100}{300 - 100}$

$\qquad = \dfrac{1}{2} < 1.$

$E = 1/2$; inelastic; a percentage change in price will result in a smaller percentage change in demand.

19. $p = (25 - q)^{1/2}$

$p^2 = 25 - q$

$q = 25 - p^2$

$\dfrac{dq}{dp} = -2p$

$\quad = -2(25 - q)^{1/2}$

$E = -\dfrac{p}{q} \dfrac{dq}{dp}$

$\quad = -\dfrac{\sqrt{25 - q}}{q}(-2\sqrt{25 - q})$

$\quad = \dfrac{2(25 - q)}{q}$

(a) When $q = 16$,

$\quad E = \dfrac{2(9)}{16}$

$\qquad = \dfrac{9}{8}.$

(b) $R = pq$

$\quad = p(25 - p^2)$

$\quad = 25p - p^3$

$R' = 25 - 3p^2$

If $R' = 0$,

$3p^2 = 25$

$p = \dfrac{5}{\sqrt{3}}$

$\quad = \dfrac{5\sqrt{3}}{3}.$

$q = 25 - p^2$

$\quad = 25 - \dfrac{25}{3}$

$\quad = \dfrac{50}{3}$

Total revenue is maximized when $q = 50/3$, $p = 5\sqrt{3}/3$.

(c) We have concluded that total revenue is maximized at the price where demand has unit elasticity. Thus, when $q = 50/3$ (or R is maximized), $E = 1$.

21. $E = -\dfrac{p}{q} \cdot \dfrac{dq}{dp}$

Since $p \neq 0$, $E = 0$ when $\dfrac{dq}{dp} = 0$. The derivative is zero, which implies that the demand function has a hori-zontal tangent line at the value of p where $E = 0$.

Section 4.3

1. $4x^2 + 3y^2 = 6$

$$\frac{d}{dx}(4x^2 + 3y^2) = \frac{d}{dx}(6)$$

$$\frac{d}{dx}(4x^2) + \frac{d}{dx}(3y^2) = \frac{d}{dx}(6)$$

$$8x + 3 \cdot 2y\frac{dy}{dx} = 0$$

$$6y\frac{dy}{dx} = -8x$$

$$\frac{dy}{dx} = -\frac{4x}{3y}$$

3. $2xy + y^2 = 8$

$$\frac{d}{dx}(2xy + y^2) = \frac{d}{dx}(8)$$

$$\frac{d}{dx}(2xy) + \frac{d}{dx}(y^2) = \frac{d}{dx}(8)$$

$$2x\frac{dy}{dx} + y(2) + 2y\frac{dy}{dx} = 0$$

$$(2x + 2y)\frac{dy}{dx} = -2y$$

$$\frac{dy}{dx} = \frac{-2y}{2x + 2y}$$

$$\frac{dy}{dx} = \frac{-y}{x + y}$$

5. $6xy^2 - 8y + 1 = 0$

$$\frac{d}{dx}(6xy^2 - 8y + 1) = \frac{d}{dx}(0)$$

$$6x\frac{d}{dx}(y^2) + y^2\frac{d}{dx}(6x) - 8\frac{dy}{dx} = 0$$

$$6x(2)y\frac{dy}{dx} + y^2(6) - 8\frac{dy}{dx} = 0$$

$$12xy\frac{dy}{dx} + 6y^2 - 8\frac{dy}{dx} = 0$$

$$(12xy - 8)\frac{dy}{dx} = -6y^2$$

$$\frac{dy}{dx} = \frac{-6y^2}{12xy - 8}$$

$$\frac{dy}{dx} = \frac{-3y^2}{6xy - 4}$$

7. $6x^2 + 8xy + y^2 = 6$

$$\frac{d}{dx}(6x^2 + 8xy + y^2) = \frac{d}{dx}(6)$$

$$12x + \frac{d}{dx}(8xy) + \frac{d}{dx}(y^2) = 0$$

$$12x + 8x\frac{dy}{dx} + y\frac{d}{dx}(8x) + 2y\frac{dy}{dx} = 0$$

$$12x + 8x\frac{dy}{dx} + 8y + 2y\frac{dy}{dx} = 0$$

$$(8x + 2y)\frac{dy}{dx} = -12x - 8y$$

$$\frac{dy}{dx} = \frac{-12x - 8y}{8x + 2y}$$

$$\frac{dy}{dx} = \frac{-6x - 4y}{4x + y}$$

9. $x^3 = y^2 + 4$

$$\frac{d}{dx}(x^3) = \frac{d}{dx}(y^2 + 4)$$

$$3x^2 = \frac{d}{dx}(y^2) + \frac{d}{dx}(4)$$

$$3x^2 = 2y\frac{dy}{dx} + 0$$

$$\frac{3x^2}{2y} = \frac{dy}{dx}$$

11. $\frac{1}{x} - \frac{1}{y} = 2$

$$\frac{d}{dx}\left(\frac{1}{x} - \frac{1}{y}\right) = \frac{d}{dx}(2)$$

$$-\frac{1}{x^2} - \frac{d}{dx}\left(\frac{1}{y}\right) = 0$$

$$-\frac{1}{x^2} + \frac{1}{y^2} \cdot \frac{dy}{dx} = 0$$

$$\frac{1}{y^2} \cdot \frac{dy}{dx} = \frac{1}{x^2}$$

$$\frac{dy}{dx} = \frac{y^2}{x^2}$$

13. $3x^2 = \dfrac{2 - y}{2 + y}$

$\dfrac{d}{dx}(3x^2) = \dfrac{d}{dx}\left(\dfrac{2 - y}{2 + y}\right)$

$6x = \dfrac{(2 + y)\dfrac{d}{dx}(2 - y) - (2 - y)\dfrac{d}{dx}(2 + y)}{(2 + y)^2}$

$6x = \dfrac{(2 + y)\left(-\dfrac{dy}{dx}\right) - (2 - y)\dfrac{dy}{dx}}{(2 + y)^2}$

$6x = \dfrac{-4\dfrac{dy}{dx}}{(2 + y)^2}$

$6x(2 + y)^2 = -4\dfrac{dy}{dx}$

$\dfrac{-3x(2 + y)^2}{2} = \dfrac{dy}{dx}$

15. $x^2y + y^3 = 4$

$\dfrac{d}{dx}(x^2y + y^3) = \dfrac{d}{dx}(4)$

$\dfrac{d}{dx}(x^2y) + \dfrac{d}{dx}(y^3) = 0$

$x^2\dfrac{dy}{dx} + y(2x) + 3y^2\dfrac{dy}{dx} = 0$

$(x^2 + 3y^2)\dfrac{dy}{dx} = -2xy$

$\dfrac{dy}{dx} = \dfrac{-2xy}{x^2 + 3y^2}$

17. $\sqrt{x} + \sqrt{y} = 4$

$\dfrac{d}{dx}(x^{1/2} + y^{1/2}) = \dfrac{d}{dx}4$

$\dfrac{1}{2}x^{-1/2} + \dfrac{1}{2}y^{-1/2}\dfrac{dy}{dx} = 0$

$\dfrac{1}{2}y^{-1/2}\dfrac{dy}{dx} = 0 - \dfrac{1}{2}x^{-1/2}$

$\dfrac{dy}{dx} = 2y^{1/2}\left(-\dfrac{1}{2}x^{-1/2}\right)$

$\dfrac{dy}{dx} = \dfrac{-y^{1/2}}{x^{1/2}}$

19. $\sqrt{xy} + y = 1$

$\dfrac{d}{dx}(\sqrt{xy} + y) = \dfrac{d}{dx}(1)$

$\dfrac{d}{dx}(x^{1/2}y^{1/2}) + \dfrac{d}{dx}(y) = 0$

$x^{1/2}\dfrac{d}{dx}(y^{1/2}) + y^{1/2}\dfrac{d}{dx}(x^{1/2}) + \dfrac{dy}{dx} = 0$

$x^{1/2} \cdot \dfrac{1}{2}y^{-1/2}\dfrac{dy}{dx} + \dfrac{1}{2}x^{-1/2}y^{1/2} + \dfrac{dy}{dx} = 0$

$\left(\dfrac{1}{2}x^{1/2}y^{-1/2} + 1\right)\dfrac{dy}{dx} = -\dfrac{1}{2}x^{-1/2}y^{1/2}$

$\dfrac{dy}{dx} = \dfrac{-\dfrac{1}{2}x^{-1/2}y^{1/2}}{\dfrac{1}{2}x^{1/2}y^{-1/2} + 1}$

$= \dfrac{-\dfrac{1}{2}x^{-1/2}y^{1/2}}{\dfrac{1}{2}(x^{1/2}y^{-1/2} + 2)}$

$= \dfrac{-x^{-1/2}y^{1/2}}{x^{1/2}y^{-1/2} + 2}$

21. $x^4y^3 + 4x^{3/2} = 6y^{3/2} + 5$

$\dfrac{d}{dx}(x^4y^3 + 4x^{3/2}) = \dfrac{d}{dx}(6y^{3/2} + 5)$

$\dfrac{d}{dx}(x^4y^3) + \dfrac{d}{dx}(4x^{3/2}) = \dfrac{d}{dx}(6y^{3/2}) + \dfrac{d}{dx}(5)$

$4x^3y^3 + x^4 \cdot 3y^2\dfrac{dy}{dx} + 6x^{1/2} = 9y^{1/2}\dfrac{dy}{dx} + 0$

$4x^3y^3 + 6x^{1/2} = 9y^{1/2}\dfrac{dy}{dx} - 3x^4y^2\dfrac{dy}{dx}$

$4x^3y^3 + 6x^{1/2} = (9y^{1/2} - 3x^4y^2)\dfrac{dy}{dx}$

$\dfrac{4x^3y^3 + 6x^{1/2}}{9y^{1/2} - 3x^4y^2} = \dfrac{dy}{dx}$

23. $(x^2 + y^3)^4 = x + 2y + 4$

$\dfrac{d}{dx}(x^2 + y^2)^4 = \dfrac{d}{dx}(x + 2y + 4)$

$4(x^2 + y^3)^3\dfrac{d}{dx}(x^2 + y^3) = \dfrac{d}{dx}(x) + \dfrac{d}{dx}(2y)$

$+ \dfrac{d}{dx}(4)$

$4(x^2 + y^3)^3\left(2x + 3y^2\dfrac{dy}{dx}\right) = 1 + 2\dfrac{dy}{dx} + 0$

$8x + 12y^2 (x^2+y^3)^3$

$$8x(x^2 + y^3)^3 + 12y^2(x^2 + y^3)^3 \frac{dy}{dx}$$

$$= 1 + 2\frac{dy}{dx}$$

$$8x(x^2 + y^3)^3 - 1 = 2\frac{dy}{dx} - 12y^2(x^2 + y^3)^3 \frac{dy}{dx}$$

$$8x(x^2 + y^3)^3 - 1 = [2 - 12y^2(x^2 + y^3)^3]\frac{dy}{dx}$$

$$\frac{8x(x^2 + y^3)^3 - 1}{2 - 12y^2(x^2 + y^3)^3} = \frac{dy}{dx}$$

25. $x^2 + y^2 = 25$; tangent at $(-3, 4)$

$$\frac{d}{dx}(x^2 + y^2) = \frac{d}{dx}(25)$$

$$2x + 2y\frac{dy}{dx} = 0$$

$$2y\frac{dy}{dx} = -2x$$

$$\frac{dy}{dx} = -\frac{x}{y}$$

$$m = -\frac{x}{y} = -\frac{-3}{4} = \frac{3}{4}$$

$$y - y_1 = m(x - x_1)$$

$$y - 4 = \frac{3}{4}[x - (-3)]$$

$$4y - 16 = 3x + 9$$

$$4y = 3x + 25$$

27. $x^2y^2 = 1$; tangent at $(-1, 1)$

$$\frac{d}{dx}(x^2y^2) = \frac{d}{dx}(1)$$

$$x^2\frac{d}{dx}(y^2) + y^2\frac{d}{dx}(x^2) = 0$$

$$x^2(2y)\frac{dy}{dx} + y^2(2x) = 0$$

$$2x^2y\frac{dy}{dx} = -2xy^2$$

$$\frac{dy}{dx} = \frac{-2xy^2}{2x^2y} = -\frac{y}{x}$$

$$m = -\frac{y}{x} = -\frac{1}{-1} = 1$$

$$y - 1 = 1[x - (-1)]$$

$$y = x + 1 + 1$$

$$y = x + 2$$

29. $x^2 + \sqrt{y} = 7$; tangent at $(2, 9)$

$$\frac{d}{dx}(x^2 + \sqrt{y}) = \frac{d}{dx}(7)$$

$$2x + \frac{1}{2}y^{-1/2}\frac{dy}{dx} = 0$$

$$\frac{1}{2}y^{-1/2}\frac{dy}{dx} = -2x$$

$$\frac{dy}{dx} = -2x(2y^{1/2})$$

$$\frac{dy}{dx} = -4xy^{1/2}$$

$$m = -4xy^{1/2} = -4(2)(9^{1/2}) = -24$$

$$y - 9 = -24(x - 2)$$

$$24x + y = 57$$

31. $y + \frac{\sqrt{x}}{y} = 3$; tangent at $(4, 2)$

$$\frac{d}{dx}\left(y + \frac{\sqrt{x}}{y}\right) = \frac{d}{dx}(3)$$

$$\frac{dy}{dx} + \frac{d}{dx}\left(\frac{\sqrt{x}}{y}\right) = 0$$

$$\frac{dy}{dx} + \frac{y(\frac{1}{2})x^{-1/2} - \sqrt{x}\frac{dy}{dx}}{y^2} = 0$$

$$\frac{dy}{dx} = \frac{-\frac{1}{2}yx^{-1/2} + \sqrt{x}\frac{dy}{dx}}{y^2}$$

$$y^2\frac{dy}{dx} = -\frac{1}{2}yx^{-1/2} + \sqrt{x}\frac{dy}{dx}$$

$$(y^2 - \sqrt{x})\frac{dy}{dx} = -\frac{1}{2}yx^{-1/2}$$

$$\frac{dy}{dx} = \frac{-y}{2x^{1/2}(y^2 - \sqrt{x})}$$

$$m = \frac{-y}{2x^{1/2}(y^2 - \sqrt{x})}$$

$$= \frac{-2}{2(2)(4 - 2)}$$

$$= -\frac{1}{4}$$

$$y - 2 = -\frac{1}{4}(x - 4)$$

$$y = -\frac{1}{4}x + 3$$

$$x + 4y = 12$$

33. $x^2 + y^2 = 100$

(a) Lines are tangent at points where
x = 6. By substituting x = 6 in
the equation, we find the points
are (6, 8) and (6, -8).

$$\frac{d}{dx}(x^2 + y^2) = \frac{d}{dx}(100)$$

$$2x + 2y\frac{dy}{dx} = 0$$

$$2y\frac{dy}{dx} = -2x$$

$$dy = -\frac{x}{y}$$

$$m_1 = -\frac{x}{y} = -\frac{6}{8} = -\frac{3}{4}$$

$$m_2 = -\frac{x}{y} = -\frac{6}{-8} = \frac{3}{4}$$

First tangent:

$$y - 8 = -\frac{3}{4}(x - 6)$$

$$y = -\frac{3}{4}x + \frac{50}{4}$$

$$3x + 4y = 50$$

Second tangent:

$$y - (-8) = \frac{3}{4}(x - 6)$$

$$y + 8 = \frac{3}{4}x - \frac{18}{4}$$

$$y = \frac{3}{4}x - \frac{50}{4}$$

$$-3x + 4y = -50$$

(b) See the answer graph in the back
of the textbook.

35. $y^5 - y - x^2 = -1$; tangent at (1, 1)

$$\frac{d}{dx}(y^5 - y - x^2) = \frac{d}{dx}(-1)$$

$$5y^4\frac{dy}{dx} - \frac{dy}{dx} - 2x = 0$$

$$(5y^4 - 1)\frac{dy}{dx} = 2x$$

$$\frac{dy}{dx} = \frac{2x}{5y^4 - 1}$$

$$m = \frac{2x}{5y^4 - 1}$$

$$= \frac{2(1)}{5(1)^4 - 1}$$

$$= \frac{2}{4}$$

$$= \frac{1}{2}$$

$$y - 1 = \frac{1}{2}(x - 1)$$

$$y = \frac{1}{2}x + \frac{1}{2}$$

$$2y = x + 1$$

37. $x^2 + y^2 + 1 = 0$

$$\frac{d}{dx}(x^2 + y^2 + 1) = \frac{d}{dx}(0)$$

$$2x + 2y\frac{dy}{dx} = 0$$

$$\frac{dy}{dx} = \frac{-2x}{2y}$$

$$= -\frac{x}{y}$$

If x and y are real numbers, x^2 and
y^2 are nonnegative. 1 + a nonnegative number cannot equal zero so
there is no function y = f(x) that
satisfies $x^2 + y^2 + 1 = 0$.

39. $\sqrt{u} + \sqrt{2v + 1} = 5$

$$\frac{dv}{du}(\sqrt{u} + \sqrt{2v + 1}) = \frac{dv}{du}(5)$$

$$\frac{1}{2}u^{-1/2} + \frac{1}{2}(2v + 1)^{-1/2}(2)\frac{dv}{du} = 0$$

$$(2v + 1)^{-1/2}\frac{dv}{du} = -\frac{1}{2}u^{-1/2}$$

$$\frac{dv}{du} = -\frac{(2v + 1)^{1/2}}{2u^{1/2}}$$

41. $C^2 = x^2 + 100\sqrt{x} + 50$

(a) $2C\frac{dC}{dx} = 2x + \frac{1}{2}(100)x^{-1/2}$

$$\frac{dC}{dx} = \frac{2x + 50x^{-1/2}}{2C}$$

$$\frac{dC}{dx} = \frac{x + 25x^{-1/2}}{C} \cdot \frac{x^{1/2}}{x^{1/2}}$$

$$\frac{dC}{dx} = \frac{x^{3/2} + 25}{Cx^{1/2}}$$

When x = 5, the approximate increase in cost of an additional unit is

$$\frac{(5)^{3/2} + 25}{(5^2 + 100\sqrt{5} + 50)^{1/2}(5)^{1/2}}$$

$$= \frac{36.18}{(17.28)\sqrt{5}}$$

$$= .94.$$

(b) $900(x - 5)^2 + 25R^2 = 22,500$

$$R^2 = 900 - 36(x - 5)^2$$

$$2R\frac{dR}{dx} = -72(x - 5)$$

$$\frac{dR}{dx} = \frac{-36(x - 5)}{R} = \frac{180 - 36x}{R}$$

When x = 5, the approximate change in revenue for a unit increase in sales is

$$\frac{180 - 36(5)}{R} = \frac{0}{R} = 0.$$

43. $2s^2 + \sqrt{st} - 4 = 3t$

$$4s\frac{ds}{dt} + \frac{1}{2}(st)^{-1/2}\left(s + t\frac{ds}{dt}\right) = 3$$

$$4s\frac{ds}{dt} + \frac{s + t\frac{ds}{dt}}{2\sqrt{st}} = 3$$

$$\frac{8s(\sqrt{st})\frac{ds}{dt} + s + t\frac{ds}{dt}}{2\sqrt{st}} = 3$$

$$\frac{(8s\sqrt{st} + t)\frac{ds}{dt} + s}{2\sqrt{st}} = 3$$

$$(8s\sqrt{st} + t)\frac{ds}{dt} = 6\sqrt{st} - s$$

$$\frac{ds}{dt} = \frac{-s + 6\sqrt{st}}{8s\sqrt{st} + t}$$

Section 4.4

1. $y = 9x^2 + 2x$; $\frac{dx}{dt} = 4$, x = 6

$$\frac{dy}{dt} = 18x\frac{dx}{dt} + 2\frac{dx}{dt}$$

$$= 18(6)(4) + 2(4) = 440$$

3. $y^2 - 5x^2 = -1$; $-\frac{dx}{dt} = -3$, x = 1, y = 2

$$2y\frac{dy}{dt} - 10x\frac{dx}{dt} = 0$$

$$y\frac{dy}{dt} = 5x\frac{dx}{dt}$$

$$2\frac{dy}{dt} = 5(-3)$$

$$\frac{dy}{dt} = -\frac{15}{2}$$

5. $xy - 5x + 2y^3 = -70$; $\frac{dx}{dt} = -5$,

   $x = 2$, $y = -3$

   $x\frac{dy}{dt} + y\frac{dx}{dt} - 5\frac{dx}{dt} + 6y^2\frac{dy}{dt} = 0$

   $(x + 6y^2)\frac{dy}{dt} + (y - 5)\frac{dx}{dt} = 0$

   $(x + 6y^2)\frac{dy}{dt} = (5 - y)\frac{dy}{dt}$

   $$\frac{dy}{dt} = \frac{(5 - y)\frac{dx}{dt}}{x + 6y^2}$$

   $$= \frac{[5 - (-3)](-5)}{2 + 6(-3)^2}$$

   $$= \frac{-40}{56} = -\frac{5}{7}$$

7. $\frac{x^2 + y}{x - y} = 9$; $\frac{dx}{dt} = 2$, $x = 4$, $y = 2$

   $$\frac{(x - y)\left(2x\frac{dx}{dt} + \frac{dy}{dt}\right) - (x^2 + y)\left(\frac{dx}{dt} - \frac{dy}{dt}\right)}{(x - y)^2} = 0$$

   $$\frac{2x(x - y)\frac{dx}{dt} + (x - y)\frac{dy}{dt} - (x^2 + y)\frac{dx}{dt} + (x^2 + y)\frac{dy}{dt}}{(x - y)^2} = 0$$

   $[2x(x - y) - (x^2 + y)]\frac{dx}{dt} + [(x - y) + (x^2 + y)]\frac{dy}{dt} = 0$

   $$\frac{dy}{dt} = \frac{[(x^2 + y) - 2x(x - y)]\frac{dx}{dt}}{(x - y) + (x^2 + y)}$$

   $$\frac{dy}{dt} = \frac{(-x^2 + y + 2xy)\frac{dx}{dt}}{x + x^2}$$

   $$= \frac{[-(4)^2 + 2 + 2(4)(2)](2)}{4 + 4^2}$$

   $$= \frac{4}{20} = \frac{1}{5}$$

9. $C = .1x^2 + 10,000$; $x = 100$, $\frac{dx}{dt} = 10$

$$\frac{dC}{dt} = .1(2x)\frac{dx}{dt}$$

$$= .1(200)(10)$$

$$= 200$$

The cost is changing at a rate of $200 per month.

11. $R = 50x - .4x^2$; $C = 5x + 15$; $x = 40$; $\frac{dx}{dt} = 10$

(a) $\frac{dR}{dt} = 50\frac{dx}{dt} - .8x\frac{dx}{dt}$

$$= 50(10) - .8(40)(10)$$

$$= 500 - 320$$

$$= 180$$

Revenue is increasing at a rate of $180 per day.

(b) $\frac{dC}{dt} = 5\frac{dx}{dt}$

$$= 5(10) = 50$$

Cost is increasing at a rate of $50 per day.

(c) Profit = Revenue − Cost, or

$$P = R - C$$

$$\frac{dP}{dt} = \frac{dR}{dt} - \frac{dC}{dt}$$

$$= 180 - 50$$

$$= 130$$

Profit is increasing at a rate of $130 per day.

13. $p = \frac{8000}{q}$; $p = 3.50$,

$$\frac{dp}{dt} = .15p$$

because price is increasing at 15% as well.

$$pq = 8000$$

$$p\frac{dq}{dt} + q\frac{dp}{dt} = 0$$

$$\frac{dq}{dt} = \frac{-q\frac{dp}{dt}}{p}$$

$$= \frac{-\left(\frac{8000}{3.50}\right)(.15)(3.50)}{3.50}$$

$$= -343$$

Demand is decreasing at a rate of approximately 343 units per unit time.

15. $V = k(R^2 - r^2)$; $r = 2$ mm, $k = 3$, $\frac{dr}{dt} = .02$ mm per minute, and R is constant.

$$V = k(R^2 - r^2)$$

$$V = 3(R^2 - r^2)$$

$$\frac{dV}{dt} = 3\left(0 - 2\frac{dr}{dt}\right)$$

$$= -6r\frac{dr}{dt}$$

$$= -6(2)(.02)$$

$$= -.24 \text{ mm/min}$$

17. $V = k(R^2 - r^2)$; $k = 555.6$, $R = .02$ mm, $\frac{dR}{dt} = .003$ mm per minute, r is constant.

$$V = k(R^2 - r^2)$$

$$V = 555.6(R^2 - r^2)$$

$$\frac{dV}{dt} = 555.6\left(2R\frac{dR}{dt} - 0\right)$$

$$= 555.6(2)(.02)(.003)$$

$$= .067 \text{ mm/min}$$

19. $T(x) = \dfrac{2 + x}{2 + x^2}$, $x = 4$, $\dfrac{dx}{dt} = 4$

Find $\dfrac{dT}{dt}$.

$$\frac{dT}{dt} = \frac{(2 + x^2)\frac{d}{dt}(2 + x) - (2 + x)\frac{d}{dt}(2 + x^2)}{(2 + x^2)^2}$$

$$= \frac{(2 + x^2)\frac{dx}{dt} - (2 + x)(2x)\frac{dx}{dt}}{(2 + x^2)^2}$$

$$= \frac{(2 + x^2)\frac{dx}{dt} - (4x + 2x^2)\frac{dx}{dt}}{(2 + x^2)^2}$$

$$= \frac{(2 - 4x - x^2)\frac{dx}{dt}}{(2 + x^2)^2}$$

$$= \frac{(2 - 16 - 16)4}{(2 + 16)^2}$$

$$= -\frac{120}{324}$$

$$= -.370$$

21. Let x = the distance of the base of the ladder from the base of the building

    y = the distance up the side of the building to the top of the ladder.

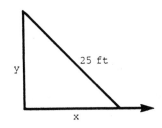

Find $\dfrac{dy}{dt}$ when x = 7 ft and $\dfrac{dx}{dt} = 4$ ft/min.

Since $y = \sqrt{25^2 - x^2}$, when x = 7,

$$y = 24 \text{ ft.}$$

$$x^2 + y^2 = 25^2$$

$$\frac{d}{dt}(x^2 + y^2) = \frac{d}{dt}(25)^2$$

$$2x\frac{dx}{dt} + 2y\frac{dy}{dt} = 0$$

$$2y\frac{dy}{dt} = -2x\frac{dx}{dt}$$

$$\frac{dy}{dt} = \frac{-2x}{2y}\frac{dx}{dt} = -\frac{x}{y}\frac{dx}{dt}$$

$$= -\frac{7}{24}(4)$$

$$= -\frac{7}{6}$$

The ladder is sliding down the building at the rate of 7/6 ft/min.

23. Let r = the radius of the circle formed by the ripple.

Find $\dfrac{dA}{dt} =$ when r = 4 ft and $\dfrac{dr}{dt} = 2$ ft/min.

$$A = \pi r^2$$

$$\frac{dA}{dt} = 2\pi r\frac{dr}{dt}$$

$$= 2\pi(4)(2)$$

$$= 16\pi \text{ ft}^2/\text{min}$$

25. Let r = the radius of the base of the conical pile.

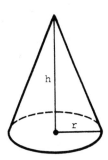

Find $\dfrac{dV}{dt}$ when r = 5 in, $\dfrac{dr}{dt}$ = 1 in/min

h = 2r for all t.

$V = \dfrac{\pi}{3}r^2 h$

$V = \dfrac{\pi}{3}r^2(2r)$

$\quad = \dfrac{2\pi}{3}r^3$

$\dfrac{dV}{dt} = \dfrac{3 \cdot 2\pi r^3}{3}\dfrac{dr}{dt}$

$\dfrac{dV}{dt} = 2\pi(5^2)(1)$

$\quad = 50\pi \ \text{in}^3/\text{min}$

$V = (xh)(16)$

$\quad = \left(\dfrac{h}{2}h\right)16$

$\quad = 8h^2$

$\dfrac{dV}{dt} = 16h\dfrac{dh}{dt}$

$\dfrac{1}{16h}\dfrac{dV}{dt} = \dfrac{dh}{dt}$

$\dfrac{1}{16(4)}(4) = \dfrac{dh}{dt}$

$\dfrac{dh}{dt} = \dfrac{1}{16} \ \text{ft/min}$

27. Let x = one-half the width of the triangular cross section

   h = the height of the water

   V = the volume of the water.

   $\dfrac{dV}{dt}$ = 4 cu ft per min

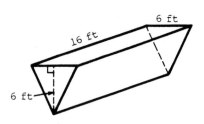

Find $\dfrac{dh}{dt}$ when h = 4.

$V = \begin{bmatrix} \text{Area of} \\ \text{triangular} \\ \text{cross section} \end{bmatrix}\begin{bmatrix} \text{length} \end{bmatrix}$

Area of triangular cross section

$= \dfrac{1}{2}(\text{base})(\text{altitude})$

$= \dfrac{1}{2}(2x)(h)$

$= xh$

By similar triangles,

$\dfrac{3}{x} = \dfrac{6}{h}$, so $x = \dfrac{h}{2}$.

29. Let x = the horizontal length

   r = the rope length.

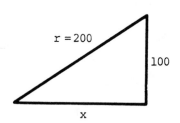

$\dfrac{dx}{dt}$ = 50 ft per min

$200^2 = x^2 + 100^2$

$173.2 = x$

$r^2 = x^2 + 100$

$2r\dfrac{dr}{dt} = 2x\dfrac{dx}{dt} + 0$

$r\dfrac{dr}{dt} = x\dfrac{dx}{dt}$

$\dfrac{dr}{dt} = \dfrac{x\dfrac{dx}{dt}}{r}$

$\dfrac{dr}{dt} = \dfrac{173.2(50)}{200}$

$\quad = 43.3 \ \text{ft/min}$

Section 4.5

1. $y = 6x^2$

$dy = 2(6x)dx$

$\quad = 12x\ dx$

3. $y = 7x^2 - 9x + 6$

$dy = [2(7x) - 9)dx$

$\quad = (14x - 9)dx$

5. $y = 2\sqrt{x}$

$dy = 2\left(\frac{1}{2}\right)x^{-1/2}\ dx$

$\quad = x^{-1/2}\ dx$

7. $y = \dfrac{8x - 2}{x - 3}$

$dy = \dfrac{(x - 3)(8) - (8x - 2)}{(x - 3)^2}\ dx$

$\quad = \dfrac{-22}{(x - 3)^2}\ dx$

9. $y = x^2\left(x - \dfrac{1}{x} + 2\right)$

$dy = \left[2x\left(x - \dfrac{1}{x} + 2\right) + x^2\left(1 + \dfrac{1}{x^2}\right)\right]dx$

$\quad = (2x^2 - 2 + 4x + x^2 + 1)dx$

$\quad = (3x^2 + 4x - 1)dx$

11. $y = \left(2 - \dfrac{3}{x}\right)\left(1 + \dfrac{1}{x}\right)$

$dy = \left[\left(2 - \dfrac{3}{x}\right)\left(-\dfrac{1}{x^2}\right) + \left(1 + \dfrac{1}{x}\right)\left(\dfrac{3}{x^2}\right)\right]dx$

$\quad = \left[\dfrac{-2}{x^2} + \dfrac{3}{x^3} + \dfrac{3}{x^2} + \dfrac{3}{x^3}\right]dx$

$\quad = \left(\dfrac{1}{x^2} + \dfrac{6}{x^3}\right)dx$ or $(x^{-2} + 6x^{-3})dx$

13. $y = 2x^2 - 5x;\ x = -2,\ \Delta x = .2$

$dy = (4x - 5)dx$

$\quad \approx (4x - 5)\Delta x$

$\quad = [4(-2) - 5](.2)$

$\quad = -2.6$

15. $y = x^3 - 2x^2 + 3,\ x = 1,\ \Delta x = -.1$

$dy = (3x^2 - 4x)dx$

$\quad \approx (3x^2 - 4x)\Delta x$

$\quad = [3(1^2) - 4(1)](-.1)$

$\quad = .1$

17. $y = \sqrt{3x},\ x = 1,\ \Delta x = .15$

$dy = \sqrt{3}\left(\dfrac{1}{2}x^{-1/2}\right)dx$

$\quad \approx \dfrac{\sqrt{3}}{2\sqrt{x}}\Delta x$

$\quad = \dfrac{1.73}{2}(.15)$

$\quad = .130$

19. $y = \dfrac{2x - 5}{x + 1};\ x = 2,\ \Delta x = -.03$

$dy = \dfrac{(x + 1)(2) - (2x - 5)(1)}{(x + 1)^2}\ dx$

$\quad = \dfrac{7}{(x + 1)^2}\ dx$

$\quad \approx \dfrac{7}{(x + 1)^2}\ \Delta x$

$\quad = \dfrac{7}{(2 + 1)^2}(-.03)$

$\quad = -.023$

21. $y = -6\left(2 - \dfrac{1}{x^2}\right),\ x = -1,\ \Delta x = .02$

$dy = -6\left(\dfrac{2}{x^3}\right)dx$

$\quad \approx -\dfrac{12}{x^3}\ \Delta x$

$\quad = \dfrac{-12}{(-1)^3}(.02)$

$\quad = .24$

23. $y = \dfrac{1 + x}{\sqrt{x}}$, $x = 9$, $\Delta x = -.03$

$dy = \dfrac{\sqrt{x}(1) - (1 + x)\left(\frac{1}{2}\right)(x^{-1/2})}{x}\, dx$

$\approx \dfrac{2\sqrt{x} - (1 + x)(x^{-1/2})}{2x}\, \Delta x$

$= \dfrac{2\sqrt{9} - (1 + 9)(9^{-1/2})}{2(9)}(-.03)$

$= -.00444$

25. Let d = the demand in thousands of pounds

x = the price in dollars.

$d(x) = -5x^3 - 2x^2 + 1500$

(a) $x = 2$, $\Delta x = .50$

$dd = (-15x^2 - 4x)\, dx$

$\approx (-15x^2 - 4x)\Delta x$

$= [-15(4) - 4(2)](.50)$

$= -34$ thousand pounds

(b) $x = 6$, $\Delta x = .30$

$dd \approx [-15(36) - 4(6)](.30)$

$= -169.2$ thousand pounds

27. $H(r) = \dfrac{300}{1 + .03r^2}$

$r = 10$, $\Delta r = .5$

$dH = \dfrac{-300(.06r)}{(1 + .03r^2)^2}\, dr$

$\approx \dfrac{-300(.06)r}{(1 + .03r^2)^2}\, \Delta r$

$= \dfrac{-300(.06)(10)}{(1 + 3)^2}(.5)$

$= \dfrac{90}{16}$

$= -5.625$ housing starts

29. $P(x) = -390 + 24x + 5x^2 - \dfrac{1}{3}x^3$

$x = 1000$, $\Delta x = 1$

$dP = (24 + 10x - x^2)\, dx$

$\approx (24 + 10x - x^2)\Delta x$

$= (24 + 10,000 - 100,000) \cdot 1$

$\approx -\$990,000$

31. Let x = the number of beach balls

V = the volume of x beach balls.

Then $\dfrac{dV}{dr} \approx$ the volume of material in beach balls since they are hollow.

$V = \dfrac{4}{3}\pi r^3 x$

$r = 6$ in, $x = 5000$, $\Delta r = .03$ in

$dV = \dfrac{4}{3}\pi(3r^2x + r^3)\Delta r$

$= \dfrac{4}{3}\pi(3 \cdot 36 \cdot 5000 + 36)(.03)$

$= 21,600\pi$

$21,600\pi$ in^3 of material would be needed.

33. $P(x) = \dfrac{25x}{8 + x^2}$

$dP = \dfrac{(8 + x^2)(25) - 25x(2x)}{(8 + x^2)^2}\, dx$

$\approx \dfrac{(8 + x^2)(25) - 25x)(2x)}{(8 + x^2)^2}\, \Delta x$

(a) $x = 2$, $\Delta x = .5$

$dP = \dfrac{[(8 + 4)(25) - (25(2)(4)](.5)}{(8 + 4)^2}$

$= .347$ million

(b) $x = 3$, $\Delta x = .25$

$dP \approx \Delta P = [(8 + 9)(25) - 25(3)(6)](.25)$

$= -.022$ million

35. r changes from 14 mm to 16 mm so
$\Delta r = 2$.

$$V = \frac{4}{3}\pi r^3$$

$$dV = \frac{4}{3}(3)\pi r^2 \; dr$$

$$\Delta V = 4\pi r^2 \Delta r$$
$$= 4\pi(14)^2(2)$$
$$= 1568\pi \text{ mm}^3$$

37. r increases from 20 mm to 22 mm so
$\Delta r = 2$.

$$A = \pi r^2$$
$$dA = 2\pi r \; dr$$
$$\Delta A \approx 2\pi r \Delta r$$
$$= 2\pi(20)(2)$$
$$= 80\pi \text{ mm}^2$$

39. r = 3 cm, $\Delta r = -.2$ cm

$$V = \frac{4}{3}\pi r^3$$
$$dV = 4\pi r^2 dr$$
$$\Delta V \approx 4\pi r^2 \Delta r$$
$$= 4\pi(9)(-.2)$$
$$= -7.2\pi \text{ cm}^3$$

41. r = 4.87 in, $\Delta r = \pm.040$

$$A = \pi r^2$$
$$dA = 2\pi r \; dr$$
$$\Delta A \approx 2\pi r \Delta r$$
$$= 2\pi(4.87)(\pm.040)$$
$$= \pm 1.224 \text{ in}^2$$

43. h = 7.284 in, r = $1.09 \pm .007$ in

$$V = \frac{1}{3}\pi r^2 h$$

$$dV = \frac{2}{3}\pi rh \; dr$$

$$\Delta V \approx \frac{2}{3}\pi rh \; \Delta r$$

$$= \frac{2}{3}\pi(1.09)(7.284)(.007)$$

$$= \pm.116 \text{ in}^3$$

Chapter 4 Review Exercises

5. $x^2y^3 + 4xy = 2$

$$\frac{d}{dx}(x^2y^3 + 4xy) = \frac{d}{dx}(2)$$

$$2xy^3 + 3y^2\left(\frac{dy}{dx}\right)x^2 + 4y + 4x\frac{dy}{dx} = 0$$

$$(3x^2y^2 + 4x)\frac{dy}{dx} = -2xy^3 - 4y$$

$$\frac{dy}{dx} = \frac{-2xy^3 - 4y}{3x^2y^2 + 4x}$$

7. $9\sqrt{x} + 4y^3 = \frac{2}{x}$

$$\frac{d}{dx}(9\sqrt{x} + 4y^3) = \frac{d}{dx}\left(\frac{2}{x}\right)$$

$$\frac{9}{2}x^{-1/2} + 12y^2\frac{dy}{dx} = \frac{-2}{x^2}$$

$$12y^2\frac{dy}{dx} = \frac{-2}{x^2} - \frac{9x^{-1/2}}{2}$$

$$12y^2\frac{dy}{dx} = \frac{-4 - 9x^{3/2}}{2x^2}$$

$$\frac{dy}{dx} = \frac{-4 - 9x^{3/2}}{24x^2y^2}$$

9. $\dfrac{x + 2y}{x - 3y} = y^{1/2}$

$x + 2y = y^{1/2}\,(x - 3y)$

$\dfrac{d}{dx}(x + 2y) = \dfrac{d}{dx}[y^{1/2}\,(x - 3y)]$

$1 + 2\dfrac{dy}{dx} = y^{1/2}\left(1 - 3\dfrac{dy}{dx}\right)$

$\qquad\qquad + \dfrac{1}{2}(x - 3y)y^{-1/2}\ \dfrac{dy}{dx}$

$1 + 2\dfrac{dy}{dx} = y^{1/2} - 3y^{1/2}\ \dfrac{dy}{dx} + \dfrac{1}{2}xy^{-1/2}\ \dfrac{dy}{dx}$

$\qquad\qquad - \dfrac{3}{2}y^{1/2}\ \dfrac{dy}{dx}$

$\left(2 + 3y^{1/2} - \dfrac{1}{2}xy^{-1/2} + \dfrac{3}{2}y^{1/2}\right)\dfrac{dy}{dx} = y^{1/2} - 1$

$\dfrac{2y^{1/2}\left(2 + \dfrac{9}{2}y^{1/2} - \dfrac{1}{2}xy^{-1/2}\right)}{2y^{1/2}}\ \dfrac{dy}{dx} = y^{1/2} - 1$

$\left(\dfrac{4y^{1/2} + 9y - x}{2y^{1/2}}\right)\dfrac{dy}{dx} = y^{1/2} - 1$

$\dfrac{dy}{dx} = \dfrac{2y - 2y^{1/2}}{4y^{1/2} + 9y - x}$

11. $(4y^2 - 3x)^{2/3} = 6x$

$\dfrac{d}{dx}[(4y^2 - 3x)^{2/3}] = \dfrac{d}{dx}(6x)$

$\dfrac{2}{3}(4y^2 - 3x)^{-1/3}\left(8y\ \dfrac{dy}{dx} - 3\right) = 6$

$8y\dfrac{dy}{dx} - 3 = 6\left(\dfrac{3}{2}\right)(4y^2 - 3x)^{1/3}$

$\dfrac{dy}{dx} = \dfrac{9(4y^2 - 3x)^{1/3} + 3}{8y}$

13. $\sqrt{2x} - 4yx = -22$, tangent line at $(2, 3)$

$\dfrac{d}{dx}(\sqrt{2x} - 4yx) = \dfrac{d}{dx}(-22)$

$\dfrac{1}{2}(2x)^{-1/2}(2) - 4y - 4x\ \dfrac{dy}{dx} = 0$

$(2x)^{-1/2} - 4y = 4x\dfrac{dy}{dx}$

$\dfrac{dy}{dx} = \dfrac{(2x)^{-1/2} - 4y}{4x} = \dfrac{\dfrac{1}{\sqrt{2x}} - 4y}{4x}$

At $(2, 3)$,

$m = \dfrac{\dfrac{1}{\sqrt{2 \cdot 2}} - 4 \cdot 3}{4 \cdot 2} = \dfrac{\dfrac{1}{2} - 12}{8}$

$\qquad = -\dfrac{23}{16}$

The equation of the tangent line is

$\qquad y - y_1 = m(x - x_1)$

$\qquad\quad y - 3 = -\dfrac{23}{16}(x - 2)$

$\qquad 23x + 16y = 94.$

15. $y = 8x^3 - 7x^2$, $\dfrac{dx}{dt} = 4$, $x = 2$

$\dfrac{dy}{dt} = \dfrac{d}{dt}(8x^3 - 7x^2)$

$\qquad = 24x^2\ \dfrac{dx}{dt} - 14x\ \dfrac{dx}{dt}$

$\qquad = 24(2)^2(4) - 14(2)(4)$

$\qquad = 272$

17. $y = \dfrac{1 + \sqrt{x}}{1 - \sqrt{x}}$, $\dfrac{dx}{dt} = -4$, $x = 4$

$\dfrac{dy}{dt} = \dfrac{d}{dt}\left[\dfrac{1 + \sqrt{x}}{1 - \sqrt{x}}\right]$

$\qquad = \dfrac{(1-\sqrt{x})\left(\dfrac{1}{2}x^{-1/2}\ \dfrac{dx}{dt}\right) - (1 + \sqrt{x})\left(-\dfrac{1}{2}\right)(x^{-1/2}\ \dfrac{dx}{dt})}{(1 - \sqrt{x})^2}$

$\qquad = \dfrac{(1 - 2)\left(\dfrac{1}{2 \cdot 2}\right)(-4) - (1 + 2)\left(\dfrac{-1}{2 \cdot 2}\right)(-4)}{(1 - 2)^2}$

$\qquad = \dfrac{1 - 3}{1}$

$\qquad = -2$

21. $y = 4(x^2 - 1)^3$

$dy = 12(x^2 - 1)^2(2x)dx$

$\qquad = 24x(x^2 - 1)^2 dx$

23. $y = \sqrt{9 + x^3} = (9 + x^3)^{1/2}$

$dy = \frac{1}{2}(9 + x^3)^{-1/2}(3x^2)dx$

$= \frac{3x^2}{2}(9 + x^3)^{-1/2} \; dx$

25. $y = \frac{3x - 7}{2x + 1}; \; x = 2, \; \Delta x = .003$

$dy = \frac{(3)(2x + 1) - (2)(3x - 7)}{(2x + 1)^2} \, dx$

$dy = \frac{17}{(2x + 1)^2} \, dx$

$\approx \frac{17}{(2x + 1)^2}\Delta x$

$= \frac{17}{(2[2] + 1)^2}(.003)$

$= .00204$

29. $x = 2 + y$

$P = xy^2 = (2 + y)y^2$

$= 2y^2 + y^3$

$P' = 4y + 3y^2 = 0$

$y(4 + 3y) = 0$

$y = 0 \quad \text{or} \quad 4 + 3y = 0$

$y = -\frac{4}{3}$

$P'' = 4 + 6y$

If $y = 0$, $P'' = 4$, which implies that $y = 0$ is location of minimum.

If $y = -\frac{4}{3}$, $P'' = -4$, which implies that $y = -\frac{4}{3}$ is location of maximum.

$x = 2 + y = 2 + 0 = 2$

So xy^2 is minimized when $x = 2$ and $y = 0$.

31.   (a) $P(x) = -x^3 + 10x^2 - 12x$

$P'(x) = -3x^2 + 20x - 12 = 0$

$3x^2 - 20x + 12 = 0$

$(3x - 2)(x - 6) = 0$

$3x - 2 = 0 \quad \text{or} \quad x - 6 = 0$

$x = \frac{2}{3} \quad \text{or} \quad x = 6$

$P''(x) = -6x + 20$

$P''\left(\frac{2}{3}\right) = 16,$

which implies that $x = 2/3$ is location of minimum.

$P''(6) = -16,$

which implies that $x = 6$ is location of maximum.

Thus, 600 boxes will produce a maximum profit.

(b) Maximum profit

$= P(6) = -(6)^3 + 10(6)^2 - 12(6)$

$P(6) = 72$

The maximum profit is $720.

33. $V = \pi r^2 h = 40$, so $h = \frac{40}{\pi r^2}$

$A = 2\pi r^2 + 2\pi rh$

$= 2\pi r^2 + 2\pi r\left(\frac{40}{\pi r^2}\right)$

$= 2\pi r^2 + \frac{80}{r}$

$\text{Cost} = C(r) = 4(2\pi r^2) + 3\left(\frac{80}{r}\right)$

$= 8\pi r^2 + \frac{240}{r}$

$C'(r) = 16\pi r - \frac{240}{r^2}$

$16\pi r - \frac{240}{r^2} = 240$

$16\pi r^3 = 240$

$r^3 = \frac{15}{\pi}$

$r \approx 1.684$

$C''(r) = 16\pi + \dfrac{240}{r^3} > 0$, so

$r = 1.684$ minimizes cost.

$h = \dfrac{40}{\pi r^2} = \dfrac{40}{\pi(1.684)^2} = 4.490$

The radius should be 1.684 in and the height should be 4.490 in.

35. $M = 980,000$ cases sold per year

$k = \$2$, cost to store 1 case for 1 yr

$f = \$20$, fixed cost for order

$x = $ the number of cases per order

$x = \sqrt{\dfrac{kM}{2f}}$

$= \sqrt{\dfrac{2(980,000)}{2(20)}}$

$= \sqrt{49,000}$

≈ 221

Each order will consist of

$\dfrac{M}{x} = \dfrac{980,000}{221} = 4434$ cases.

37. $M = 240,000$ cases per year

$k = \$2$, cost to store 1 case for 1 yr

$f = \$15$, fixed cost for 1 batch

$x = $ the number of batches that should be produced annually

$x = \sqrt{\dfrac{kM}{2f}}$

$= \sqrt{\dfrac{2(240,000)}{2(15)}}$

$= \sqrt{16,000} \approx 126$ batches

39. $q = \dfrac{A}{p^k}$

$\dfrac{dq}{dp} = -k\dfrac{A}{p^{k+1}}$

$E = -\dfrac{p}{q} \cdot \dfrac{dq}{dp}$

$= -\dfrac{p}{\dfrac{A}{p^k}}\left(-k\dfrac{A}{p^{k+1}}\right)$

$= \left(-\dfrac{p^{k+1}}{A}\right)\left(-k\dfrac{A}{p^{k+1}}\right)$

$= k$

The demand is elastic when $k > 1$ and inelastic when $k < 1$.

41. Let $x = $ the distance from the base of the ladder to the building

$y = $ the height on the building at the top of the ladder.

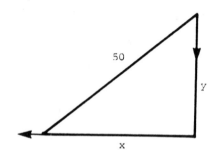

$\dfrac{dy}{dt} = 2$ ft/min;

$(50)^2 = x^2 + y^2$

$0 = 2x\dfrac{dx}{dt} + 2y\dfrac{dy}{dt}$

$\dfrac{dx}{dt} = -\dfrac{y}{x}\dfrac{dy}{dt}$

When $x = 30$, $y = \sqrt{2500 - (30)^2} = 40$.

So $\dfrac{dx}{dt} = -\dfrac{40}{30}(2) = -\dfrac{80}{30} = -\dfrac{8}{3}$.

The base of the ladder is slipping away from the building at a rate of 8/3 ft/min.

43. Let x = one-half the width of the triangular cross section

h = the height of the water

V = the volume of the water.

$$\frac{dV}{dt} = 3.5 \text{ ft}^3/\text{min}$$

Find $\frac{dV}{dt}$ when $h = \frac{1}{3}$.

$$V = \begin{bmatrix} \text{Area of} \\ \text{triangular} \\ \text{side} \end{bmatrix} \begin{bmatrix} \text{length} \end{bmatrix}$$

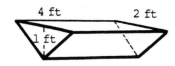

Area of triangular cross section

$$= \frac{1}{2}(\text{base})(\text{altitude})$$

$$= \frac{1}{2}(2x)(h)$$

$$= xh$$

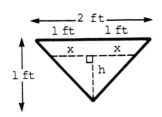

By similar triangles, $\frac{1}{x} = \frac{1}{h}$,

so x = h.

$$V = (xh)(4)$$

$$= h^2 \cdot 4$$

$$= 4h^2$$

$$\frac{dV}{dt} = 8h\frac{dh}{dt}$$

$$\frac{1}{8h}\frac{dV}{dt} = \frac{dh}{dt}$$

$$\frac{1}{8\left(\frac{1}{3}\right)}(3.5) = \frac{dh}{dt}$$

$$\frac{dh}{dt} = \frac{21}{16} = 1.3125 \text{ ft/min}$$

45. $A = s^2$; $s = 9.2''$, $\Delta s = \pm.04$

$$ds = 2sds$$

$$\Delta A \approx 2s\Delta s$$

$$= 2(9.2)(\pm.04)$$

$$= \pm.736 \text{ in}^2$$

47. $-12x + x^3 + y + y^2 = 4$

$$\frac{d}{dx}(-12x + x^3 + y + y^2) = \frac{d}{dx}(4)$$

$$-12 + 3x^2 + \frac{dy}{dx} + 2y\frac{dy}{dx} = 0$$

$$(1 + 2y)\frac{dy}{dx} = 12 - 3x^2$$

$$\frac{dy}{dx} = \frac{12 - 3x^2}{1 + 2y}$$

(a) If $\frac{dy}{dx} = 0$, $12 - 3x^2 = 0$

$$12 = 3x^2$$

$$\pm 2 = x$$

$$x = 2: -24 + 8 + y + y^2 = 4$$

$$y + y^2 = 20$$

$$y^2 + y - 20 = 0$$

$$(y + 5)(y - 4) = 0$$

$$y = -5, \ y = 4$$

(2, -5) and (2, 4) are critical points.

x = -2:

$$24 - 8 + y + y^2 = 4$$

$$y + y^2 = -12$$

$$y^2 + y + 12 = 0$$

$$y = \frac{-1 \pm \sqrt{1^2 - 48}}{2}$$

This leads to imaginary roots.
x = -2 does not produce critical
points.

(b)

x	y_1	y_2
1.9	-4.99	3.99
2	-5	4
2.1	-4.99	3.99

The point (2, -5) is a relative
minimum.
The point (2, 4) is a relative
maximum.

Extended Application

1. $Z(m) = \dfrac{C_1}{m} + DtC_2 + DC_2\left(\dfrac{m-1}{2}\right)$

$Z'(m) = -\dfrac{C_1}{m^2} + 0 + \dfrac{DC_3}{2}$

$\quad\quad = -\dfrac{C_1}{m^2} + \dfrac{DC_3}{2}$

2. $Z'(m) = 0$ when

$\dfrac{DC_3}{2} = \dfrac{C_1}{m^2}$

$m^2 = \dfrac{2C_1}{DC_3}$

$m = \sqrt{\dfrac{2C_1}{DC_3}}$

3. $D = 3$, $t = 12$, $C_1 = \$15,000$,
$C_3 = 900$

$m = \sqrt{\dfrac{2(15,000)}{3(900)}}$

$\quad = \sqrt{\dfrac{30,000}{2700}}$

$\quad = \sqrt{\dfrac{100}{9}}$

$\quad \approx 3.33$

4. $3 < 3.33 < 4$
$m^+ = 4$ and $m^- = 3$

5. $Z(m) = \dfrac{C_1 + mDtC_2}{m} + \dfrac{DC_3\left[\dfrac{m(m-1)}{2}\right]}{m}$

$\quad\quad = \dfrac{C_1}{m} + DtC_2 + DC_3\left(\dfrac{m-1}{2}\right)$

$C_1 = 15,000$; $D = 3$, $t = 12$,
$C_2 = 100$; $C_3 = 900$

$Z(m^+) = Z(4)$

$\quad = \dfrac{15,000}{4} + 3(12)(100)$

$\quad\quad + 3(900)\left(\dfrac{4-1}{2}\right)$

$\quad = 3750 + 3600 + 4050$

$\quad = \$11,400$

$Z(m^-) = Z(3)$

$\quad = \dfrac{15,000}{3} + 3600$

$\quad\quad + 3\dfrac{(900)(3-1)}{2}$

$\quad = 5000 + 3600 + 2700$

$\quad = \$11,300$

6. Since $Z(3) < Z(4)$, the optimal time
interval is 3 months.

$N = mD$

$\quad = 3 \cdot 3$

$\quad = 9$

There should be 9 trainees per batch

CHAPTER 4 TEST

[4.1]   **1.**   Find two nonnegative numbers x and y such that x + y = 50 and P = xy is maximized.

[4.1]   **2.**   A travel agency offers a tour to the Bahamas for 12 people at $800 each. For each person more than the original 12, up to a total of 30 people, who signs up for the cruise, the fare is reduced $20. What tour group size produces the greatest revenue for the travel agency?

[4.1]   **3.** $320 is available for fencing for a rectangular garden. The fencing for the two sides parallel to the back of the house costs $6 per linear foot, while the fencing for the other sides costs $2 per linear foot. Find the dimensions that will maximize the area of the garden.

[4.2]   **4.**   A small beauty supply store sells 600 hair dryers each year. Each hair dryer costs the store $5. It costs the store $3 a year to store one hair dryer for one year. The fixed cost of placing the order is $36. Find the number of orders that should be placed each year. Find the number of dryers that should be ordered each time.

[4.2]   **5.**   A demand function is given by

$$q = 100 - 2p$$

where q is the number of units produced and p is the price in dollars.

(a)   Find E.   (b)   Find the price that maximizes the total revenue.

[4.2]   **6.**   What does elasticity of demand measure? What does it mean for demand to be elastic? What does it mean for demand to be inelastic?

[4.3]   **7.**   Find $\frac{dy}{dx}$ for $x^3 - x^2y + y^2 = 0$.

[4.3]   **8.**   Find an equation for the tangent line to the graph of $2x^2 - y^2 = 2$ at (3, 4).

[4.3]   9. Suppose $3x^2 + 4y^2 + 6 = 0$.

Use implicit differentiation to find dy/dx. Explain why your result is meaningless.

[4.4]   10. Find $\frac{dy}{dt}$ if $y = 3x^3 - 2x^2$, $\frac{dx}{dt} = -2$, and $x = 3$.

[4.4]   11. Find $\frac{dy}{dt}$ if $2xy - y^2 = x$, $x = -1$, $y = 3$, and $\frac{dx}{dt} = 2$.   1.25

[4.4]   12. When solving related rates problems, how should you interpret a negative derivative?

[4.4]   13. A ladder 25 ft long is leaning against a vertical wall. If the bottom of the ladder is pulled horizontally away from the wall at 3 ft/sec, how fast is the top of ladder sliding down the wall when the bottom is 15 ft from the wall?

[4.4]   14. A real estate developer estimates that his monthly sales are given by

$$S = 30y - xy - \frac{y^2}{3000},$$

where y is the average cost of a new house in the development and x percent is the current interest rate for home mortages. If the current rate is 11% and is rising at a rate of 1/4% per month and the current average price of a new home is $90,000 and is increasing at a rate of $500 per month, how fast are his expected sales changing?

[4.4]   Given the revenue and cost functions $R = 30q - .2q^2$ and $C = 12q + 20$, where q is the daily production (and sales), answer Problems 15–17 when 50 units are produced and the rate of change of production is 8 units per day.

15. Find the rate of change of revenue with respect to time.

16. Find the rate of change of cost with respect to time.

17. Find the rate of change of profit with respect to time.

[4.5]   18.   If $y = \frac{2x - 3}{x^2 - 7}$, find dy.

[4.5]   19.   Find the value of dy if $y = \sqrt{5 - x^2}$, $x = 1$, $\Delta x = .01$.

[4.5]   20.   The radius of a sphere is claimed to be 5 cm with a possible error of .01 cm. Estimate the possible error in the volume of the sphere.

CHAPTER 4   TEST ANSWERS

1. 25 and 25

2. 26 people

3. $13\frac{1}{3}$ ft on each side parallel to the back, 40 ft on each of the other sides

4. 5 orders, 120 dryers per order

5. (a) $E = \dfrac{2p}{100 - 2p}$ (b) $25

6. The instantaneous responsiveness of demand to price; the relative change in demand is greater than the relative change in price; the relative change in demand is less than the relative change in price.

7. $\dfrac{dy}{dx} = \dfrac{2xy - 3x^2}{2y - x^2}$

8. $3x - 2y = 1$

9. $-\dfrac{3x}{4y}$; no function exists such that $3x^2 + 4y^2 + 6 = 0$

10. -138

11. 1.25

12. A decrease in rate or travel in a negative direction

13. $-\dfrac{9}{4}$ ft/sec

14. Decreasing $43,000/mo

15. Increasing at a rate of $80/day

16. Increasing at a rate of $96/day

17. Decreasing at a rate of $16/day

18. $dy = \dfrac{-2x^2 + 6x - 14}{(x^2 - 7)^2} dx$

19. $-.005$

20. 3.14 cm^3

CHAPTER 5 EXPONENTIAL AND LOGARITHMIC FUNCTIONS

Section 5.1

For Exercises 1–15, see the graphs in the answer section at the back of your textbook.

1. $y = 3^x$

x	-2	-1	0	1	2
y	1/9	1/3	1	3	9

3. $y = 3^{-x/2}$

x	-4	-2	0	2	4
y	9	3	1	1/3	1/9

5. $y = \left(\frac{1}{4}\right)^x + 1$

x	-2	-1	0	1	2
y	17	5	2	5/4	17/16

7. $y = -3^{2x-1}$

x	-1	-1/2	0	1/2	1
2x - 1	-3	-2	-1	0	1
y	-1/27	-1/9	-1/3	-1	-3

9. $y = e^{-x^2+1}$

x	-2	-1	0	1	2
y	.05	1	$e \approx 2.72$	1	.05

11. $y = 10 - 5e^{-x}$

x	-2	-1	0	1	2	3	4
y	-26.9	-3.6	5	8.2	9.3	9.8	9.9

13. $y = x \cdot 2^x$

x	-3	-2	-1	-.5	0	1	2
y	-.38	-.5	-.5	-.35	0	2	8

15. $y = \dfrac{e^x - e^{-x}}{2}$

x	-3	-2	-1	0	1	2
y	-10.02	-3.6	-1.2	0	1.2	3.6

17. To solve the equation $4^x = 6$ using the method described in this section, 4 and 6 must be written as powers of the same base. This is not easily done.

19. $4^x = 64$
$4^x = 4^3$
$x = 3$

21. $e^x = \dfrac{1}{e^2}$
$e^x = e^{-2}$
$x = -2$

23. $25^x = 125^{x-2}$
$(5^2)^x = (5^3)^{x-2}$
$5^{2x} = 5^{3x-6}$
$2x = 3x - 6$
$6 = x$

25. $16^{x+2} = 64^{2x-1}$

$(2^4)^{x+2} = (2^6)^{2x-1}$

$2^{4x+8} = 2^{12x-6}$

$4x + 8 = 12x - 6$

$14 = 8x$

$\dfrac{7}{4} = x$

27. $e^{-x} = (e^2)^{x+3}$

$e^{-x} = e^{2x+6}$

$-x = 2x + 6$

$-3x = 6$

$x = -2$

29. $5^{-|x|} = \dfrac{1}{25}$

$5^{-|x|} = 5^{-2}$

$-|x| = -2$

$|x| = 2$

$x = 2 \quad \text{or} \quad x = -2$

31. $5^{x^2+x} = 1$

$5^{x^2+x} = 5^0$

$x^2 + x = 0$

$x(x + 1) = 0$

$x = 0 \quad \text{or} \quad x + 1 = 0$

$x = 0 \quad \text{or} \qquad x = -1$

33. $9^{x+4} = 3^{x^2}$

$(3^2)^{x+4} = 3^{x^2}$

$3^{2x+8} = 3^{x^2}$

$2x + 8 = x^2$

$0 = x^2 - 2x - 8$

$0 = (x + 2)(x - 4)$

$x = -2 \quad \text{or} \quad x = 4$

35. $f(x) = e^x$

(a) $\dfrac{f(1) - f(0)}{1 - 0} = \dfrac{e^1 - e^0}{1} = 1.718$

(b) $\dfrac{f(.1) - f(0)}{.1 - 0} = \dfrac{e^{.1} - e^0}{.1} = 1.052$

(c) $\dfrac{f(.01) - f(0)}{.01 - 0} = \dfrac{e^{.01} - e^0}{.01}$

$= 1.005$

(d) $\dfrac{f(.001) - f(0)}{.001 - 0} = \dfrac{e^{.001} - e^0}{.001}$

$= 1.001$

(e) The slope of the graph of $f(x)$ at $x = 0$ is about 1.

39. $A = P\left(1 + \dfrac{r}{m}\right)^{tm}$, $P = 26,000$, $r = .12$, $t = 3$

(a) annually, $m = 1$

$A = 26,000\left(1 + \dfrac{.12}{1}\right)^{3 \cdot 1}$

$= 26,000(1.12)^3$

$= 36,528.13$

Interest $= \$36,528.13 - \$26,000$

$= \$10,528.13$

(b) semiannually, $m = 2$

$A = 26,000\left(1 + \dfrac{.12}{2}\right)^{3 \cdot 2}$

$= 26,000(1.06)^6$

$= 36,881.50$

Interest $= \$36,881.50 - \$26,000$

$= \$10,881.50$

(c) quarterly, m = 4

$$A = 26{,}000\left(1 + \frac{.12}{4}\right)^{3 \cdot 4}$$

$$= 26{,}000(1.03)^{12}$$

$$= 37{,}069.78$$

Interest = $37,069.78 - $26,000

$$= \$11{,}069.78$$

(d) monthly, m = 12

$$A = 26{,}000\left(1 + \frac{.12}{12}\right)^{3 \cdot 12}$$

$$= 26{,}000(1.01)^{36}$$

$$= 37{,}199.99$$

Interest = $37,199.99 - $26,000

$$= \$11{,}199.99$$

41. $A = P\left(1 + \frac{r}{m}\right)^{tm}$, P = 17,500, r = .10,

m = 2, t = 6

$$A = 17{,}500\left(1 + \frac{.10}{2}\right)^{2 \cdot 6}$$

$$= 17{,}500(1.05)^{12}$$

$$= 31{,}427.49$$

She will owe $31,427.49.

43. $A = P\left(1 + \frac{r}{m}\right)^{tm}$, P = 5000, A = 8000,

t = 4

(a) m = 1

$$8000 = 5000\left(1 + \frac{r}{1}\right)^{4 \cdot 1}$$

$$\frac{8}{5} = (1 + r)^4$$

$$\left(\frac{8}{5}\right)^{1/4} - 1 = r$$

$$.125 = r$$

The interest rate is 12.5%.

(b) m = 4

$$8000 = 5000\left(1 + \frac{r}{4}\right)^{4 \cdot 4}$$

$$\frac{8}{5} = \left(1 + \frac{r}{4}\right)^{16}$$

$$\left(\frac{8}{5}\right)^{1/16} - 1 = \frac{r}{4}$$

$$4\left[\left(\frac{8}{5}\right)^{1/16} - 1\right] = r$$

$$.119 = r$$

The interest rate is 11.9%.

45. $p(t) = 250 - 120(2.8)^{-.5t}$

(a) $p(2) = 250 - 120(2.8)^{-.5(2)}$

$$\approx 207$$

(b) $p(4) = 240 - 120(2.8)^{-.5(4)}$

$$\approx 235$$

(c) $p(10) = 250 - 120(2.8)^{-.5(10)}$

$$\approx 249$$

(d) See the answer graph in the back of your textbook.

(e) The number of symbols typed per minute gets very close to 250.

(f) The limit appears to be 250 symbols per minute.

47. $A(t) = 2600e^{.018t}$

(a) 1970: t = 20

$$A(20) = 2600e^{.018(20)}$$

$$= 2600e^{.36}$$

$$= 3727 \text{ million}$$

This is very close to the actual population of 3700 million.

(b) 1990:   t = 40

$A(40) = 2600e^{.018(40)}$

$= 2600e^{.72}$

$= 5341$ million

(c) 2000:   t = 50

$A(50) = 2600e^{.018(50)}$

$= 2600e^{.9}$

$= 6395$ million

49. $y = 6 \cdot 2^t$

(a) t = 0

$y = 6 \cdot 2^0$

$= 6$

6 animals were introduced originally.

(b) $24 = 6 \cdot 2^t$

$4 = 2^t$

$2^2 = 2^t$

$2 = t$

It will take 2 yr.

(c) $384 = 6 \cdot 2^t$

$64 = 2^t$

$2^6 = 2^t$

$6 = t$

It will take 6 yr.

51. $Q(t) = 1000(5^{-.3t})$

(a) $Q(6) = 1000[5^{-.3(6)}]$

$= 1000(5^{-1.8})$

$= 55$ g

(b) $8 = 1000(5^{-.3t})$

$\frac{1}{125} = 5^{-.3t}$

$5^{-3} = 5^{-.3t}$

$-3 = -.3t$

$10 = t$

It will take 10 mo.

53. This exercise should be solved by a graphing calculator or computer methods. The solution will vary according to the method that is used. See the answer graph at the back of the textbook.

55. This exercise should be solved by a graphing calculator or computer methods. The solution will vary according to the method that is used. See the answer graph at the back of the textbook.

Section 5.2

1. $2^3 = 8$

Since $a^y = x$ means $y = \log_a x$, the equation in logarithmic form is

$\log_2 8 = 3.$

3. $3^4 = 81$

$\log_3 81 = 4$

5. $\left(\frac{1}{3}\right)^{-2} = 9$

$\log_{1/3} 9 = -2$

7. $\log_2 128 = 7$

Since $y = \log_a x$ means $a^y = x$, the equation in exponential form is

$$2^7 = 128.$$

9. $\log_{25} \frac{1}{25} = -1$

$$25^{-1} = \frac{1}{25}$$

11. $\log 10{,}000 = 4$
$\log_{10} 10{,}000 = 4$
$$10^4 = 10{,}000$$

When no base is written, $\log_{10}$ is understood.

13. Let $\log_5 25 = x$.

Then, $5^x = 25$.
$$5^x = 5^2$$
$$x = 2$$

Thus, $\log_5 25 = 2$.

15. $\log_4 64 = x$
$$4^x = 64$$
$$4^x = 4^3$$
$$x = 3$$

17. $\log_2 \frac{1}{4} = x$
$$2^x = \frac{1}{4}$$
$$2^x = 2^{-2}$$
$$x = -2$$

19. $\log_2 \sqrt[3]{1/4} = x$
$$2^x = \left(\frac{1}{4}\right)^{1/3}$$
$$2^x = \left(\frac{1}{2^2}\right)^{1/3}$$
$$2^x = 2^{-2/3}$$
$$x = -\frac{2}{3}$$

21. $\ln e = x$

Recall that $\ln$ is $\log_e$.
$$e^x = e$$
$$x = 1$$

23. $\ln e^{5/3} = x$
$$e^x = e^{5/3}$$
$$x = \frac{5}{3}$$

25. The logarithm to the base 3 of 4 is written $\log_3 4$. The subscript denotes the base.

For Exercises 27–31, see the answer graphs in the textbook.

27. $y = \log_4 x$
$4^y = x$

To find plotting points, use values of y to find x.

x	1/16	1/4	1	4	16
y	-2	-1	0	1	2

29. $y = \log_{1/3}(x - 1)$

$\left(\frac{1}{3}\right)^y = x - 1$

$\left(\frac{1}{3}\right)^y + 1 = x$

x	10	4	2	4/3	10/9
y	-2	-1	0	1	2

31. $y = \ln x^2$

Use a calculator to find plotting points.

x	-2	-1	0	1	2
y	1.39	0	does not exist	0	1.39

x cannot be zero because there is no value of y such that $e^y = 0$.

33. $\log_9 7m = \log_9 7 + \log_9 m$

35. $\log_3 \frac{3p}{5k}$

$= \log_3 3p - \log_3 5k$

$= (\log_3 3 + \log_3 p) - (\log_3 5 + \log_3 k)$

$= 1 + \log_3 p - \log_3 5 - \log_3 k$

37. $\log_3 \frac{5\sqrt{2}}{\sqrt[4]{7}}$

$= \log_3 5\sqrt{2} - \log_3 \sqrt[4]{7}$

$= \log_3 5 \cdot 2^{1/2} - \log_3 7^{1/4}$

$= \log_3 5 + \log_3 2^{1/2} - \log_3 7^{1/4}$

$= \log_3 5 + \frac{1}{2}\log_3 2 - \frac{1}{4}\log_3 7$

For Exercises 39–41, $\log_b 2 = a$ and $\log_b 3 = c$.

39. $\log_b 8 = \log_b 2^3$

$= 3 \log_b 2$

$= 3a$

41. $\log_b 72b = \log_b 72 + \log_b b$

$= \log_b 72 + 1$

$= \log_b 2^3 \cdot 3^2 + 1$

$= \log_b 2^3 + \log_b 3^2 + 1$

$= 3 \log_b 2 + 2 \log_b 3 + 1$

$= 3a + 2c + 1$

43. $\log_5 20 = \frac{\ln 20}{\ln 5}$

$\approx \frac{3}{1.61}$

≈ 1.86

45. $\log_{1.2} 5.5 = \frac{\ln 5.5}{\ln 1.2}$

≈ 9.35

47. $\log_3 1.1^{-2.4} = -2.4(\log_3 1.1)$

$= -2.4\left(\frac{\ln 1.1}{\ln 3}\right)$

$\approx -.21$

49. $\log_x 25 = -2$

$x^{-2} = 25$

$(x^{-2})^{-1/2} = 25^{-1/2}$

$x = \frac{1}{5}$

51. $\log_8 4 = z$

$8^z = 4$

$(2^3)^z = 2^2$

$2^{3z} = 2^2$

$3z = 2$

$z = \dfrac{2}{3}$

53. $\log_r 7 = \dfrac{1}{2}$

$r^{1/2} = 7$

$(r^{1/2})^2 = 7^2$

$r = 49$

55. $\log_5 (9x - 4) = 1$

$5^1 = 9x - 4$

$9 = 9x$

$1 = x$

57. $\log_9 m - \log_9 (m - 4) = -2$

$\log_9 \dfrac{m}{m - 4} = -2$

$9^{-2} = \dfrac{m}{m - 4}$

$\dfrac{1}{81} = \dfrac{m}{m - 4}$

$m - 4 = 81m$

$-4 = 80m$

$-.05 = m$

This value is not possible since $\log_9 (-.05)$ does not exist.
Thus, there is no solution to the original equation.

59. $4^x = 12$

$x \log 4 = \log 12$

$x = \dfrac{\log 12}{\log 4}$

≈ 1.79

61. $e^{2y} = 12$

$\ln e^{2y} = \ln 12$

$2y \ln e = \ln 12$

$2y(1) = \ln 12$

$y = \dfrac{\ln 12}{2}$

≈ 1.24

63. $10e^{3z-7} = 5$

$\ln 10e^{3z-7} = \ln 5$

$\ln 10 + \ln e^{3z-7} = \ln 5$

$\ln 10 + (3z - 7) \ln e = \ln 5$

$3z - 7 = \ln 5 - \ln 10$

$3z = \ln 5 - \ln 10 + 7$

$z = \dfrac{\ln 5 - \ln 10 + 7}{3}$

≈ 2.10

65. Prove: $\log_a x^r = r \log_a x$.

Let $s = \log_a x$.

The equivalent exponential equation is $a^s = x$.

Now $\log_a x^r = \log_a (a^s)^r$

$= \log_a a^{rs}$

So, $\log_a x^r = rs$

or $\log_a x^r = r \log_a x$.

67. $P = 15,000$, $r = .06$, $m = 2$

$A = P\left(1 + \dfrac{r}{m}\right)^{mt}$

$= 15,000\left(1 + \dfrac{.06}{2}\right)^{2t}$

(a) $A = 2P$

Notice that the doubling time is the same for all values of P.

$$2P = P(1 + .03)^{2t}$$
$$2 = 1.03^{2t}$$
$$\ln 2 = \ln 1.03^{2t}$$
$$= 2t \ln 1.03$$
$$t = \frac{\ln 2}{2 \ln 1.03}$$
$$\approx 11.7 \text{ yr}$$

(b) $A = 3P$
$$3P = P(1 + .03)^{2t}$$
$$3 = 1.03^{2t}$$
$$\ln 3 = 1.03^{2t}$$
$$= 2t \ln 1.03$$
$$t = \frac{\ln 3}{2 \ln 1.03}$$
$$t \approx 18.6 \text{ yr}$$

(c) Since $.05 \le .06 \le .12$, the rule of 72 approximates the doubling time in (a).

$$\frac{72}{100r} = \frac{72}{100(.06)} = 12 \text{ yr}$$

69. $H = \frac{-1}{\ln 2} [P_1 \ln P_1 + P_2 \ln P_2$
$$+ P_3 \ln P_3 + P_4 \ln P_4]$$

Let $P_1 = .521$, $P_2 = .324$, $P_3 = .081$, $P_4 = .074$.

$H = \frac{-1}{\ln 2} [.521 \ln .521$
$$+ .324 \ln .324$$
$$+ .081 \ln .081$$
$$+ .074 \ln .074$$
$$= 1.589$$

71. From the graph, a body mass of .3 kg (on horizontal axis) corresponds to oxygen consumption of 4.3 ml/min. A body mass of .7 kg has oxygen consumption of 7.8 ml/min.

73. Decibal rating: $10 \log\frac{I}{I_0}$

(a) Intensity, $I = 115I_0$
$$10 \log \left(\frac{115I_0}{I_0}\right)$$
$$= 10 \cdot \log 115$$
$$\approx 21$$

(b) $I = 9,500,000I_0$
$$10 \log \left(\frac{9.5 \times 10^6 I_0}{I_0}\right)$$
$$= 10 \log 9.5 \times 10^6$$
$$\approx 70$$

(c) $I = 1,200,000,000I_0$
$$10 \log \left(\frac{1.2 \times 10^9 I_0}{I_0}\right)$$
$$= 10 \log 1.2 \times 10^9$$
$$\approx 91$$

(d) $I = 895,000,000,000I_0$
$$10 \log \left(\frac{8.95 \times 10^{11} I_0}{I_0}\right)$$
$$= 10 \log 8.95 \times 10^{11}$$
$$\approx 120$$

(e) $I = 109,000,000,000,000I_0$
$$10 \log \left(\frac{1.09 \times 10^{14} I_0}{I_0}\right)$$
$$= 10 \log 1.09 \times 10^{14}$$
$$\approx 140$$

(f) Let I_1 = the intensity of 96 decibel sound

I_2 = the intensity of 98 decibel sound.

$x_1 = \frac{I_1}{I_0}$ and $x_2 = \frac{I_2}{I_0}$

Then $I_1 = x_1 I_0$ and $I_2 = x_2 I_0$.
$$96 = 10 \log x_1$$
$$9.6 = \log x_1$$
$$10^{9.6} = x_1$$
$$4,000,000,000 \approx x_1$$

$I_1 \approx 4,000,000,000 I_0$

$$98 = 10 \log x_2$$

$$9.8 = \log x_2$$

$$10^{9.8} = x_2$$

$$6,300,000,000 \approx x_2$$

$I_2 \approx 6,300,000,000 I_0$

The difference is about $2,300,000,000 I_0$.

75. $pH = -\log [H^+]$

(a) For pure water:

$$7 = -\log [H^+]$$

$$-7 = \log [H^+]$$

$$10^{-7} = [H^+]$$

$$.0000001 = [H^+]$$

For acid rain:

$$4 = -\log [H^+]$$

$$-4 = \log [H^+]$$

$$10^{-4} = [H^+]$$

$$.0001 = [H^+]$$

$$\frac{.0001}{.0000001} = 1000$$

The acid rain has a hydrogen ion concentration 1000 times greater than pure water.

(b) For laundry solution:

$$11 = -\log [H^+]$$

$$10^{-11} = [H^+]$$

For black coffee:

$$5 = -\log [H^+]$$

$$10^{-5} = [H^+]$$

$$\frac{10^{-5}}{10^{-11}} = 10^6 \quad \text{or} \quad 1,000,000$$

The black coffee has a hydrogen ion concentration 1,000,000 times greater than laundry solution.

Exercises 77 and 79 should be solved by graphing calculator or computer methods. The solutions will vary according to the method that is used. See the answer graphs in the back of the textbook.

Section 5.3

1. $r = 5\%$ compounded monthly

$m = 12$

$$r_E = \left(1 + \frac{r}{m}\right)^m - 1$$

$$= \left(1 + \frac{.05}{12}\right)^{12} - 1$$

$$\approx .0512$$

$$= 5.12\%$$

3. $r = 10\%$ compounded semiannually

$m = 2$

$$r_E = \left(1 + \frac{.10}{2}\right)^2 - 1$$

$$= .1025$$

$$= 10.25\%$$

5. $r = 11\%$ compounded continuously

$$r_E = e^r - 1$$

$$= e^{.11} - 1$$

$$\approx .1163$$

$$= 11.63\%$$

7. A = $2000, r = 6%, m = 2, t = 11

$$P = A\left(1 + \frac{r}{m}\right)^{-tm}$$

$$= 2000\left(1 + \frac{.06}{2}\right)^{-11(2)}$$

$$\approx \$1043.79$$

9. A = $10,000, r = 10%, m = 4, t = 8

$$P = A\left(1 + \frac{r}{m}\right)^{-tm}$$

$$= 10,000\left(1 + \frac{.10}{4}\right)^{-8(4)}$$

$$\approx \$4537.71$$

11. A = $7300, r = 11% compounded continuously, t = 3

$$A = Pe^{rt}$$

$$P = \frac{A}{e^{rt}}$$

$$= \frac{7300}{e^{.11(3)}}$$

$$\approx \$5248.14$$

17. $A(t) = A_0\,e^{kt}$, t = T, $A(T) = \frac{1}{2}\,A_0$

$$\frac{1}{2}\,A_0 = A_0 e^{kT}$$

$$\frac{1}{2} = e^{kT}$$

$$\ln\frac{1}{2} = \ln e^{kT}$$

$$\ln 1 - \ln 2 = kT\,\ln e$$

Since $\ln 1 = 0$ and $\ln e = 1$,

$$0 - \ln 2 = kT$$

$$-\frac{\ln 2}{T} = k.$$

19. A = Pe^{rt}

(a) r = 3%

$$A = 10e^{.03(3)}$$

$$= \$10.94$$

(b) r = 4%

$$A = 10e^{.04(3)}$$

$$= \$11.27$$

(c) r = 5%

$$A = 10e^{.05(3)}$$

$$= \$11.62$$

21. P = $60,000

(a) r = 10% compounded quarterly:

$$A = P\left(1 + \frac{r}{m}\right)^{tm}$$

$$= 60,000\left(1 + \frac{.10}{4}\right)^{5\cdot4}$$

$$\approx \$98,316.99$$

r = 9.75% compounded

continuously:

$$A = Pe^{rt}$$

$$= 60,000\,e^{.0975(5)}$$

$$\approx \$97,694.43$$

Linda will earn more money at 10% compounded quarterly.

(b) She will earn $622.56 more.

(c) r = 10%, m = 4:

$$r_E = \left(1 + \frac{r}{m}\right)^{m} - 1$$

$$= \left(1 + \frac{.10}{4}\right)^{4} - 1$$

$$\approx .1038$$

$$= 10.38\%$$

r = 9.75% compounded continously:

$$r_E = e^r - 1$$

$$= e^{.0975} - 1$$

$$\approx .1024$$

$$= 10.24\%$$

(d) A = $80,000

$$A = Pe^{rt}$$

$$80,000 = 60,000e^{.0975t}$$

$$\frac{4}{3} = e^{.0975t}$$

$$\ln \frac{4}{3} = \ln e^{.0975t}$$

$$\ln 4 - \ln 3 = .0975t$$

$$\frac{\ln 4 - \ln 3}{.0975} = t$$

$$2.95 \approx t$$

$60,000 will grow to $80,000 in about 2.95 years.

23. r = 7.2%, m = 4

$$r_E = \left(1 + \frac{.072}{4}\right)^4 - 1$$

$$\approx .0740$$

$$= 7.40\%$$

25. A = $20,000, t = 4, r = 8%, m = 1

$$A = P\left(1 + \frac{r}{m}\right)^{mt}$$

$$20,000 = P\left(1 + \frac{.08}{1}\right)^{1 \cdot 4}$$

$$\frac{20,000}{(1.08)^4} = P$$

$$\$14,700.60 = P$$

27. A = $20,000, t = 5

(a) r = .08, m = 4

$$A = P\left(1 + \frac{r}{m}\right)^{mt}$$

$$20,000 = P\left(1 + \frac{.08}{4}\right)^{5(4)}$$

$$\frac{20,000}{(1.02)^{20}} = P$$

$$\$13,459.43 = P$$

(b) Interest = 20,000 - 13,459.43

$$= \$6540.57$$

(c) P = $10,000

$$A = 10,000\left(1 + \frac{.08}{4}\right)^{5(4)}$$

$$= \$14,859.47$$

The amount needed will be

$20,000 - $14,859.47 = $5140.53

29. $S(x) = 1000 - 800e^{-x}$

(a) $S(0) = 1000 - 800e^0$

$$= 1000 - 800$$

$$= 200$$

(b) S(x) = 500

$$500 = 1000 - 800e^{-x}$$

$$-500 = -800e^{-x}$$

$$\frac{5}{8} = e^{-x}$$

$$\ln \frac{5}{8} = \ln e^{-x}$$

$$-\ln \frac{5}{8} = x$$

$$.47 \approx x$$

Sales reach 500 in about 1/2 yr.

(c) Since $800e^{-x}$ will never actually be zero, $S(x) = 1000 - 800e^{-x}$ will never be 1000.

(d) $\lim_{x \to \infty} (1000 - 800e^{-x})$

$\qquad = \lim_{x \to \infty} 1000 - \lim_{x \to \infty} 800e^{-x}$

$\qquad = 1000 - 0 = 1000$

31. $y = y_0 e^{kt}$

(a) $y = 125$, $t = 2$

$\qquad 125 = 100e^{k(2)}$

$\qquad 1.25 = e^{2k}$

$\qquad \ln 1.25 = 2k$

$\qquad .11 = k$

The equation is

$\qquad y = 100e^{.11t}$

(b) $y = 500$

$\qquad 500 = 100e^{.11t}$

$\qquad 5 = e^{.11t}$

$\qquad \ln 5 = .11t$

$\qquad 14.6 = t$

There will be 500 lice in about 15 mo.

33. $y = y_0 e^{kt}$

(a) $y = 20{,}000$, $y_0 = 50{,}000$, $t = 9$

$\qquad 20{,}000 = 50{,}000e^{9k}$

$\qquad .4 = e^{9k}$

$\qquad \ln .4 = 9k$

$\qquad -.102 = k$

The equation is

$\qquad y = 50{,}000e^{-.102t}.$

(b) $\frac{1}{2}(50{,}000) = 25{,}000$

$\qquad 25{,}000 = 50{,}000e^{-.102t}$

$\qquad .5 = e^{-.102t}$

$\qquad \ln .5 = -.102t$

$\qquad 6.8 = t$

Half the bacteria remain after about 6.8 hours.

35. $G(t)$

$\quad = \dfrac{(2500)(1000)}{1000 + (2500 - 1000)e^{-(.0004)(2500)t}}$

$\quad = \dfrac{2{,}500{,}000}{1000 + 1500\,e^{-t}}$

(a) $G(.2) = \dfrac{2{,}500{,}000}{1000 + 1500e^{-.2}}$

$\qquad \approx 1100$

(b) $G(1) = \dfrac{2{,}500{,}000}{1000 + 1500e^{-1}}$

$\qquad \approx 1600$

(c) $G(3) = \dfrac{2{,}500{,}000}{1000 + 1500e^{-3}}$

$\qquad \approx 2300$

(d) $2000 = \dfrac{2{,}500{,}000}{1000 + 1500e^{-t}}$

$\quad 2000(1000 + 1550e^{-t}) = 2{,}500{,}000$

$\quad 2{,}000{,}000 + 3{,}000{,}000e^{-t} = 2{,}500{,}000$

$\qquad 3{,}000{,}000e^{-t} = 500{,}000$

$\qquad\qquad e^{-t} = .167$

$\qquad\qquad -t = \ln(.167)$

$\qquad\qquad t \approx 1.8 \text{ decades}$

37. $y = y_0 e^{kt}$

Let $y = .37y_0$ and $t = 5$; that is the number of survivors in 5 yr is 37% of the total number.

$\qquad .37y_0 = y_0 e^{k(5)}$

$\qquad .37 = e^{5k}$

$\qquad \ln .37 = 5k$

$\qquad \dfrac{\ln .37}{5} = k$

$\qquad -.1989 = k$

39. $P(x) = 500 - 500e^{-x}$

(a) $P_0 = P(0) = 500 - 500e^0$
$$= 0$$

(b) $P(2) = 500 - 500e^{-2}$
$$\approx 432$$

(c) $P(5) = 500 - 500e^{-5}$
$$\approx 497$$

(d) $P = 400$
$$400 = 500 - 500e^{-x}$$
$$-100 = -500e^{-x}$$
$$e^{-x} = \frac{1}{5}$$
$$e^x = 5$$
$$x = \ln 5 \approx 1.6$$

400 items per day are produced after about 1.6 days.

(e) $\lim_{x \to \infty} (500 - 500e^{-x})$
$$= \lim_{x \to \infty} 500 - \lim_{x \to \infty} 500e^{-x}$$
$$= 500 - 0$$
$$= 500$$

(f) See the answer graph in your textbook.

41.
$$A(t) = A_0 e^{kt}$$
$$.60 A_0 = A_0 e^{(-\ln 2/5600)t}$$
$$.60 = e^{(-\ln/5600)t}$$
$$\ln .60 = -\frac{\ln 2}{5600}t$$
$$\frac{5600(\ln .60)}{-\ln 2} = t$$
$$4127 \approx t$$

The sample was about 4100 yr old.

43.
$$\frac{1}{2} A_0 = A_0 e^{-.00043t}$$
$$\frac{1}{2} = e^{-.00043t}$$
$$\ln \frac{1}{2} = -.00043t$$
$$\ln 1 - \ln 2 = -.00043t$$
$$\frac{0 - \ln 2}{-.00043} = t$$
$$1612 \approx t$$

The half-life of radium 226 is about 1600 yr.

45. $y = 40e^{-.004t}$

(a) $t = 180$
$$y = 40e^{-.004(180)}$$
$$= 40e^{-.72}$$
$$\approx 19.5 \text{ watts}$$

(b)
$$20 = 40e^{-.004t}$$
$$\frac{1}{2} = e^{-.004t}$$
$$\ln \frac{1}{2} = -.004t$$
$$\frac{\ln 1 - \ln 2}{-.004} = t$$
$$173 \approx t$$

(c) The power will never be completely gone. $\lim_{t \to \infty} 40e^{-.004t} = 0$, but it will never actually reach zero.

47. (a) Let t = the number of degrees Celsius.

$y = y_0 \cdot e^{kt}$, $y_0 = 10$ when $t = 0°$.
To find k, let $y = 11$, when $t = 10°$.

$$11 = 10e^{10k}$$

$$e^{10k} = \frac{11}{10}$$

$$10k = \ln 1.1$$

$$k = \frac{\ln 1.1}{10}$$

$$\approx .0095$$

$$y = 10\, e^{.0095t}$$

(b) Let $y = 15$,

Solve for t.

$$15 = 10e^{.0095t}$$

$$\ln 1.5 = .0095t$$

$$t = \frac{\ln 1.5}{.0095}$$

$$\approx 42.7°\ C$$

49. $f(t) = T_0 + Ce^{-kt}$

$$25 = 20 + 100e^{-.1t}$$

$$5 = 100e^{-.1t}$$

$$e^{-.1t} = .05$$

$$-.1t = \ln .05$$

$$t = \frac{\ln .05}{-.1}$$

$$\approx 30$$

It will take about 30 min.

Section 5.4

1. $y = \ln (8x)$

$$y' = \frac{d}{dx}(\ln 8x)$$

$$= \frac{d}{dx}(\ln 8 + \ln x)$$

$$= \frac{d}{dx}(\ln 8) + \frac{d}{dx}(\ln x)$$

$$= 0 + \frac{1}{x}$$

$$= \frac{1}{x}$$

3. $y = \ln (3 - x)$

$$g(x) = 3 - x$$

$$g'(x) = -1$$

$$y' = \frac{g'(x)}{g(x)}$$

$$= \frac{-1}{3 - x} \quad \text{or} \quad \frac{1}{x - 3}$$

5. $y = \ln |2x^2 - 7x|$

$$g(x) = 2x^2 - 7x$$

$$g'(x) = 4x - 7$$

$$y' = \frac{4x - 7}{2x^2 - 7x}$$

7. $y = \sqrt{x + 5}$

$$g(x) = \sqrt{x + 5}$$

$$= (x + 5)^{1/2}$$

$$g'(x) = \frac{1}{2}(x + 5)^{-1/2}$$

$$y' = \frac{\frac{1}{2}(x + 5)^{-1/2}}{(x + 5)^{1/2}}$$

$$= \frac{1}{2(x + 5)}$$

9. $y = \ln (x^4 + 5x^2)^{3/2}$

$$= \frac{3}{2} \ln (x^4 + 5x^2)$$

$$y' = \frac{3}{2}D_x [\ln (x^4 + 5x^2)]$$

$$g(x) = x^4 + 5x^2$$

$$g'(x) = 4x^3 + 10x$$

$$y' = \frac{3}{2}\left(\frac{4x^3 + 10x}{x^4 + 5x^2}\right)$$

$$= \frac{3}{2}\left[\frac{2x(2x^2 + 5)}{x^2(x^2 + 5)}\right]$$

$$= \frac{3(2x^2 + 5)}{x(x^2 + 5)}$$

11. $y = -3x \ln (x + 2)$

Use the product rule.

$$y' = -3x\left[\frac{d}{dx} \ln (x + 2)\right]$$

$$+ \ln (x + 2)\left[\frac{d}{dx}(-3x)\right]$$

$$= -3x\left(\frac{1}{x + 2}\right) + \left[\ln (x + 2)\right](-3)$$

$$= -\frac{3x}{x + 2} - 3 \ln (x + 2)$$

13. $y = x^2 \ln |x|$

Use the product rule.

$$y' = x^2\left(\frac{1}{x}\right) + 2x \ln |x|$$

$$= x + 2x \ln |x|$$

15. $y = \frac{2 \ln (x + 3)}{x^2}$

Use the quotient rule.

$$y' = \frac{x^2\left(\frac{2}{x + 3}\right) - 2 \ln (x + 3) \cdot 2x}{(x^2)^2}$$

$$= \frac{\frac{2x^2}{x + 3} - 4x \ln (x + 3)}{x^4}$$

$$= \frac{2x^2 - 4x(x + 3) \ln (x + 3)}{x^4(x + 3)}$$

$$= \frac{x[2x - 4(x + 3) \ln (x + 3)]}{x^4(x + 3)}$$

$$= \frac{2x - 4(x + 3) \ln (x + 3)}{x^3(x + 3)}$$

17. $y = \frac{\ln x}{4x + 7}$

Use the quotient rule.

$$y' = \frac{(4x + 7)\left(\frac{1}{x}\right) - (\ln x)(4)}{(4x + 7)^2}$$

$$= \frac{\frac{4x + 7}{x} - 4 \ln x}{(4x + 7)^2}$$

$$= \frac{4x + 7 - 4x \ln x}{x(4x + 7)^2}$$

19. $y = \frac{3x^2}{\ln x}$

$$y' = \frac{(\ln x)(6x) - 3x^2\left(\frac{1}{x}\right)}{(\ln x)^2}$$

$$= \frac{6x \ln x - 3x}{(\ln x)^2}$$

21. $y = (\ln |x + 1|)^4$

$$y' = 4(\ln |x + 1|^3\left(\frac{1}{x + 1}\right)$$

$$= \frac{4(\ln |x + 1|)^3}{x + 1}$$

23. $y = \ln |\ln x|$

$g(x) = \ln x$

$g'(x) = \frac{1}{x}$

$$y' = \frac{g'(x)}{g(x)}$$

$$= \frac{\frac{1}{x}}{\ln x}$$

$$= \frac{1}{x \ln x}$$

27. $y = \log (6x)$

$$= \frac{\ln 6x}{\ln 10}$$

$$= \frac{1}{\ln 10}(\ln 6x)$$

$$y' = \frac{1}{\ln 10}\left(\frac{1}{x}\right)$$

$$= \frac{1}{x \ln 10}$$

29. $y = \log |1 - x| = \dfrac{\ln |1 - x|}{\ln 10}$

$$y' = \frac{1}{\ln 10} \cdot \frac{d}{dx}[\ln |1 - x|]$$

$$= \frac{1}{\ln 10} \cdot \frac{-1}{1 - x}$$

$$= -\frac{1}{(\ln 10)(1 - x)}$$

or $\dfrac{1}{(\ln 10)(x - 1)}$

31. $y = \log_5 \sqrt{5x + 2}$

$$= \log_5 (5x + 2)^{1/2}$$

$$= \frac{\ln (5x + 2)^{1/2}}{\ln 5}$$

$$y' = \frac{1}{\ln 5} \cdot \frac{\frac{1}{2}(5x + 2)^{-1/2}(5)}{(5x + 2)^{1/2}}$$

$$= \frac{5}{2 \ln 5 \, (5x + 2)}$$

33. $y = \log_3 (x^2 + 2x)^{3/2}$

$$= \frac{\ln (x^2 + 2x)^{3/2}}{\ln 3}$$

$$= \frac{1}{\ln 3} \cdot \frac{3}{2} \ln (x^2 + 2x)$$

$$y' = \frac{3}{2 \ln 3} \cdot \frac{2x + 2}{x^2 + 2x}$$

$$= \frac{3}{2 \ln 3} \cdot \frac{2(x + 1)}{x^2 + 2x}$$

$$= \frac{3(x + 1)}{(\ln 3)(x^2 + 2x)}$$

For Exercises 35–39, see the answer graphs in the textbook.

35. $y = x \ln x$, $x > 0$

To find any maxima or minima, find the first derivative.

$$y = x \cdot \ln x$$

$$y' = x\left(\frac{1}{x}\right) + \ln x$$

$$= 1 + \ln x$$

Set the derivative equal to 0.

$$1 + \ln x = 0$$

$$\ln x = -1$$

$$e^{-1} = x$$

$$x = \frac{1}{e} \approx .3679$$

$$y = \frac{1}{e} \ln \left(\frac{1}{e}\right)$$

$$= \frac{1}{e} \ln e^{-1}$$

$$= -\frac{1}{e} \approx -.3679$$

Find the second derivative to determine whether y is a maximum or minimum at x = 1/e.

$$y'' = 0 + \frac{1}{x} = \frac{1}{x} > 0 \text{ since } x > 0.$$

There is a minimum value of

$$y = -\frac{1}{e} \approx -.3679 \text{ at}$$

$$x = \frac{1}{e} \approx .3679.$$

37. $y = x \ln |x|$

Find the first derivative.

$$y' = x\left(\frac{1}{x}\right) + \ln |x| = 1 + \ln |x|$$

Set $y' = 0$.

$$\ln |x| = -1$$
$$e^{-1} = |x|$$

If $x > 0$, $x = 1/e \approx .3679$.

If $x < 0$, $x = -1/e \approx -.3679$.

$y'' = \dfrac{1}{x} > 0$ if $x > 0$,

which indicates a minimum.

At $x = 1/e \approx .3679$, there is a

minimum:

$$y = \frac{1}{e} \ln \frac{1}{e} = \frac{1}{e} \ln e^{-1}$$

$$= -\frac{1}{e} \approx -.3679.$$

$y'' = \dfrac{1}{x} < 0$ if $x < 0$,

which indicates a maximum.

At $x = -1/e \approx -.3679$, there is a

maximum.

$$y = -\frac{1}{e} \ln \left|-\frac{1}{e}\right| = -\frac{1}{e} \ln e^{-1}$$

$$= \frac{1}{e} \approx .3679.$$

39. $y = \dfrac{\ln x}{x}$, $x > 0$

$x \leq 0$ is not in the domain of the

function.

$$y' = \frac{x\left(\frac{1}{x}\right) - \ln x}{x^2}$$

$$= \frac{1 - \ln x}{x^2}$$

Set $y' = 0$.

$$\ln x = 1$$
$$x = e$$

$$y'' = \frac{x^2(0) - (\ln x)2x}{x^4} < 0$$

$$= -\frac{2 \ln x}{x^3} < 0$$

At $x = e$,

$y'' = -2e^{-3} < 0$.

Therefore, at $x = e \approx 2.718$, there

is a maximum of

$$y = \frac{\ln e}{e}$$

$$= \frac{1}{e} \approx .3679.$$

41. $f(x) = \ln x$

The slope of the tangent line is

$f'(x) = \dfrac{1}{x}$.

$$\lim_{x \to \infty} \left(\frac{1}{x}\right) = 0$$

$$\lim_{x \to \infty} \left(\frac{1}{x}\right) = \infty$$

As $x \to \infty$ the slope approaches 0; as

$x \to \infty$, the slope becomes infinitely

large.

43. $\dfrac{d}{dx} \ln |ax|$ Note: a is a constant.

$$= \frac{d}{dx}(\ln |a| + \ln |x|)$$

$$= \frac{d}{dx} \ln |a| + \frac{d}{dx} \ln |x|$$

$$= 0 + \frac{d}{dx} \ln |x|$$

$$= \frac{d}{dx} \ln |x|$$

Therefore,

$$\frac{d}{dx} \ln |ax| = \frac{d}{dx} \ln |x|.$$

45. $p = 100 + \dfrac{50}{\ln x}$

(a) $R = px$

$$R = 100x + \frac{50x}{\ln x}$$

The marginal revenue is

$$\frac{dR}{dx} = 100 + \frac{(\ln x)(50) - 50x\left(\frac{1}{x}\right)}{(\ln x)^2}$$

$$= 100 + \frac{50(\ln x - 1)}{(\ln x)^2}.$$

(b) The revenue from one more unit is dR/dx for x = 8.

$$100 + \frac{50(\ln 8 - 1)}{(\ln 8)^2} = \$112.48$$

(c) The manager can use the information from (b) to decide if it is reasonable to sell additional items. If the revenue does not exceed the cost, there will be no profit.

47. p = 100 - 10 ln x, 1 < x ≤ 20,000

x = 6n, n = number of employees

(a) The revenue function is

$$R(x) = px$$
$$= (100 - 10 \ln x)x$$
$$= 100x - 10x \ln x,$$
$$1 < x < 20,000.$$

(b) The marginal revenue product function is

$$\frac{dR}{dn} = \frac{dR}{dx} \cdot \frac{dx}{dn}$$

$$= \left[100 - 10\left(x \cdot \frac{1}{x} + \ln x\right)\right]6$$

$$= (90 - 10 \ln x)6$$

$$= 60(9 - \ln x)$$

$$= 60(9 - \ln 6n).$$

(c) When x = 20, the marginal revenue product is

$$60(9 - \ln 20) \approx 360.$$

Hiring an additional worker will produce an increase of about $360.

49.   P(t) = (t + 100) ln (t + 2)

P'(t) = $(t + 100)\left(\frac{1}{t + 2}\right)$ + ln (t + 2)

P'(2) = $(102)\left(\frac{1}{4}\right)$ + ln (4)

≈ 26.9

P'(8) = $(108)\left(\frac{1}{10}\right)$ + ln (10)

≈ 13.1

51. This exercise should be solved by a graphing calculator or by computer methods. The solution will vary according to the method that is used. See the answer graph at the back of the textbook. The minimum of about .85 should occur at about .7.

Section 5.5

1. $y = e^{4x}$

Let   g(x) = 4x,

with g'(x) = 4.

$y' = 4e^{4x}$

3. $y = -8e^{2x}$

$y' = -8(2e^{2x})$

$= -16e^{2x}$

5. $y = -16e^{x+1}$

g(x) = x + 1

g'(x) = 1

$y' = -16[(1)e^{x+1}]$

$= -16e^{x+1}$

7. $y = e^{x^2}$

$g(x) = x^2$

$g'(x) = 2x$

$y' = 2xe^{x^2}$

9. $y = 3e^{2x^2}$

$g(x) = 2x^2$

$g'(x) = 4x$

$y' = 3(4xe^{2x^2})$

$\quad = 12xe^{2x^2}$

11. $y = 4e^{2x^2-4}$

$g(x) = 2x^2 - 4$

$g'(x) = 4x$

$y' = 4[(4x)e^{2x^2-4}]$

$\quad = 16xe^{2x^2-4}$

13. $y = xe^x$

Use the product rule.

$y' = xe^x + e^x$

$\quad = e^x(x + 1)$

15. $y = (x - 3)^2 e^{2x}$

Use the product rule.

$y' = (x - 3)^2(2)e^{2x} + e^{2x}(2)(x - 3)$

$\quad = 2(x - 3)^2 e^{2x} + 2(x - 3)e^{2x}$

$\quad = 2(x - 3)e^{2x}[(x - 3) + 1]$

$\quad = 2(x - 3)(x - 2)e^{2x}$

17. $y = e^{x^2} \ln x, \; x > 0$

$y' = e^{x^2}\left(\frac{1}{x}\right) + (\ln x)(2x)e^{x^2}$

$\quad = \dfrac{e^{x^2}}{x} + 2xe^{x^2} \ln x$

19. $y = \dfrac{e^x}{\ln x}, \; x > 0$

Use the quotient rule.

$y' = \dfrac{(\ln x)e^x - e^x\left(\frac{1}{x}\right)}{(\ln x)^2} \cdot \dfrac{x}{x}$

$\quad = \dfrac{xe^x \ln x - e^x}{x(\ln x)^2}$

21. $y = \dfrac{x^2}{e^x}$

Use the quotient rule.

$y' = \dfrac{e^x(2x) - x^2 e^x}{(e^x)^2}$

$\quad = \dfrac{xe^x(2 - x)}{e^{2x}}$

$\quad = \dfrac{x(2 - x)}{e^x}$

23. $y = \dfrac{e^x + e^{-x}}{x}$

$y' = \dfrac{x(e^x - e^{-x}) - (e^x + e^{-x})}{x^2}$

25. $y = \dfrac{5000}{1 + 10e^{.4x}}$

$y' = \dfrac{(1 + 10e^{.4x}) \cdot 0 - 5000[0 + 10(.4)e^{.4x}]}{(1 + 10e^{.4x})^2}$

$\quad = \dfrac{-20{,}000e^{.4x}}{(1 + 10e^{.4x})^2}$

27. $y = \dfrac{10{,}000}{9 + 4e^{-.2x}}$

$y' = \dfrac{(9 + 4e^{-.2x}) \cdot 0 - 10{,}000[0 + 4(.-2)e^{-.2x}]}{(9 + 4e^{-.2x})^2}$

$\quad = \dfrac{8000e^{-.2x}}{(9 + 4e^{-.2x})^2}$

29. $y = (2x + e^{-x^2})^2$

Use the chain rule.

$y' = 2(2x + e^{-x^2})(2 - 2xe^{-x^2})$

31. $y = 8^{5x}$

$= e^{\ln 8^{5x}}$

$= e^{5x \ln 8}$

$y' = 5(\ln 8)e^{5x \ln 8}$

$= 5(\ln 8)8^{5x}$

33. $y = 3 \cdot 4^{x^2+2} = 3 \cdot e^{\ln 4^{x^2+2}} = 3 \cdot e^{(x^2+2)\ln 4}$

$y' = 3(\ln 4)(2x)e^{(x^2+2)\ln 4} = 6x(\ln 4)e^{(x^2+2)\ln 4} = 6x(\ln 4)4^{x^2+2}$

35. $y = 2 \cdot 3^{\sqrt{x}} = 2 \cdot e^{\ln 3^{\sqrt{x}}} = 2 \cdot e^{\sqrt{x} \ln 3}$

$y' = 2 \cdot \ln 3 \, \dfrac{1}{2}x^{-1/2} e^{\sqrt{x} \ln 3} = \dfrac{\ln 3 e^{\sqrt{x} \ln 3}}{\sqrt{x}} = \dfrac{(\ln 3)(3^{\sqrt{x}})}{\sqrt{x}}$

37. $S'(x) = \dfrac{3,000,000e^{-.3x}}{(1 + 100e^{-.3x})^2} = (3,000,000e^{-.3x})(1 + 100e^{-.3x})^{-2}$

$S''(x) = (3,000,000e^{-.3x})[-2(1 + 100e^{-.3x})^{-3}(-30e^{-.3x})]$

$\quad + (1 + 100e^{-.3x})^{-2}(-900,000e^{-.3x})$

$= (3,000,000e^{-.3x})(60e^{-.3x})(1 + 100e^{-.3x})^{-3} + (1 + 100e^{-.3x})^{-2}(-900,000e^{-.3x})$

$= (1 + 100e^{-.3x})^{-3}[180,000,000e^{-.6x} + (1 + 100e^{-.3x})(-900,000e^{-.3x})]$

$= (1 + 100e^{-.3x})^{-3}(90,000,000e^{-.6x} - 900,000e^{-.3x})$

$= (1 + 100e^{-.3x})(900,000e^{-.3x})(100e^{-.3x} - 1)$

$S''(x) = 0$ when $100e^{-.3x} - 1 = 0$.

$100e^{-.3x} = 1$

$e^{-.3x} = .01$

$-.3x = \ln .01$

$x = \dfrac{\ln .01}{-.3} \approx 15.4$

$S(15.4) = \dfrac{100,000}{1 + 100e^{-.3(15.4)}}$

$\approx 50,370.7$

The inflection point is approximately (15.4, 50,000).

R′(4) > 0 and R′(6) < 0, so there is a relative maximum when x = 5.

53. $A(t) = 500e^{-.25t}$

$A'(t) = 500(-.25)e^{-.25t}$

$= -125e^{-.25t}$

(a) $A'(4) = -125e^{-.25(4)}$

$= -125e^{-1}$

≈ -46.0

(b) $A'(6) = -125e^{-.25(6)}$

$= -125e^{-1.5}$

≈ -27.9

(c) $A'(10) = -125e^{-.25(10)}$

$= -125e^{-2.5}$

≈ -10.3

(d) $\lim_{t \to \infty} A'(t) = \lim_{t \to \infty} -125e^{-.25t} = 0$

As the number of years increase, the rate of change approaches zero.

(e)   A′(t) will never equal zero, although as t increases, it approaches zero. Powers of e are always positive.

55.   This exercise should be solved by a graphing calculator or computer methods. The solution will vary according to the method that is used. See the answer graph at the back of the textbook. There are no relative extrema. An inflection point occurs at (1, 0).

Chapter 5   Review Exercises

3. $2^{3x} = \frac{1}{8}$

$2^{3x} = 2^{-3}$

$3x = -3$

$x = -1$

5. $9^{2y-1} = 27^y$

$3^{2(2y-1)} = 3^{3y}$

$2(2y - 1) = 3y$

$4y - 2 = 3y$

$y = 2$

For Exercises 7–13, see the answer graphs in your textbook.

7. $y = 5^x$

x	-2	-1	0	1	2
y	1/25	1/5	1	5	25

9. $y = \left(\frac{1}{5}\right)^{2x-3}$

x	0	1	2
y	125	5	1/5

11. $y = \log_2 (x - 1)$

$2^y = x - 1$

$x = 1 + 2^y$

x	2	3	5	9
y	0	1	2	3

13. $y = -\log_3 x$

$-y = \log_3 x$

$3^{-y} = x$

x	1/3	1	3	9
y	1	0	-1	-2

15. $2^6 = 64$

$\log_2 64 = 6$

17. $e^{.09} = 1.09417$

$\ln 1.09417 = .09$

19. $\log_2 32 = 5$

$2^5 = 32$

21. $\ln 82.9 = 4.41763$

$e^{4.41763} = 82.9$

23. $\log_3 81 = x$

$3^x = 81$

$3^x = 3^4$

$x = 4$

25. $\log_{32} 16 = x$

$32^x = 16$

$2^{5x} = 2^4$

$5x = 4$

$x = \dfrac{4}{5}$

27. $\log_{100} 1000 = x$

$100^x = 1000$

$(10^2)^x = 10^3$

$2x = 3$

$x = \dfrac{3}{2}$

29. $\log_5 3k + \log_5 7k^3$

$= \log_5 3k(7k^3)$

$= \log_5 (21k^4)$

31. $2 \log_2 x - 3 \log_2 m$

$= \log_2 x^2 - \log_2 m^3$

$= \log_2 \left(\dfrac{x^2}{m^3}\right)$

33. $8^p = 19$

$\ln 8^p = \ln 19$

$p \ln 8 = \ln 19$

$p = \dfrac{\ln 19}{\ln 8}$

≈ 1.416

35. $2^{1-m} = 7$

$\ln 2^{1-m} = \ln 7$

$(1 - m) \ln 2 = \ln 7$

$1 - m = \dfrac{\ln 7}{\ln 2}$

$-m = \dfrac{\ln 7}{\ln 2} - 1$

$m = 1 - \dfrac{\ln 7}{\ln 2}$

≈ -1.807

37. $e^{-5-2x} = 5$

$\ln e^{-5-2x} = \ln 5$

$(-5 - 2x) \ln e = \ln 5$

$(-5 - 2x) \cdot 1 = \ln 5$

$-2x = \ln 5 + 5$

$x = \dfrac{\ln 5 + 5}{-2}$

≈ -3.305

39. $\left(1 + \frac{m}{3}\right)^5 = 10$

$\left[\left(1 + \frac{m}{3}\right)^5\right]^{1/5} = 10^{1/5}$

$1 + \frac{m}{3} = 10^{1/5}$

$\frac{m}{3} = 10^{1/5} - 1$

$m = 3(10^{1/5} - 1)$

≈ 1.7547

41. $y = -6e^{2x}$
$y' = -6(2e^{2x}) = -12\ e^{2x}$

43. $y = e^{-2x^3}$
$g(x) = -2x^3$
$g'(x) = -6x^2$
$y' = -6x^2\ e^{-2x^3}$

45. $y = 5x \cdot e^{2x}$

Use the product rule.

$y' = 5x(2e^{2x}) + e^{2x(5)}$

$= 10xe^{2x} + 5e^{2x}$

$= 5e^{2x}(2x + 1)$

47. $y = \ln\ (2 + x^2)$
$g(x) = 2 + x^2$
$g'(x) = 2x$
$y' = \frac{2x}{2 + x^2}$

49. $y = \frac{\ln\ |3x|}{x - 3}$

$y' = \dfrac{(x - 3)\left(\frac{1}{3x}\right)(3) - (\ln\ |3x|)1}{(x - 3)^2}$

$= \dfrac{\frac{x - 3}{x} - \ln\ |3x|}{(x - 3)^2} \cdot \frac{x}{x}$

$= \dfrac{x - 3 - x \ln\ |3x|}{x(x - 3)^2}$

51. $y = \dfrac{xe^x}{\ln\ (x^2 - 1)}$

$y' = \dfrac{\ln\ (x^2 - 1)\,[xe^x + e^x] - xe^x\left(\frac{1}{x^2 - 1}\right)(2x)}{[\ln\ (x^2 - 1)]^2}$

$= \dfrac{e^x(x + 1)\ \ln\ (x^2 - 1) - \frac{2x^2 e^x}{x^2 - 1}}{[\ln\ (x^2 - 1)]^2} \cdot \dfrac{x^2 - 1}{x^2 - 1}$

$= \dfrac{e^x(x + 1)(x^2 - 1)\ \ln\ (x^2 - 1) - 2x^2 e^x}{(x^2 - 1)[\ln\ (x^2 - 1)]^2}$

53. $y = (x^2 + e^x)^2$

Use the chain rule.

$y' = 2(x^2 + e^x)(2x + e^x)$

55. $y = x \cdot e^x$

Find the first derivative and set it equal to 0.

$y' = xe^x + e^x = 0$

$e^x(x + 1) = 0$

$x = -1$

To determine minimum or maximum, check the second derivative.

$y'' = xe^x + e^x + e^x = xe^x + 2e^x$

At $x = -1$, $y'' = -1e^{-1} + 2e^{-1}$

$= -\frac{1}{e} + \frac{2}{e} = \frac{1}{e} > 0$

y is concave up, minimum.

$y = -1e^{-1}$

$= -\frac{1}{e} \approx -.368$

Relative minimum of $-e^{-1} \approx -.368$
at $x = -1$

Set y" = 0 to find inflection points.

$$xe^x + 2e^x = 0$$
$$e^x(x + 2) = 0$$
$$x = -2$$
$$y = -2e^{-2} \approx .27$$

Inflection point at $(-2, -.27)$
See the answer graph in your
textbook.

57. $y = \dfrac{e^x}{x - 1}$

Find the first derivative and set it
equal to zero.

$$y' = \frac{(x - 1)e^x - e^x}{(x - 1)^2} = 0$$
$$\frac{e^x(x - 1 - 1)}{(x - 1)^2} = 0$$
$$\frac{e^x(x - 2)}{(x - 1)^2} = 0$$
$$x - 2 = 0$$
$$x = 2$$

To determine minimum or maximum,
check the second derivative.

$$y'' = \frac{(x-1)^2[e^x+(x-2)e^x]-e^x(x-2)(2)(x-1)}{(x - 1)^4}$$
$$= \frac{e^x(x^2 - 4x + 5)}{(x - 1)^3}$$

At $x = 2$,

$$y'' = \frac{1^2(e^2) - 0}{1^4} = e^2 > 0$$

y is concave up, minimum at x = 2.

$$y = \frac{e^2}{1} = e^2 \approx 7.39$$

Relative minimum at $e^2 \approx 7.39$ at
$x = 2$

Set y" = 0 to find inflection points.

$$\frac{e^x(x^2 - 4x + 5)}{(x - 1)^3} = 0$$
$$x^2 - 4x + 5 = 0$$
$$x = \frac{4 \pm \sqrt{16 - 20}}{2}$$

This quadratic equation has no real
roots. There are no inflection
points. See the answer graph in your
textbook.

59. P = $6902, r = 12%, t = 8, m = 2

$$A = P\left(1 + \frac{r}{m}\right)^{tm}$$
$$A = 6902\left(1 + \frac{.12}{2}\right)^{8(2)}$$
$$= 6902(1.06)^{16}$$
$$= \$17,533.51$$

Interest = A - P
$$= \$17,533.51 - \$6902$$
$$= \$10,631.51$$

For Exercises 61–65, use $A = Pe^{rt}$.

61. P = $12,104, r = 8%, t = 2

$$A = 12,104e^{.08(2)}$$
$$= \$14,204.18$$

63. P = $12,104, r = 8%, t = 7

$$A = 12,104e^{.08(7)}$$
$$= \$21,190.14$$

65. P = $12,000, r = .05, t = 8

$$A = 12,000e^{.05(8)}$$
$$= 12,000e^{.40}$$
$$= \$17,901.90$$

67. $r = 9\%$, $m = 12$

$$r_E = \left(1 + \frac{r}{m}\right)^m - 1$$

$$= \left(1 + \frac{.09}{12}\right)^{12} - 1$$

$$= .0938 = 9.38\%$$

69. $r = 9\%$ compounded continuously

$$r_E = e^r - 1$$

$$= e^{.09} - 1$$

$$= .0942 = 9.42\%$$

71. $A = \$2000$, $r = 6\%$, $t = 5$, $m = 1$

$$P = A\left(1 + \frac{r}{m}\right)^{-tm}$$

$$= 2000\left(1 + \frac{.06}{1}\right)^{-5(1)}$$

$$= 2000(1.06)^{-5}$$

$$= \$1494.52$$

73. $A = \$43,200$, $r = 8\%$, $t = 4$, $m = 4$

$$P = 43,200\left(1 + \frac{.08}{4}\right)^{-4(4)}$$

$$= 43,200(1.02)^{-16}$$

$$= \$31,468.86$$

75. (a) $P(x) = 100 - 100e^{-.8x}$

$$\lim_{x \to \infty} P(x) = \lim_{x \to \infty} \left(100 - 100e^{-.8x}\right)$$

$$= \lim_{x \to \infty} 100 - \lim_{x \to \infty} 100e^{-.8x}$$

$$= 100 - 0$$

$$= 100$$

(b) An experienced worker, who has had many days on the job should produce 100 items per day.

(c) $$50 = 100 - 100e^{-.8x}$$

$$-50 = -100e^{-.8x}$$

$$\frac{1}{2} = e^{-.8x}$$

$$\ln\left(\frac{1}{2}\right) = -.8x$$

$$\frac{-\ln 2}{-.8} = x$$

$$1 \approx x$$

A new employee will produce 50 items after 1 day on the job.

77. $P = \$1$, $r = .08$

$$A = Pe^{rt}, \quad A = 3(1)$$

$$3 = 1e^{.08t}$$

$$\ln 3 = .08t$$

$$\frac{\ln 3}{.08} = t$$

$$13.7 = t$$

It would take about 13.7 yr.

79. $$P = A\left(1 + \frac{r}{m}\right)^{-tm}$$

$$P = 25,000\left(1 + \frac{.06}{12}\right)^{-3(12)}$$

$$= 25,000(1.005)^{-36}$$

$$= \$20,891.12$$

81. $I(x) \geq 1$

$$I(x) = 10e^{-.3x}$$

$$10e^{-.3x} \geq 1$$

$$e^{-.3x} \geq .1$$

$$-.3x \geq \ln .1$$

$$x \leq \frac{\ln .1}{-.3} \approx 7.7$$

The greatest depth is about 7.7 m.

83. $g(t) = \frac{c}{a} + \left(g_0 - \frac{c}{a}\right)e^{-at}$

(a) $g'(t) = -a\left(g_0 - \frac{c}{a}\right)e^{-at}$

$= (-ag_0 + c)e^{-at}$

$g_0 = .08$, $c = .1$ and $a = 1.3$

$g'(t) = [(-1.3)(.08) + .1]e^{-at}$

$= -.004e^{-1.3t}$

$g'(t)$ cannot equal zero. The maximum amount occurs when the glucose is first infused. The maximum amount is $g_0 = .08$ g.

(b) $g(t) = \frac{c}{a} + \left(g_0 - \frac{c}{a}\right)e^{-at}$

$.1 = \frac{.1}{1.3} + \left(.08 - \frac{.1}{1.3}\right)e^{-1.3t}$

$.1 - \frac{.1}{1.3} = \left(.08 - \frac{.1}{1.3}\right)e^{-1.3t}$

$7.5 = e^{-1.3t}$

$\ln 7.5 = -1.3t$

$-\frac{1}{1.3}\ln 7.5 = t$

$t = -1.55$

Since time is never negative, the amount of glucose is never .1 g. (As shown in part (a), the maximum amount is .08 g.)

(c) If the glucose is continuously infused into the bloodstream, it continuously decreases at a slower and slower rate over time.

85. $t = (1.26 \times 10^9)\dfrac{\ln\left[1 + 8.33\left(\frac{A}{K}\right)\right]}{\ln 2}$

(a) $A = 0$, $K > 0$

$t = (1.26 \times 10^9)\dfrac{\ln[1 + 8.33(0)]}{\ln 2}$

$= (1.26 \times 10^9)(0) = 0$ years

(b) $t = (1.26 \times 10^9)\dfrac{\ln[1 + 8.33(.212)]}{\ln 2}$

$= (1.26 \times 10^9)\dfrac{\ln 2.76596}{\ln 2}$

$= 1,849,403,169$

or about 1.85×10^9 years.

(c) $t = (1.26 \times 10^9)\dfrac{\ln(1 + 8.33r)}{\ln 2}$

$\dfrac{dt}{dr} = \dfrac{(1.26 \times 10^9)}{\ln 2} \cdot \dfrac{1}{1 + 8.33r}(8.33)$

$= \dfrac{10.4958 \times 10^9}{\ln 2(1 + 8.33r)}$

(d) As r increases, t increases, but at a slower and slower rate. As r decreases, t decreases at a faster and faster rate.

87. $L(t) = 9 + 2e^{.15t}$

where $t = 0$ corresponds to 1982.

(a) $L(0) = 9 + 2e^{.15(0)}$

$= 9 + 2$

$= 11$

(b) $L(4) = 9 + 2e^{.15(4)}$

≈ 12.6

(c) $L(10) = 9 + 2e^{.15(10)}$

≈ 18.0

(d) See the answer graph in the textbook.

89. $f(x) = a^x$; $a > 0$, $a \neq 1$.

(a) The domain is $(-\infty, \infty)$.

(b) The range is $(0, \infty)$.

(c) The y-intercept is 1.

(d) There are no discontinuities.

(e) $y = 0$ is an asymptote.

(f) f(x) is increasing when a is, greater than 1.

(g) f(x) is decreasing when a is between 0 and 1.

Extended Application

1. (a) P = 2000, i = .1, n = 20, t = .37

$$A = P[(1 + i)^n(1 - t) + t]$$
$$= 2000[(1 + .1)^{20}(1 - .37) + .37]$$
$$= \$9216.65$$
$$M = \frac{9216.65}{2000}$$
$$= 4.6$$

(b) $A = 2000[1 + (1 - t)i]^n$
$$= 2000[1 + .63(.1)]^{20}$$
$$= 2000(1.063)^{20}$$
$$= \$6787.27$$
$$m = \frac{6787.27}{2000}$$
$$= 3.4$$

(c) $\frac{M}{m} = \frac{4.6}{3.4}$
$$= 1.35$$

Investment (a) will yield approximately 35% more after-tax dollars than investment (b).

2. (a) P = 2000, i = .1, n = 10, t = .37

$$A = P[(1 + i)^n(1 - t) + t]$$
$$= 2000[(1 + .1)^{10}(1 - .37) + .37]$$
$$= \$4008.12$$
$$M = \frac{4008.12}{2000}$$
$$= 2.0$$

(b) $A = 2000[1 + (1 - t)i]^n$
$$= 2000[1 + (1 - .37).1]^{10}$$
$$= \$3684.37$$
$$M = \frac{3684.37}{2.0}$$
$$= 1.8$$

(c) $\frac{M}{m} = \frac{2.0}{1.8}$
$$= 1.11$$

Investment (a) will yield approximately 11% more after-tax dollars than investment (b).

3. M is an increasing function of t. The longer you leave your money in the account, the greater yield in after-tax dollars.

4. As n increases, the ratio $\frac{M}{m}$ increases for given values of i and t; that is, m is an increasing function of n.

5. The multiplier function is an increasing function of i. The advantage of the IRA over a regular account widens as the interest rate i increases and is particularly dramatic for high income tax rates.

CHAPTER 5   TEST

[5.1]   **1.** Solve $8^{2y-1} = 4^{y+1}$ [5.1] **2.** Graph the function $y = 4^{x-1}$.

[5.1]   **3.** $315 is deposited in an account paying 6% compounded quarterly for 3 yr. Find the following.

   (a) The amount in the account after 3 yr.

   (b) The amount of interest earned by this deposit

[5.3]   **4.** **(a)** Explain the concept of effective rate.

   (b) Find the effective rate for the situation in Exercise 3.

[5.2]   **5.** Use exponents to write the equation $\ln 42.8 = 3.75654$.

[5.2]   **6.** Evaluate $\log_8 16$.   [5.2] **7.** Evaluate $\ln 543$.

[5.2]   **Use properties of logarithms to simplify the following.**

   8. $\log_2 3k + \log_2 4k^2$ **9.** $4 \log_3 r - 3 \log_3 m$

[5.2]   **Solve each equation. Round to the nearest thousandth.**

   10. $2^{x+1} = 10$ **11.** $\left(1 + \dfrac{2m}{3}\right)^5 = 8$

[5.3]   **12.** Find the interest rate needed for $5000 to grow to $10,000 in 8 yr with continuous compounding.

[5.3]   **13.** How long will it take for $1 to triple at an average rate of 7% compounded continuously?

[5.3]   **14.** Suppose sales of a certain item are given by $S(x) = 2500 - 1500e^{-2x}$ where x represents the number of years that the item has been on the market and $S(x)$ represents sales in thousands. Find the limit on sales.

[5.4]   **15.** Find the derivative of $y = \dfrac{\ln |3x - 1|}{x - 2}$.

[5.4]   **16.** Find all relative maxima or minima of $y = 2x \ln x$, $x > 0$. Sketch the graph of the function.

[5.5]   **Find the derivative of each function.**

17. $y = 3x^2 e^{2x}$ **18.** $y = (e^{2x} - \ln |x|)^3$

[5.5]   **19.** Find all relative maxima or minima of $y = 4xe^{-x}$ and sketch the graph of the function.

[5.5]   **20.** The concentration of a certain drug in the bloodstream at time t in minutes is given by

$$c(t) = e^{-t} - e^{-3t}.$$

Find the concentration at each of the following times.

(a) $t = 0$ **(b)** $t = 1$ **(c)** $t = 2$

(d) Find the maximum concentration and when it occurs.

[5.5]   **21.** The population of Smalltown has grown exponentially from 14,000 in 1990 to 16,500 in 1993. At this rate, in what year will the population reach 17,400?

[5.5]   **22.** Potassium 42 decays exponentially. A sample which contained 1000 g 5 hr ago has decreased to 758 g at present.

(a) Write an exponential equation to express the amount, y, present after t hr.

(b) What is the half-life of potassium 42?

[5.5]   **23.** What is meant by the half-life of a substance?

[5.5]   **24.** Find the present value of $15,000 at 6% compounded quarterly for 4 yr.

[5.5]   **25.** Mr. Jones needs $20,000 for a down payment on a house in 5 yr. How much must he deposit now at 5.8% compounded quarterly in order to have $20,000 in 5 yr?

CHAPTER 5   TEST ANSWERS

1. 5/4

2.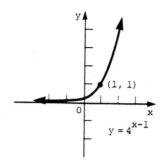
$y = 4^{x-1}$
(1, 1)

3. (a) $376.62   (b) $61.62

4. (a) The effective interest rate is the actual annual rate yielded when compounding occurs.

    (b) 6.14%

5. $e^{3.75654} = 42.8$

6. 4/3

7. 6.29711

8. $\log_2 12k^3$

9. $\log_3 \dfrac{r^4}{m^3}$

10. 2.322

11. .774

12. 8.66%

13. About 15.7 yr

14. 2,500,000

15. $\dfrac{3x - 6 - (3x - 1) \ln |3x - 1|}{(3x - 1)(x - 2)^2}$

16. Relative minimum of $-2/e$ at $x = 1/e$

17. $6x^2e^{2x} + 6xe^{2x}$

18. $\dfrac{3(e^{2x} - \ln |x|)^2(2xe^{2x} - 1)}{x}$

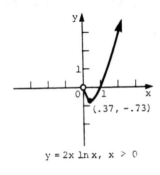
(.37, −.73)
$y = 2x \ln x,\ x > 0$

19. Relative maximum of $4/e$ at $x = 1$

20. (a) 0   (b) .318   (c) .133

     (d) Maximum of .38 at $t = .55$

21. 1994

22. (a) $y = 1000e^{-.0554t}$

     (b) About 12.5 hr

(1, 1.47)
$y = 4xe^{-x}$

23. The half-life is the amount of time needed for half the quantity to decay.

24. $11,820.47

25. $14,996.47

CHAPTER 6   INTEGRATION

Section 6.1

1. $\displaystyle\int 4x\ dx = 4\int x\ dx$

$\displaystyle = 4\cdot\frac{1}{1+1}x^{1+1} + C$

$\displaystyle = \frac{4x^2}{2} + C$

$= 2x^2 + C$

3. $\displaystyle\int 5t^2\ dt = 5\int t^2\ dt$

$\displaystyle = 5\cdot\frac{1}{2+1}t^{2+1} + C$

$\displaystyle = \frac{5t^3}{3} + C$

5. $\displaystyle\int 6\ dk = 6\int 1\ dk = 6\int k^0\ dy$

$\displaystyle = 6\cdot\frac{1}{1}k^{0+1} + C$

$= 6k + C$

7. $\displaystyle\int (2z + 3)\ dz$

$\displaystyle = 2\int z\ dz + 3\int z^0\ dz$

$\displaystyle = 2\cdot\frac{1}{1+1}z^{1+1} + 3\cdot\frac{1}{0+1}z^{0+1} + C$

$= z^2 + 3z + C$

9. $\displaystyle\int (x^2 + 6x)dx = \int x^2\ dx + 6\int x\ dx$

$\displaystyle = \frac{x^3}{3} + \frac{6x^2}{2} + C$

$\displaystyle = \frac{x^3}{3} + 3x^2 + C$

11. $\displaystyle\int (t^2 - 4t + 5)dt$

$\displaystyle = \int t^2\ dt - 4\int t\ dt + 5\int t^0\ dt$

$\displaystyle = \frac{t^3}{3} - \frac{4t^2}{2} + 5t + C$

$\displaystyle = \frac{t^3}{3} - 2t^2 + 5t + C$

13. $\displaystyle\int (4z^3 + 3z^2 + 2z - 6)dz$

$\displaystyle = 4\int z^3\ dz + 3\int z^2\ dz + 2\int z\ dz$

$\displaystyle \qquad - 6\int z^0\ dz$

$\displaystyle = \frac{4z^4}{4} + \frac{3z^3}{3} + \frac{2z^2}{2} - 6z + C$

$= z^4 + z^3 + z^2 - 6z + C$

15. $\displaystyle\int 5\sqrt{z}\ dz = 5\int z^{1/2}\ dz$

$\displaystyle = \frac{5z^{3/2}}{3/2} + C$

$\displaystyle = 5\left(\frac{2}{3}\right)z^{3/2} + C$

$\displaystyle = \frac{10z^{3/2}}{3} + C$

17. $\displaystyle\int (u^{1/2} + u^{3/2})du$

$\displaystyle = \int u^{1/2}\ du + \int u^{3/2}\ du$

$\displaystyle = \frac{u^{3/2}}{3/2} + \frac{u^{5/2}}{5/2} + C$

$\displaystyle = \frac{2u^{3/2}}{3} + \frac{2u^{5/2}}{5} + C$

19. $\displaystyle\int (15x\sqrt{x} + 2\sqrt{x})dx$

$\displaystyle = 15\int x(x^{1/2})\ dx + 2\int x^{1/2}\ dx$

$\displaystyle = 15\int x^{3/2}\ dx + 2\int x^{1/2}\ dx$

$$= \frac{15x^{5/2}}{5/2} + \frac{2x^{3/2}}{3/2} + C$$

$$= 15\left(\frac{2}{5}\right)x^{5/2} + 2\left(\frac{2}{3}\right)x^{3/2} + C$$

$$= 6x^{5/2} + \frac{4x^{3/2}}{3} + C$$

21. $\displaystyle\int (10u^{3/2} - 14u^{5/2})\,du$

$$= 10\int u^{3/2}\,du - 14\int u^{5/2}\,du$$

$$= \frac{10u^{5/2}}{5/2} - \frac{14u^{7/2}}{7/2} + C$$

$$= 10\left(\frac{2}{5}\right)u^{5/2} - 14\left(\frac{2}{7}\right)u^{7/2} + C$$

$$= 4u^{5/2} - 4u^{7/2} + C$$

23. $\displaystyle\int \left(\frac{1}{z^2}\right)dz = \int z^{-2}\,dz$

$$= \frac{z^{-2+1}}{-2+1} + C$$

$$= \frac{z^{-1}}{-1} + C$$

$$= -\frac{1}{z} + C$$

25. $\displaystyle\int \left(\frac{1}{y^3} - \frac{1}{\sqrt{y}}\right)dy$

$$= \int y^{-3}\,dy - \int y^{-1/2}\,dy$$

$$= \frac{y^{-2}}{-2} - \frac{y^{1/2}}{1/2} + C$$

$$= \frac{-1}{2y^2} - 2y^{1/2} + C$$

27. $\displaystyle\int (-9t^{-2} - 2t^{-1})\,dt$

$$= -9\int t^{-2}\,dt - 2\int t^{-1}\,dt$$

$$= \frac{-9t^{-1}}{-1} - 2\int \frac{dt}{t}$$

$$= \frac{9}{t} - 2\ln|t| + C$$

29. $\displaystyle\int e^{2t}\,dt = \frac{e^{2t}}{2} + C$

31. $\displaystyle\int 3e^{-.2x}\,dx = 3\int e^{-.2x}\,dx$

$$= 3\left(\frac{1}{-.2}\right)e^{-.2x} + C$$

$$= \frac{3(e^{-.2x})}{-.2} + C$$

$$= -15e^{-.2x} + C$$

33. $\displaystyle\int \left(\frac{3}{x} + 4e^{-.5x}\right)dx$

$$= 3\int \frac{dx}{x} + 4\int e^{-.5x}\,dx$$

$$= 3\ln|x| + \frac{4e^{-.5x}}{-.5} + C$$

$$= 3\ln|x| - 8e^{-.5x} + C$$

35. $\displaystyle\int \frac{1+2t^3}{t}\,dt = \int \left(\frac{1}{t} + 2t^2\right)dt$

$$= \int \frac{1}{t}\,dt + 2\int t^2\,dt$$

$$= \ln|t| + \frac{2t^3}{3} + C$$

37. $\displaystyle\int (e^{2u} + 4u)\,du = \frac{e^{2u}}{2} + \frac{4u^2}{2} + C$

$$= \frac{e^{2u}}{2} + 2u^2 + C$$

39. $\displaystyle\int (x+1)^2\,dx = \int (x^2 + 2x + 1)\,dx$

$$= \frac{x^3}{3} + \frac{2x^2}{2} + x + C$$

$$= \frac{x^3}{3} + x^2 + x + C$$

41. $\int \dfrac{\sqrt{x} + 1}{\sqrt[3]{x}}\, dx = \int \left(\dfrac{\sqrt{x}}{\sqrt[3]{x}} + \dfrac{1}{\sqrt[3]{x}} \right) dx$

$\qquad = \int (x^{(1/2 - 1/3)} + x^{-1/3})\, dx$

$\qquad = \int x^{1/6}\, dx + \int x^{-1/3}\, dx$

$\qquad = \dfrac{x^{7/6}}{7/6} + \dfrac{x^{2/3}}{2/3} + C$

$\qquad = \dfrac{6x^{7/6}}{7} + \dfrac{3x^{2/3}}{2} + C$

43. Find $f(x)$ such that $f'(x) = x^{2/3}$, and $(1, 3/5)$ is on the curve.

$\int x^{2/3} = \dfrac{x^{5/3}}{5/3} + C$

$\quad f(x) = \dfrac{3x^{5/3}}{5} + C$

Since $(1, 3/5)$ is on the curve,

$f(1) = \dfrac{3}{5}$.

$f(1) = \dfrac{3(1)^{5/3}}{5} + C = \dfrac{3}{5}$

$\qquad\qquad \dfrac{3}{5} + C = \dfrac{3}{5}$

$\qquad\qquad\qquad C = 0$

Thus,

$f(x) = \dfrac{3x^{5/3}}{5}$.

45. $C'(x) = 4x - 5$, fixed cost is \$8.

$C(x) = \int (4x - 5)\, dx$

$\qquad = \dfrac{4x^2}{2} - 5x + k$

$\qquad = 2x^2 - 5x + k$

$C(0) = 2(0)^2 - 5(0) + k = k$

Since $C(0) = 8$, $k = 8$.

$C(x) = 2x^2 - 5x + 8$

47. $C'(x) = .2x^2 + 5x$, fixed cost is \$10.

$C(x) = \int (.2x^2 + 5x)\, dx$

$\qquad = \dfrac{.2x^3}{3} + \dfrac{5x^2}{2} + k$

$\qquad = \dfrac{.2x^3}{3} + \dfrac{5x^2}{2} + k$

$C(0) = \dfrac{.2(0)^3}{3} + \dfrac{5(0)^2}{2} + k = k$

Since $C(0) = 10$, $k = 10$.

$C(x) = \dfrac{.2x^3}{3} + \dfrac{5x^2}{2} + 10$

49. $C'(x) = .03e^{.01x}$, fixed cost is \$8.

$C(x) = \int .03e^{.01x}\, dx$

$\qquad = .03 \int e^{.01x}\, dx$

$\qquad = .03\left(\dfrac{1}{.01}e^{.01x}\right) + k$

$\qquad = 3e^{.01x} + k$

$C(0) = 3e^{.01(0)} + k = 3(1) + k$

$\qquad = 3 + k$

Since $C(0) = 8$, $3 + k = 8$, and $k = 5$.

So, $C(x) = 3e^{.01x} + 5$.

51. $C'(x) = x^{2/3} + 2$, 8 units costs \$58, so

$C(x) = \int (x^{2/3} + 2)\, dx$

$\qquad = \dfrac{3x^{5/3}}{5} + 2x + k$

$C(8) = \dfrac{3(8)^{5/3}}{5} + 2(8) + k$

$\qquad = \dfrac{3(32)}{5} + 16 + k$

Since $C(8) = 58$,

$58 - 16 - \dfrac{96}{5} = k$

$\qquad \dfrac{114}{5} = k.$

So, $C(x) = \dfrac{3x^{5/3}}{5} + 2x + \dfrac{114}{5}.$

53. $C'(x) = x + \dfrac{1}{x^2}$, 2 units cost \$5.50,

so

$$C(2) = 5.50.$$

$C(x) = \displaystyle\int \left(x + \dfrac{1}{x^2}\right) dx$

$\qquad = \displaystyle\int (x + x^{-2}) dx$

$\qquad = \dfrac{x^2}{2} + \dfrac{x^{-1}}{-1} + k$

$C(x) = \dfrac{x^2}{2} - \dfrac{1}{x} + k$

$C(2) = \dfrac{(2)^2}{2} - \dfrac{1}{2} + k$

$\qquad = 2 - \dfrac{1}{2} + k$

Since $C(2) = 5.50$,

$5.50 - 1.5 = k$

$\qquad 4 = k.$

So, $C(x) = \dfrac{x^2}{2} - \dfrac{1}{x} + 4.$

55. $C'(x) = 5x - \dfrac{1}{x}$, 10 units cost

\$94.20, so

$$C(10) = 94.20.$$

$C(x) = \displaystyle\int \left(5x - \dfrac{1}{x}\right) dx$

$\qquad = \dfrac{5x^2}{2} - \ln x + k$

$C(10) = \dfrac{5(10)^2}{2} - \ln (10) + k$

$\qquad = 250 - 2.30 + k$

Since $C(10) = 94.20$,

$94.20 = 247.70 + k$

$-153.50 = k$

So, $C(x) = \dfrac{5x^2}{2} - \ln x - 153.50.$

57. $P'(x) = 2x + 20$, profit is -50 when

0 hamburgers are sold.

$P(x) = \displaystyle\int (2x + 20) dx$

$\qquad = \dfrac{2x^2}{2} + 20x + k$

$\qquad = x^2 + 20x + k$

$P(0) = 0^2 + 20(0) + k$

Since $P(0) = -50$,

$k = -50.$

$P(x) = x^2 + 20x - 50$

59. **(a)** $f'(t) = .01e^{-.01t}$

$f(t) = \displaystyle\int .01e^{-.01t} \, dt$

$\qquad = -\dfrac{.01e^{-.01t}}{.01} + k$

$\qquad = -e^{-.01t} + k$

(b) $f(0) = -e^{-.01(0)} + k$

$\qquad = -e^0 + k$

$\qquad = -1 + k$

Since $f(0) = 0$,

$0 = -1 + k$

$k = 1.$

$f(t) = -e^{-.01t} + 1$

$$f(10) = -e^{-.01(10)} + 1$$
$$= -e^{-.1} + 1$$
$$= -.905 + 1$$
$$= .095 \text{ unit}$$

61. $a(t) = t^2 + 1$

$$v(t) = \int (t^2 + 1)dt$$

$$= \frac{t^3}{3} + t + C$$

$$v(0) = \frac{0^3}{3} + 0 + C$$

Since $v(0) = 6$,

$$C = 6.$$

$$v(t) = \frac{t^3}{3} + t + 6$$

63. $a(t) = -32$

$$v(t) = \int -32 \, dt = -32t + C_1$$

$$v(0) = -32(0) + C_1$$

Since $v(0) = 0$,

$$C_1 = 0.$$

$$v(t) = -32t$$

$$s(t) = \int -32t \, dt$$

$$= \frac{-32t^2}{2} + C_2$$

$$= -16t^2 + C_2$$

At $t = 0$, the plane is at 6400 ft.
That is, $s(0) = 6400$.

$$s(0) = -16(0)^2 + C_2$$
$$6400 = 0 + C_2$$
$$C_2 = 6400$$
$$s(t) = -16t^2 + 6400$$

When the object hits the ground,
$S(t) = 0$.

$$-16t^2 + 6400 = 0$$
$$-16t^2 = -6400$$
$$t^2 = 400$$
$$t = \pm 20$$

Discard -20 since time must be
positive in this problem.
The object hits the ground in
20 sec.

65. $a(t) = \frac{15}{2}\sqrt{t} = 3e^{-t}$

$$v(t) = \int \left(\frac{15}{2}\sqrt{t} + 3e^{-t}\right)dt$$

$$= \int \left(\frac{15}{2}t^{1/2} + 3e^{-t}\right)dt$$

$$= \frac{15}{2}\left(\frac{t^{3/2}}{3/2}\right) + 3\left(\frac{1}{-1}e^{-t}\right) + C_1$$

$$= 5t^{3/2} - 3e^{-t} + C_1$$

$$v(0) = 5(0)^{3/2} - 3e^{-0} + C_1 = -3 + C_1$$

Since $v(0) = -3$, $C_1 = 0$.

$$v(t) = 5t^{3/2} - 3e^{-t}$$

$$s(t) = \int (5t^{3/2} - 3e^{-t})dt$$

$$= 5\left(\frac{t^{5/2}}{5/2}\right) - 3\left(-\frac{1}{1}e^{-t}\right) + C_2$$

$$= 2t^{5/2} + 3e^{-t} + C_2$$

$$s(0) = 2(0)^{5/2} + 3e^{-0} + C_2 = 3 + C_2$$

Since $s(0) = 4$, $C_2 = 1$.

So $s(t) = 2t^{5/2} + 3e^{-t} + 1$.

Section 6.2

1. $\displaystyle\int 4(2x + 3)^4 \, dx = 2 \int 2(2x + 3)^4 \, dx$

Let $u = 2x + 3$, so that

$du = 2 \, dx.$

$$= 2 \int u^4 \, du$$

$$= \frac{2 \cdot u^5}{5} + C$$

$$= \frac{2(2x + 3)^5}{5} + C$$

3. $\displaystyle\int \frac{2 \, dm}{(2m + 1)^3} = \int 2(2m + 1)^{-3} \, dm$

Let $u = 2m + 1$, so that

$du = 2 \, dm.$

$$= \int u^{-3} \, du$$

$$= \frac{u^{-2}}{-2} + C$$

$$= \frac{-(2m + 1)^{-2}}{2} + C$$

5. $\displaystyle\int \frac{2x + 2}{(x^2 + 2x - 4)^4} \, dx$

$$= \int (2x + 2)(x^2 + 2x - 4)^{-4} \, dx$$

Let $u = x^2 + 2x - 4$, so that

$du = (2x + 2) \, dx.$

$$= \int u^{-4} \, du$$

$$= \frac{u^{-3}}{-3} + C$$

$$= \frac{-(x^2 + 2x - 4)^{-3}}{3} + C$$

7. $\displaystyle\int z\sqrt{z^2 - 5} \, dz = \int z(z^2 - 5)^{1/2} \, dz$

$$= \frac{1}{2} \int 2z(z^2 - 5)^{1/2} \, dz$$

Let $u = z^2 - 5$, so that

$du = 2z \, dz.$

$$= \frac{1}{2} \int u^{1/2} \, du$$

$$= \frac{1}{2} \cdot \frac{u^{3/2}}{3/2} + C$$

$$= \frac{1}{2}\left(\frac{2}{3}\right)u^{3/2} + C$$

$$= \frac{(z^2 - 5)^{3/2}}{3} + C$$

9. $\displaystyle\int (-4e^{2p}) \, dp = -2 \int 2e^{2p} \, dp$

Let $u = 2p$, so that

$du = 2 \, dp.$

$$= -2 \int e^u \, du$$

$$= -2e^u + C$$

$$= -2e^{2p} + C$$

11. $\displaystyle\int 3x^2 \, e^{2x^3} \, dx = \frac{1}{2} \int 2 \cdot 3x^2 \, e^{2x^3} \, dx$

Let $u = 2x^3$, so that

$du = 6x^2 \, dx.$

$$= \frac{1}{2} \int e^u \, du$$

$$= \frac{1}{2}e^u + C$$

$$= \frac{e^{2x^3}}{2} + C$$

13. $\displaystyle\int (1 - t)e^{2t - t^2} \, dt$

$$= \frac{1}{2} \int 2(1 - t)e^{2t - t^2} \, dt$$
$\qquad\qquad\qquad$ *Multiply by 1/2 · 2*

Let $u = 2t - t^2$, so that

$du = (2 - 2t)dt$

$\qquad = 2(1 - t)dt.$

$$= \frac{1}{2} \int e^u \, du$$

$$= \frac{e^u}{2} + C$$

$$= \frac{e^{2t-t^2}}{2} + C$$

15. $\int \frac{e^{1/z}}{z^2} \, dz = -\int e^{1/z} \cdot \frac{-1}{z^2} \, dz$

Let $u = 1/z$, so that

$$du = \frac{-1}{z^2} \, dx.$$

$$= -\int e^u \, du$$

$$= -e^u + C$$

$$= -e^{1/z} + C$$

17. $\int \frac{-8}{1 + 3x} \, dx = -8 \int \frac{1}{1 + 3x} \, dx$

$$= -8\left(\frac{1}{3}\right) \int \frac{3}{1 + 3x} \, dx$$

Let $u = 1 + 3x$, so that

$$du = 3 \, dx.$$

$$= \frac{-8}{3} \int \frac{du}{u}$$

$$= \frac{-8}{3} \ln |u| + C$$

$$= \frac{-8 \ln |1 + 3x|}{3} + C$$

19. $\int \frac{dt}{2t + 1} = \frac{1}{2} \int \frac{2 \, dt}{2t + 1}$

Let $u = 2t + 1$, so that

$$du = 2 \, dt.$$

$$= \frac{1}{2} \int \frac{du}{u}$$

$$= \frac{1}{2} \ln |u| + C$$

$$= \frac{\ln |2t + 1|}{2} + C$$

21. $\int \frac{v \, dv}{(3v^2 + 2)^4} = \frac{1}{6} \int \frac{6v \, dv}{(3v^2 + 2)^4}$

Let $u = 3v^2 + 2$, so that

$$du = 6v \, dv.$$

$$= \frac{1}{6} \int \frac{du}{u^4}$$

$$= \frac{1}{6} \int u^{-4} \, du$$

$$= \left(\frac{1}{6}\right)\frac{u^{-3}}{-3} + C$$

$$= -\frac{1}{18}(3v^2 + 2)^{-3} + C$$

$$= \frac{-(3v^2 + 2)^{-3}}{18} + C$$

23. $\int \frac{x - 1}{(2x^2 - 4x)^2} \, dx = \frac{1}{4} \int \frac{4(x - 1) \, dx}{(2x^2 - 4x)^2}$

Let $u = 2x^2 - 4x$, so that

$$du = (4x - 4) \, dx$$

$$= 4(x - 1) \, dx.$$

$$= \frac{1}{4} \int \frac{du}{u^2}$$

$$= \frac{1}{4} \int u^{-2} \, du$$

$$= \frac{1}{4}\left(\frac{u^{-1}}{-1}\right) + C$$

$$= \frac{-(2x^2 - 4x)^{-1}}{4} + C$$

25. $\int \left(\frac{1}{r} + r\right)\left(1 - \frac{1}{r^2}\right) dr = \int u \, du$

Let $u = 1/r + r$, so that

$$du = \left(-\frac{1}{r^2} + 1\right) dr$$

$$= \left(1 - \frac{1}{r^2}\right) dr.$$

$$= \frac{u^2}{2} + C$$

$$= \frac{1}{2}\left(\frac{1}{r} + r\right)^2 + C$$

$$= \frac{\left(\frac{1}{r} + r\right)^2}{2} + C$$

27. $\int \dfrac{x^2 + 1}{(x^3 + 3x)^{2/3}} \, dx$

$= \dfrac{1}{3} \int \dfrac{3(x^2 + 1)\,dx}{(x^3 + 3x)^{2/3}}$

Let $u = x^3 + 3x$, so that

$\quad du = (3x^2 + 3)\,dx.$

$\quad = 3(x^2 + 1)\,dx$

$= \dfrac{1}{3} \int \dfrac{du}{u^{2/3}}$

$= \dfrac{1}{3} \int u^{-2/3} \, du$

$= \dfrac{1}{3}\left(\dfrac{u^{1/3}}{1/3}\right) + C$

$= u^{1/3} + C$

$= (x^3 + 3x)^{1/3} + C$

29. $\int p(p + 1)^5 \, dp$

Let $u = p + 1$, so that

$\quad du = dp;$ also, $p = u - 1.$

$= \int (u - 1)u^5 \, du$

$= \int (u^6 - u^5)\,du$

$= \dfrac{u^7}{u} - \dfrac{u^6}{6} + C$

$= \dfrac{(p + 1)^7}{7} - \dfrac{(p + 1)^6}{6} + C$

31. $\int t\sqrt{5t - 1} \, dt$

$= \dfrac{1}{5} \int 5t(5t - 1)^{1/2} \, dt$

Let $u = 5t - 1$, so that

$\quad du = 5 \, dt;$ also,

$\quad t = \dfrac{u + 1}{5}.$

$= \dfrac{1}{5} \int \left(\dfrac{u + 1}{5}\right)u^{1/2} \, du$

$= \dfrac{1}{25} \int (u^{3/2} + u^{1/2})\,du$

$= \dfrac{1}{25}\left(\dfrac{u^{5/2}}{5/2} + \dfrac{u^{3/2}}{3/2}\right) = C$

$= \dfrac{1}{25}\left[\dfrac{2}{5}(5t - 1)^{5/2} + \dfrac{2}{3}(5t - 1)^{3/2}\right] + C$

$= \dfrac{2(5t - 1)^{5/2}}{125} + \dfrac{2(5t - 1)^{3/2}}{75} + C$

33. $\int \dfrac{u}{\sqrt{u - 1}} \, du$

$= \int u(u - 1)^{-1/2} \, du$

Let $w = u - 1$, so that

$\quad dw = du$ and

$\quad u = w + 1.$

$= \int (w + 1)w^{-1/2} \, dw$

$= \int (w^{1/2} + w^{-1/2})\,dw$

$= \dfrac{w^{3/2}}{3/2} + \dfrac{w^{1/2}}{1/2} + C$

$= \dfrac{2(u - 1)^{3/2}}{3} + 2(u - 1)^{1/2} + C$

35. $\int (\sqrt{x^2 + 12x})(x + 6)\,dx$

$= \int (x^2 + 12x)^{1/2}(x + 6)\,dx$

Let $x^2 + 12x = u$, so that

$\quad (2x + 12)\,dx = du$

$\quad 2(x + 6)\,dx = du.$

$= \dfrac{1}{2} \int u^{1/2} \, du$

$= \dfrac{1}{2}\left(\dfrac{2}{3}\right)u^{3/2} + C$

$= \dfrac{(x^2 + 12x)^{3/2}}{3} + C$

37. $\int \frac{t}{t^2 + 2} \, dt$

Let $t^2 + 2 = u$, so that

$$2t \, dt = du.$$

$$= \frac{1}{2} \int \frac{du}{u}$$

$$= \frac{1}{2} \ln |u| + C$$

$$= \frac{\ln (t^2 + 2)}{2} + C$$

39. $\int z e^{2z^2} \, dz$

Let $2z^2 = u$, so that

$$4z \, dz = du.$$

$$= \frac{1}{4} \int e^u \, du$$

$$= \frac{1}{4} e^u + C$$

$$= \frac{e^{2z^2}}{4} + C$$

41. $\int \frac{(1 + \ln x)^2}{x} \, dx$

Let $1 + \ln x = u$, so that

$$\frac{1}{x} \, dx = du,$$

$$= \int u^2 \, du$$

$$= \frac{1}{3} u^3 + C$$

$$= \frac{(1 + \ln x)^3}{3} + C$$

43. $\int x^{3/2} \sqrt{x^{5/2} + 4} \, dx$

Let $x^{5/2} + 4 = u$, so that

$$\frac{5}{2} x^{3/2} \, dx = du$$

$$x^{3/2} \, dx = \frac{2}{5} \, du.$$

$$= \frac{2}{5} \int \sqrt{u} \, du$$

$$= \frac{2}{5} \int u^{1/2} \, du$$

$$= \frac{2}{5} \left(\frac{u^{3/2}}{3/2}\right) + C$$

$$= \frac{4}{15} u^{3/2} + C$$

$$= \frac{4}{15} (x^{5/2} + 4)^{3/2} + C$$

45. **(a)** $R'(x) = 2x(x^2 + 50)^2$

$$R(x) = \int 2x(x^2 + 50)^2 \, dx$$

Let $u = x^2 + 50$, so that

$$du = 2x \, dx.$$

$$R = \int u^2 \, du$$

$$= \frac{u^3}{3} + C$$

$$= \frac{1}{3}(x^2 + 50)^3 + C$$

$$R(3) = \frac{1}{3}(9 + 50)^3 + C$$

Since $R(3) = \$206,379$,

$$\frac{1}{3}(59)^3 + C = 206,379$$

$$\frac{205,379}{3} + C = 206,379$$

$$C = 137,919.33.$$

$$R(x) = \frac{(x^2 + 50)^3}{3} + 137,919.33$$

(b)

$$R(x) = \frac{(x^2 + 50)^3}{3} + 137,919.33 \geq 450,000$$

$$\frac{(x^2 + 50)^3}{3} \geq 312,080.67$$

$$(x^2 + 50)^3 \geq 936,242.01$$

$$x^2 + 50 \geq 97.83$$

$$x^2 \geq 47.83$$

$$x \geq 6.92$$

At least 7 planes must be sold.

47. **(a)** $p'(x) = xe^{-x^2}$

Let $-x^2 = u$, so that

$-2x\,dx = du$

$x\,dx = -\dfrac{du}{2}$

$p = -\dfrac{1}{2}\displaystyle\int e^u\,du$

$= -\dfrac{1}{2}e^u$

$= -\dfrac{e^{-x^2}}{2} + C$

$p(3) = -\dfrac{e^{-9}}{2} + C$

Since $10,000 = .01$ millions and $p(3) = .01$,

$-\dfrac{e^{-9}}{2} + C = .01$

$C = .01 + \dfrac{e^{-9}}{2}$

$= .01006$

$\approx .01$.

$p(x) = \dfrac{-e^{-x^2}}{2} + .01$

(b) $\displaystyle\lim_{x\to\infty}(x) = \lim_{x\to\infty}\left(\dfrac{-e^{-x^2}}{2} + .01\right)$

$= \displaystyle\lim_{x\to\infty}\left(-\dfrac{1}{2e^{x^2}} + .01\right)$

$= .01$

Since profit is expressed in millions of dollars, the profit approaches $.01(1,000,000) = \$10,000$.

49. **(a)** $D'(x) = \dfrac{2}{x + 9}$

$D(x) = \displaystyle\int D'(x)\,dx$

$= \displaystyle\int \dfrac{2}{x + 9}\,dx = 2\int \dfrac{dx}{x + 9}$

$D(x) = 2\ln|x + 9| + C$

If $x = 1$, $D(x) = 2.5$.

So

$2.5 = 2\ln|1 + 9| + C$

$2.5 = 2\ln 10 + C$

$2.5 - 2\ln 10 = C$

$-2.11 \approx C.$

Thus

$D(x) = 2\ln|x + 9| - 2.11.$

(b)

$D(x)$

$= 2\ln|x + 9| - 2.11 = 3$

$2\ln|x + 9| = 5.11$

$\ln|x + 9| = 2.555$

$|x + 9| = e^{2.555}$

$|x + 9| = 12.87$

$x + 9 = \pm 12.87$

$x = -9 \pm 12.87$

$x = 3.87 \quad\text{or}\quad x = -21.87$

Disregard negative solution. Thus, about 3.9 mg of the drug is necessary.

Section 6.3

1. $\displaystyle\sum_{i=1}^{3} 3i = 3(1) + 3(2) + 3(3) = 18$

3. $\displaystyle\sum_{i=1}^{5} (2i + 7)$

$= [2(1) + 7] + [2(2) + 7] + [2(3) + 7]$

$\quad + [2(4) + 7] + [2(5) + 7]$

$= 9 + 11 + 13 + 15 + 17$

$= 65$

The sum of these rectangles approximates $\int_0^8 (2x + 1)dx$.

5. $\sum\limits_{i=1}^{4} x_i = x_1 + x_2 + x_3 + x_4$

$= -5 + 8 + 7 + 10$

$= 20$

7. $f(x) = x - 3$ and $x_1 = 4$, $x_2 = 6$, $x_3 = 7$

$\sum\limits_{i=1}^{3} f(x_i)$

$= \sum\limits_{i=1}^{3} (x_i - 3)$

$= (x_1 - 3) + (x_2 - 3) + (x_3 - 3)$

$= (4 - 3) + (6 - 3) + (7 - 3)$

$= 1 + 3 + 4$

$= 8$

9. $f(x) = 2x + 1$, $x_1 = 0$, $x_2 = 2$, $x_3 = 4$, $x_4 = 6$, and $\Delta x = 2$

(a) $\sum\limits_{i=1}^{4} f(x_i)\Delta x$

$= f(x_1)\Delta x + f(x_2)\Delta x + f(x_3)\Delta x + f(x_4)\Delta x$

$= f(0)(2) + f(2)(2) + f(4)(2) + f(6)(2)$

$= [2(0) + 1](2) + [2(2) + 1](2) + [2(4) + 1](2) + [2(6) + 1](2)$

$= 2 + 5(2) + 9(2) + 13(2)$

$= 56$

(b)

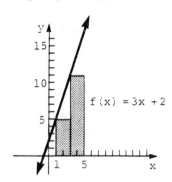

$f(x) = 2x + 1$

11. $f(x) = 3x + 2$ from $x = 1$ to $x = 5$

For $n = 2$ rectangles:

$$\Delta x = \frac{5 - 1}{2} = 2$$

i	x_i	$f(x_i)$
1	1	5
2	3	11

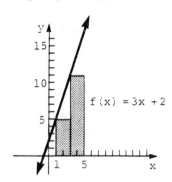

$A = \sum\limits_{i=1}^{2} f(x_i)\Delta x = f(x_1)\Delta x + f(x_2)\Delta x$

$= 5(2) + 11(2)$

$= 32$

For $n = 4$ rectangles:

$$\Delta x = \frac{5 - 1}{4} = 1$$

i	x_i	$f(x_i)$
1	1	5
2	2	8
3	3	11
4	4	14

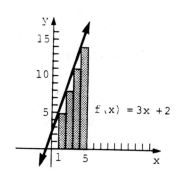

$f(x) = 3x + 2$

$$A = \sum_{i=1}^{4} f(x_i)\Delta x$$

$$= 5(1) + 8(1) + 11(1) + 14(1)$$

$$= 38$$

13. $f(x) = x + 5$ from $x = 2$ to $x = 4$

For n = 2 rectangles:

$$\Delta x = \frac{4 - 2}{2} = 1$$

i	x_i	$f(x_i)$
1	2	7
2	3	8

$$A = \sum_{i=1}^{2} f(x_i)\Delta x = 7(1) + 8(1)$$
$$= 15$$

For n = 4 rectangles:

$$\Delta x = \frac{4 - 2}{4} = \frac{1}{2}$$

i	x_i	$f(x_i)$
1	2	7
2	5/2	15/2
3	3	8
4	7/2	17/2

$$A = \sum_{i=1}^{4} f(x_i)\Delta x$$

$$= 7\left(\frac{1}{2}\right) + \frac{15}{2}\left(\frac{1}{2}\right) + 8\left(\frac{1}{2}\right) + \frac{17}{2}\left(\frac{1}{2}\right)$$

$$= \frac{31}{2}$$

15. $f(x) = x^2$ from $x = 1$ to 5

For n = 2 rectangles:

$$\Delta x = \frac{5 - 1}{2} = 2$$

i	x_i	$f(x_i)$
1	1	1
2	3	9

$$A = \sum_{i=1}^{2} f(x_i)\Delta x = 1(2) + 9(2)$$
$$= 20$$

For n = 4 rectangles:

$$\Delta x = \frac{5 - 1}{4} = 1$$

i	x_i	$f(x_i)$
1	1	1
2	2	4
3	3	9
4	4	16

$$A = \sum_{i=1}^{4} f(x_i)\Delta x$$

$$= 1(1) + 4(1) + 9(1) + 16(1)$$

$$= 30$$

17. $f(x) = x^2 + 2$ from $x = -2$ to $x = 2$

For n = 2 rectangles:

$$\Delta x = \frac{2 - (-2)}{2} = 2$$

i	x_i	$f(x_i)$
1	-2	6
2	0	2

$$A = \sum_{i=1}^{2} f(x_i)\Delta x = 6(2) + 2(2)$$
$$= 16$$

For n = 4 rectangles:

$$\Delta x = \frac{2 - (-2)}{4} = 1$$

i	x_i	$f(x_i)$
1	-2	6
2	-1	3
3	0	2
4	1	3

$$A = \sum_{i=1}^{4} f(x_i)\Delta x = 6 + 3 + 2 + 3$$
$$= 14$$

19. $f(x) = e^x - 1$ from x = 0 to x = 4

For n = 2 rectangles:

$$\Delta x = \frac{4 - 0}{2} = 2$$

i	x_i	$f(x_i)$
1	0	0
2	2	6.39

$$A = \sum_{i=1}^{2} f(x_i)\Delta x = 0(2) + 6.39(2)$$
$$= 12.8$$

For n = 4 rectangles:

$$\Delta x = \frac{4 - 0}{4} = 1$$

i	x_i	$f(x_i)$
1	0	0
2	1	1.72
3	2	6.39
4	3	19.09

$$A = \sum_{i=1}^{4} f(x_i)\Delta x$$
$$= 0 + 1.72 + 6.39 + 10.09$$
$$= 27.2$$

21. $f(x) = \frac{1}{x}$ from x = 1 to 5

For n = 2 rectangles:

$$\Delta x = \frac{5 - 1}{2} = 2$$

i	x_i	$f(x_i)$
1	1	1
2	3	.333

$$A = \sum_{i=1}^{2} f(x_i)\Delta x = 1(2) + .333(2)$$
$$= 2.67$$

For n = 4 rectangles:

$$\Delta x = \frac{5 - 1}{4} = 1$$

i	x_i	$f(x_i)$
1	1	1
2	2	.5
3	3	.333
4	4	.25

$$A = \sum_{i=1}^{4} f(x_i)\Delta x = 1 + .5 + .333 + .25$$
$$= 2.08$$

23. $f(x) = \frac{x}{2}$ between x = 0 and x = 4

(a) For n = 4 rectangles:

$$\Delta x = \frac{4 - 0}{4} = 1$$

i	x_i	$f(x_i)$
1	0	0
2	1	.5
3	2	1
4	3	1.5

$$A = \sum_{i=1}^{4} f(x_i)\Delta x = 0 + .5 + 1 + 1.5$$
$$= 3$$

(b) For n = 8 rectangles:

$$\Delta x = \frac{4 - 0}{8} = .5$$

i	x_i	$f(x_i)$
1	0	0
2	.5	.25
3	1	.5
4	1.5	.75
5	2	1
6	2.5	1.25
7	3	1.5
8	3.5	1.75

$$A = \sum_{i=1}^{8} f(x_i)\Delta x$$

$$= 0(.5) + .25(.5) + .5(.5)$$
$$+ .75(.5) + 1(.5) + 1.25(.5)$$
$$+ 1.5(.5) + 1.75$$
$$= 3.5$$

(c)

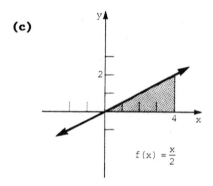

$f(x) = \frac{x}{2}$

$$\int_0^4 f(x)\, dx = \int_0^4 \frac{x}{2}\, dx$$

$$= \frac{1}{2}(\text{base})(\text{height})$$

$$= \frac{1}{2}(4)(2)$$

$$= 4$$

25. $\displaystyle\int_0^3 2x\, dx$

Graph y = 2x.

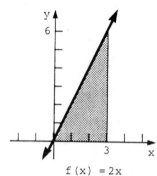

$f(x) = 2x$

$\displaystyle\int_0^3 2x\, dx$ is the area of a triangle

with base = 3 − 0 = 3 and altitude 6.

$$\text{Area} = \frac{1}{2}(\text{base})(\text{altitude})$$

$$= \frac{1}{2}(3)(6)$$

$$= 9$$

27. $\displaystyle\int_{-3}^3 \sqrt{9 - x^2}\, dx$

Graph $y = \sqrt{9 - x^2}$.

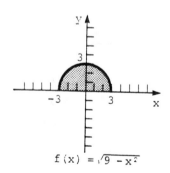

$f(x) = \sqrt{9 - x^2}$

$\displaystyle\int_{-3}^3 \sqrt{9 - x^2}\, dx$ is the area of a semi-

circle with radius 3 centered at the origin.

Area $= \frac{1}{2}\pi r^2$

$= \frac{1}{2}\pi(3)^2$

$= \frac{9}{2}\pi$

29. $\int_1^3 (5 - x)\, dx$

Graph $y = 5 - x$.

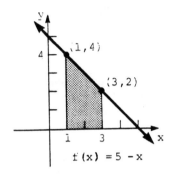

$\int_1^3 (5 - x)\, dx$ is the area of a

trapezoid with bases of length 4
and 2 and height of length 2.

Area $= \frac{1}{2}(\text{height})(\text{base}_1 + \text{base}_2)$

$= \frac{1}{2}(2)(4 + 2)$

$= 6$

For Exercises 31–39, readings on the
graphs and answers may vary.

31. **(a)** Read values for estimated production per day, current policy
base, on the graph for every 2 yr
from 1990 to 2008. These are the
left sides of rectangles with width
$\Delta x = 2$.

$\Sigma = 8.5(2) + 8(2) + 7.7(2) + 7.5(2)$
$+ 7.2(2) + 7.0(2) + 6.7(2)$
$+ 6.5(2) + 6.3(2) + 6.1(2)$
≈ 143 million barrels

This sum must be multiplied by 365
since there are 365 days in a year.
The total production is $365(143) \approx$
52 billion barrels.

(b) Read values for estimated production, with strategy, on the graph
for every 2 yr from 1990 to 2008.
These are the left sides of rectangles with width $\Delta x = 2$.

$\Sigma = 8.5(2) + 8(2) + 7.7(2) + 7.7(2)$
$+ 8.4(2) + 9(2) + 9.5(2)$
$+ 9.8(2) + 10.2(2) + 10(2)$
≈ 177

Total production is $177(365) \approx 64$
billion barrels.

33. **(a)** Read the value for United States
for every year from $x = 1987$ to
$x = 1990$. These are the left sides
of rectangles with width $\Delta x = 1$.

$\Sigma = 9.9(1) + 10.2(1) + 10.5(1)$
$+ 10.8(1)$
≈ 41.4

This sum must be multiplied by 2000
hr/yr, so $41.4(2000) \approx \$83,000$.

(b) Read the value for Canada for
every year from $x = 1987$ to $x =$
1990. Thee are the left sides of
rectangles with width $\Delta x = 1$.

$\Sigma = 9.5(1) + 10.8(1) + 11.8(1)$
$+ 12.2(1)$
≈ 44.3

This sum must be multiplied by 2000 hr/yr, so $44.3(2000) \approx \$89,000$.

35. Read the value of the function for every minute from $x = 0$ to $x = 19$. These are the left sides of rectangles with width $\Delta x = 1$.

$$\sum_{i=1}^{20} f(x_i)\Delta x$$

$$\approx 0(1) + 2.5(1) + 2.8(1) + 3(1)$$
$$+ 3(1) + 3(1) + 3(1) + 3(1)$$
$$+ 3(1) + 3(1) + 3(1) + 1(1)$$
$$+ .8(1) + .6(1) + .4(1) + .3(1)$$
$$+ .2(1) + .2(1) + .2(1) + .2(1)$$
$$\approx 33.2$$

The total volume of oxygen inhaled is 33.2 liters.

37. Read the value for the speed every 3 sec from $x = 3$ to $x = 27$. These are the right sides of rectangles with width $\Delta x = 3$. Then read the speed for $x = 28$, which is the right side of a rectangle with width $\Delta x = 1$.

$$\sum_{i=1}^{9} f(x_i)\Delta x$$

$$\approx 30(3) + 45(3) + 60(3) + 68(3)$$
$$+ 76(3) + 85(3) + 90(3) + 94(3)$$
$$= 1742$$

$$\frac{1742}{3600}(5280) \approx 2600$$

The BMW 733i traveled about 2600 ft.

39. **(a)** Read the value for a plain glass window facing south for every 2 hr from 6 to 6. These are the heights of rectangles with width $\Delta x = 2$.

$$\Sigma = 10(2) + 28(2) + 80(2) + 108(2)$$
$$+ 80(2) + 28(2) + 10(2)$$
$$\approx 690 \text{ BTUs}$$

(b) Read the value for a window with Shadescreen facing south for every 2 hr from 6 to 6. These are the heights of rectangles with width $\Delta x = 2$.

$$\Sigma = 4(2) + 10(2) + 20(2) + 22(2)$$
$$+ 20(2) + 10(2) + 4(2)$$
$$\approx 180 \text{ BTUs}$$

For Exercises 41 and 43 use computer methods. Answers will vary depending upon the computer program that is used. The answers are given.

41. $f(x) = x^2 e^{-x}$; $[-1, 3]$

The area is 1.91837.

43. $f(x) = \dfrac{e^x - e^{-x}}{2}$; $[0, 4]$

The area is 25.7659.

Section 6.4

1. $\displaystyle\int_{-2}^{4} (-1) \ dp = -1 \int_{-2}^{4} dp$

$$= -1 \cdot p \ \Big|_{-2}^{4}$$

$$= -1[4 - (-2)]$$

$$= -6$$

3. $\int_{-1}^{2} (3t - 1)dt$

$= 3\int_{-1}^{2} t\,dt - \int_{-1}^{2} dt$

$= \frac{3}{2}t^2 \Big|_{-1}^{2} - t\Big|_{-1}^{2}$

$= \frac{3}{2}[2^2 - (-1)^2] - [2 - (-1)]$

$= \frac{3}{2}(4 - 1) - (2 + 1)$

$= \frac{9}{2} - 3 = \frac{9 - 6}{2}$

$= \frac{3}{2}$

$\int_{0}^{2} 3\sqrt{4u + 1}\, du$

$= \frac{3}{4}\int_{0}^{2} \sqrt{4u + 1}\,(4\,du)$

$= \frac{3}{4}\int_{1}^{9} x^{1/2}\,dx$

$= \frac{3}{4}\cdot\frac{x^{3/2}}{3/2}\Big|_{1}^{9}$

$= \frac{3}{4}\cdot\frac{2}{3}(9^{2/3} - 1^{3/2})$

$= \frac{1}{2}(27 - 1) = \frac{26}{2}$

$= 13$

5. $\int_{0}^{2} (5x^2 - 4x + 2)dx$

$= 5\int_{0}^{2} x^2\,dx - 4\int_{0}^{2} x\,dx + 2\int_{0}^{2} dx$

$= \frac{5x^3}{3}\Big|_{0}^{2} - 2x^2\Big|_{0}^{2} + 2x\Big|_{0}^{2}$

$= \frac{5}{3}(2^3 - 0^3) - 2(2^2 - 0^2) + 2(2 - 0)$

$= \frac{5}{3}(8) - 2(4) + 2(2)$

$= \frac{40 - 24 + 12}{3}$

$= \frac{28}{3}$

7. $\int_{0}^{2} 3\sqrt{4u + 1}\, du$

Let $4u + 1 = x$, so that
$4\,du = dx$.
When $u = 0$, $x = 4(0) + 1 = 1$.
When $u = 2$, $x = 4(2) + 1 = 9$.

9. $\int_{0}^{1} 2(t^{1/2} - t)dt$

$= 2\int_{0}^{1} t^{1/2} - 2\int_{0}^{1} t\,dt$

$= 2\frac{t^{3/2}}{3/2}\Big|_{0}^{1} - 2\frac{t^2}{2}\Big|_{0}^{1}$

$= \frac{4}{3}(1^{3/2} - 0^{3/2}) - (1^2 - 0^2)$

$= \frac{4}{3} - 1 = \frac{1}{3}$

11. $\int_{1}^{4} (5y\sqrt{y} + 3\sqrt{y})dy$

$= 5\int_{1}^{4} y^{3/2}\,dy + 3\int_{1}^{4} y^{1/2}\,dy$

$= 5\left(\frac{y^{5/2}}{5/2}\right)\Big|_{1}^{4} + 3\left(\frac{y^{3/2}}{3/2}\right)\Big|_{1}^{4}$

$= 2y^{5/2}\Big|_{1}^{4} + 2y^{3/2}\Big|_{1}^{4}$

$= 2(4^{5/2} - 1) + 2(4^{3/2} - 1)$

$= 2(32 - 1) + 2(8 - 1)$

$= 62 + 14$

$= 76$

13. $\displaystyle\int_{4}^{6} \frac{2}{(x - 3)^2} \, dx$

Let $x - 3 = u$, so that

$$dx = du$$
$$x = u + 3.$$

When $x = 6$, $u = 6 - 3 = 3$.

When $x = 4$, $u = 4 - 3 = 1$.

$$\int_{4}^{6} \frac{2}{(x - 3)^2} = 2 \int_{4}^{6} (x - 3)^{-2} \, dx$$

$$= 2 \int_{1}^{3} u^{-2} \, du$$

$$= 2 \cdot \frac{u^{-1}}{-1} \Big|_{1}^{3} = -2 \cdot u^{-1} \Big|_{1}^{3}$$

$$= -2\left(\frac{1}{3} - 1\right) = -2\left(-\frac{2}{3}\right)$$

$$= \frac{4}{3}$$

15. $\displaystyle\int_{1}^{5} (5n^{-2} + n^{-3}) \, dn$

$$= \int_{1}^{5} 5n^{-2} \, dn + \int_{1}^{5} n^{-3} \, dn$$

$$= \frac{5n^{-1}}{-1} \Big|_{1}^{5} + \frac{n^{-2}}{-2} \Big|_{1}^{5}$$

$$= \frac{-5}{n} \Big|_{1}^{5} + \frac{-1}{2n^2} \Big|_{1}^{5}$$

$$= \frac{-5}{5} - \frac{-5}{1} + \frac{-1}{2(25)} - \frac{-1}{2(1)}$$

$$= \frac{-5}{5} + \frac{5}{1} - \frac{1}{50} + \frac{1}{2}$$

$$= \frac{-50 + 250 - 1 + 25}{50} = \frac{224}{50}$$

$$= \frac{112}{25}$$

17. $\displaystyle\int_{2}^{3} \left(2e^{-.1A} + \frac{3}{A}\right) dA$

$$= 2 \int_{2}^{3} e^{-.1A} \, dA + 3 \int_{2}^{3} \frac{1}{A} \, dA$$

$$= 2\frac{e^{-.1A}}{-.1} \Big|_{2}^{3} + 3 \ln |A| \Big|_{2}^{3}$$

$$= -20e^{-.1A} \Big|_{2}^{3} + 3 \ln |A| \Big|_{2}^{3}$$

$$= 20e^{-.2} - 20e^{-.3} + 3 \ln 3$$
$$\quad - 3 \ln 2$$

$$\approx 2.775$$

19. $\displaystyle\int_{1}^{2} \left(e^{5u} - \frac{1}{u^2}\right) du$

$$= \int_{1}^{2} e^{5u} \, du - \int_{1}^{2} \frac{1}{u^2} \, du$$

$$= \frac{e^{5u}}{5} \Big|_{1}^{2} + \frac{1}{u} \Big|_{1}^{2}$$

$$= \frac{e^{10}}{5} - \frac{e^5}{5} + \frac{1}{2} - 1$$

$$= \frac{e^{10}}{5} - \frac{e^5}{5} - \frac{1}{2}$$

$$\approx 4375.1$$

21. $\displaystyle\int_{-1}^{0} y(2y^2 - 3)^5 \, dy$

Let $u = 2y^2 - 3$, so that

$$du = 4y \, dy \text{ and } \frac{1}{4} \, du = y \, dy.$$

When $y = -1$, $u = 2(-1)^2 - 3 = -1$.
When $y = 0$, $u = 2(0)^2 - 3 = -3$.

$$\frac{1}{4}\int_{-1}^{-3} u^5 \, du = \frac{1}{4} \cdot \frac{u^6}{6}\Big|_{-1}^{-3}$$

$$= \frac{1}{24}u^6\Big|_{-1}^{-3}$$

$$= \frac{1}{24}(-3)^6 - \frac{1}{24}(-1)^6$$

$$= \frac{729}{24} - \frac{1}{24}$$

$$= \frac{728}{24}$$

$$= \frac{91}{3}$$

23. $\displaystyle\int_{1}^{64} \frac{\sqrt{z} - 2}{\sqrt[3]{z}} \, dz$

$$= \int_{1}^{64} \left(\frac{z^{1/2}}{z^{1/3}} - 2z^{-1/3}\right) dz$$

$$= \int_{1}^{64} z^{1/6} \, dz - 2\int_{1}^{64} z^{-1/3} \, dz$$

$$= \frac{z^{7/6}}{7/6}\Big|_{1}^{64} - 2\frac{z^{2/3}}{2/3}\Big|_{1}^{64}$$

$$= \frac{6z^{7/6}}{7}\Big|_{1}^{64} - 3z^{2/3}\Big|_{1}^{64}$$

$$= \frac{6(64)^{7/6}}{7} - \frac{6(1)^{7/6}}{7}$$

$$\quad - 3(64^{2/3} - 1^{2/3})$$

$$= \frac{6(128)}{7} - \frac{6}{7} - 3(16 - 1)$$

$$= \frac{768 - 6 - 315}{7}$$

$$= \frac{447}{7} \approx 63.857$$

25. $\displaystyle\int_{1}^{2} \frac{\ln x}{x} \, dx$

Let $u = \ln x$, so that

$$du = \frac{1}{x} \, dx.$$

When $x = 1$, $u = \ln 1 = 0$.
When $x = 2$, $u = \ln 2$.

$$\int_{0}^{\ln 2} u \, du = \frac{u^2}{2}\Big|_{0}^{\ln 2}$$

$$= \frac{(\ln 2)^2}{2} - 0$$

$$\approx .24023$$

27. $\displaystyle\int_{0}^{8} x^{1/3} \sqrt{x^{4/3} + 9} \, dx$

Let $u = x^{4/3} + 9$, so that

$$du = \frac{4}{3}x^{1/3} \, dx \text{ and } \frac{3}{4} \, du = x^{1/3} \, dx.$$

When $x = 0$, $u = 0^{4/3} + 9 = 9$.
When $x = 8$, $u = 8^{4/3} + 9 = 25$.

$$\frac{3}{4}\int_{9}^{25} \sqrt{u} \, du = \frac{3}{4}\int_{9}^{25} u^{1/2} \, du$$

$$= \frac{3}{4} \cdot \frac{u^{3/2}}{3/2}\Big|_{9}^{25}$$

$$= \frac{1}{2}u^{3/2}\Big|_{9}^{25}$$

$$= \frac{1}{2}(25)^{3/2} - \frac{1}{2}(9)^{3/2}$$

$$= \frac{125}{2} - \frac{27}{2}$$

$$= 49$$

29. $\displaystyle\int_0^1 \frac{e^t}{(3 + e^t)^2}\, dt$

Let $u = 3 + e^t$, so that

$du = e^t\, dt.$

When $t = 0$, $u = 3 + e^0 = 4$.

When $t = 1$, $u = 3 + e$.

$\displaystyle\int_4^{3+e} \frac{1}{u^2}\, du = \int_4^{3+e} u^{-2}\, du$

$\displaystyle = \frac{u^{-1}}{-1}\Big|_4^{3+e}$

$\displaystyle = \frac{-1}{u}\Big|_4^{3+e}$

$\displaystyle = -\frac{1}{3 + e} + \frac{1}{4}$

$\approx .075122$

31. $\displaystyle\int_1^{49} \frac{(1 + \sqrt{x})^{4/3}}{\sqrt{x}}\, dx$

Let $u = 1 + \sqrt{x}$, so that

$du = \frac{1}{2}x^{-1/2}\, dx$ and $2\, du = \frac{1}{\sqrt{x}}\, dx$.

When $x = 1$, $u = 1 + \sqrt{1} = 2$.

When $x = 49$, $u = 1 + \sqrt{49} = 8$.

$\displaystyle 2\int_2^8 u^{4/3}\, du = 2 \cdot \frac{u^{7/3}}{7/3}\Big|_2^8$

$\displaystyle = \frac{6}{7}u^{7/3}\Big|_2^8$

$\displaystyle = \frac{6}{7}(8)^{7/3} - \frac{6}{7}(2)^{7/3}$

$\displaystyle = \frac{6}{7}(128 - 2^{7/3})$

≈ 105.39

33. $f(x) = 2x + 3;\ [8, 10]$

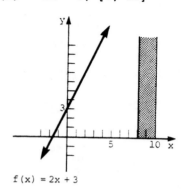

$f(x) = 2x + 3$

Graph does not cross x-axis in given interval [8, 10].

$\displaystyle\int_8^{10} (2x + 3)\, dx$

$\displaystyle = \left(\frac{2x^2}{2} + 3x\right)\Big|_8^{10}$

$= (10^2 + 30) - (8^2 + 24)$

$= 42$

35. $f(x) = 2 - 2x^2;\ [0, 5]$

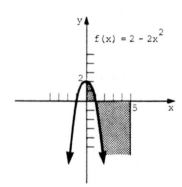

$f(x) = 2 - 2x^2$

Find the points where the graph crosses the x-axis by solving

$2 - 2x^2 = 0$

$2x^2 = 2$

$x^2 = 1$

$x = \pm 1.$

The only solution in the interval [0, 5] is 1.

The total area is

$$\int_0^1 (2 - 2x^2)\,dx + \left|\int_1^5 (2 - 2x^2)\,dx\right|$$

$$= \left(2x - \frac{2x^3}{3}\right)\Big|_0^1 + \left|\left(2x - \frac{2x^3}{3}\right)\Big|_1^5\right|$$

$$= 2 - \frac{2}{3} + \left|10 - \frac{2(5^3)}{3} - 2 + \frac{2}{3}\right|$$

$$= \frac{4}{3} + \left|\frac{-224}{3}\right|$$

$$= \frac{228}{3}$$

$$= 76.$$

37. $f(x) = x^2 + 4x - 5$; $[-6, 3]$

$f(x) = (x + 5)(x - 1)$

$f(x) = 0$ when $x = -5$ or $x = 1$ which means the graph crosses the x-axis at $x = -5$ and $x = 1$.

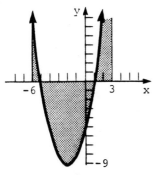

$$f(x) = x^2 + 4x - 5$$

The total area is

$$\int_{-6}^{-5} (x^2 + 4x - 5)\,dx + \left|\int_{-5}^1 (x^2 + 4x - 5)\,dx\right|$$

$$+ \int_1^3 (x^2 + 4x - 5)\,dx.$$

$$= \frac{x^3}{3} + 2x^2 - 5x\Big|_{-6}^{-5} + \left|\frac{x^3}{3} + 2x^2 - 5x\Big|_{-5}^{1}\right|$$

$$+ \frac{x^3}{3} + 2x^2 - 5x\Big|_1^3$$

$$= \left(\frac{100}{3} - 30\right) + \left|\left(-\frac{8}{3} - \frac{100}{3}\right| + \left(12 + \frac{8}{3}\right)\right.$$

$$= \frac{10}{3} + 36 + \frac{44}{3}$$

$$= 54$$

39. $f(x) = x^3$; $[-1, 3]$

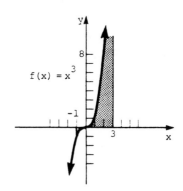

$$f(x) = x^3$$

The solution

$$x^3 = 0$$

$$x = 0$$

indicates that the graph crosses the x-axis at 0 in the given interval $[-1, 3]$.

The total area is

$$\left|\int_{-1}^0 x^3\,dx\right| + \int_0^3 x^3\,dx$$

$$= \left|\frac{x^4}{4}\Big|_{-1}^0\right| + \frac{x^4}{4}\Big|_0^3$$

$$= \left|\left(0 - \frac{1}{4}\right)\right| + \left(\frac{3^4}{4} - 0\right)$$

$$= \frac{1}{4} + \frac{81}{4} = \frac{82}{4}$$

$$= \frac{41}{2}$$

41. $f(x) = e^x - 1$; $[-1, 2]$

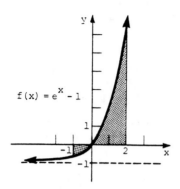

Solve

$$e^x - 1 = 0.$$
$$e^x = 1$$
$$x \ln e = \ln 1$$
$$x = 0$$

The graph crosses the x-axis at 0 in the given interval $[-1, 2]$. The total area is

$$\left| \int_{-1}^{0} (e^x - 1)dx \right| + \int_{0}^{2} (e^x - 1)dx$$

$$= \left| (e^x - x) \Big|_{-1}^{0} \right| + (e^x - x) \Big|_{0}^{2}$$

$$= |(1 - 0) - (e^{-1} + 1)|$$
$$+ (e^2 - 2) - (1 - 0)$$

$$= |1 - e^{-1} - 1| + e^2 - 2 - 1$$

$$= \frac{1}{e} + e^2 - 3$$

$$\approx 4.757$$

43. $f(x) = \frac{1}{x}$; $[1, e]$

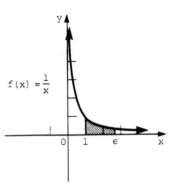

$\frac{1}{x} = 0$ has no solution so the graph does not cross the x-axis in the given interval $[1, e]$.

$$\int_{1}^{e} \frac{1}{x} dx = \ln x \Big|_{1}^{e}$$

$$= \ln e - \ln 1$$
$$= 1$$

45. $y = 4 - x^2$; $[0, 3]$

From the graph, we see that the total area is

$$\int_{0}^{2} (4 - x^2)dx + \left| \int_{2}^{3} (4 - x^2)dx \right|$$

$$= \left(4x - \frac{x^3}{3} \right) \Big|_{0}^{2} + \left| \left(4x - \frac{x^3}{3} \right) \Big|_{2}^{3} \right|$$

$$= \left[\left(8 - \frac{8}{3} \right) - 0 \right]$$

$$+ \left| \left[(12 - 9) - \left(8 - \frac{8}{3} \right) \right] \right|$$

$$= \frac{16}{3} + \left| 3 - \frac{16}{3} \right|$$

$$= \frac{16}{3} + \frac{7}{3}$$

$$= \frac{23}{3}$$

47. $y = e^2 - e^x$; [1, 3]

From the graph, we see that the total area is

$$\int_1^2 (e^2 - e^x)\,dx + \left|\int_2^3 (e^2 - e^x)\,dx\right|$$

$$= (e^2 x - e^x)\Big|_1^2 + \left|(e^2 x - e^x)\Big|_2^3\right|$$

$$= (2e^2 - e^2) - (e^2 - e)$$
$$\quad + \left|(3e^2 - e^3) - (2e^2 - e^2)\right|$$

$$= e^2 - e^2 + e$$
$$\quad + \left|3e^2 - e^3 - 2e^2 + e^2\right|$$

$$= e + \left|2e^2 - e^3\right|$$

$$= e + e^3 - 2e^2 \quad e^3 > 2e^2$$

$$\approx 8.026$$

49. Let $F'(x) = f(x)$, and hence $F(x)$ is an antiderivative of $f(x)$.

$$\int_a^a f(x)\,dx = F(x)\Big|_a^a$$

$$= F(a) - F(a)$$

$$= 0$$

51. Let $F'(x) = f(x)$, and hence $F(x)$ is an antiderivative of $f(x)$.

$$\int_a^c f(x)\,dx + \int_c^b f(x)\,dx$$

$$= F(x)\Big|_a^c + F(x)\Big|_c^b$$

$$= [F(c) - F(a)] + [F(b) - F(a)]$$

$$= -F(a) + F(b)$$

$$= F(b) - F(a)$$

$$= f(x)\Big|_a^b$$

$$= \int_a^b f(x)\,dx$$

Therefore,

$$\int_a^b f(x)\,dx = \int_a^c f(x)\,dx + \int_c^b f(x)\,dx.$$

53. $E'(x) = 4x + 2$ is the rate of expenditure per day

(a) The total expenditure in hundreds of dollars in 10 days is

$$\int_0^{10} (4x + 2)\,dx$$

$$= \left(\frac{4x^2}{2} + 2x\right)\Big|_0^{10}$$

$$= 2(100) + 20 - 0$$

$$= 220.$$

Therefore, since 220(100) = 22,000, the total expenditure is $22,000.

(b) From the tenth to the twentieth day:

$$\int_{10}^{20} (4x + 2)\,dx$$

$$= \left(\frac{4x^2}{2} + 2x\right)\Big|_{10}^{20}$$

$$= [2(400) + 40] - [2(100) + 20]$$

$$= 620$$

That is, $62,000 is spent.

(c) Let a = the number of days on the job.

The total expenditure in a days is

$$\int_0^a (4x + 2)dx$$

$$= (2x^2 + 2x) \Big|_0^a$$

$$= 2a^2 + 2a.$$

If no more than $5000, or 50(100) is spent, $50 \geq 2a^2 + 2a$.

Solve $50 = 2a^2 + 2a$ by the quadratic formula.

$$2a^2 + 2a - 50 = 0$$

$$a^2 + a - 25 = 0$$

$$a = \frac{-1 \pm \sqrt{1 - 4(-25)}}{2}$$

$$a = \frac{-1 + \sqrt{101}}{2} \quad \text{or} \quad a = \frac{-1 - \sqrt{101}}{2}$$

$$\approx 4.5 \qquad\qquad = -5.5$$

To spend less than $5000, the company must complete the job within 4.5 days.

55. $P'(t) = (3t + 3)(t^2 + 2t + 2)^{1/3}$

(a) $\int_0^3 3(t + 1)(t^2 + 2t + 2)^{1/3} \ dt$

Let $u = t^2 + 2t + 2$, so that

$du = (2t + 2)dt$ and $\frac{1}{2}du = (t + 1)dt$.

When $t = 0$, $u = 0^2 + 2 \cdot 0 + 2 = 2$.
When $t = 3$, $u = 3^2 + 2 \cdot 3 + 2 = 17$.

$$\frac{3}{2}\int_2^{17} u^{1/3} \ du = \frac{3}{2} \cdot \frac{u^{4/3}}{4/3} \Big|_2^{17}$$

$$= \frac{9}{8}u^{4/3} \Big|_2^{17}$$

$$= \frac{9}{8}(17)^{4/3} - \frac{9}{8}(2)^{4/3}$$

$$\approx 46.341$$

Total profits for the first 3 yr were

$$\frac{9000}{8}(17^{4/3} - 2^{4/3}) \approx \$46,341.$$

(b) $\int_3^4 3(t + 1)(t^2 + 2t + 2)^{1/3} \ dt$

Let $u = t^2 + 2t + 2$, so that

$du = (2t + 2)dt = 2(t + 1)dt$ and $\frac{3}{2} du = 3(t + 1)dt$.

When $t = 3$, $u = 3^2 + 2 \cdot 3 + 2 = 17$.
When $t = 4$, $u = 4^2 + 2 \cdot 4 + 2 = 26$.

$$\frac{3}{2}\int_{17}^{26} u^{1/3} \ du = \frac{9}{8}u^{4/3} \Big|_{17}^{26}$$

$$= \frac{9}{8}(26)^{4/3} - \frac{9}{8}(17)^{4/3}$$

$$\approx 37.477$$

Profit in the fourth year was

$$\frac{9000}{8}(26^{4/3} - 17^{4/3}) \approx \$37,477.$$

(c) $\lim_{t \to \infty} P'(t)$

$$= \lim_{t \to \infty} (3t + 3)(t^2 + 2t + 2)^{1/3}$$

$$= \infty$$

The annual profits increase slowly without bound.

57. $P'(t) = 140t^{5/2}$

The total concentration of polu-
tants in 4 yr is

$$\int_0^4 140t^{5/2}\,dt = \frac{140t^{7/2}}{7/2}\Big|_0^4$$

$$= 40t^{7/2}\Big|_0^4$$

$$= 40(4^{7/2}) - 0$$

$$= 5120.$$

Since 5120 > 4850, the factory cannot operate for 4 yr without killing
all the fish.

59. Growth rate is $.2 + 4t^{-4}$ ft/yr.

(a) Total growth in the second year
is

$$\int_1^2 (.2 + 4t^{-4})\,dt$$

$$= \left(.2t + \frac{4t^{-3}}{-3}\right)\Big|_1^2$$

$$= \left[.2(2) - \frac{4}{3}(2)^{-3}\right] - \left[.2 - \frac{4}{3}\right]$$

$$= .2 - \frac{1}{6} + \frac{4}{3}$$

$$\approx 1.37 \text{ ft.}$$

(b) Total growth in the third year
is

$$\int_2^3 (.2 + 4t^{-4})\,dt$$

$$= \left(.2t - \frac{4}{3}t^{-3}\right)\Big|_2^3$$

$$= \left[.2(3) - \frac{4}{81}\right] - \left[.2(2) - \frac{1}{6}\right]$$

$$= .2 - \frac{4}{81} + \frac{1}{6} \approx .32 \text{ ft.}$$

61. $R'(t) = \frac{5}{t} + \frac{2}{t^2}$

(a) Total reaction from t = 1 to
t = 12 is

$$\int_1^{12} \left(\frac{5}{t} + 2t^{-2}\right)dt$$

$$= (5 \ln t - 2t^{-1})\Big|_1^{12}$$

$$= \left(5 \ln 12 - \frac{1}{6}\right) - (5 \ln 1 - 2)$$

$$\approx 14.26.$$

(b) Total reaction from t = 12 to
t = 24 is

$$\int_{12}^{24} \left(\frac{5}{t} + 2t^{-2}\right)dt$$

$$= \left(5 \ln t - \frac{2}{t}\right)\Big|_{12}^{24}$$

$$= \left(5 \ln 24 - \frac{1}{12}\right) - \left(5 \ln 12 - \frac{1}{6}\right)$$

$$\approx 3.55.$$

63. $c'(t) = ke^{rt}$

(a) $c(t) = 1.2e^{.04t}$

(b) $\int_0^{10} 1.2e^{.04t}\,dt$

(c) $= \frac{1.2e^{.04t}}{.04}\Big|_0^{10}$

$$= 30e^{.04t}\Big|_0^{10}$$

$$= 30e^{.4} - 30$$

$$\approx 14.75 \text{ billion}$$

(d) $\int_0^T 1.2e^{.04t}\ dt = 30e^{.04t}\Big|_0^T$

$$= 30e^{.04T} - 30$$

Solve

$$20 = 30e^{.04T} - 30.$$

$$50 = 30e^{.04T}$$

$$\frac{5}{3} = e^{.04T}$$

$$\ln\frac{5}{3} = .04T \ln e$$

$$T = \frac{\ln\frac{5}{3}}{.04}$$

$$\approx 12.8 \text{ yr}$$

(e) $\int_0^T 1/2e^{.02t}\ dt = 60e^{.02t}\Big|_0^T$

$$= 60e^{.02T} - 60$$

Solve

$$20 = 60e^{.02T} - 60.$$

$$80 = 60e^{.02T}$$

$$\frac{4}{3} = e^{.02T}$$

$$\ln\frac{4}{3} = .02T \ln e$$

$$T = \frac{\ln\frac{4}{3}}{.02}$$

$$\approx 14.4 \text{ yr}$$

65. $C'(t) = 72e^{.014t}$

$$\int_0^T 72e^{.014t}\ dt = \frac{72e^{.014t}}{.014}\Big|_0^T$$

$$= \frac{72e^{.014T}}{.014} - \frac{72e^0}{.014}$$

$$= \frac{72}{.014}(e^{.014T} - 1)$$

$$\approx 5142.9(e^{.014T} - 1)$$

Section 6.5

1. $x = -2$, $x = 1$, $y = x^2 + 4$, $y = 0$

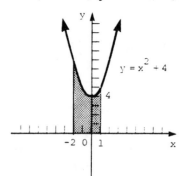

$$\int_{-2}^1 [(x^2 + 4) - 0]dx$$

$$= \left(\frac{x^3}{3} + 4x\right)\Big|_{-2}^1$$

$$= \left(\frac{1}{3} + 4\right) - \left(-\frac{8}{3} - 8\right)$$

$$= 3 + 12$$

$$= 15$$

3. $x = -3$, $x = 1$, $y = x + 1$, $y = 0$

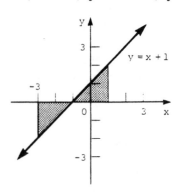

The region is composed of two separate regions because $y = x + 1$ intersects $y = 0$ at $x = -1$.
Let $f(x) = x + 1$, $g(x) = 0$.
In interval $[-3, -1]$, $g(x) \geq f(x)$.
In interval $[-1, 1]$, $f(x) \geq g(x)$.

$$\int_{-3}^{-1} [0 - (x + 1)]dx + \int_{-1}^{1} [(x + 1) - 0]dx$$

$$= \left(\frac{-x^2}{2} - x\right)\Big|_{-3}^{-1} + \left(\frac{x^2}{2} + x\right)\Big|_{-1}^{1}$$

$$= \left(-\frac{1}{2} + 1\right) - \left(-\frac{9}{2} + 3\right)$$

$$+ \left(\frac{1}{2} + 1\right) - \left(\frac{1}{2} - 1\right)$$

$$= 4$$

$$= \left(\frac{x^3}{3} - 3x - x^2\right)\Big|_{-2}^{-1} + \left(x^2 - \frac{x^3}{3} + 3x\right)\Big|_{-1}^{1}$$

$$= -\frac{1}{3} + 3 - 1 - \left(-\frac{8}{3} + 6 - 4\right) + 1 - \frac{1}{3} + 3$$

$$- \left(1 + \frac{1}{3} - 3\right)$$

$$= \frac{5}{3} + 6 = \frac{23}{3}$$

5. x = -2, x = 1, y = 2x, y = x² - 3

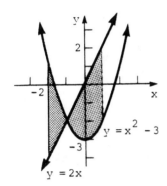

Find the intersections of y = 2x and
y = x² - 3 by substituting for y.

$$2x = x^2 - 3$$
$$0 = x^2 - 2x - 3$$
$$0 = (x - 3)(x + 1)$$

The only intersection in [-2, 1] is
at x = -1.
In interval [-2, -1], (x² - 3) ≥ 2x.
In interval [-1, 1], 2x ≥ (x² - 3).

$$\int_{-2}^{-1} [(x^2 - 3) - (2x)]dx$$

$$+ \int_{-1}^{1} [(2x) - (x^2 - 3)]dx$$

$$= \int_{-2}^{-1} (x^2 - 3 - 2x)dx + \int_{-1}^{1} (2x - x^2 + 3)dx$$

7. y = x² - 30
y = 10 - 3x

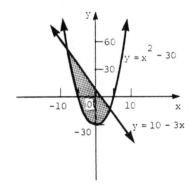

Find points of intersection:

$$x^2 - 30 = 10 - 3x$$
$$x^2 + 3x - 40 = 0$$
$$(x + 8)(x - 5) = 0$$
$$x = -8 \quad \text{or} \quad x = 5$$

Let f(x) = 10 - 3x and g(x) = x² - 30.
The area between the curves is given
by

$$\int_{-8}^{5} [f(x) - g(x)]dx$$

$$= \int_{-8}^{5} [(10 - 3x) - (x^2 - 30)]dx$$

$$= \int_{-8}^{5} (-x^2 - 3x + 40)dx$$

$$= \left(\frac{-x^3}{3} - \frac{3x^2}{2} + 40x\right)\Big|_{-8}^{5}$$

$$= \frac{-5^3}{3} - \frac{3(5)^2}{2} + 40(5)$$

$$- \left[\frac{-(-8)^3}{3} - \frac{3(-8)^2}{2} + 40(-8)\right]$$

$$= \frac{-125}{3} - \frac{75}{2} + 200 - \frac{512}{3}$$

$$+ \frac{192}{2} + 320$$

$$\approx 366.1667.$$

9. $y = x^2$, $y = 2x$

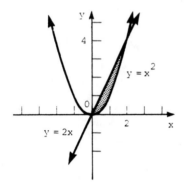

Find points of intersection:

$$x^2 = 2x$$
$$x^2 - 2x = 0$$
$$x(x - 2) = 0$$
$$x = 0 \quad \text{or} \quad x = 2$$

Let $f(x) = 2x$ and $g(x) = x^2$.
The area between the curves is given by

$$\int_0^2 [f(x) - g(x)]dx$$

$$= \int_0^2 (2x - x^2)dx$$

$$= \left(\frac{2x^2}{2} - \frac{x^3}{3}\right)\Big|_0^2$$

$$= 4 - \frac{8}{3} = \frac{4}{3}.$$

11. $x = 1$, $x = 6$, $y = \frac{1}{x}$, $y = -1$

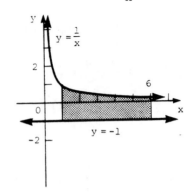

Find the points of intersection.

$$-1 = \frac{1}{x}$$

$$x = -1$$

The curves do not intersect in the interval $[1, 6]$.

If $f(x) = \frac{1}{x}$ and $g(x) = -1$, the area between the curves is

$$\int_1^6 [f(x) - g(x)]dx$$

$$= \int_1^6 \left[\frac{1}{x} - (-1)\right]dx$$

$$= \int_1^6 \left(\frac{1}{x} + 1\right)dx$$

$$= (\ln|x| + x)\Big|_1^6$$

$$= \ln 6 + 6 - \ln 1 - 1$$
$$= 5 + \ln 6$$
$$\approx 6.792.$$

13. $x = -1$, $x = 1$, $y = e^x$, $y = 3 - e^x$.

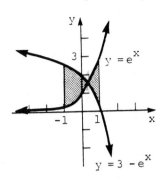

To find the point of intersection, set $e^x = 3 - e^x$ and solve for x.

$$e^x = 3 - e^x$$
$$2e^x = 3$$
$$e^x = \frac{3}{2}$$
$$\ln e^x = \ln \frac{3}{2}$$
$$x \ln e = \ln \frac{3}{2}$$
$$x = \ln \frac{3}{2}$$

The area of the region between the curves from $x = -1$ to $x = 1$ is

$$\int_{-1}^{\ln 3/2} [(3 - e^x) - e^x]dx$$

$$+ \int_{\ln 3/2}^{1} [e^x - (3 - e^x)]dx$$

$$= \int_{-1}^{\ln 3/2} (3 - 2e^x)dx + \int_{\ln 3/2}^{1} (2e^x - 3)dx$$

$$= (3x - 2e^x)\Big|_{-1}^{\ln 3/2} + (2e^x - 3x)\Big|_{\ln 3/2}^{1}$$

$$= \left[\left(3 \ln \frac{3}{2} - 2e^{\ln 3/2}\right) - (3(-1) - 2e^{-1})\right]$$
$$+ \left[2e^1 - 3(1) - \left(2e^{\ln 3/2} - 3 \ln \frac{3}{2}\right)\right]$$

$$= \left[(3 \ln \frac{3}{2} - 3) - (-3 - \frac{2}{e})\right]$$

$$+ \left[2e - 3 - (3 - 3 \ln \frac{3}{2})\right]$$

$$= 6 \ln \frac{3}{2} + \frac{2}{e} + 2e - 6$$

$$\approx 2.6051.$$

15. $x = 1$, $x = 2$, $y = e^x$, $y = \frac{1}{x}$

Sketch the graph of $y = e^x$ and $y = \frac{1}{x}$.

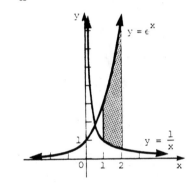

Let $f(x) = e^x$ and $g(x) = \frac{1}{x}$. For all x in [1, 2] $f(x) \geq g(x)$. The area between the curves is

$$\int_{1}^{2} [f(x) - g(x)]dx$$

$$= \int_{1}^{2} \left(e^x - \frac{1}{x}\right)dx$$

$$= (e^x - \ln |x|)\Big|_{1}^{2}$$

$$= e^2 - \ln 2 - e + \ln 1$$
$$= e^2 - e - \ln 2$$
$$\approx 3.978.$$

17. $y = x^3 - x^2 + x + 1$, $y = 2x^2 - x + 1$

Find the points of intersection.

$$x^3 - x^2 + x + 1 = 2x^2 - x + 1$$
$$x^3 - 3x^2 + 2x = 0$$
$$x(x^2 - 3x + 2) = 0$$
$$x(x - 2)(x - 1) = 0$$

So the points of intersection are at
$x = 0$, $x = 1$, and $x = 2$.

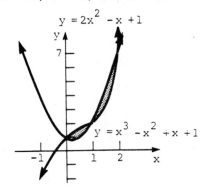

Area between the curves is

$$\int_0^1 [(x^3 - x^2 + x + 1) - (2x^2 - x + 1)]\,dx$$

$$+ \int_1^2 [(2x^2 - x + 1) - (x^3 - x^2 + x + 1)]\,dx$$

$$= \int_0^1 (x^3 - 3x^2 + 2x)\,dx + \int_1^2 (-x^3 + 3x^2 - 2x)\,dx$$

$$= \left(\frac{x^4}{4} - x^3 + x^2\right)\Big|_0^1 + \left(\frac{-x^4}{4} + x^3 - x^2\right)\Big|_1^2$$

$$= \left[\left(\frac{1}{4} - 1 + 1\right) - (0)\right]$$

$$\quad + \left[(-4 + 8 - 4) - \left(-\frac{1}{4} + 1 - 1\right)\right]$$

$$= \frac{1}{4} + \frac{1}{4}$$

$$= \frac{1}{2}.$$

19. $y = x^4 + \ln (x + 10)$,
$y = x^3 + \ln (x + 10)$

Find the points of intersection.

$$x^4 + \ln (x + 10) = x^3 + \ln (x + 10)$$
$$x^4 - x^3 = 0$$
$$x^3(x - 1) = 0$$
$$x = 0 \quad \text{or} \quad x = 1$$

The points of intersection are at
$x = 0$ and $x = 1$.
The area between the curves is

$$\int_0^1 [(x^3 + \ln (x + 10)) - (x^4 + \ln (x + 10))]\,dx$$

$$= \int_0^1 (x^3 - x^4)\,dx$$

$$= \left(\frac{x^4}{4} - \frac{x^5}{5}\right)\Big|_0^1$$

$$= \left(\frac{1}{4} - \frac{1}{5}\right) - (0)$$

$$= \frac{1}{20}.$$

21. $y = x^{4/3}$, $y = 2x^{1/3}$

Find the points of intersection.

$$x^{4/3} = 2x^{1/3}$$
$$x^{4/3} - 2x^{1/3} = 0$$
$$x^{1/3}(x - 2) = 0$$
$$x = 0 \quad \text{or} \quad x = 2$$

The points of intersection are at
$x = 0$ and $x = 2$.

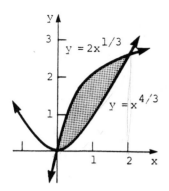

The area between the curves is

$$\int_0^2 (2x^{1/3} - x^{4/3})\,dx$$

$$= 2\frac{x^{4/3}}{4/3} - \frac{x^{7/3}}{7/3}\Big|_0^2$$

$$= \frac{3}{2}x^{4/3} - \frac{3}{7}x^{7/3}\Big|_0^2$$

$$= \left[\frac{3}{2}(2)^{4/3} - \frac{3}{7}(2)^{7/3}\right] - 0$$

$$= \frac{3(2^{4/3})}{2} - \frac{3(2^{7/3})}{7}$$

$$\approx 1.6199.$$

23. **(a)** It is profitable to use the machine until $S(x) = C(x)$.

$$150 - x^2 = x^2 + \frac{11}{4}x$$

$$2x^2 + \frac{11}{4}x - 150 = 0$$

$$8x^2 + 11x - 600 = 0$$

$$x = \frac{-11 \pm \sqrt{121 - 4(8)(-600)}}{16}$$

$$= \frac{-11 \pm 139}{16}\quad \text{\textit{Discard negative}}$$
$$\text{\textit{solution}}$$

$$= 8 \text{ yr}$$

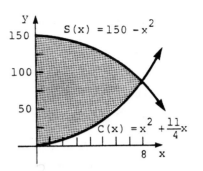

(b) Since $150 - x^2 > x^2 + \frac{11}{4}x$, in the interval $[0, 8]$, the net total savings are in the first year.

$$\int_0^1 [(150 - x^2) - (x^2 + \frac{11}{4}x)]\,dx$$

$$= \int_0^1 (-2x^2 - \frac{11}{4}x + 150)\,dx$$

$$= \left(\frac{-2x^3}{3} - \frac{11x^2}{8} + 150x\right)\Big|_0^1$$

$$= -\frac{2}{3} - \frac{11}{8} + 150$$

$$\approx \$148.$$

(c) The net total savings over the entire period of use are

$$\int_0^8 [(150 - x^2) - (x^2 + \frac{11}{4}x)]\,dx$$

$$= \left(\frac{-2x^3}{3} - \frac{11x^2}{8} + 150x\right)\Big|_0^8$$

$$= \frac{-2(8^3)}{3} - \frac{11(8^2)}{8} + 150(8)$$

$$= \frac{-1024}{3} - \frac{704}{8} + 1200$$

$$\approx \$771.$$

25. (a) $E(x) = e^{.1x}$ and $I(x) = 98.8 - e^{.1x}$

To find the point of intersection, where profit will be maximized, set the functions equal to each other and solve for x.

$$e^{.1x} = 98.8 - e^{.1x}$$
$$2e^{.1x} = 98.8$$
$$e^{.1x} = 49.4$$
$$.1x = \ln 49.4$$
$$x = \frac{\ln 49.4}{.1}$$
$$x \approx 39 \text{ days}$$

(b) The total income for 39 days is

$$\int_0^{39} (98.8 - e^{.1x})dx$$

$$= \left(98.8x - \frac{e^{.1x}}{.1}\right)\Big|_0^{39}$$

$$= (98.8x - 10e^{.1x})\Big|_0^{39}$$

$$= [98.8(39) - 10e^{3.9}] - (0 - 10)$$

$$= \$3369.18$$

(c) The total expenditures for 39 days is

$$\int_0^{39} e^{.1x} \, dx = \frac{e^{.1x}}{.1}\Big|_0^{39}$$

$$= 10e^{.1x}\Big|_0^{39}$$

$$= 10e^{3.9} - 10$$

$$= \$484.02$$

(d) Profit = Income - Expense
$$= 3369.18 - 484.02$$
$$= \$2885.16$$

27. $S(q) = q^{5/2} + 2q^{3/2} + 50$, $q = 16$ is the equilibrium quantity.

$$\text{Producers' surplus} = \int_0^{q_0} [p_0 - S(q)]dq$$

where p_0 is the equilibrium price and q_0 is equilibrium supply.

$$p_0 = S(16) = (16)^{5/2} + 2(16)^{3/2} + 50$$
$$= 1202$$

Therefore, the producers' surplus is

$$\int_0^{16} [1202 - (q^{5/2} + 2q^{3/2} + 50]dq$$

$$= \int_0^{16} (1152 - q^{5/2} - 2q^{3/2})dq$$

$$= \left(1152q - \frac{2}{7}q^{7/2} - \frac{4}{5}q^{5/2}\right)\Big|_0^{16}$$

$$= 1152(16) - \frac{2}{7}(16)^{7/2} - \frac{4}{5}(16)^{5/2}$$

$$= 18,432 - \frac{32,768}{7} - \frac{4096}{5}$$

$$= \$12,931.66.$$

29. $D(q) = \frac{100}{(3q + 1)^2}$, $q = 3$ is the equilibrium quantity.

$$\text{Consumers' surplus} = \int_0^{q_0} [D(q) - p_0]dq$$

$$p_0 = D(3) = 1$$

Therefore, the consumers' surplus is

$$\int_0^3 \left[\frac{100}{(3q + 1)^2} - 1\right]dq$$

$$\int_0^3 \frac{100}{(3q + 1)^2} \, dq - \int 1 \, dq.$$

Let u = 3q + 1 so that

du = 3 dq and $\frac{1}{3}$ du = dq.

$= \frac{1}{3} \int_{1}^{10} \frac{100}{u^2} \, du - \int_{0}^{3} 1 \, dq$

$= \frac{100}{3} \int_{1}^{10} u^{-2} \, du - \int_{0}^{3} 1 \, dq$

$= \frac{100}{3} \cdot \frac{u^{-1}}{-1} \Big|_{1}^{10} - q \Big|_{0}^{3}$

$= -\frac{100}{3u} \Big|_{1}^{10} - 3$

$= \frac{-100}{30} + \frac{100}{3} - 3$

$= 27$

31. $S(q) = q^2 + 10q$

$D(q) = 900 - 20q - q^2$

(a) The graphs of the supply and demand functions are parabolas with vertices at $(-5, -25)$ and $(-10, 1900)$, respectively. See the answer graph at the back of the textbook.

(b) The graphs intersect at the point where the y-coordinates are equal.

$q^2 + 10q = 900 - 20q - q^2$

$2q^2 + 30q - 900 = 0$

$q^2 + 15q - 450 = 0$

$(q + 30)(q - 15) = 0$

$q = -30 \quad \text{or} \quad q = 15.$

Disregard the negative solution. The supply and demand functions are in equilibrium when $q = 15$.

$S(15) = 15^2 + 10(15) = 375$

The point is $(15, 375)$.

(c) The consumers' surplus is

$$\int_{0}^{q_0} [(D(q) - p_0)] \, dq$$

$p_0 = D(15) = 375$

$$\int_{0}^{15} [(900 - 20q - q^2) - 375] \, dq$$

$$\int_{0}^{15} (525 - 20q - q^2) \, dq$$

$= \left(525q - 10q^2 - \frac{1}{3}q^3\right) \Big|_{0}^{15}$

$= \left[525(15) - 10(15)^2 - \frac{1}{3}(15)^3\right] - 0$

$= 4500.$

(d) The producers' surplus is

$$\int_{0}^{q_0} [p_0 - S(q)] \, dq$$

$p_0 = S(15) = 375$

$$\int_{0}^{15} [375 - (q^2 + 10q)] \, dq$$

$$\int_{0}^{15} (375 - q^2 - 10q) \, dq$$

$= 375q - \frac{1}{3}q^3 - 5q^2 \Big|_{0}^{15}$

$= \left[375(15) - \frac{1}{3}(15)^3 - 5(15)^2\right] - 0$

$= 3375.$

33. $I(x) = .9x^2 + .1x$

(a) $I(.1) = .9(.1)^2 + .1(.1)$

$= .019$

The lower 10% of income producers earn 1.9% of total income of the population.

(b) $I(.4) = .9(.4)^2 + .1(.4) = .184$

The lower 40% of income producers earn 18.4% of income of the population.

(c) $I(.6) = .9(.6)^2 + .1(.6) = .384$

The lower 60% of income producers earn 38.4% of total income of the population.

(d) $I(.9) = .9(.9)^2 + .1(.9) = .819$

The lower 90% of income producers earn 81.9% of total income of the population.

(e) See the answer graph in your textbook.

(f) To find points of intersection, solve

$$x = .9x^2 + .1x$$
$$.9x^2 - .9x = 0$$
$$.9x(x - 1) = 0$$
$$x = 0 \quad \text{or} \quad x = 1$$

The area between the curves is given by

$$\int_0^1 [x - (.9x^2 + .1x)]\,dx$$

$$= \int_0^1 (.9x - .9x^2)\,dx$$

$$= \left(\frac{.9x^2}{2} - \frac{.9x^3}{3}\right)\Big|_0^1$$

$$= \frac{.9}{2} - \frac{.9}{3}$$

$$= .15.$$

For Exercises 35 and 37, use computer methods. The solutions will vary depending on the computer program used. The answers are given.

35. $y = \ln x$ and $y = xe^x$; $[1, 4]$

The area is 161.2.

37. $y = \sqrt{9 - x^2}$ and $y = \sqrt{x + 1}$; $[-1, 3]$

The area is 5.516.

Chapter 6   Review Exercises

5. $\displaystyle\int 6\,dx = 6x + C$

7. $\displaystyle\int (2x + 3)\,dx = \frac{2x^2}{2} + 3x + C$

$$= x^2 + 3x + C$$

9. $\displaystyle\int (x^2 - 3x + 2)\,dx$

$$= \frac{x^3}{3} - \frac{3x^2}{2} + 2x + C$$

11. $\displaystyle\int 3\sqrt{x}\,dx = 3\int x^{1/2}\,dx$

$$= \frac{3x^{3/2}}{3/2} + C$$

$$= 2x^{3/2} + C$$

13. $\displaystyle\int (x^{1/2} + 3x^{-2/3})\,dx$

$$= \frac{x^{3/2}}{3/2} + \frac{3x^{1/3}}{1/3} + C$$

$$= \frac{2x^{3/2}}{3} + 9x^{1/3} + C$$

15. $\displaystyle\int \frac{-4}{x^3}\, dx = \int -4x^{-3}\, dx$

$$= \frac{-4x^{-2}}{-2} + C$$

$$= 2x^{-2} + C$$

17. $\displaystyle\int -3e^{2x}\, dx = \frac{-3e^{2x}}{2} + C$

19. $\displaystyle\int \frac{2}{x-1}\, dx = 2 \int \frac{dx}{x-1}$

Let $u = x - 1$, so that

$du = dx$.

$$= 2 \int \frac{du}{u}$$

$$= 2 \ln |u| + C$$

$$= 2 \ln |x - 1| + C$$

21. $\displaystyle\int xe^{3x^2}\, dx = \frac{1}{6} \int 6xe^{3x^2}\, dx$

Let $u = 3x^2$, so that

$du = 6x\, dx$.

$$= \frac{1}{6} \int e^u\, du$$

$$= \frac{1}{6}e^u + C$$

$$= \frac{e^{3x^2}}{6} + C$$

23. $\displaystyle\int \frac{3x}{x^2-1}\, dx = 3\left(\frac{1}{2}\right) \int \frac{2x\, dx}{x^2-1}$

Let $u = x^2 - 1$, so that

$du = 2x\, dx$.

$$= \frac{3}{2} \int \frac{du}{u}$$

$$= \frac{3}{2} \ln |u| + C$$

$$= \frac{3 \ln |x^2 - 1|}{2} + C$$

25. $\displaystyle\int \frac{x^2\, dx}{(x^3+5)^4} = \frac{1}{3} \int \frac{3x^2\, dx}{(x^3+5)^4}$

Let $u = x^3 + 5$, so that

$du = 3x^2\, dx$.

$$= \frac{1}{3} \int \frac{du}{u^4}$$

$$= \frac{1}{3} \int u^{-4}\, du$$

$$= \frac{1}{3}\left(\frac{u^{-3}}{-3}\right) + C$$

$$= \frac{-(x^3+5)^{-3}}{6} + C$$

27. $\displaystyle\int \frac{4x-5}{2x^2-5x}\, dx$

Let $u = 2x^2 - 5x$, so that

$du = (4x - 5)dx$.

$$= \int \frac{du}{u}$$

$$= \ln |u| + C$$

$$= \ln |2x^2 - 5x| + C$$

29. $\displaystyle\int \frac{x^3}{e^{3x^4}}\, dx = \int x^3 e^{-3x^4}$

$$= -\frac{1}{12} \int -12x^3 e^{-3x^4}\, dx$$

Let $u = -3x^4$, so that

$du = -12x^3\, dx$.

$$= -\frac{1}{12} \int e^u\, du$$

$$= -\frac{1}{12}e^u + C$$

$$= \frac{-e^{-3x^4}}{12} + C$$

31. $\int -2e^{-5x} \, dx = \frac{-2}{-5} \int -5e^{-5x} \, dx$

Let $u = 5x$, so that

   $du = -5 \, dx$.

   $= \frac{2}{5} \int e^u \, du$

   $= \frac{2}{5}e^u + C$

   $= \frac{2e^{-5x}}{5} + C$

33. $\sum_{i=1}^{4} (i^2 - i)$

   $= (1^2 - 1) + (2^2 - 2) + (3^2 - 3)$

   $\quad + (4^2 - 4)$

   $= 0 + 2 + 6 + 12$

   $= 20$

35. $f(x) = 2x + 3$, from $x = 0$ to $x = 4$

   $\Delta x = \frac{4 - 0}{4} = 1$

i	x_i	$f(x_i)$
1	0	3
2	1	5
3	2	7
4	3	9

   $A = \sum_{i=1}^{4} f(x_i)\Delta x$

   $= 3(1) + 5(1) + 7(1) + 9(1)$

   $= 24$

39. $\int_{1}^{6} (2x^2 + x) \, dx$

   $= \left(\frac{2x^3}{3} + \frac{x^2}{2} \right) \Big|_{1}^{6}$

   $= \left[\frac{2(6)^3}{3} + \frac{(6)^2}{2} \right] - \left[\frac{2(1)^3}{3} + \frac{(1)^2}{2} \right]$

   $= 144 + 18 - \frac{2}{3} - \frac{1}{2}$

   $= 162 - \frac{2}{3} - \frac{1}{2}$

   $= \frac{965}{6}$

   ≈ 160.83

41. $\int_{2}^{3} (5x^{-2} + x^{-4}) \, dx$

   $= \left(\frac{5x^{-1}}{-1} + \frac{x^{-3}}{-3} \right) \Big|_{2}^{3}$

   $= \left(\frac{-5}{x} - \frac{1}{3x^3} \right) \Big|_{2}^{3}$

   $= \left(\frac{-5}{3} - \frac{1}{81} \right) - \left(\frac{-5}{2} - \frac{1}{24} \right)$

   $= \frac{-5}{3} - \frac{1}{81} + \frac{5}{2} + \frac{1}{24}$

   $= \frac{-1080 - 8 + 1620 + 27}{648}$

   $= \frac{559}{648}$

   $\approx .863$

43. $\int_{1}^{6} 8x^{-1} \, dx = \int_{1}^{6} \frac{8}{x} \, dx$

   $= 8(\ln x) \Big|_{1}^{6}$

   $= 8(\ln 6 - \ln 1)$

   $= 8 \ln 6$

   ≈ 14.334

45.
$$\int_1^6 \frac{5}{2}e^{4x}\,dx = \frac{1}{4} \cdot \frac{5}{2}\int_1^6 4e^{4x}\,dx$$

$$= \frac{5e^{4x}}{8}\bigg|_1^6$$

$$= \frac{5(e^{24} - e^4)}{8}$$

$$\approx 1.656 \times 10^{10}$$

47.
$$\int_0^1 x\sqrt{5x^2 + 4}\,dx$$

Let $u = 5x^2 + 4$, so that

$du = 10x\,dx$ and $\frac{1}{10}\,du = x\,dx$.

When $x = 0$, $u = 5(0^2) + 4 = 4$.

When $x = 1$, $u = 5(1^2) + 4 = 9$.

$$= \frac{1}{10}\int_4^9 \sqrt{u}\,du = \frac{1}{10}\int_4^9 u^{1/2}\,du$$

$$= \frac{1}{10} \cdot \frac{u^{3/2}}{3/2}\bigg|_4^9 = \frac{1}{15}u^{3/2}\bigg|_4^9$$

$$= \frac{1}{15}(9)^{3/2} - \frac{1}{15}(4)^{3/2}$$

$$= \frac{27}{15} - \frac{8}{15}$$

$$= \frac{19}{15}$$

49. $f(x) = x(x + 2)^6$; $[-2, 0]$

$$\text{Area} = \int_{-2}^0 [0 - x(x + 2)^6]\,dx$$

$$= -\int_{-2}^0 x(x + 2)^6\,dx$$

Let $u = x + 2$, so that

$du = dx$ and $x = u - 2$.

When $x = -2$, $u = 0$.

When $x = 0$, $u = 2$.

$$= -\int_0^2 (u - 2)u^6\,du$$

$$= -\int_0^2 u^7\,du + 2\int_0^2 u^6\,du$$

$$= -\frac{u^8}{8}\bigg|_0^2 + 2 \cdot \frac{u^7}{7}\bigg|_0^2$$

$$= -\frac{256}{8} + \frac{256}{7}$$

$$= -32 + \frac{256}{7}$$

$$= \frac{32}{7}$$

51. $f(x) = 1 + e^{-x}$; $[0, 4]$

$$\int_0^4 (1 + e^{-x})\,dx = (x - e^{-x})\bigg|_0^4$$

$$= (4 - e^{-4}) - (0 - e^0)$$

$$= 5 - e^{-4}$$

$$\approx 4.982$$

53. $f(x) = x^2 - 4x$, $g(x) = x - 6$

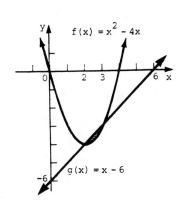

Points of intersection:

$$x^2 - 4x = x - 6$$
$$x^2 - 5x + 6 = 0$$
$$(x - 3)(x - 2) = 0$$
$$x = 2 \text{ or } x = 3$$

Since $g(x) \geq f(x)$ in the interval [2, 3], the area between the graphs is

$$\int_{2}^{3} [g(x) - f(x)]dx$$

$$= \int_{2}^{3} [(x - 6) - (x^2 - 4x)]dx$$

$$= \int_{2}^{3} (-x^2 + 5x - 6)dx$$

$$= \left(\frac{-x^3}{3} + \frac{5x^2}{2} - 6x\right)\Big|_{2}^{3}$$

$$= \frac{-27}{3} + \frac{5(9)}{2} - 6(3) - \frac{-8}{3}$$

$$\quad - \frac{5(4)}{2} + 6(2)$$

$$= -\frac{19}{3} + \frac{25}{2} - 6$$

$$= \frac{1}{6}$$

55. $f(x) = 5 - x^2$, $g(x) = x^2 - 3$, $x = 0$, $x = 4$

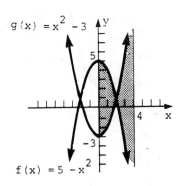

Find the points of intersection.

$$5 - x^2 = x^2 - 3$$
$$8 = 2x^2$$
$$4 = x^2$$
$$\pm 2 = x$$

So the curves intersect at $x = 2$, -2.

Thus, the area

$$= \int_{0}^{2} [(5 - x^2) - (x^2 - 3)]dx$$

$$\quad + \int_{2}^{4} [(x^2 - 3) - (5 - x^2)]dx$$

$$= \int_{0}^{2} (-2x^2 + 8)dx + \int_{2}^{4} (2x^2 - 8)dx$$

$$= \left(\frac{-2x^3}{3} + 8x\right)\Big|_{0}^{2} + \left(\frac{2x^3}{3} - 8x\right)\Big|_{2}^{4}$$

$$= \frac{-16}{3} + 16 + \left(\frac{128}{3} - 32\right) - \left(\frac{16}{3} - 16\right)$$

$$= \frac{32}{3} + \frac{128}{3} - 32 - \frac{16}{3} + 16$$

$$= 32.$$

57. $C'(x) = 10 - 2x$; fixed cost is $4

$$C(x) = \int(10 - 2x)dx$$

$$= 10x - \frac{2x^2}{2} + C$$

$$= 10x - x^2 + C$$

$$C(0) = 10(0) - 0^2 + C$$

Since $C(0) = 4$,

$$C = 4.$$

$$C(x) = 10x - x^2 + 4$$

59. $C(x) = \int 3(2x - 1)^{1/2} \, dx$

$= \dfrac{3}{2} \int 2(2x - 1)^{1/2} \, dx$

Let $u = 2x - 1$, so that

$du = 2 \, dx.$

$= \dfrac{3}{2} \int u^{1/2} \, du$

$= \dfrac{3}{2} \left(\dfrac{u^{3/2}}{3/2} \right) + C$

$= (2x - 1)^{3/2} + C$

$C(13) = [2(13) - 1]^{3/2} + C$

Since $C(13) = 270$,

$270 = 25^{3/2} + C$

$270 = 125 + C$

$C = 145.$

$C(x) = (2x - 1)^{3/2} + 145$

61. Read values for the rate of investment income accumulation for every 2 year from year 1 to yr 9. These are the heights of rectangles with width $\Delta x = 2$.

Total accumulated income

$= 11{,}000(2) + 9000(2) + 12{,}000(2)$

$+ 10{,}000(2) = 6000(2)$

$\approx \$96{,}000$

63. $S' = \sqrt{x} + 2$

$S(x) = \int_{0}^{9} (x^{1/2} + 2) \, dx$

$= \left(\dfrac{x^{3/2}}{3/2} + 2x \right) \Big|_{0}^{9}$

$= \dfrac{2}{3}(9)^{3/2} + 18$

$= 36$

Total sales are $36(1000)$, or $36{,}000$ units.

65. $S'(x) = 225 - x^2$, $C'(x) = x^2 + 25x + 150$

$S'(x) = C'(x)$

$225 - x^2 = x^2 + 25x + 150$

$2x^2 + 25x - 75 = 0$

$(2x - 5)(x + 15) = 0$

$x = \dfrac{5}{2} = 2.5 \text{ yr}$

$\displaystyle\int_{0}^{2.5} [225 - x^2) - (x^2 + 25x + 150)] \, dx$

$= \displaystyle\int_{0}^{2.5} (-2x^2 - 25x + 75) \, dx$

$= \left(\dfrac{-2x^3}{3} - \dfrac{25x^2}{2} + 75x \right) \Big|_{0}^{2.5}$

$= \dfrac{-2(2.5^3)}{3} - \dfrac{25(2.5^2)}{2} + 75(2.5)$

≈ 98.95833

Net savings are $98{,}958.33$, or about $\$99{,}000$.

67. The total number of infected people over the first four months is

$\displaystyle\int_{0}^{4} \dfrac{100t}{t^2 + 1} \, dt$, where t is time in

months.

Let $u = t^2 + 1$, so that

$du = 2t \, dt$ and $50 \, du = 100t \, dt$

$$= 50 \int_{1}^{17} \frac{1}{u} \, du = 50 \ln |u| \Big|_{1}^{17}$$

$$= 50 \ln 17 - 50 \ln |1|$$

$$= 50 \ln 17$$

$$\approx 141.66$$

Approximately 142 people are infected.

69. For each month, subtract the average temperature from 65° if it falls below 65° F, then multiply this number times the number of days in the month. The sum is the total number of heating degree days. Readings may vary, but the sum is approximately 4500 degree–days. (The actual value is 4868 degree–days.)

Extended Application

1. $2,000,000 \div 19,600 \approx 102$ yr

2.

$$2,000,000 = \frac{19,600}{.02}(e^{.02t_1} - 1)$$

$$\frac{2,000,000(.02)}{19,600} = e^{.02t_1} - 1$$

$$2.0408 + 1 = e^{.02t_1}$$

$$3.0408 = e^{.02t_1}$$

$$\ln 3.0408 = .02t_1$$

$$t_1 = \frac{\ln 3.0408}{.02}$$

$$t_1 \approx 55.6,$$

or about 55.6 yr

3.

$$15,000,000 = \frac{63,000}{.06}(e^{.06t_1} - 1)$$

$$\frac{15,000,000(.06)}{63,000} + 1 = e^{.06t_1}$$

$$15.286 \approx e^{.06t_1}$$

$$\ln 15.286 \approx .06t_1 \ln e$$

$$t_1 \approx \frac{\ln 15.286}{.06}$$

$$\approx 45.4$$

or about 45.4 yr

4.

$$2,000,000 = \frac{2200}{.04}(e^{.04t_1} - 1)$$

$$\frac{2,000,000(.04)}{2200} + 1 = e^{.04t_1}$$

$$37.36 = e^{.04t_1}$$

$$\ln 37.36 = .04t_1$$

$$t_1 \approx 90.5$$

or about 90 yr

CHAPTER 6   TEST

[6.1]   1.   Explain what an antiderivative of a function f(x) is.

[6.1]   **Find each indefinite integral.**

2. $\int (3x^3 - 5x^2 + x + 1)\,dx$ 3. $\int \left(\frac{5}{x} + e^{\cdot 5x}\right)dx$

4. $\int \dfrac{2x^3 - 3x^2}{\sqrt{x}}\,dx$

5.   Find the cost function if $C'(x) = 200 + 2x^{-1/4}$ and 16 units cost $4000.

6.   A ball is thrown upward at time t = 0 with initial velocity of 64 ft/sec from a height of 100 ft. Assume that $a(t) = -32$ ft/sec^2. Find v(t) and s(t).

[6.2]   **Use substitution to find each indefinite integral.**

7. $\int 6x^2(3x^3 - 5)^8\,dx$ 8. $\int \dfrac{6x + 5}{3x^2 + 5x}\,dx$

9. $\int \sqrt[3]{2x^2 - 8x}\,(x - 2)\,dx$ 10. $\int 4x^3 e^{-x^4}\,dx$

11.   A city's population is predicted to grow at a rate of

$$P'(x) = \frac{400e^{10t}}{1 + e^{10t}}$$

people per year where t is the time in years from the present. Find the total population 3 yr from now if P(0) = 100,000.

[6.3]   12.   Evaluate $\displaystyle\sum_{i=1}^{4} \frac{2}{i^2}$.

[6.3]   13.   Approximate the area under the graph of $f(x) = x^2 + x$ and above the x-axis from x = 0 to x = 2 using 4 rectangles. Let the height of each rectangle be the function value of the left side.

[6.3]   14.   Approximate the value of $\int_1^5 x^2\ dx$ by using the formula for the area

of a trapezoid, $A = \frac{1}{2}(B + b)h$, where B and b are the lengths of the

parallel sides and h is the distance between them.

[6.4]   Evaluate the following definite integrals.

15. $\int_1^5 (4x^3 - 5x)\,dx$

16. $\int_1^5 \left(\frac{3}{x^2} + \frac{2}{x}\right)dx$

17. $\int_1^2 \sqrt{3r - 2}\ dr$

18. $\int_0^1 4xe^{x^2+1}\ dx$

19. $\int_{-2}^1 3x(x^2 - 4)^5\ dx$

[6.4]   20.   Find the area of the region between the x-axis and $f(x) = e^{x/2}$ on
the interval [0, 2].

[6.4]   21.   The rate at which a substance grows is given by $R(x) = 500e^{.5x}$, where
x is the time in days. What is the total accumulated growth after 4
days?

[6.5]   22.   Find the area of the region enclosed by $f(x) = -x + 4$,
$g(x) = -x^2 + 6x - 6$, $x = 2$, and $x = 4$.

[6.5]   23.   Find the area of the region enclosed by $f(x) = 5x$ and $g(x) = x^3 - 4x$.

[6.5]   24.   A company has determined that the use of a new process would produce
a savings rate (in thousands of dollars) of

$$S(x) = 2x + 7,$$

where x is the number of years the process is used. However, the use
of this process also creates additional costs (in thousands of dollars)
according to the rate-of-cost function

$$C(x) = x^2 + 2x + 3.$$

(a) For how many years does the new process save the company money?

(b) Find the net savings in thousands of dollars over this period.

[6.5] 25. Suppose that the supply function of a commodity is p = .05q + 5 and the demand function is p = 12 - .02q.

(a) Find the producers' surplus.

(b) Find the consumers' surplus.

CHAPTER 6   TEST ANSWERS

1. An antiderivative of $f(x)$ is $F(x)$ such that $F'(x) = f(x)$.

2. $\frac{3}{4}x^4 - \frac{5}{3}x^3 + \frac{x^2}{2} + x + C$

3. $5 \ln |x| + 2e^{.5x} + C$

4. $\frac{4}{7}x^{7/2} - \frac{6}{5}x^{5/2} + C$

5. $C(x) = 200x + \frac{8}{x}x^{3/4} + 778.67$

6. $v(t) = -32t + 64;\ s(t) = -16t^2 + 64t + 100$

7. $\frac{2}{27}(3x^3 - 5)^9 + C$

8. $\ln |3x^2 + 5x| + C$

9. $\frac{3}{16}(2x^2 - 8x)^{4/3} + C$

10. $-e^{-x^4} + C$

11. 101,172

12. $2\frac{61}{72}$ or 2.85

13. 3.25   14. 52   15. 564   16. 5.62   17. $\frac{14}{9}$ or 1.56

18. $2e^2 - 2e$ or 9.34

19. 182.25

20. $2e - 2$ or 3.44

21. 6389

22. 3.33

23. 40.5

24. (a) 2 yr   (b) $5.33 thousand

25. (a) 250   (b) 100

**CHAPTER 7   FURTHER TECHNIQUES AND
APPLICATIONS OF INTEGRATION**

Section 7.1

1. $\int xe^x \, dx$

   Let $dv = e^x \, dx$ and $u = x$.

   Then $v = \int e^x \, dx$ and $du = dx$.

   $v = e^x + C$

   Use the formula

   $$\int u \, dv = uv - \int v \, du.$$

   $\int xe^x \, dx = xe^x - \int e^x \, dx$

   $\qquad = xe^x - e^x + C$

3. $\int (5x - 9)e^{-3x} \, dx$

   Let $dv = e^{-3x} \, dx$ and $u = 5x - 9$.

   Then $v = \int e^{-3x} \, dx$ and $du = 5 \, dx$.

   $v = \dfrac{e^{-3x}}{-3}$

   $\int (5x - 9)e^{-3x} \, dx$

   $\quad = \dfrac{-(5x - 9)e^{-3x}}{3} + \int \dfrac{5}{3} e^{-3x} \, dx$

   $\quad = \dfrac{-5xe^{-3x}}{3} + 3e^{-3x} + \dfrac{5e^{-3x}}{-9} + C$

   $\quad = \dfrac{-5xe^{-3x}}{3} - \dfrac{5e^{-3x}}{9} + 3e^{-3x} + C$

   or $\dfrac{-5xe^{-3x}}{3} + \dfrac{22e^{-3x}}{9} + C$

5. $\int_0^1 \dfrac{2x + 1}{e^x} \, dx$

   $= \int_0^1 (2x + 1)e^{-x} \, dx$

Let $dv = e^{-x} \, dx$ and $u = 2x + 1$.

Then $v = \int e^{-x} \, dx$ and $du = 2 \, dx$

$\qquad v = -e^{-x}$

$\int \dfrac{2x + 1}{e^x} \, dx$

$\quad = -(2x + 1)e^{-x} + \int 2e^{-x} \, dx$

$\quad = -(2x + 1)e^{-x} - 2e^{-x}$

$\int_0^1 \dfrac{2x + 1}{e^x} \, dx$

$\quad = \left[-(2x + 1)e^{-x} - 2e^{-x} \right] \Big|_0^1$

$\quad = [-(3)e^{-1} - 2e^{-1}] - (-1 - 2)$

$\quad = -5e^{-1} + 3$

$\quad \approx 1.1606$

7. $\int_1^4 \ln 2x \, dx$

   Let $dv = dx$ and $u = \ln 2x$.

   Then $v = x$ and $du = \dfrac{1}{x}$.

   $\int \ln 2x \, dx = x \ln 2x - \int dx$

   $\int_1^4 \ln 2x \, dx$

   $\quad = (x \ln 2x - x) \Big|_1^4$

   $\quad = (4 \ln 8 - 4) - (\ln 2 - 1)$

   $\quad = 4 \ln 2^3 - 4 - \ln 2 + 1$

   $\quad = 12 \ln 2 - \ln 2 - 3$

   $\quad = 11 \ln 2 - 3$

   $\quad \approx 4.6246$

9. $\int x \ln x \, dx$

Let $dv = x \, dx$ and $u = \ln x$.

Then $v = \dfrac{x^2}{2}$ and $du = \dfrac{1}{x} \, dx$.

$\int x \ln x \, dx = \dfrac{x^2}{2} \ln x - \int \dfrac{x}{2} \, dx$

$\quad = \dfrac{x^2 \ln x}{2} - \dfrac{x^2}{4} + C$

11. The area is $\displaystyle\int_2^4 (x - 2)e^x \, dx$.

Let $dv = e^x \, dx$ and $u = x - 2$.

Then $v = e^x$ and $du = dx$.

$\int (x - 2)e^x \, dx$

$\quad = (x - 2)e^x \int e^x \, dx$

$\int_2^4 (x - 2)e^x \, dx$

$\quad = [(x - 2)e^x - e^x]\Big|_2^4$

$\quad = (2e^4 - e^4) - (0 - e^2)$

$\quad = e^4 + e^2$

$\quad \approx 61.9872$

13. $\int x^2 e^{2x} \, dx$

Let $u = x^2$ and $dv = e^{2x} \, dx$.

Use column integration.

D	I
x^2	e^{2x}
$2x$	$e^{2x}/2$
2	$e^{2x}/4$
0	$e^{2x}/8$

$\int x^2 e^{2x} \, dx$

$\quad = x^2(e^{2x}/2) - 2x(e^{2x}/4)$

$\qquad + 2e^{2x}/8 + C$

$\quad = \dfrac{x^2 e^{2x}}{2} - \dfrac{xe^{2x}}{2} + \dfrac{e^{2x}}{4} + C$

15. $\displaystyle\int_0^5 x\sqrt[3]{x^2 + 2} \, dx$

$\quad = \displaystyle\int_0^5 x(x^2 + 2)^{1/3} \, dx$

$\quad = \dfrac{1}{2} \displaystyle\int_0^5 2x(x^2 + 2)^{1/3} \, dx$

Let $u = x^2 + 2$. Then $du = 2x \, dx$.

If $x = 5$, $u = 27$

$\quad x = 0$, $u = 2$.

$\quad = \dfrac{1}{2} \displaystyle\int_2^{27} u^{1/3} \, du$

$\quad = \dfrac{1}{2}\left(\dfrac{u^{4/3}}{1}\right)\left(\dfrac{3}{4}\right)\Big|_2^{27}$

$\quad = \dfrac{3}{8}(27)^{4/3} - \dfrac{3}{8}(2)^{4/3}$

$\quad = \dfrac{243}{8} - \dfrac{3(2^{4/3})}{8}$

$\quad = \dfrac{243}{8} - \dfrac{3\sqrt[3]{2}}{4} \approx 29.4301$

17. $\int (8x + 7) \ln (5x) \, dx$

Let $dv = (8x + 7)dx$ and $u = \ln 5x$.

Then $v = \dfrac{8x^2}{2} + 7x$ and $du = \dfrac{1}{5x}(5)dx$

$\qquad = 4x^2 + 7x$ $\qquad\qquad = \dfrac{1}{x} \, dx$.

$$\int (8x + 7) \ln (5x)\,dx$$

$$= (4x^2 + 7x) \ln 5x - \int (4x + 7)\,dx$$

$$= (4x^2 + 7x) \ln 5x - \left(\frac{4x^2}{2} + 7x\right) + C$$

$$= 4x^2 \ln (5x) + 7x \ln (5x)$$
$$\quad - 2x^2 - 7x + C$$

19. $\displaystyle\int x^2\sqrt{x + 2}\,dx$

Let $u = x^2$ and $dv = (x + 2)^{1/2}\,dx$.
Use column integration.

D	I
x^2	$(x + 2)^{1/2}$
$2x$	$(x + 2)^{3/2}(2/3)$
2	$(x + 2)^{5/2}(2/3)(2/5)$
0	$(x + 2)^{7/2}(2/3)(2/5)(2/7)$

$$\int x^2\sqrt{x + 2}\,dx$$

$$= x^2(x + 2)^{3/2}(2/3)$$
$$\quad - 2x(x + 2)^{5/2}(2/3)(2/5)$$
$$\quad + 2(x + 2)^{7/2}(2/3)(2/5)(2/7) + C$$
$$= \frac{2x^2(x + 2)^{3/2}}{3} - \frac{8x(x + 2)^{5/2}}{15}$$
$$\quad + \frac{16(x + 2)^{7/2}}{105} + C$$

21. $\displaystyle\int_0^1 \frac{x^3\,dx}{\sqrt{3 + x^2}}$

$$= \int_0^1 x^3(3x + x^2)^{-1/2}\,dx$$

Let $dv = x(3 + x^2)^{-1/2}\,dx$ and $u = x^2$.

Then $v = \dfrac{2(3 + x^2)^{1/2}}{2}$ and $du = 2x\,dx$.

$$v = (3 + x^2)^{1/2}$$

$$\int \frac{x^3\,dx}{\sqrt{3 + x^2}}$$

$$= x^2(3 + x^2)^{1/2} - \int 2x(3 + x^2)^{1/2}\,dx$$

$$= x^2(3 + x^2)^{1/2} - \frac{2}{3}(3 + x^2)^{3/2}$$

$$\int_0^1 \frac{x^3\,dx}{\sqrt{3 + x^2}}$$

$$= \left[x^2(3 + x^2)^{1/2} - \frac{2}{3}(3 + x^2)^{3/2}\right]\Bigg|_0^1$$

$$= 4^{1/2} - \frac{2}{3}(4^{3/2}) - 0 + \frac{2}{3}(3^{3/2})$$

$$= 2 - \frac{2}{3}(8) + \frac{2}{3}(3^{3/2})$$

$$\approx .13077$$

23. $\displaystyle\int \frac{-4}{\sqrt{x^2 + 36}}\,dx$

$$= -4 \int \frac{dx}{\sqrt{x^2 + 36}}$$

If $a = 6$, this integral matches
entry 5 in the table.

$$= -4 \ln \left|x + \sqrt{x^2 + 36}\right| + C$$

25. $\displaystyle\int \frac{6}{x^2 - 9}\,dx$

$$= 6 \int \frac{1}{x^2 - 9}\,dx$$

Use entry 8 from table with $a = 3$

$$= 6\left[\frac{1}{2(3)} \ln \left|\frac{x - 3}{x + 3}\right|\right] + C$$

$$= \ln \left|\frac{x - 3}{x + 3}\right| + C$$

27. $\displaystyle\int \frac{-4}{x\sqrt{9-x^2}}\,dx$

$\displaystyle = -4\int \frac{dx}{x\sqrt{9-x^2}}$

Use entry 9 from table with a = 3.

$\displaystyle = -4\left(-\frac{1}{3}\ln\left|\frac{3+\sqrt{9-x^2}}{x}\right|\right)+C$

$\displaystyle = \frac{4}{3}\ln\left|\frac{3+\sqrt{9-x^2}}{x}\right|+C$

29. $\displaystyle\int \frac{-2x}{3x+1}\,dx$

$\displaystyle = -2\int \frac{x\,dx}{3x+1}$

Use entry 11 from table with a = 3, b = 1.

$\displaystyle = -2\left(\frac{x}{3}-\frac{1}{9}\ln\left|3x+1\right|\right)+C$

$\displaystyle = -\frac{2x}{3}+\frac{2}{9}\ln\left|3x+1\right|+C$

31. $\displaystyle\int \frac{2}{3x(3x-5)}\,dx$

$\displaystyle = \frac{2}{3}\int \frac{1}{x(3x-5)}\,dx$

Use entry 13 from table with a = 3, b = -5.

$\displaystyle = \frac{2}{3}\left(-\frac{1}{5}\ln\left|\frac{x}{3x-5}\right|\right)+C$

$\displaystyle = -\frac{2}{15}\ln\left|\frac{x}{3x-5}\right|+C$

33. $\displaystyle\int \frac{4}{4x^2-1}\,dx$

$\displaystyle = 4\int \frac{dx}{4x^2-1}$

$\displaystyle = 2\int \frac{2\,dx}{4x^2-1}$

Let u = 2x.
Then du = 2 dx.

$\displaystyle = 2\int \frac{du}{u^2-1}$

Use entry 8 from table with a = 1.

$\displaystyle = 2\cdot\frac{1}{2}\ln\left|\frac{u-1}{u+1}\right|+C$

Substitute 2x for u.

$\displaystyle = \ln\left|\frac{2x-1}{2x+1}\right|+C$

35. $\displaystyle\int \frac{3}{x\sqrt{1-9x^2}}\,dx$

$\displaystyle = 3\int \frac{3\,dx}{3x\sqrt{1-9x^2}}$

Let   u = 3x.
Then du = 3 dx.

$\displaystyle = 3\int \frac{du}{u\sqrt{1-u^2}}$

Use entry 9 from table with a = 1.

$\displaystyle = -3\ln\left|\frac{1+\sqrt{1-u^2}}{u}\right|+C$

$\displaystyle = -3\ln\left|\frac{1+\sqrt{1-9x^2}}{3x}\right|+C$

37. $\displaystyle\int \frac{4x}{2x+3}\,dx$

$\displaystyle = 4\int \frac{x}{2x+3}\,dx$

Use entry 11 from table with a = 2, b = 3.

$\displaystyle = 4\left(\frac{x}{2}-\frac{3}{4}\ln\left|2x+3\right|\right)+C$

$\displaystyle = 2x - 3\ln\left|2x+3\right|+C$

39. $\int \dfrac{-x}{(5x-1)^2}\,dx$

$\quad = -\int \dfrac{x\,dx}{(5x-1)^2}$

Use entry 12 from table with $a = 5$, $b = -1$.

$\quad = -\left[\dfrac{-1}{25(5x-1)} + \dfrac{1}{25}\ln|5x-11|\right] + C$

$\quad = \dfrac{1}{25(5x-1)} - \dfrac{\ln|5x-1|}{25} + C$

41. $\int x^n \cdot \ln|x|\,dx$

Let $u = \ln|x|$ and $dv = x^n dx$.

Use column integration.

D	I
$\ln|x|$ $\quad +$	x^n
$\dfrac{1}{x}$ $\quad -$	$\dfrac{1}{n+1}x^{n+1}$

$\int x^n \cdot \ln|x|\,dx$

$\quad = \dfrac{1}{n+1}x^{n+1}\ln|x| - \int\left[\dfrac{1}{x}\cdot\dfrac{1}{n+1}x^{n+1}\right]dx$

$\quad = \dfrac{1}{n+1}x^{n+1}\ln|x| - \int \dfrac{1}{n+1}x^n\,dx$

$\quad = \dfrac{1}{n+1}x^{n+1}\ln|x| - \dfrac{1}{(n+1)^2}x^{n+1} + C$

$\quad = x^{n+1}\left[\dfrac{\ln|x|}{n+1} - \dfrac{1}{(n+1)^2}\right] + C$

Notice that $n \ne -1$.

43. $\int x\sqrt{x+1}\,dx$

(a) Let $u = x$ and $dv = \sqrt{x+1}\,dx$.
Use column integration.

D	I
x $\quad +$	$\sqrt{x+1}$
1 $\quad -$	$\left(\dfrac{2}{3}\right)(x+1)^{3/2}$
0	$\left(\dfrac{4}{15}\right)(x+1)^{5/2}$

$\int x\sqrt{x+1}\,dx$

$\quad = \left(\dfrac{2}{3}\right)x(x+1)^{3/2} - \left(\dfrac{4}{15}\right)(x+1)^{5/2} + C$

(b) Let $u = x+1$; then $u - 1 = x$
and $\quad du = dx$.

$\int x\sqrt{x+1}\,dx$

$\quad = \int (u-1)u^{1/2}\,du$

$\quad = \int (u^{3/2} - u^{1/2})\,du$

$\quad = \left(\dfrac{2}{5}\right)u^{5/2} - \left(\dfrac{2}{3}\right)u^{3/2} + C$

$\quad = \left(\dfrac{2}{5}\right)(x+1)^{5/2} - \left(\dfrac{2}{3}\right)(x+1)^{3/2} + C$

45. $r(x) = \displaystyle\int_1^6 2x^2 e^{-x}\,dx$

Let $dv = e^{-x}dx$ and $u = 2x^2$.
Then $v = -e^{-x}$ and $du = 4x\,dx$.

Use column integration.

D	I
$2x^2$	e^{-x}
$4x$	$-e^{-x}$
4	e^{-x}
0	$-e^{-x}$

$\displaystyle\int 2x^2 e^{-x}\ dx$

$= 2x^2(-e^{-x}) - 4x(e^{-x}) + 4(-e^{-x})$

$= -2x^2 e^{-x} - 4xe^{-x} - 4e^{-x}$

$= -2e^{-x}(x^2 + 2x + 2)$

$\displaystyle\int_1^6 2x^2 e^{-x}dx$

$= -2e^{-x}(x^2 + 2x + 2)\Big|_1^6$

$= -100e^{-6} + 10e^{-1}$

≈ 3.431

47. $h'(x) = \sqrt{x^2 + 16}$

$h(x) = \displaystyle\int_0^7 \sqrt{x^2 + 16}\ dx$

If $a = 4$, this matches entry 15 in the table.

$\displaystyle\int_0^7 \sqrt{x^2 + 16}\ dx$

$= \left(\dfrac{x}{2}\sqrt{x^2 + 16} + 8\ \ln\ |x + \sqrt{x^2 + 16}|\right)\Big|_0^7$

$= \dfrac{7\sqrt{65}}{2} + 8\ \ln\ (7 + \sqrt{65}) - 8\ \ln\ 4$

≈ 38.8

Section 7.2

1. $\displaystyle\int_0^2 x^2\ dx$

$n = 4,\ b = 2,\ a = 0,\ f(x) = x^2$

i	x_i	$f(x_i)$
0	0	0
1	1/2	.25
2	1	1
3	3/2	2.25
4	2	4

(a) Trapezoidal rule

$\displaystyle\int_0^2 x^2\ dx$

$\approx \dfrac{2 - 0}{4}\left[\dfrac{1}{2}(0) + .25 + 1 + 2.25 + \dfrac{1}{2}(4)\right]$

$= .5(5.5) = 2.7500$

(b) Simpson's rule

$\displaystyle\int_0^2 x^2\ dx$

$\approx \dfrac{2}{3(4)}[0 + 4(.25) + 2(1) + 4(2.25) + 4]$

$= \dfrac{2}{12}(16) \approx 2.6667$

(c) Exact value

$\displaystyle\int_0^2 x^2\ dx = \dfrac{x^3}{3}\Big|_0^2 = \dfrac{8}{3} \approx 2.6667$

3. $\displaystyle\int_{-1}^{3} \frac{1}{4-x}$

$n = 4,\ b = 3,\ a = -1,\ f(x) = \dfrac{1}{4-x}$

i	x_i	$f(x_i)$
0	-1	.2
1	0	.25
2	1	.3333
3	2	.5
4	3	1

(a) Trapezoidal rule

$\displaystyle\int_{-1}^{3} \frac{1}{4-x}\ dx$

$\approx \dfrac{3-(-1)}{4}$

$\cdot \left[\dfrac{1}{2}(.2) + .25 + .3333 + .5 + \dfrac{1}{2}(1)\right]$

≈ 1.6833

(b) Simpson's rule

$\displaystyle\int_{-1}^{3} \frac{1}{4-x}\ dx$

$\approx \dfrac{3-(-1)}{12}$

$\cdot [.2 + 4(.25) + 2(.3333) + 4(.5) + 1]$

≈ 1.6222

(c) Exact value

$\displaystyle\int_{-1}^{3} \frac{1}{4-x}\ dx = -\int_{-1}^{3} \frac{-dx}{4-x}$

$= -(\ln |4-x|)\ \Big|_{-1}^{3}$

$= -\ln 1 + \ln 5$

$= \ln 5 \approx 1.6094$

5. $\displaystyle\int_{-2}^{2} (2x^2 + 1)\ dx$

$n = 4,\ b = 2,\ a = -2,\ f(x) = 2x^2 + 1$

i	x	f(x)
0	-2	9
1	-1	3
2	0	1
3	1	3
4	2	9

(a) Trapezoidal rule

$\displaystyle\int_{-2}^{2} (2x^2 + 1)\ dx$

$\approx \dfrac{2-(-2)}{4}$

$\cdot \left[\dfrac{1}{2}(9) + 3 + 1 + 3 + \dfrac{1}{2}(9)\right]$

≈ 16

(b) Simpson's rule

$\displaystyle\int_{-2}^{2} (2x^2 + 1)\ dx$

$\approx \dfrac{2-(-2)}{12}$

$\cdot [9 + 4(3) + 2(1) + 4(3) + 9]$

≈ 14.6667

(c) Exact value

$\displaystyle\int_{-2}^{2} (2x^2 + 1)\ dx$

$= \left(\dfrac{2x^3}{3} + x\right)\ \Big|_{-2}^{2}$

$= \left(\dfrac{16}{3} + 2\right) - \left(-\dfrac{16}{3} - 2\right)$

$= \dfrac{32}{3} + 4$

$= \dfrac{44}{3}$

≈ 14.6667

7. $\displaystyle\int_{1}^{5} \frac{1}{x^2}\ dx$

$n = 4,\ b = 5,\ a = 1,\ f(x) = \dfrac{1}{x^2}$

i	x_i	$f(x_i)$
0	1	1
1	2	.25
2	3	.1111
3	4	.0625
4	5	.04

(a) Trapezoidal rule

$\displaystyle\int_{1}^{5} \frac{1}{x^2}\ dx$

$\approx \dfrac{5-1}{4}$

$\cdot \left[\dfrac{1}{2}(1) + .25 + .1111 + .0625 + \dfrac{1}{2}(.04)\right]$

$\approx .9436$

(b) Simpson's rule

$\displaystyle\int_{1}^{5} \frac{1}{x^2}\ dx$

$\approx \dfrac{5-1}{12}$

$\cdot [1 + 4(.25) + 2(.1111)$

$+ 4(.0625) + .04)]$

$\approx .8374$

(c) Exact value

$\displaystyle\int_{1}^{5} x^{-2}\ dx = -x^{-1}\Big|_{1}^{5}$

$= -\dfrac{1}{5} + 1$

$= \dfrac{4}{5} = .8$

9. $\displaystyle\int_{0}^{4} \sqrt{x^2 + 1}\ dx$

$n = 4,\ b = 4,\ a = 0,\ f(x) = \sqrt{x^2 + 1}$

i	x_i	$f(x_i)$
0	0	1
1	1	1.4142
2	2	2.2361
3	3	3.1623
4	4	4.1231

(a) Trapezoidal rule

$\displaystyle\int_{0}^{4} \sqrt{x^2 + 1}\ dx$

$\approx \dfrac{4-0}{4}\left[\dfrac{1}{2}(1) + 1.4142 + 2.2361 + 3.1623\right.$

$\left. + \dfrac{1}{2}(4.1231)\right]$

≈ 9.3741

(b) Simpson's rule

$\displaystyle\int_{0}^{4} \sqrt{x^2 + 1}\ dx$

$\approx \dfrac{4-0}{12}[1 + 4(1.4142) + 2(2.2361)$

$+ 4(3.1623) + 4.1231]$

≈ 9.3004

(c) Exact value

$\displaystyle\int_{0}^{4} \sqrt{x^2 + 1}\ dx$

If $a = 1$, this integral matches entry 15 in the table.

$$\int_0^4 \sqrt{x^2 + 1}\ dx$$

$$= \left(\frac{x}{2}\sqrt{x^2 + 1} + \frac{1}{2}\ln\left|x + \sqrt{x^2 + 1}\right|\right)\Bigg|_0^4$$

$$= 2\sqrt{17} + \frac{1}{2}\ln(4 + \sqrt{17}) - \frac{1}{2}\ln 1$$

$$= 2\sqrt{17} + \frac{1}{2}\ln(4 + \sqrt{17})$$

$$\approx 9.2936$$

11. $y = \sqrt{4 - x^2}$

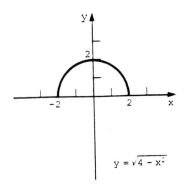

$$y = \sqrt{4 - x^2}$$

$n = 8$, $b = 2$, $a = -2$, $f(x) = \sqrt{4 - x^2}$

i	x_i	y
0	-2.0	0
1	-1.5	1.32289
2	-1.0	1.73205
3	-.5	1.93649
4	0	2
5	.5	1.93649
6	1.0	1.73205
7	1.5	1.32289
8	2.0	0

(a) Trapezoidal rule

$$\int_{-2}^2 \sqrt{4 - x^2}\ dx$$

$$\approx \frac{2 - (-2)}{8}$$

$$\cdot \left[\frac{1}{2}(0) + 1.32289 + 1.73205 + \cdots\right.$$

$$\left. + \frac{1}{2}(0)\right]$$

$$\approx 5.9914$$

(b) Simpson's rule

$$\int_{-2}^2 \sqrt{4 - x^2}\ dx$$

$$\approx \frac{2 - (-2)}{8}$$

$$\cdot [0 + 4(1.32289) + 2(1.73205)$$

$$+ 4(1.93649) + 2(2) + 4(1.93649)$$

$$+ 2(1.73205) + 4(1.32289) + 0]$$

$$\approx 6.1672$$

(c) Area of semicircle $= \frac{1}{2}\pi r^2$

$$= \frac{1}{2}\pi(2)^2$$

$$\approx 6.2832$$

Simpson's rule is more accurate.

13. (a) See the answer graph in your textbook.

(b) $A = \dfrac{7 - 1}{6}\left[\frac{1}{2}(.4) + .6 + .9 + 1.1\right.$

$$\left. + 1.3 + 1.4 + \frac{1}{2}(1.6)\right]$$

$$= 6.3$$

(c) $A = \dfrac{7 - 1}{3(6)}[.4 + 4(.6) + 2(.9)$

$$+ 4(1.1) + 2(1.3) + 4(1.4)$$

$$+ 1.6]$$

$$\approx 6.27$$

15. $y = e^{-t^2} + \dfrac{1}{t}$

The total reaction is

$$\int_1^9 \left(e^{-t^2} + \tfrac{1}{t}\right)\, dt.$$

$n = 8$, $b = 9$, $a = 1$, $f(t) = e^{-t^2} + \dfrac{1}{t}$

i	x_i	$f(x_i)$
0	1	1.3679
1	2	.5183
2	3	.3335
3	4	.2500
4	5	.2000
5	6	.1667
6	7	.1429
7	8	.1250
8	9	.1111

(a) Trapezoidal rule

$$\int_1^9 \left(e^{-t^2} + \tfrac{1}{t}\right) dt$$

$$\approx \frac{9-1}{8}\left[\tfrac{1}{2}(1.3679) + .5183 + .3335\right.$$

$$\left. + \cdots + \tfrac{1}{2}(.1111)\right]$$

$$\approx 2.4759$$

(b) Simpson's rule

$$\int_1^9 \left(e^{-t^2} + \tfrac{1}{t}\right) dt$$

$$\approx \frac{9-1}{3(8)}[1.3679 + 4(.5183) + 2(.3334)$$

$$+ 4(.2500) + 2(.2000) + 4(.1667)$$

$$+ 2(.1429) + 4(.1250) + .1111]$$

$$\approx 2.3572$$

17. Note that heights may differ depending on the readings of the graph. Thus, answers may vary.

$n = 10$, $b = 20$, $a = 0$

i	x_i	y
0	0	0
1	2	5
2	4	3
3	6	2
4	8	1.5
5	10	1.2
6	12	1
7	14	.5
8	16	.3
9	18	.2
10	20	.2

Area under curve for Formulation A

$$= \frac{20-0}{10}\left[\tfrac{1}{2}(0) + 5 + 3 + 2 + 1.5 + 1.2\right.$$

$$\left. + 1 + .5 + .3 + .2 + \tfrac{1}{2}(.2)\right]$$

$$= 2(14.8)$$

$$\approx 30 \text{ mcg/ml}$$

This represents the total amount of drug available to the patient.

19. As in Exercise 17, readings on the graph may vary, so answers may vary. The area both under the curve for Formulation A and above the minimum effective concentration line is on the interval [1/2, 6].

Area under curve for Formulation A on [1/2, 1], with n = 1,

$$= \frac{1 - \frac{1}{2}}{1}\left[\frac{1}{2}(2 + 6)\right]$$

$$= \frac{1}{2}(4) = 2$$

Area under curve for Formulation A on [1, 6], with n = 5,

$$= \frac{6 - 1}{5}\left[\frac{1}{2}(6) + 5 + 4 + 3 + 2.4 + \frac{1}{2}(2)\right]$$

$$= 18.4$$

Area under minimum effective concentration line on [1/2, 6]

$$= 5.5(2)$$

$$= 11.0$$

Area under the curve for Formulation A and above minimum effective concentration line

$$= 2 + 18.4 - 11.0$$

$$\approx 9 \text{ mcg/ml}$$

This represents the total effective amount of drug available to the patient.

21. **(a)** See the answer graph in textbook.

(b) $\frac{(7 - 1)}{6}\left[\frac{1}{2}(4) + 7 + 11 + 9 + 15 + 16 + \frac{1}{2}(23)\right]$

$$= 71.5$$

(c) $\frac{(7 - 1)}{3(6)}[4 + 4(7) + 2(11) + 4(9) + 2(15) + 4(16) + 23]$

$$= 69.0$$

Exercises 23–31 should be solved by computer methods. The solutions will vary according to the computer program that is used. The answers are given.

23. $\displaystyle\int_{4}^{8} \ln(x^2 - 10)\,dx$

Trapezoidal rule: 12.6027
Simpson's rule: 12.6029

25. $\displaystyle\int_{-2}^{2} \sqrt{9 - 2x^2}\,dx$

Trapezoidal rule: 9.83271
Simpson's rule: 9.83377

27. $\displaystyle\int_{1}^{5} (2x^2 + 3x - 1)^{2/5}\,dx$

Trapezoidal rule: 14.5192
Simpson's rule: 14.5193

29. Trapezoidal rule: 3979.24
Simpson's rule: 3979.24

31. **(a)** Trapezoidal rule: .682673
Simpson's rule: .682689

(b) Trapezoidal rule: .954471
Simpson's rule: .954500

(c) Trapezoidal rule: .997292
Simpson's rule: .997300

33. As n changes from 4 to 8, for example, the error changes from .020703 to .005200.

$$.020703a = .005200$$

$$a \approx \frac{1}{4}$$

Similar results would be obtained using other values for n.

The error is multiplied by 1/4.

35. As n changes form 4 to 8, the error changes from .0005208 to .0000326.

$$.0005208a = .0000326$$

$$a \approx \frac{1}{16}$$

Similar results would be obtained using other values for n.

The error is multiplied by 1/16.

Section 7.3

1. $f(x) = x$, $y = 0$, $x = 0$, $x = 2$

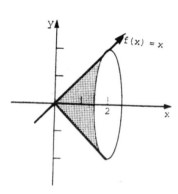

$$V = \pi \int_0^2 x^2 \, dx = \frac{\pi x^3}{2} \Big|_0^2$$

$$= \frac{\pi(8)}{3} - 0$$

$$= \frac{8\pi}{3}$$

3. $f(x) = 2x + 1$, $y = 0$, $x = 4$, $x = 4$

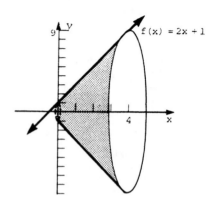

$$V = \pi \int_0^4 (2x + 1)^2 \, dx$$

Let $u = 2x + 1$.

Then $du = 2 \, dx$.

If $x = 4$, $u = 9$.

If $x = 0$, $u = 1$.

$$V = \frac{1}{2}\pi \int_0^4 2(2x + 1)^2 \, dx$$

$$= \frac{1}{2}\pi \int_1^9 u^2 \, du$$

$$= \frac{\pi}{2}\left(\frac{u^3}{3}\right)\Big|_1^9$$

$$= \frac{729\pi}{6} - \frac{\pi}{6}$$

$$= \frac{728\pi}{6}$$

$$= \frac{364\pi}{3}$$

5. $f(x) = \frac{1}{3}x + 2$, $y = 0$, $x = 1$, $x = 3$

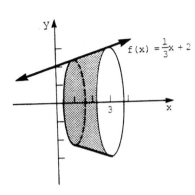

$$V = \pi \int_1^3 \left(\frac{1}{3}x + 2\right)^2 dx$$

$$= 3\pi \int_1^3 \frac{1}{3}\left(\frac{1}{3}x + 2\right)^2 dx$$

$$= 3\pi \frac{\left(\frac{1}{3}x + 2\right)^3}{3}\Bigg|_1^3$$

$$= \pi\left(\frac{1}{3}x + 2\right)^3\Bigg|_1^3$$

$$= 27\pi - \frac{343\pi}{27}$$

$$= \frac{386\pi}{27}$$

7. $f(x) = \sqrt{x}$, $y = 0$, $x = 1$, $x = 2$

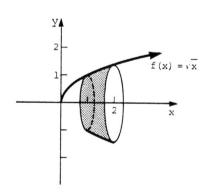

$$V = \pi \int_1^2 (\sqrt{x})^2 \, dx = \pi \int_1^2 x \, dx$$

$$= \frac{\pi x^2}{2}\Bigg|_1^2$$

$$= 2\pi - \frac{\pi}{2}$$

$$= \frac{3\pi}{2}$$

9. $f(x) = \sqrt{2x + 1}$, $y = 0$, $x = 1$, $x = 4$

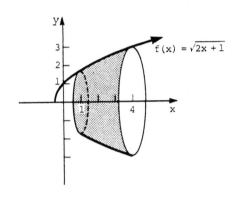

$$V = \pi \int_1^4 (\sqrt{2x + 1})^2 \, dx$$

$$= \pi \int_1^4 (2x + 1) \, dx$$

$$= \pi\left(\frac{2x^2}{2} + x\right)\Bigg|_1^4$$

$$= \pi[(16 + 4) - 2]$$

$$= 18\pi$$

11. $f(x) = e^x$; $y = 0$, $x = 0$, $x = 2$

$$V = \pi \int_0^2 e^{2x}\ dx = \frac{\pi e^{2x}}{2}\Big|_0^2$$

$$= \frac{\pi e^4}{2} - \frac{\pi}{2}$$

$$= \frac{\pi}{2}(e^4 - 1)$$

$$\approx 84.19$$

13. $f(x) = \dfrac{1}{\sqrt{x}}$, $y = 0$, $x = 1$, $x = 4$

$$V = \pi \int_1^4 \frac{1}{(\sqrt{x})^2}\ dx$$

$$= \pi \int_1^4 \frac{1}{x}\ dx$$

$$= \pi \ln |x|\ \Big|_1^4$$

$$= \pi \ln 4 - \pi \ln 1$$

$$= \pi \ln 4 \approx 4.36$$

15. $f(x) = x^2$, $y = 0$, $x = 1$, $x = 5$

$$V = \pi \int_1^5 x^4\ dx = \frac{\pi x^5}{5}\Big|_1^5$$

$$= 625\pi - \frac{\pi}{5}$$

$$= \frac{3124\pi}{5}$$

17. $f(x) = 1 - x^2$, $y = 0$

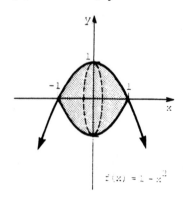

$$f(x) = 1 - x^2$$

Since $f(x) = 1 - x^2$ intersects $y = 0$ where

$$1 - x^2 = 0$$

$$x = \pm 1,$$

$$a = -1 \quad \text{and} \quad b = 1.$$

$$V = \pi \int_{-1}^{1} (1 - x^2)^2\ dx$$

$$= \pi \int_{-1}^{1} (1 - 2x^2 + x^4)\ dx$$

$$= \pi \left(x - \frac{2x^3}{3} + \frac{x^5}{5}\right)\Big|_{-1}^{1}$$

$$= \pi \left(1 - \frac{2}{3} + \frac{1}{5}\right) - \pi \left(-1 + \frac{2}{3} - \frac{1}{5}\right)$$

$$= 2\pi - \frac{4\pi}{3} + \frac{2\pi}{5}$$

$$= \frac{16\pi}{15}$$

19. $f(x) = \sqrt{1 - x^2}$, $r = \sqrt{1} = 1$

$$V = \pi \int_{-1}^{1} (\sqrt{1 - x^2})^2\ dx$$

$$= \pi \int_{-1}^{1} (1 - x^2)\ dx$$

$$= \pi \left(x - \frac{x^3}{3}\right)\Big|_{-1}^{1}$$

$$= \pi \left(1 - \frac{1}{3}\right) - \pi \left(-1 + \frac{1}{3}\right)$$

$$= 2\pi - \frac{2}{3}\pi$$

$$= \frac{4\pi}{3}$$

21. $f(x) = \sqrt{r^2 - x^2}$

$$V = \pi \int_{-r}^{r} (\sqrt{r^2 - x^2})^2 \, dx$$

$$= \pi \int_{-r}^{r} (r^2 - x^2) \, dx$$

$$= \pi \left(r^2 x - \frac{x^3}{3}\right)\Big|_{-r}^{r}$$

$$= \pi \left(r^3 - \frac{r^3}{3}\right) - \pi\left(-r^3 + \frac{r^3}{3}\right)$$

$$= 2r^3\pi - \frac{2r^3\pi}{3})$$

$$= \frac{4r^3\pi}{3}$$

23. $f(x) = r,\ x = 0,\ x = h$
Graph $f(x) = r$; then show solid of revolution formed by rotating about the x-axis the region bounded by $f(x)$, $x = 0$, $x = h$.

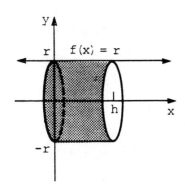

$$\int_{0}^{h} \pi r^2 \, dx = \pi r^2 x \Big|_{0}^{h}$$

$$= \pi r^2 h - 0$$

$$= \pi r^2 h$$

25. $f(x) = x^2 - 2;\ [0, 5]$

Average value

$$= \frac{1}{5 - 0}\int_{0}^{5} (x^2 - 2) \, dx$$

$$= \frac{1}{5}\left(\frac{x^3}{3} - 2x\right)\Big|_{0}^{5}$$

$$= \frac{1}{5}\left(\frac{125}{3} - 10\right)$$

$$= \frac{1}{5}\left(\frac{95}{3}\right)$$

$$= \frac{19}{3}$$

27. $f(x) = \sqrt{x + 1};\ [3, 8]$

Average value

$$= \frac{1}{8 - 3} \int_{3}^{8} \sqrt{x + 1} \, dx$$

$$= \frac{1}{5}\int_{3}^{8} (x + 1)^{1/2} \, dx$$

$$= \frac{1}{5} \cdot \frac{2}{3}(x + 1)^{3/2} \Big|_{3}^{8}$$

$$= \frac{2}{15}(9^{3/2} - 4^{3/2})$$

$$= \frac{2}{15}(27 - 8)$$

$$= \frac{38}{15}$$

29. $f(x) = e^{x/5};\ [0, 5]$

Average value

$$= \frac{1}{5 - 0} \int_{0}^{e} e^{x/5} \, dx$$

$$= \int_{0}^{5} \frac{1}{5}e^{x/5} \, dx$$

$$= e^{x/5} \Big|_{0}^{5} = e^1 - e^0$$

$$= e - 1 \approx 1.718$$

31. $f(x) = x^2 e^{2x}$; $[0, 2]$

Average value

$$= \frac{1}{2 - 0} \int_0^2 x^2 e^{2x}\, dx$$

Let $u = x^2$ and $dv = e^{2x}dx$.

Use column integration.

D	I
x^2	e^{2x}
$2x$	$(1/2)e^{2x}$
2	$(1/4)e^{2x}$
0	$(1/8)e^{2x}$

$$\frac{1}{2-0} \int_0^2 x^2 e^{2x}\, dx$$

$$= \frac{1}{2}\left(\left[(x^2)\left(\frac{1}{2}\right)e^{2x} - (2x)\left(\frac{1}{4}\right)e^{2x}\right.\right.$$

$$\left.\left. + 2\left(\frac{1}{8}\right)e^{2x}\right]\Big|_0^2\right)$$

$$= \frac{1}{2}\left(2e^4 - e^4 + \frac{1}{4}e^4 - \frac{1}{4}\right)$$

$$= \frac{(5e^4 - 1)}{8} \approx 33.999$$

33. $R(t) = te^{-.1t}$

During the nth hour is the interval $(n - 1, n)$.

Average intensity during nth hour

$$= \frac{1}{n - (n - 1)} \int_{n-1}^n te^{-.1t}\, dt$$

$$= \int_{n-1}^n te^{-.1t}\, dt$$

Let $u = t$ and $dv = e^{-.1t}\, dt$.

D	I
t	$e^{-.1t}$
1	$-10e^{-.1t}$
0	$100e^{-.1t}$

$$\int_{n-1}^n te^{-.1t}\, dt$$

$$= (-10te^{-.1t} - 100e^{-.1t})\Big|_{n-1}^n$$

(a) Second hour, $n = 2$

Average intensity

$$= -10e^{-.2}(12) + 10e^{-.1}(11)$$

$$= 110e^{-.1} - 120e^{-.2}$$

$$\approx 1.2844$$

(b) Twelfth hour, $n = 12$

Average intensity

$$= -10e^{-1.2}(12 + 10) + 10e^{-1.1}(11 + 10)$$

$$= 210e^{-1.1} - 220e^{-1.2}$$

$$\approx 3.6402$$

(c) Twenty-fourth hour, $n = 24$

Average intensity

$$= -10e^{-2.4}(24 + 10) + 10e^{-2.3}(23 + 10)$$

$$= 330e^{-2.3} - 340e^{-2.4}$$

$$\approx 2.2413$$

35. $W(t) = -6t^2 + 10t + 80$

(a) $W(0) = -6(0)^2 + 10(0) + 80$

$W(0) = 80$

(b) W′(t) = -12t + 10

If W′(t) = 0, t = $\frac{5}{6}$.

$W\left(\frac{5}{6}\right) = -6\left(\frac{5}{6}\right)^2 + 10\left(\frac{5}{6}\right) + 80$

$= -\frac{25}{6} + \frac{50}{6} + 80$

$= \frac{505}{6}$

≈ 84 words per minute
when t = 5/6

(c) Average speed

$= \frac{1}{5-0} \int_0^5 (-6t^2 + 10t + 80)dt$

$= \frac{1}{5}(-2t^3 + 5t^2 + 80t)\Big|_0^5$

$= \frac{1}{5}(-250 + 125 + 400)$

$= \frac{275}{5}$

= 55 words per minute

Section 7.4

1. f(x) = 1000

(a) $P = \int_0^{10} 1000e^{-.12x}\ dx$

$= \frac{1000}{-.12}\ e^{-.12x}\Big|_0^{10}$

$= \frac{1000}{-.12}\ e^{-1.2} + \frac{1000}{.12}$

= \$5823.38

Store the value for P without round-
ing in your calculator.

(b) $A = e^{.12(10)} \int_0^{10} 1000e^{-.12x}\ dx$

$= e^{.12(10)}P$

= \$19,334.31

3. f(x) = 500

(a) $P = \int_0^{10} 500e^{-.12x}\ dx$

$= \frac{500}{.12}\ e^{-.12x}\Big|_0^{10}$

$= \frac{500}{-.12}e^{-1.2} + \frac{500}{.12}$

= \$2911.69

(b) $A = e^{.12(10)} \int_0^{10} 500e^{-.12x}\ dx$

$= e^{1.2}\ P$

= \$9667.16

5. f(x) = 400e^{.03x}

(a) $\int_0^{10} 400e^{.03x} \cdot e^{-.12x}\ dx$

$= \int_0^{10} 400e^{-.09}dx$

$= \frac{400}{-.09}\ e^{-.09x}\Big|_0^{10}$

$= \frac{400}{-.09}\ e^{-.9} + \frac{400}{.09}$

= \$2637.47

(b) $e^{.12(10)} \int_0^{10} 400e^{-.09x}dx$

$= e^{1.2}\left(\frac{400}{-.09}\ e^{-.9} + \frac{400}{.09}\right)$

= \$8756.70

7. (a) $\displaystyle\int_0^{10} 5000e^{-.01x} \cdot e^{-.12x}\ dx$

$= 1000 \displaystyle\int_0^{10} 5000e^{-.13x}\ dx$

$= \dfrac{5000}{-.13}\ e^{-.13x}\ \Big|_0^{10}$

$= \dfrac{5000}{-.13}\ e^{-1.3} + \dfrac{5000}{.13}$

$= \$27{,}979.55$

(b) $e^{.12(10)} \displaystyle\int_0^{10} 5000e^{-.13x}\ dx$

$= e^{1.2}\left(\dfrac{5000}{-.13}\ e^{-1.3} + \dfrac{5000}{.13}\right)$

$= \$92{,}895.37$

9. f(x) = .1x

(a) $P = \displaystyle\int_0^{10} .1xe^{-.12x}\ dx$

Let $u = .1x$ and $dv = e^{-.12x}$.

Then $du = .1dx$ and $v = \dfrac{e^{-.12x}}{-.12}$.

$\displaystyle\int .1xe^{-.12x}\ dx$

$= .1x\dfrac{e^{-.12x}}{-.12} - \displaystyle\int \dfrac{.1}{-.12}\ e^{-.12x}dx$

$= \dfrac{.1xe^{-.12x}}{-.12} + \dfrac{.1}{.12}\left(\dfrac{e^{-.12x}}{-.12}\right)$

$\displaystyle\int_0^{10} .1xe^{-.12x}dx$

$= \left[\dfrac{.1xe^{-.12x}}{-.12} + \dfrac{.1}{.12}\left(\dfrac{e^{-.12x}}{-.12}\right)\right]\Big|_0^{10}$

$= \dfrac{1}{-.12}\ e^{-1.2} + \dfrac{.1e^{-1.2}}{-.12(.12)}$

$\quad + \dfrac{.1}{(.12)^2}$

$= \$2.34$

(b) $A = e^{.12(10)} \displaystyle\int_0^{10} .1xe^{-.12x}\ dx$

$\approx \$7.78$

$= e^{.12(10)}P$

$= \$7.78$

11. f(x) = .01x + 100

(a) $P = \displaystyle\int_0^{10} (.01x + 100)e^{-.12x}\ dx$

$= \displaystyle\int_0^{10} .01xe^{-.12x}\ dx$

$\quad + \displaystyle\int_0^{10} 100e^{-.12x}\ dx$

Let $u = .01x$ and $dv = e^{-.12x}$.

Then $du = .01dx$ and $v = \dfrac{e^{-.12x}}{-.12}$.

$\displaystyle\int .01xe^{-.12x}\ dx$

$= \dfrac{.01xe^{-.12x}}{-.12} - \displaystyle\int \dfrac{.01e^{-.12x}}{-.12}\ dx$

$= \dfrac{.01xe^{-.12x}}{-.12} - \dfrac{.01e^{-.12x}}{(-.12)^2}$

$\displaystyle\int_0^{10} .01xe^{-.12x}dx + \displaystyle\int_0^{10} 100e^{-.12x}\ dx$

$= \left(\dfrac{.01xe^{-.12x}}{-.12} - \dfrac{.01e^{-.12x}}{(-.12)^2}\right)\Big|_0^{10}$

$\quad + \dfrac{100e^{-.12x}}{-.12}\Big|_0^{10}$

$= \dfrac{.1e^{-1.2}}{-.12} - \dfrac{.01e^{-1.2}}{(-.12)^2} - 0 + \dfrac{.01}{(-.12)^2}$

$\quad + \dfrac{100e^{-1.2}}{-.12} - \dfrac{100}{-.12}$

$P = \$582.57$

(b) $e^{.12(10)} \int_0^{10} (.01x + 100)e^{-.12x}dx$

$= e^{1.2}P$

$\approx \$1934.20$

13. $f(x) = 1000x - 100x^2$

(a) $P = \int_0^{10} (1000x - 100x^2)e^{-.12x}\ dx$

$= \int_0^{10} 1000xe^{-.12x}\ dx$

$\qquad - \int_0^{10} 100x^2e^{-.12x}\ dx$

Let $u = 1000x$ and $dv = e^{-.12x}\ dx$;
then $du = 1000\ dx$ and

$\qquad v = \frac{-1}{.12}e^{-.12x}\ dx.$

$\int 1000xe^{-.12x}\ dx$

$= \frac{1000xe^{-.12x}}{(-.12)} - \int \frac{1000e^{-.12x}}{-.12}\ dx$

$= \frac{1000xe^{-.12x}}{(-.12)} - \frac{1000e^{-.12x}}{(-.12)^2} + C$

For $\int 100x^2e^{-.12x}\ dx$, let $u = 100x^2$
and $dv = e^{-.12x}\ dx$.

Use column integration

D	I
$100x^2$	$e^{-.12x}$
$200x$	$\frac{1}{(-.12)}e^{-.12x}$
200	$\frac{1}{(-.12)^2}e^{-.12x}$
0	$\frac{1}{(-.12)^3}e^{-.12x}$

$\int 100x^2e^{-.12x}\ dx$

$= \left(\frac{100}{-.12}\right)x^2e^{-.12x} - \frac{200}{(-.12)^2}xe^{-.12x}$

$\qquad + \frac{200}{(-.12)^3}e^{-.12x} + C$

$\int_0^{10} 1000xe^{-.12x}\ dx - \int_0^{10} 100x^2e^{-.12x}\ dx$

$= \left(\frac{1000xe^{-.12x}}{(-.12)} - \frac{1000e^{-.12x}}{(-.12)^2}\right.$

$\qquad - \frac{100}{(-.12)}x^2e^{-.12x} + \frac{200}{(-.12)^2}xe^{-.12x}$

$\qquad \left. - \frac{200}{(-.12)^3}e^{-.12x}\right)\Big|_0^{10}$

$= \frac{10{,}000e^{-1.2}}{(-.12)} - \frac{1000}{(-.12)^2}e^{-1.2} - \frac{10{,}000}{(-.12)}e^{-1.2}$

$\qquad + \frac{2000}{(-.12)^2}e^{-1.2} - \frac{200}{(-.12)^3}e^{-1.2}$

$\qquad + \frac{1000}{(-.12)^2} + \frac{200}{(-.12)^3}$

$= \frac{1000}{(-.12)^2}(e^{-1.2} + 1)$

$\qquad - \frac{200}{(-.12)^3}(e^{-1.2} - 1)$

$= \$9480.41$

(b)

$e^{.12(10)} \int_0^{10} (1000x - 100x^2)e^{-.12x}\ dx$

$= e^{1.2}(P)$

$= \$31{,}476.07$

15. $A = e^{.14(3)} \int_0^3 20{,}000e^{-.14x}\ dx$

$= e^{.42}\left(\frac{20{,}000}{-.14}e^{-.14x}\right)\Big|_0^3$

$= e^{.42}\left(\frac{60{,}000}{-.14}e^{-.42} + \frac{20{,}000}{.14}\right)$

$\approx \$74{,}565.94$

17. **(a)** Present value

$$= \int_0^8 5000e^{-.01x}e^{-.08x}\ dx$$

$$= \int_0^8 5000e^{-.09x}\ dx$$

$$= \left(\frac{5000}{-.09}\ e^{-.09x}\right)\Big|_0^8$$

$$= \frac{5000e^{-.72}}{-.09} + \frac{5000}{.09}$$

$$= \$28,513.76$$

(b) Final amount

$$= e^{.08(8)} \int_0^8 5000e^{-.01x}e^{-.08x}dx$$

$$= e^{.64}(28,513.76)$$

$$= \$54,075.80$$

19. $P = \int_0^5 (1500 - 60x^2)e^{-.10x}\ dx$

$$= \int_0^5 1500e^{-.10x}\ dx$$

$$- \int_0^5 60x^2e^{-.10x}\ dx$$

$$= 1500\int_0^5 e^{-.10x}\ dx$$

$$- 60\int_0^5 x^2e^{-.10x}\ dx$$

Second integral, by column integration.

D	I
x^2	$e^{-.10x}$
$2x$	$e^{-.10x}/-.10$
2	$e^{-.10x}/.01$
0	$e^{-.10x}/-.001$

$$-60 \int x^2e^{-.10x}dx$$

$$= \frac{x^2e^{-.10x}}{-.10} - \frac{2xe^{-.10x}}{.01} + \frac{2e^{-.10x}}{-.001} + C$$

$$1500 \int_0^5 e^{-.10x}dx - 60 \int_0^5 x^2e^{-.10x}dx$$

$$= \frac{1500}{-.10}\ e^{-.10x}\Big|_0^5$$

$$- 60(-\frac{x^2e^{-.10x}}{.10} - \frac{2xe^{-.10x}}{.01}$$

$$- \frac{2e^{-.10x}}{.001})\Big|_0^5$$

$$= -15,000e^{-.5} + 15,000$$

$$+ 60(\frac{25e^{-.5}}{.1} + \frac{10e^{-.5}}{.01} + \frac{2e^{-.5}}{.001}$$

$$- 0 - \frac{2}{.001})$$

$$= -15,000e^{-.5} + 15,000 + 15,000e^{-.5}$$

$$+ 60,000e^{-.5} + 120,000e^{-.5}$$

$$- 120,000$$

$$= 180,000e^{-.5} - 105,000$$

$$= \$4175.52$$

Section 7.5

1. $\displaystyle\int_2^\infty \frac{1}{x^2}\,dx \;=\; \lim_{b\to\infty}\int_2^b x^{-2}\,dx$

$\displaystyle = \lim_{b\to\infty}\int (-x^{-1})\Big|_2^b$

$\displaystyle = \lim_{b\to\infty}\left(-\frac{1}{b} + \frac{1}{2}\right)$

$\displaystyle = \lim_{b\to\infty}\left(-\frac{1}{b}\right) + \lim_{b\to\infty}\frac{1}{2}$

As $b\to\infty$, $-\dfrac{1}{b}\to 0$. The integral is convergent.

$\displaystyle\int_2^\infty \frac{1}{x^2}\,dx = 0 + \frac{1}{2} = \frac{1}{2}$

3. $\displaystyle\int_1^\infty \frac{1}{\sqrt{x}}\,dx = \lim_{b\to\infty}\int_1^b x^{-1/2}\,dx$

$\displaystyle = \lim_{b\to\infty}(2x^{1/2})\Big|_1^b$

$\displaystyle = \lim_{b\to\infty}(2\sqrt{b} - 2)$

$\displaystyle = \lim_{b\to\infty} 2\sqrt{b} - \lim_{b\to\infty} 2$

As $b\to\infty$, $2\sqrt{b}\to\infty$.
The integral is divergent.

5. $\displaystyle\int_{-\infty}^{-1} \frac{2}{x^3}\,dx$

$\displaystyle = \int_{-\infty}^{-1} 2x^{-3}\,dx$

$\displaystyle = \lim_{a\to-\infty}\int_a^{-1} 2x^{-3}\,dx$

$\displaystyle = \lim_{a\to-\infty}\left(\frac{2x^{-2}}{-2}\right)\Big|_a^{-1}$

$\displaystyle = \lim_{a\to-\infty}\left(-1 + \frac{1}{a^2}\right)$

As $a\to-\infty$, $\dfrac{1}{a^2}\to 0$. The integral is convergent.

$\displaystyle\int_{-\infty}^{-1} \frac{2}{x^3}\,dx = -1 + 0 = -1$

7. $\displaystyle\int_1^\infty \frac{1}{x^{1.001}}\,dx$

$\displaystyle = \int_1^\infty x^{-1.001}\,dx$

$\displaystyle = \lim_{b\to\infty}\int_1^b x^{-1.001}\,dx$

$\displaystyle = \lim_{b\to\infty}\left(\frac{x^{-.001}}{-.001}\right)\Big|_1^b$

$\displaystyle = \lim_{b\to\infty}\left(-\frac{1}{(.001)b^{.001}} + \frac{1}{.001}\right)$

As $b\to\infty$, $-\dfrac{1}{.001\,b^{.001}}\to 0$.
The integral is convergent.

$\displaystyle\int_1^\infty \frac{1}{x^{1.001}}\,dx = 0 + \frac{1}{.001} = 1000$

9. $\displaystyle\int_{-\infty}^{-1} x^{-2}\,dx = \lim_{a\to-\infty}\int_a^{-1} x^{-2}\,dx$

$\displaystyle = \lim_{a\to-\infty}(-x^{-1})\Big|_a^{-1}$

$\displaystyle = \lim_{a\to-\infty}\left(1 + \frac{1}{a}\right)$

$\displaystyle = 1 + 0 = 1$

The integral is convergent.

11. $\displaystyle\int_{-\infty}^{-1} x^{-8/3}\,dx = \lim_{a\to-\infty}\int_{a}^{-1} x^{-8/3}\,dx$

$$= \lim_{a\to-\infty}\left(-\tfrac{3}{5}x^{-5/3}\right)\Big|_{a}^{-1}$$

$$= \lim_{a\to-\infty}\left(\tfrac{3}{5} + \tfrac{3}{5a^{5/3}}\right)$$

$$= \tfrac{3}{5} + 0 = \tfrac{3}{5}$$

The integral is convergent.

13. $\displaystyle\int_{0}^{\infty} 4e^{-4x}\,dx = \lim_{b\to\infty}\int_{0}^{b} 4e^{-4x}\,dx$

$$= \lim_{b\to\infty}\left(\frac{4e^{-4x}}{-4}\right)\Big|_{0}^{b}$$

$$= \lim_{b\to\infty}\left(-e^{-4b} + 1\right)$$

$$= \lim_{b\to\infty}\left(-\frac{1}{e^{4b}} + 1\right)$$

$$= 0 + 1 = 1$$

The integral is convergent.

15. $\displaystyle\int_{-\infty}^{0} 4e^{x}\,dx = \lim_{a\to-\infty}\int_{a}^{0} 4e^{x}\,dx$

$$= \lim_{a\to-\infty}\left(4e^{x}\right)\Big|_{a}^{0}$$

$$= \lim_{a\to-\infty}(4 - 4e^{a})$$

Since a approaches $-\infty$, e^{a} is in the denominator of a fraction.

As $a\to-\infty$, $-4e^{a}\to 0$. The integral is convergent.

$$\int_{-\infty}^{0} 4e^{x}\,dx = 4 - 0 = 4$$

17. $\displaystyle\int_{-\infty}^{-1} \ln|x|\,dx = \lim_{a\to-\infty}\int_{a}^{-1} \ln|x|\,dx$

Let $\quad u = \ln|x|$ and $dv = dx$.

Then $du = \dfrac{1}{x}\,dx \quad$ and $\quad v = x$.

$$\int \ln|x|\,dx = x\ln|x| - \int \frac{x}{x}\,dx$$

$$= x\ln|x| - x + C$$

$$\int_{-\infty}^{-1} \ln|x|\,dx$$

$$= \lim_{a\to-\infty}\left(x\ln|x| - x\right)\Big|_{a}^{-1}$$

$$= \lim_{a\to-\infty}\left(-\ln 1 + 1 - a\ln|a| + a\right)$$

$$= \lim_{a\to-\infty}\left(1 + a - a\ln|a|\right)$$

The integral is divergent, since as $a\to-\infty$,

$$(a - a\ln|a|)\to\infty.$$

19. $\displaystyle\int_{1}^{\infty} \frac{dx}{(x+1)^{2}} = \lim_{b\to\infty}\int_{0}^{b} \frac{dx}{(x+1)^{2}}$

(Ue substitution.)

$$= \lim_{b\to\infty} -(x+1)^{-1}\Big|_{0}^{b}$$

$$= \lim_{b\to\infty}\left(\frac{-1}{b+1} + 1\right)$$

As $b\to\infty$, $-\dfrac{1}{b+1}\to 0$. The integral is convergent.

$$\int_{0}^{\infty} \frac{dx}{(x+1)^{2}} = 0 + 1 = 1$$

21. $\displaystyle\int_{-\infty}^{-1} \frac{2x - 1}{x^2 - x}\ dx$

$\displaystyle = \lim_{a \to -\infty} \int_a^{-1} \frac{2x - 1}{x^2 - x}\ dx$

(Use substitution.)

$\displaystyle = \lim_{a \to -\infty} \ln |x^2 - x|\ \Big|_a^{-1}$

$\displaystyle = \lim_{a \to -\infty} (\ln 2 - \ln |a^2 - a|)$

As $a \to -\infty$, $\ln |a^2 - a) \to \infty$. The integral is divergent.

23. $\displaystyle\int_2^{\infty} \frac{1}{x \ln x}\ dx$

$\displaystyle = \lim_{b \to \infty} \int_2^b \frac{1}{x \ln x}\ dx$

(Use substitution.)

$\displaystyle = \lim_{b \to \infty} \left[\ln (\ln x)\ \Big|_2^b\ \right]$

$\displaystyle = \lim_{b \to \infty} [\ln (\ln b) - \ln (\ln 2)]$

As $b \to \infty$, $\ln (\ln b) \to \infty$. The integral is divergent.

25. $\displaystyle\int_0^{\infty} xe^{2x}dx = \lim_{b \to \infty} \int_0^b xe^{2x}\ dx$

Let $dv = e^{2x}dx$ and $u = x$.

Then $v = \dfrac{1}{2} e^{2x}$ and $du = dx$.

$\displaystyle\int xe^{2x}dx = \frac{x}{2}e^{2x} - \int \frac{1}{2}e^{2x}dx$

$\displaystyle = \frac{x}{2}e^{2x} - \frac{1}{4}e^{2x}$

$\displaystyle = \frac{1}{4}(2x - 1)e^{2x}$

$\displaystyle\int_0^{\infty} xe^{2x}dx$

$\displaystyle = \lim_{b \to \infty} \left[\frac{1}{4}(2x - 1)e^{2x}\right]\Big|_0^b$

$\displaystyle = \lim_{b \to \infty} \left[\frac{1}{4}(2b - 1)e^{2b} - \frac{1}{4}(-1)(1)\right]$

$\displaystyle = \lim_{b \to \infty} \left[\frac{1}{4}(2b - 1)e^{2b} + \frac{1}{4}\right]$

As $b \to \infty$, $\frac{1}{4}(2b - 1)e^{2b} \to \infty$. The integral is divergent.

27. $\displaystyle\int_{-\infty}^{-1} \frac{2}{3x(2x - 7)}\ dx$

$\displaystyle = \lim_{a \to -\infty} \int_a^{-1} \left[\frac{2}{3} \cdot \frac{1}{x(2x - 7)}\right]\ dx$

Use entry 13 from the table.

$\displaystyle = \lim_{a \to -\infty} \left(\frac{2}{3}\left[-\frac{1}{7} \ln \left|\frac{x}{2x - 7}\right|\right]\right)\Big|_a^{-1}$

$\displaystyle = \lim_{a \to -\infty} \frac{2}{21}\left(-\ln \frac{1}{9} + \ln \left|\frac{a}{2a - 7}\right|\right)$

As $a \to -\infty$, $\ln \left|\dfrac{a}{2a - 7}\right| \to \ln \dfrac{1}{2}$.

$\displaystyle = \frac{2}{21}\left(\ln \frac{1}{2} - \ln \frac{1}{9}\right)$

$\displaystyle = \frac{2}{21}\left(\ln \frac{9}{2}\right)$

$\displaystyle = \frac{2 \ln 4.5}{21} \approx .143$

29. $\displaystyle\int_1^{\infty} \frac{4}{9x(x + 1)^2}\ dx$

$\displaystyle = \lim_{b \to \infty} \frac{4}{9} \int_1^b \frac{dx}{x(x + 1)^2}$

Use entry 14 from the table.

$$= \frac{4}{9} \lim_{b \to \infty} \left(\frac{1}{x+1} + \ln \left| \frac{x}{x+1} \right| \right) \Bigg|_{1}^{b}$$

$$= \frac{4}{9} \lim_{b \to \infty} \left(\frac{1}{b+1} + \ln \frac{b}{b+1} \right.$$

$$\left. - \frac{1}{2} - \ln \frac{1}{2} \right)$$

As $b \to \infty$, $\frac{1}{b+1} \to 0$ and $\ln \frac{b}{b+1} \to \ln 1 = 0$. The integral is convergent.

$$\int_{1}^{\infty} \frac{4}{9x(x+1)^2} \, dx$$

$$= \frac{4}{9} \left(0 + 0 - \frac{1}{2} - \ln \frac{1}{2} \right)$$

$$= \frac{4}{9} \left(-\frac{1}{2} - \ln 1 + \ln 2 \right)$$

$$= \frac{4}{9} \left(\ln 2 - \frac{1}{2} \right) \approx .086$$

31. $\displaystyle\int_{0}^{\infty} \frac{1}{\sqrt{1+x^2}} \, dx$

$$= \lim_{b \to \infty} \int_{0}^{b} \frac{1}{\sqrt{1+x^2}} \, dx$$

Use entry 5 from the table.

$$= \lim_{b \to \infty} \left(\ln \left| x + \sqrt{x^2+1} \right| \Bigg|_{0}^{b} \right)$$

$$= \lim_{b \to \infty} \left(\ln \left| b + \sqrt{b^2+1} \right| + \ln 1 \right)$$

$$= \lim_{b \to \infty} \ln \left| b + \sqrt{b^2+1} \right|$$

As $b \to \infty$, $\ln \left| b + \sqrt{b^2+1} \right| \to \infty$.
The integral is divergent.

33. $f(x) = \dfrac{1}{x-1}$ for $(-\infty, 0]$

$$\int_{-\infty}^{0} \frac{1}{x-1} \, dx$$

$$= \lim_{a \to -\infty} \int_{a}^{0} \frac{dx}{x-1}$$

$$= \lim_{a \to -\infty} \left(\ln |x-1) \right) \Bigg|_{a}^{0}$$

$$= \lim_{a \to -\infty} \left(\ln |-1| - \ln |a-1| \right)$$

But $\displaystyle\lim_{a \to -\infty} (\ln |a-1|) = \infty$.

The integral is divergent, so the area cannot be found.

35. $f(x) = \dfrac{1}{(x-1)^2}$ for $(-\infty, 0]$

$$\int_{-\infty}^{0} \frac{1}{(x-1)^2}$$

$$= \lim_{a \to -\infty} \int_{a}^{0} \frac{1}{(x-1)^2}$$
(Use substitution.)

$$= \lim_{a \to -\infty} -(x-1)^{-1} \Bigg|_{a}^{0}$$

$$= \lim_{a \to -\infty} \left(-\frac{1}{-1} + \frac{1}{a-1} \right)$$

As $a \to -\infty$, $\dfrac{1}{a-1} \to 0$. The integral is convergent.

$$= 1 + 0$$

$$= 1$$

Therefore, the area is 1.

37. $\displaystyle\int_{-\infty}^{\infty} xe^{-x^2}\, dx$

$$= \int_{-\infty}^{\infty} xe^{-x^2}\, dx + \int_{0}^{\infty} xe^{-x^2} dx$$

Let $u = -x^2$, so that

du = -2x dx.

$$= \lim_{a\to -\infty} \left(-\frac{1}{2}\int_{a}^{0} -2xe^{-x^2} dx\right.$$

$$+ \lim_{b\to\infty}\left(-\frac{1}{2}\int_{0}^{b} -2xe^{-x^2} dx\right)$$

$$= \lim_{a\to -\infty}\left(-\frac{1}{2}e^{-x^2}\right)\Big|_{a}^{0}$$

$$+ \lim_{b\to\infty}\left(-\frac{1}{2}e^{-x^2}\right)\Big|_{0}^{b}$$

$$= \lim_{a\to -\infty}\left(-\frac{1}{2} + \frac{1}{2e^{a^2}}\right)$$

$$+ \lim_{b\to\infty}\left(-\frac{1}{2e^{b^2}} + \frac{1}{2}\right)$$

$$= -\frac{1}{2} + \frac{1}{2}$$

$$= 0$$

The area is 0.

39. $\displaystyle\int_{1}^{\infty} \frac{1}{x^p}\, dx$

Case $p < 1$:

$$\int_{1}^{\infty} \frac{1}{x^p}\, dx$$

$$= \int_{1}^{\infty} x^{-p}\, dx$$

$$= \lim_{a\to\infty}\int_{1}^{a} x^{-p}\, dx$$

$$= \lim_{a\to\infty}\left[\frac{x^{-p+1}}{(-p+1)}\Big|_{1}^{a}\right]$$

$$= \lim_{a\to\infty}\left[\frac{1}{(-p+1)}(a^{-p+1} - 1)\right]$$

$$= \lim_{a\to\infty}\left[\frac{1}{(-p+1)}a^{1-p} - \frac{1}{(-p+1)}\right]$$

Since $p < 1$, $1 - p$ is positive and, as $a\to\infty$, $a^{1-p}\to\infty$. The integral diverges.

$p = 1$:

$$\int_{1}^{\infty}\frac{1}{x^p}\, dx = \int_{1}^{\infty}\frac{1}{x}\, dx$$

$$= \lim_{a\to\infty}\int_{1}^{a}\frac{1}{x}\, dx$$

$$= \lim_{a\to\infty}\left(\ln|x|\Big|_{1}^{a}\right)$$

$$= \lim_{a\to\infty}(\ln|a| - \ln 1)$$

$$= \lim_{a\to\infty}\ln|a|$$

As $a\to\infty$, $\ln|a|\to\infty$. The integral diverges.

Case $p > 1$:

$$\int_{1}^{\infty}\frac{1}{x^p}\, dx$$

$$= \lim_{a\to\infty}\int_{1}^{a} x^{-p}\, dx$$

$$= \lim_{a\to\infty}\left(\frac{x^{-p+1}}{-p+1}\Big|_{1}^{a}\right)$$

$$= \lim_{a\to\infty}\left[\frac{a^{-p+1}}{(-p+1)} - \frac{1}{(-p+1)}\right]$$

Since $p > 1$, $-p + 1 < 0$; thus as

$a \to \infty$, $\dfrac{a^{-p+1}}{(-p + 1)} \to 0$.

Hence, $\lim\limits_{a \to \infty} \left[\dfrac{a^{-p+1}}{(-p + 1)} - \dfrac{1}{(-p + 1)} \right]$

$= 0 - \dfrac{1}{(-p + 1)}$

$= \dfrac{-1}{-p + 1}$

$= \dfrac{1}{p + 1}$

The integral converges.

41. Capital value is

$\displaystyle\int_0^\infty 60,000e^{-.08t}\ dt$

$= \lim\limits_{b \to \infty} \displaystyle\int_0^b 60,000e^{-.08t}\ dt$

$= \lim\limits_{b \to \infty} \left. \left(\dfrac{60,000}{-.08}\ e^{-.08t} \right) \right|_0^b$

$= -750,000 \lim\limits_{b \to \infty} (e^{-.08b} - 1)$

$= -750,000 \lim\limits_{b \to \infty} \left(\dfrac{1}{e^{.08b}} - 1 \right)$

As $b \to \infty$, $\dfrac{1}{e^{.08b}} \to 0$.

$= -750,000(-1)$

$= \$750,000$

43. (a) $\displaystyle\int_0^\infty 6000e^{-.08t}\ dt$

$= \lim\limits_{b \to \infty} \displaystyle\int_0^b 6000e^{-.08t}\ dt$

$= \lim\limits_{b \to \infty} \left. \dfrac{6000e^{-.08t}}{-.08} \right|_0^b$

$= \lim\limits_{b \to \infty} \left(\dfrac{6000e^{-.08b}}{-.08} - \dfrac{6000}{-.08} \right)$

$= 0 + 75,000 = \$75,000$

(b) $\displaystyle\int_0^\infty 6000e^{-.1t}\ dt$

$= \lim\limits_{b \to \infty} \displaystyle\int_0^b 6000e^{-.1t}\ dt$

$= \lim\limits_{b \to \infty} \left. \dfrac{6000e^{-.1t}}{-.1} \right|_0^b$

$= \lim\limits_{b \to \infty} \left(\dfrac{6000e^{-.1b}}{-.1} - \dfrac{6000}{-.1} \right)$

$= 0 + 60,000$

$= \$60,000$

45. $\displaystyle\int_0^\infty 3000e^{-.1t}\ dt$

$= \lim\limits_{b \to \infty} \displaystyle\int_0^b 3000e^{-.1t}\ dt$

$= \lim\limits_{b \to \infty} \left. \dfrac{3000e^{-.1b}}{-.1} \right|_0^b$

$= \lim\limits_{b \to \infty} \left(\dfrac{3000e^{-.1b}}{-.1} + \dfrac{3000}{.1} \right)$

$= 0 + 30,000$

$= \$30,000$

47. $\displaystyle\int_0^\infty 50e^{-.06t}\ dt$

$= 50 \lim\limits_{b \to \infty} \displaystyle\int_0^b e^{-.06t}\ dt$

$= 50 \lim\limits_{b \to \infty} \left. \dfrac{e^{-.06t}}{-.06} \right|_0^b$

$= \dfrac{50}{-.06} \lim\limits_{b \to \infty} (e^{-.06b} - e^0)$

$= -\dfrac{50}{.06}(0 - 1)$

$= \dfrac{50}{.06}$

$= 833.33$

Chapter 7   Review Exercises

5. $\int x(8-x)^{3/2}\,dx$

Let $u = x$ and $dv = (8-x)^{3/2}$.

Then $du = dx$ and $v = -\frac{2}{5}(8-x)^{5/2}$.

$\int x(8-x)^{3/2}\,dx$

$= -\frac{2}{5}x(8-x)^{5/2} + \int \frac{2}{5}(8-x)^{5/2}\,dx$

$= -\frac{2}{5}x(8-x)^{5/2} - \frac{2}{5}\left(\frac{2}{7}\right)(8-x)^{7/2} + C$

$= \frac{-2x}{5}(8-x)^{5/2} - \frac{4}{35}(8-x)^{7/2} + C$

7. $\int xe^x\,dx$

Let $u = x$ and $dv = e^x\,dx$.

Then $du = dx$ and $v = e^x$.

$\int xe^x\,dx = xe^x - \int e^x\,dx$

$= xe^x - e^x + C$

9. $\int \ln|2x+3|\,dx$

First, use substitution.

Let $a = 2x + 3$.

Then $da = 2\,dx$.

$\int \ln|2x+3|\,dx = \frac{1}{2}\int \ln|a|\,da$

Second, integrate by parts.

Let $u = \ln|a|$ and $dv = da$.

Then $du = \frac{1}{a}\,da$ and $v = a$.

$\frac{1}{2}\int \ln|a|\,da$

$= \frac{1}{2}\left(a\ln|a| - \int da\right)$

$= \frac{1}{2}(a\ln|a| - a) + C$

Finally, substitute $2x + 3$ for a.

$\int \ln|2x+3|\,dx$

$= \frac{1}{2}\Big[(2x+3)\ln|2x+3| - (2x+3) + C\Big]$

$= \frac{1}{2}(2x+3)\big[\ln|2x+3| - 1\big] + C$

11. $\int \frac{x}{9-4x^2}\,dx$

Use substitution.

Let $u = 9 - 4x^2$.

Then $du = -8x\,dx$.

$\int \frac{x}{9-4x^2}\,dx = -\frac{1}{8}\int \frac{-8x\,dx}{9-4x^2}$

$= -\frac{1}{8}\int \frac{du}{u}$

$= -\frac{1}{8}\ln|u| + C$

$= -\frac{1}{8}\ln|9-4x^2| + C$

13. $\int \frac{1}{9-4x^2}\,dx$

First use substitution.

Let $u = 2x$.

Then $du = 2\,dx$.

$\int \frac{1}{9-4x^2}\,dx$

$= \frac{1}{2}\int \frac{2}{9-4x^2}\,dx$

$= \frac{1}{2}\int \frac{1}{9-u^2}\,du$

Now use entry 7 in the table.

$$= \frac{1}{2}\left[\frac{1}{6}\ln\left|\frac{3+u}{3-u}\right|\right] + C$$

$$= \frac{1}{12}\ln\left|\frac{3+u}{3-u}\right| + C$$

$$= \frac{1}{12}\ln\left|\frac{3+2x}{3-2x}\right| + C$$

15. $\displaystyle\int_{1}^{2} \frac{1}{x\sqrt{9-x^2}}\, dx$

(Use entry 9 in the table.)

$$= \left(-\frac{1}{3}\ln\left|\frac{3+\sqrt{9-x^2}}{x}\right|\right)\Bigg|_{1}^{2}$$

$$= -\frac{1}{3}\ln\left|\frac{3+\sqrt{5}}{2}\right| + \frac{1}{3}\ln\left|\frac{3+\sqrt{8}}{1}\right|$$

$$= \frac{1}{3}\left[\ln(3+\sqrt{8}) - \ln\left(\frac{3+\sqrt{5}}{2}\right)\right]$$

$$= \frac{1}{3}\ln\left[\frac{(3+2\sqrt{2})2}{3+\sqrt{5}}\right]$$

$$= \frac{1}{3}\ln\left(\frac{6+4\sqrt{2}}{3+\sqrt{5}}\right) \approx .26677$$

17. $\displaystyle\int_{1}^{e} x^3 \ln x\, dx$

Let $u = \ln x$ and $dv = x^3\, dx$.
Use column integration.

D	I
$\ln x$	x^3
$\dfrac{1}{x}$	$\dfrac{1}{4}x^4$

$$\int x^3 \ln x\, dx$$

$$= \frac{1}{4}x^4 \ln x - \int\left(\frac{1}{4}x^4 \cdot \frac{1}{x}\right)dx$$

$$= \frac{1}{4}x^4 \ln x - \frac{1}{4}\int x^3\, dx$$

$$= \frac{1}{4}x^4 \ln x - \frac{1}{16}x^4 + C$$

$$\int_{1}^{e} x^3 \ln x\, dx$$

$$= \left(\frac{1}{4}x^4 \ln x - \frac{1}{16}x^4\right)\Bigg|_{1}^{e}$$

$$= \left(\frac{e^4}{4}\right)(1) - \frac{e^4}{16} - 0 + \frac{1}{16}$$

$$= \frac{e^4}{4} - \frac{e^4}{16} + \frac{1}{16}$$

$$= \frac{3e^4 + 1}{16}$$

$$\approx 10.300$$

19. $A = \displaystyle\int_{0}^{1} (3 + x^2)e^{2x}\, dx$

$$= \int_{0}^{1} 3e^{2x}\, dx + \int_{0}^{1} x^2 e^{2x}\, dx$$

$$\int 3e^{2x}\, dx = \frac{3}{2}e^{2x} + C$$

For the second integral, $\displaystyle\int x^2 e^{2x}\, dx$,
let $u = x^2$ and $dv = e^{2x}\, dx$
Use column integration.

D	I
x^2	e^{2x}
$2x$	$e^{2x}/2$
2	$e^{2x}/4$
0	$e^{2x}/8$

$$\int x^2 \, e^{2x} dx$$

$$= x^2 \frac{e^{2x}}{2} - 2x \frac{e^{2x}}{4} + 2 \frac{e^{2x}}{8}$$

$$= \frac{x^2 e^{2x}}{2} - \frac{xe^{2x}}{2} + \frac{e^{2x}}{4}$$

$$A = \left(\frac{3}{2} e^{2x} + \frac{x^2 e^{2x}}{2} - \frac{xe^{2x}}{2} + \frac{e^{2x}}{4} \right) \Big|_0^1$$

$$= \frac{3}{2} e^2 + \frac{e^2}{2} - \frac{e^2}{2} + \frac{e^2}{4} - \left(\frac{3}{2} + \frac{1}{4} \right)$$

$$= \left(\frac{6 + 2 - 2 + 1}{4} \right) e^2 - \left(\frac{7}{4} \right)$$

$$= \frac{7}{4} (e^2 - 1)$$

$$\approx 11.181$$

21. $\displaystyle\int_2^6 \frac{dx}{x^2 - 1}$

Trapezoidal rule:

$n = 4$, $b = 6$, $a = 2$, $f(x) = \dfrac{1}{x^2 - 1}$

i	x_i	$f(x_i)$
0	2	.3333
1	3	.125
2	4	.0667
3	5	.0417
4	6	.0286

$$\int_2^6 \frac{dx}{x^2 - 1}$$

$$\approx \frac{6 - 2}{4} \left[\frac{1}{2}(.3333) + .125 + .0667 \right.$$

$$\left. + .0417 + \frac{1}{2}(.0286) \right]$$

$$\approx .4143$$

Exact value:

Use entry 8 in the table.

$$\int_2^6 \frac{dx}{x^2 - 1} = \frac{1}{2} \cdot \ln \left| \frac{x - 1}{x + 1} \right| \Big|_2^6$$

$$= \frac{1}{2} \ln \frac{5}{7} - \frac{1}{2} \ln \frac{1}{3}$$

$$= \frac{1}{2} \ln \frac{15}{7}$$

$$\approx .3811$$

23. $\displaystyle\int_1^5 \ln x \, dx$

Trapezoidal rule:

$n = 4$, $b = 5$, $a = 1$, $f(x) = \ln x$

i	x_i	$f(x_i)$
0	1	0
1	2	.6931
2	3	1.0986
3	4	1.3863
4	5	1.6094

$$\int_1^5 \ln x \, dx$$

$$\approx \frac{5 - 1}{4} \left[\frac{1}{2}(0) + .6931 + 1.0986 \right.$$

$$\left. + 1.3863 + \frac{1}{2}(1.6094) \right]$$

$$\approx 3.983$$

Exact value:

Use entry 4 in the table.

$$\int_1^5 \ln x \, dx$$

$$= x(\ln |x| - 1) \Big|_1^5$$

$$= 5(\ln 5 - 1) - (\ln 1 - 1)$$

$$= 5 \ln 5 - 5 - 0 + 1$$

$$= 5 \ln 5 - 4$$

$$\approx 4.047$$

25. $\displaystyle\int_{2}^{10} \frac{x\ dx}{x - 1}$

$n = 4,\ b = 10,\ a = 2,\ f(x) = \dfrac{x}{x - 1}$

i	x_i	$f(x_i)$
0	2	2
1	4	1.3333
2	6	1.2
3	8	1.1429
4	10	1.1111

$\displaystyle\int_{2}^{10} \frac{x}{x - 1}\ dx$

$\approx \dfrac{10 - 2}{12}[2 + 4(1.3333) + 2(1.2)$

$+ 4(1.1429) + 1.1111]$

≈ 10.28

27. $A = \displaystyle\int_{-1}^{1} \sqrt{1 - x^2}\ dx$

$n = 6,\ b = 1,\ a = -1,\ f(x) = \sqrt{1 - x^2}$

i	x_i	$f(x_i)$
0	-1	0
1	$-\dfrac{2}{3}$	.7454
2	$-\dfrac{1}{3}$	.9428
3	0	1
4	$\dfrac{1}{3}$	.9428
5	$\dfrac{2}{3}$	.7454
6	1	0

$A \approx \dfrac{1 - (-1)}{6}[\frac{1}{2}(0) + .7454 + .9428$

$+ 1 + .9428 + .7454 + \frac{1}{2}(0)]$

≈ 1.459

29. $f(x) = 2x - 1;\ y = 0,\ x = 3$

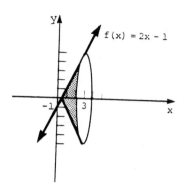

Since $f(x) = 2x - 1$ intersects $y = 0$ at $x = 1/2$, the integral has a lower bound $a = 1/2$.

$V = \pi \displaystyle\int_{1/2}^{3} (2x - 1)^2\ dx$

$= \pi \displaystyle\int_{1/2}^{3} (4x^2 - 4x + 1)\ dx$

$= \pi\left(\dfrac{4x^3}{3} - \dfrac{4x^2}{2} + x\right)\Big|_{1/2}^{3}$

$= \pi\left(36 - 18 + 3 - \dfrac{1}{6} + \dfrac{1}{2} - \dfrac{1}{2}\right)$

$= \pi\left(21 - \dfrac{1}{6}\right)$

$= \dfrac{125}{6}\pi \approx 65.45$

31. $f(x) = e^{-x},\ y = 0,\ x = -2,\ x = 1$

$V = \pi \displaystyle\int_{-2}^{1} e^{-2x}\ dx = \dfrac{\pi e^{-2x}}{-2}\Big|_{-2}^{1}$

$= \dfrac{\pi e^{-2}}{-2} + \dfrac{\pi e^4}{2}$

$= \dfrac{\pi(e^4 - e^{-2})}{2}$

≈ 85.55

33. $f(x) = 4 - x^2$, $y = 0$, $x = -1$, $x = 1$

$$V = \pi \int_{-1}^{1} (4 - x^2)^2 \, dx$$

$$= \pi \int_{-1}^{1} (16 - 8x^2 + x^4) \, dx$$

$$= \pi \left(16x - \frac{8x^3}{3} + \frac{x^5}{5} \right) \Big|_{-1}^{1}$$

$$= \pi \left(16 - \frac{8}{3} + \frac{1}{5} + 16 - \frac{8}{3} + \frac{1}{5} \right)$$

$$= \pi \left(32 - \frac{16}{3} + \frac{2}{5} \right)$$

$$= \frac{406\pi}{15} \approx 85.03$$

35. The frustum may be shown as follows.

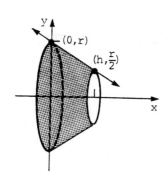

Use the two points given to find

$$f(x) = -\frac{r}{2h}x + r.$$

$$V = \pi \int_{0}^{h} \left(-\frac{r}{2h}x + r \right)^2 \, dx$$

$$= -\frac{2\pi h}{3r} \left(-\frac{r}{2h}x + r \right)^3 \Big|_{0}^{h}$$

$$= -\frac{2\pi h}{3r} \left[\left(-\frac{r}{2} + r \right)^3 - (0 + r)^3 \right]$$

$$= -\frac{2\pi h}{3r} \left[\left(\frac{r}{2} \right)^3 - r^3 \right]$$

$$-\frac{2\pi h}{3r} \left(\frac{r^3}{8} - r^3 \right)$$

$$= -\frac{2\pi h}{3r} \left(-\frac{7r^3}{8} \right)$$

$$= \frac{7\pi r^2 h}{12}$$

37. $f(x) = \sqrt{x + 1}$

$$\frac{1}{b - a} \int_{a}^{b} f(x) \, dx$$

$$= \frac{1}{8 - 0} \int_{0}^{8} \sqrt{x + 1} \, dx$$

$$= \frac{1}{8} \int_{0}^{8} (x + 1)^{1/2} \, dx$$

$$= \frac{1}{8} \left(\frac{2}{3} \right) (x + 1)^{3/2} \Big|_{0}^{8}$$

$$= \frac{1}{12} (9)^{3/2} - \frac{1}{12}(1)$$

$$= \frac{27}{12} - \frac{1}{12} = \frac{26}{12}$$

$$= \frac{13}{6}$$

39. $\int_{1}^{\infty} x^{-1} \, dx$

$$= \lim_{b \to \infty} \int_{1}^{b} \frac{dx}{x}$$

$$= \lim_{b \to \infty} \ln x \Big|_{1}^{b}$$

$$= \lim_{b \to \infty} \ln b$$

As $b \to \infty$, $\ln b \to \infty$. The integral is divergent.

41. $\displaystyle\int_0^\infty \frac{dx}{(5x+2)^2}$

$\displaystyle= \lim_{b\to\infty} \int_0^b (5x+2)^{-2}\ dx$

$\displaystyle= \lim_{b\to\infty} \left[-\frac{1}{5}(5x+2)^{-1}\right]\Big|_0^b$

$\displaystyle= \lim_{b\to\infty} \left[\frac{-1}{5(5x+2)}\right]\Big|_0^b$

$\displaystyle= \lim_{b\to\infty} \left[\frac{-1}{5(5b+2)} + \frac{1}{10}\right]$

$\displaystyle= \frac{1}{10}$

43. $\displaystyle\int_{-\infty}^0 \frac{x}{x^2+3}\ dx$

$\displaystyle= \lim_{a\to-\infty} \frac{1}{2}\int_a^0 \frac{2x\ dx}{x^2+3}$

$\displaystyle= \lim_{a\to-\infty} \frac{1}{2}(\ln|x^2+3|)\Big|_a^0$

$\displaystyle= \lim_{a\to-\infty} \frac{1}{2}(\ln 3 - \ln|a^2+3|)$

As $a\to-\infty$, $\frac{1}{2}(\ln 3 - \ln|a^2+3|)\to\infty$.

The integral is divergent.

45. $\displaystyle A = \int_{-\infty}^1 \frac{3}{(x-2)^2}\ dx$

$\displaystyle= \lim_{a\to-\infty} \int_a^1 3(x-2)^{-2}dx$

$\displaystyle= \lim_{a\to-\infty} \frac{3(x-2)^{-1}}{-1}\Big|_a^1$

$\displaystyle= \lim_{a\to-\infty} \left(-\frac{3}{x-2}\right)\Big|_a^1$

$\displaystyle= \lim_{a\to-\infty} \left(-\frac{3}{-1} + \frac{3}{a-2}\right)$

$\displaystyle= 3$

49. Sales

$\displaystyle= \frac{7-1}{6}[\frac{1}{2}(.7) + 1.2 + 1.5 + 1.9$

$\displaystyle\qquad + 2.2 + 2.4 + \frac{1}{2}(2.0)]$

≈ 10.55

Total sales are \$10.55 million.

51. $f(x) = 5000$; 8 yr; 9%

$\displaystyle P = \int_0^8 5000e^{-.09x}\ dx$

$\displaystyle= \frac{5000}{-.09}\ e^{-.09x}\Big|_0^8$

$\displaystyle= \frac{-5000}{.09}\ e^{-.72} + \frac{5000}{.09}$

$\approx \$28{,}513.76$

53. $f(x) = 100e^{.02x}$; 5 yr; 11%

$\displaystyle P = \int_0^5 100e^{.02x}\cdot e^{-.11x}\ dx$

$\displaystyle= \int_0^5 100e^{-.09x}\ dx$

$\displaystyle= \frac{100}{-.09}\ e^{-.09x}\Big|_0^5$

$\displaystyle= \frac{-100}{.09}\ e^{-.45} + \frac{100}{.09}$

$\approx \$402.64$

55. $f(x) = 2000$; 5 mo; 1% per month

$\displaystyle e^{.01(5)} \int_0^5 2000e^{-.01t}\ dt$

$\displaystyle= e^{.05}\left(\frac{2000}{-.01}\ e^{-.01t}\right)\Big|_0^5$

$\displaystyle= e^{.05}(-2000000e^{-.05} + 200{,}000)$

$\approx \$10{,}254.22$

57. $f(x) = 20x$; 6 yr; 12% per year

$$e^{.12(6)} \int_0^6 20xe^{-.12x}\, dx$$

$$= 20e^{.72} \int_0^6 xe^{-.12x}\, dx$$

Let $u = x\, dx$ and $dv = e^{-.12x}$.
Then $du = dx$ and $v = \dfrac{e^{-.12x}}{-.12}$.

$$\int xe^{-.12x}\, dx$$
$$= \frac{xe^{-.12x}}{-.12} - \int \frac{e^{-.12x}}{-.12}\, dx$$
$$= \frac{-xe^{-.12x}}{.12} - \frac{e^{-.12x}}{(.12)^2}$$

$$20e^{.72} \int_0^6 xe^{-.12x}\, dx$$

$$= 20e^{.72}\left[\frac{-xe^{-.12x}}{.12} - \frac{e^{-.12x}}{(.12)^2}\right]\Big|_0^6$$

$$= 20e^{.72}\left[\frac{-6e^{-.72}}{.12} - \frac{e^{-.72}}{(.12)^2}\right.$$
$$\left. + 0 + \frac{1}{(.12)^2}\right]$$

$$\approx \$464.49$$

59. $f(x) = Ce^{kx}$ where $C = 1000$, $k = .05$
$f(x) = 1000e^{.05x}$

Use $P = \displaystyle\int_0^t f(x)e^{-rx}\, dx$ with $r = .11$

and $t = 7$.

$$P = \int_0^7 1000e^{.05x} \cdot e^{-.11x}\, dx$$

$$= 1000 \int_0^7 e^{-.06x}\, dx$$

$$= 1000\left(\frac{1}{-.06}\, e^{-.06x}\right)\Big|_0^7$$

$$= \frac{1000}{-.06}(e^{-.42} - 1)$$

$$= \$5715.89$$

61. $\displaystyle\int_0^b Re^{-kt}\, dt$

$$= \int_0^\infty 50{,}000e^{-.09t}\, dt$$

$$= \lim_{b\to\infty} \left(\frac{50{,}000e^{-.09t}}{-.09}\right)\Big|_0^b$$

$$= \lim_{b\to\infty} \left[\frac{50{,}000}{-.09}\, e^{-.09(b)} + \frac{50{,}000}{.09}(1)\right]$$

$$= \lim_{b\to\infty} \left[\frac{-50{,}000}{.09e^{.09b}} + \frac{50{,}000}{.09}\right]$$

$$= \$555{,}555.56$$

As $b \to \infty$, $e^{.09b} \to \infty$, and so

$$\frac{-50{,}000}{.09e^{.09b}} \to 0.$$

63. $f(x) = 100e^{-.05x}$

The total amount of oil is

$$\int_0^\infty 100e^{-.05x}\, dx$$

$$= \lim_{b\to\infty} \int_0^b 100e^{-.05x}\, dx$$

$$= \lim_{b\to\infty}\left(\frac{100e^{-.05x}}{-.05}\right)\Big|_0^b$$

$$= \lim_{b\to\infty}\left(\frac{-2000}{e^{.05b}} + 2000\right)$$

$$= 0 + 2000$$

$$= 2000 \text{ gal.}$$

65. This exercise should be solved by computer or programmable calculator methods. The solutions will vary according to the method that is used.

The answers are given.

(a) .68270

(b) .95450

(c) .99730

(d) .99994

(e) The values approach 1 as n increases. The value of the integral over the interval $(-\infty, \infty)$ is 1.

Extended Application:

1. $\int_0^W c\left(1 - \dfrac{t}{w}\right)\left(\dfrac{N}{m}\right)e^{-t/m}\ dt$

$= \dfrac{Nc}{m}\left[\int_0^W \left(1 - \dfrac{t}{w}\right)e^{-t/m}\ dt\right]$

$= \dfrac{Nc}{m}\left[\int_0^W e^{-t/m}\ dt - \int_0^W \dfrac{t}{w}e^{-t/m}\ dt\right]$

Integrate $\int \dfrac{t}{w}e^{-t/m}\,dt$ by parts.

Let $u = \dfrac{t}{w}$ and $dv = e^{-t/m}dt$.

Use column integration.

D	I
$\dfrac{t}{w}$	$e^{-t/m}$
$\dfrac{1}{w}$	$-me^{-t/m}$
0	$m^2 e^{-t/m}$

$\int \dfrac{t}{w}\,e^{-t/m}\ dt$

$= -\dfrac{mt}{w}\,e^{-t/m} - \dfrac{m^2}{w}\,e^{-t/m} + C$

$\dfrac{Nc}{m}\left[\int_0^W e^{-t/m}\,dt - \int_0^W \dfrac{t}{w}\,e^{-t/m}\,dt\right]$

$= \dfrac{Nc}{m}\Big[-me^{-t/m}$

$\quad - \left(-\dfrac{mt}{w}\,e^{-t/m} - \dfrac{m^2}{w}\,e^{-t/m}\right)\Big]\Big|_0^W$

$= \dfrac{Nc}{m}\left(-me^{-t/m} + \dfrac{mt}{w}\,e^{-t/m} + \dfrac{m^2}{w}\,e^{-t/m}\right)\Big|_0^W$

$= \dfrac{Nc}{m}\left(-me^{-w/m} + me^{-w/m} + \dfrac{m^2}{w}\,e^{-w/m}\right.$

$\quad\left. + m - 0 - \dfrac{m^2}{w}\right)$

$= \dfrac{Nc}{m}\left(\dfrac{m^2}{w}\,e^{-w/m} + m - \dfrac{m^2}{w}\right)$

$= Nc\left(\dfrac{m}{w}\,e^{-w/m} + 1 - \dfrac{m}{w}\right)$

$= Nc\left(1 - \dfrac{m}{w} + \dfrac{m}{w}\,e^{-w/m}\right)$

2. $r = 50\left(1 - \dfrac{48}{24} + \dfrac{48}{24}\,e^{-24/48}\right)$

$= 50(1 - 2 + 2e^{-.5})$

$= 10.65$

3. $r = 1000\left(1 - \dfrac{60}{24} + \dfrac{60}{24}\,e^{-24/60}\right)$

$= 1000(1 - 2.5 + 2.5e^{-.4})$

≈ 175.8

4. $r = 1200\left(1 - \dfrac{30}{30} + \dfrac{30}{30}\,e^{-1}\right)$

$= 1200(e^{-1})$

≈ 441.66

CHAPTER 7 TEST

[7.1] **Use integration by parts or column integration to find the integrals.**

1. $\displaystyle\int x \ln |5x| \; dx$

2. $\displaystyle\int x\sqrt{2x - 1} \; dx$

3. $\displaystyle\int (2x - 1)e^x \; dx$

4. $\displaystyle\int \frac{x + 3}{(3x - 1)^4} \; dx$

[7.1] 5. Find the area between the x-axis and the graph of $f(x) = x\sqrt[3]{x - 3}$ from $x = 4$ to $x = 11$.

[7.1] 6. Given that the marginal profit in dollars earned from the sale of x computers is

$$P'(x) = xe^{.001x} - 100,$$

find the total profit from the sale of the first 500 computers.

[7.2] 7. Use the trapezoidal rule with $n = 4$ to approximate the value of

$$\int_0^2 \sqrt{1 + x^3} \; dx.$$

[7.2] 8. Use Simpson's rule with $n = 6$ to approximate the value of $\displaystyle\int_0^3 \frac{1}{x^2 + 1} \; dx$.

[7.2] 9. Find the area between the curve $y = e^{-x^2}$ and the x-axis from $x = 0$ to $x = 3$, using the trapezoidal rule with $n = 6$.

[7.2] 10. Find the area between the curve $y = \dfrac{x}{x^2 + 1}$ and the x-axis from $x = 1$ to $x = 4$, using Simpson's rule with $n = 6$.

[7.3] **Find the volume of the solid of revolution formed by rotating each of the following bounded regions about the x-axis.**

11. $f(x) = 3x - 1$, $y = 0$, $x = 4$

12. $f(x) = \dfrac{1}{\sqrt{3x + 1}}$, $y = 0$, $x = 0$, $x = 5$

13. $f(x) = e^x$, $y = 0$, $x = -1$, $x = 2$

[7.3]   14. Find the average value of $f(x) = x^3 - x^2$ from $a = -1$ to $b = 1$.

[7.3]   15. The rate of depreciation t years after purchase of a certain machine is

$$D = 10,000(t - 7), \quad 0 \le t \le 6.$$

What is the average depreciation over the first 3 yr?

[7.4]   16. The rate of flow of an investment is $2000 per year for 10 yr. Find the present value if the annual interest rate is 11% compounded continuously.

[7.4]   17. The rate of flow of an investment is given by

$$f(x) = 200e^{-.1x}$$

for 10 yr. Find the present value if the annual interest rate is 8% compounded continuously.

[7.4]   18. The rate of flow of an investment is given by the function $f(x) = 1000$. Find the final amount at the end of 15 yr at an interest rate of 11% compounded continuously.

[7.4]   19. An investment scheme is expected to produce a continuous flow of money, starting at $2000 and increasing exponentially at 4% per year for 8 yr. Find the present value at an interest rate of 12% compounded continuously.

[7.5]   **Find the value of each interval that converges.**

20. $\displaystyle\int_{2}^{\infty} \frac{14}{(x + 1)^2} \, dx$ 21. $\displaystyle\int_{-\infty}^{2} \frac{4}{x + 1} \, dx$

22. $\displaystyle\int_{1}^{\infty} e^{-3x} \, dx$

[7.5]   **Find the area between the graph of the function and the x-axis over the given interval, if possible.**

23. $f(x) = 8e^{-x}$ on $[1, \infty)$ 24. $f(x) = \dfrac{1}{(2x - 1)^2}$ on $[2, \infty)$

[7.5]   25. Find the capital value of an asset that generates income at an annual rate of $4000 if the interest rate is 8% compounded continuously.

CHAPTER 7 TEST ANSWERS

1. $\frac{1}{2}x^2 \ln |5x| - \frac{1}{4}x^2 + C$

2. $\frac{x}{3}(2x - 1)^{3/2} - \frac{1}{15}(2x - 1)^{5/2} + C$

3. $(2x - 3)e^x + C$

4. $-\frac{1}{9}\left[\frac{x + 3}{(3x - 1)^3}\right] - \frac{1}{54(3x - 1)^2} + C$ or $-\frac{9x + 17}{54(3x - 1)^3} + C$

5. 88.18 6. $\$125,639.36$ 7. 3.28 8. 1.25 9. $.89$ 10. 1.07

11. $\frac{1331}{9}\pi$ or 464.61 12. 2.90 13. 85.55 14. $-\frac{1}{3}$

15. $55,000$ per year 16. $\$12,129.62$ 17. $\$927.45$ 18. $\$38,245.27$

19. $\$11,817.69$ 20. $\frac{14}{3}$ 21. Divergent 22. $.017$

23. $\frac{8}{e}$ or 2.943 24. $\frac{1}{6}$ 25. $\$50,000$

CHAPTER 8   MULTIVARIABLE CALCULUS

Section 8.1

1. $f(x, y) = 4x + 5y + 3$

    (a) $f(2, -1) = 4(2) + 5(-1) + 3$
    $$= 6$$

    (b) $f(-4, 1) = 4(-4) + 5(1) + 3$
    $$= -8$$

    (c) $f(-2, -3) = 4(-2) + 5(-3) + 3$
    $$= -20$$

    (d) $f(0, 8) = 4(0) + 5(8) + 3$
    $$= 43$$

3. $h(x, y) = = \sqrt{x^2 + 2y^2}$

    (a) $h(5, 3) = \sqrt{25 + 2(9)} = \sqrt{43}$

    (b) $h(2, 4) = \sqrt{4 + 32} = 6$

    (c) $h(-1, -3) = \sqrt{1 + 18} = \sqrt{19}$

    (d) $h(-3, -1) = \sqrt{9 + 2} = \sqrt{11}$

For Exercises 5–15, see the graphs in the answer section at the back of your textbook.

5. $x + y + z = 6$

    If $x = 0$ and $y = 0$, $z = 6$.
    If $x = 0$ and $z = 0$, $y = 6$.
    If $y = 0$ and $z = 0$, $x = 6$.

7. $2x + 3y + 4z = 12$

    If $x = 0$ and $y = 0$, $z = 3$.
    If $x = 0$ and $z = 0$, $y = 4$.
    If $y = 0$ and $z = 0$, $x = 6$.

9. $x + y = 4$

    If $x = 0$, $y = 4$.
    If $y = 0$, $x = 4$.

    There is no z-intercept.

11. $x = 2$

    The point $(2, 0, 0)$ is on the graph.
    There are no y- or z-intercepts.
    The plane is parallel to the yz-plane.

13. $3x + 2y + z = 18$

    For $z = 0$, $3x + 2y = 18$. Graph the line $3x + 2y = 18$ in the xy-plane.
    For $z = 2$, $3x + 2y = 16$. Graph the line $3x + 2y = 16$ in the plane $z = 2$.
    For $z = 4$, $3x + 2y = 14$. Graph the line $3x + 2y = 14$ in the plane $z = 4$.

15. $y^2 - x = -z$

    For $z = 0$, $x = y^2$. Graph $x = y^2$ in the xy-plane.
    For $z = 2$, $x = y^2 + 2$. Graph $x = y^2 + 2$ in the plane $z = 2$.
    For $z = 4$, $x = y^2 + 4$. Graph $x = y^2 + 4$ in the plane $z = 4$.

17. $P(x, y) = 100\left[\frac{3}{5}x^{-2/5} + \frac{2}{5}y^{-2/5}\right]^{-5}$

    (a) $P(32, 1)$
    $$= 100\left[\frac{3}{5}(32)^{-2/5} + \frac{2}{5}(1)^{-2/5}\right]^{-5}$$
    $$= 100\left[\frac{3}{5}\left(\frac{1}{4}\right) + \frac{2}{5}(1)\right]^{-5}$$

$$= 100\left[\frac{11}{20}\right]^{-5}$$

$$= 100\left(\frac{20}{11}\right)^{5}$$

$$\approx 1986.95$$

The production is approximately 1987 cameras.

(b) P(1, 32)

$$= 100\left[\frac{3}{5}(1)^{-2/5} + \frac{2}{5}(32)^{-2/5}\right]^{-5}$$

$$= 100\left[\frac{3}{5}(1) + \frac{2}{5}\left(\frac{1}{4}\right)\right]^{-5}$$

$$= 100\left(\frac{7}{10}\right)^{-5}$$

$$= 100\left(\frac{10}{7}\right)^{5}$$

$$\approx 595$$

The production is approximately 595 cameras.

(c) 32 work hours means that x = 32. 243 units of capital means that y = 243.

P(32, 243)

$$= 100\left[\frac{3}{5}(32)^{-2/5} + \frac{2}{5}(243)^{-2/5}\right]^{-5}$$

$$= 100\left[\frac{3}{5}\left(\frac{1}{4}\right) + \frac{2}{5}\left(\frac{1}{9}\right)\right]^{-5}$$

$$= 100\left(\frac{7}{36}\right)^{-5}$$

$$= 100\left(\frac{36}{7}\right)^{5}$$

$$\approx 359,767.81$$

The production is approximately 359,768 cameras.

19. M = f(n, i, t)

$$= \frac{(1 + i)^n(1 - t) + t}{[1 + (1 - t)i]^n}$$

Therefore,

f(25, .06, .33)

$$= \frac{(1 + .06)^{25}(1 - .33) + .33}{[1 + (1 - .33)(.06)]^{25}}$$

$$= \frac{(1.06)^{25}(.67) + .33}{[1 + (.67)(.06)]^{25}}$$

$$\approx 1.197$$

Since the value of M ≈ 1.197, which is greater than 1, the IRA account grows faster.

21. $z = x^{.4}y^{.6}$ where z = 500

$$500 = x^{2/5}\,y^{3/5}$$

$$\frac{500}{x^{2/5}} = y^{3/5}$$

$$\left(\frac{500}{x^{2/5}}\right)^{5/3} = (y^{3/5})^{5/3}$$

$$y = \frac{(500)^{5/3}}{x^{2/3}}$$

$$y \approx \frac{31,498}{x^{2/3}}$$

See answer graph in the back of your textbook.

23. C(x, y, z) = 200x + 100y + 50z

25. $A = .202W^{.425}H^{.725}$

(a) $A = .202(72)^{.425}(1.78)^{.725}$

$$\approx 1.89 \text{ m}^2$$

(b) $A = .202(65)^{.425}(1.40)^{.725}$

$$\approx 1.52 \text{ m}^2$$

(c) $A = .202(70)^{.425}(1.60)^{.725}$

$$\approx 1.73 \text{ m}^2$$

(d) Answers vary.

27. Let the area be given by $g(L, W, H)$. Then,

$$g(L, W, H) = 2LW + 2WH + 2LH \text{ ft}^2$$

29. $z = x^2 + y^2$

The xy-trace is

$$z = x^2 + 0 = x^2.$$

The yz-trace is

$$z = 0 + y^2 = y^2.$$

Both are parabolas with vertices at the origin that open upward.
The xy-trace is

$$0 = x^2 + y^2.$$

This is a point, the origin.
The equation is represented by a paraboloid as shown in (c).

31. $x^2 - y^2 = z$

The xz-trace is

$$x^2 = z,$$

which is a parabola with vertex at the origin that opens upward.
The yz-trace is

$$-y^2 = z,$$

which is a parabola with vertex at the origin that opens downward.
The xy-trace is

$$x^2 - y^2 = 0$$
$$x^2 = y^2$$
$$x = y \quad \text{or} \quad x = -y,$$

which are two lines that intersect at the origin.

The equation is represented by a hyperbolic paraboloid, as shown in (e).

33. $\dfrac{x^2}{16} + \dfrac{y^2}{25} + \dfrac{z^2}{4} = 1$

xz-trace:

$\dfrac{x^2}{16} + \dfrac{z^2}{4} = 1$, an ellipse

yz-trace:

$\dfrac{y^2}{25} + \dfrac{z^2}{4} = 1$, and ellipse

xy-trace:

$\dfrac{x^2}{16} + \dfrac{y^2}{25} = 1$, an ellipse

The graph is an ellipsoid as shown in (b).

35. **(a)** If $f(x, y) = 9x^2 - 3y^2$, then

$$\frac{f(x + h, y) - f(x, y)}{h}$$

$$= \frac{[9(x + h)^2 - 3y^2] - [9x^2 - 3y^2]}{h}$$

$$= \frac{9x^2 + 18xh + 9h^2 - 3y^2 - 9x^2 + 3y^2}{h}$$

$$= \frac{h(18x + 9h)}{h}$$

$$= 18x + 9h.$$

(b) If $f(x, y) = 9x^2 - 3y^2$, then

$$\frac{f(x, y + h) - f(x, y)}{h}$$

$$= \frac{[9x^2 - 3(y + h)^2] - [9x^2 - 3y^2]}{h}$$

$$= \frac{9x^2 - 3y^2 - 6yh - 3h^2 - 9x^2 + 3y^2}{h}$$

$$= \frac{-h(6y + 3h)}{h}$$

$$= -(6y + 3h)$$

$$= -6y - 3h.$$

Section 8.2

1. $z = f(x, y) = 12x^2 - 8xy + 3y^2$

(a) $\dfrac{\partial z}{\partial x} = 24x - 8y$

(b) $\dfrac{\partial z}{\partial y} = -8x + 6y$

(c) $\dfrac{\partial f}{\partial x}(2, 3) = 24(2) - 8(3) = 24$

(d) $f_y(1, -2) = -8(1) + 6(-2)$
$$= -20$$

3. $f(x, y) = -2xy + 6y^3 + 2$

$f_x = -2y$

$f_y = -2x + 18y^2$

$f_x(2, -1) = -2(-1) = 2$

$f_y(-4, 3) = -2(-4) + 18(3)^2$
$$= 8 + 18(9)$$
$$= 170$$

5. $f(x, y) = 3x^3y^2$

$f_x = 9x^2y^2$

$f_y = 6x^3y$

$f_x(2, -1) = 9(4)(1) = 36$

$f_y(-4, 3) = 6(-64)3 = -1152$

7. $f(x, y) = e^{x+y}$

$f_x = e^{x+y}$

$f_y = e^{x+y}$

$f_x(2, 1) = e^{2-1} = e^1 = e$

$f_y(-4, 3) = e^{-4+3} = e^{-1} = \dfrac{1}{e}$

9. $f(x, y) = -5e^{3x-4y}$

$f_x = -15e^{3x-4y}$

$f_y = 20e^{3x-4y}$

$f_x(2, -1) = -15e^{10}$

$f_y(-4, 3) = 20e^{-12-12} = 20e^{-24}$

11. $f(x, y) = \dfrac{x^2 + y^3}{x^3 - y^2}$

$f_x = \dfrac{2x(x^3 - y^2) - 3x^2(x^2 + y^3)}{(x^3 - y^2)^2}$

$\quad = \dfrac{2x^4 - 2xy^2 - 3x^4 - 3x^2y^3}{(x^3 - y^2)^2}$

$\quad = \dfrac{-x^4 - 2xy^2 - 3x^2y^3}{(x^3 - y^2)^2}$

$f_y = \dfrac{3y^2(x^3 - y^2) - (-2y)(x^2 + y^3)}{(x^3 - y^2)^2}$

$\quad = \dfrac{3x^3y^2 - 3y^4 + 2x^2y + 2y^4}{(x^3 - y^2)^2}$

$\quad = \dfrac{3x^3y^2 - y^4 + 2x^2y}{(x^3 - y^2)^2}$

$f_x(2, -1)$

$\quad = \dfrac{-2^4 - 2(2)(-1)^2 - 3(2^2)(-1)^3}{[2^3 - (-1)^2]^2}$

$\quad = -\dfrac{8}{49}$

$f_y(-4, 3)$

$\quad = \dfrac{3(-4)^3(3)^2 - 3^4 + 2(-4)^2(3)}{[(-4)^3 - 3^2]^2}$

$\quad = -\dfrac{1713}{5329}$

13. $f(x, y) = \ln|1 + 3x^2y^3|$

$f_x = \dfrac{1}{1 + 3x^2y^3} \cdot 6xy^3$

$\quad = \dfrac{6xy^3}{1 + 3x^2y^3}$

$f_y = \dfrac{9x^2y^2}{1 + 3x^2y^3}$

$f_x(2, -1) = \dfrac{6(2)(-1)}{1 + 3(4)(-1)}$

$\qquad\quad = \dfrac{-12}{1 - 12} = \dfrac{12}{11}$

$f_y(-4, 3) = \dfrac{9(16)(9)}{1 + 3(16)(27)}$

$\qquad\quad = \dfrac{1296}{1297}$

15. $f(x, y) = xe^{x^2 y}$

$f_x = e^{x^2 y} \cdot 1 + x(2xy)(e^{x^2 y})$

$\quad = e^{x^2 y}(1 + 2x^2 y)$

$f_y = x^3 e^{x^2 y}$

$f_x(2, -1) = e^{-4}(1 - 8) = -7e^{-4}$

$f_y(-4, 3) = -64e^{48}$

17. $f(x, y) = 6x^3 y - 9y^2 + 2x$

$f_x = 18x^2 y + 2$

$f_y = 6x^3 - 18y$

$f_{xx} = 36xy$

$f_{yy} = -18$

$f_{xy} = 18x^2 = f_{yx}$

19. $R(x, y) = 4x^2 - 5xy^3 + 12y^2 x^2$

$R_x = 8x - 5y^3 + 24y^2 x$

$R_y = -15xy^2 + 24yx^2$

$R_{xx} = 8 + 24y^2$

$R_{yy} = -30xy + 24x^2$

$R_{xy} = -15y^2 + 48xy = R_{yx}$

21. $r(x, y) = \dfrac{4x}{x + y}$

$r_x = \dfrac{4(x + y) - 4x}{(x + y)^2}$

$\quad = 4y(x + y)^{-2}$

$r_y = \dfrac{-4x}{(x + y)^2}$

$r_{xx} = -8y(x + y)^{-3} = \dfrac{-8y}{(x + y)^3}$

$r_{yy} = 8x(x + y)^{-3} = \dfrac{8x}{(x + y)^3}$

$r_{xy} = 4(x - y)(x + y)^{-3}$

$\quad = \dfrac{4x - 4y}{(x + y)^3} = r_{yx}$

23. $z = 4xe^y$

$z_x = 4e^y$

$z_y = 4xe^y$

$z_{xx} = 0$

$z_{yy} = 4xe^y$

$z_{xy} = 4e^y = z_{yx}$

25. $r = \ln |x + y|$

$r_x = \dfrac{1}{x + y}$

$r_y = \dfrac{1}{x + y}$

$r_{xx} = \dfrac{-1}{(x + y)^2}$

$r_{yy} = \dfrac{-1}{(x + y)^2}$

$r_{xy} = \dfrac{-1}{(x + y)^2} = r_{yx}$

27. $z = x \ln |xy|$

$z_x = \ln |xy| + 1$

$z_y = \dfrac{x}{y}$

$z_{xx} = \dfrac{1}{x}$

$z_{yy} = -xy^{-2} = \dfrac{-x}{y^2}$

$z_{xy} = \dfrac{1}{y} = z_{yx}$

29. $f(x, y) = 6x^2 + 6y^2 + 6xy + 36x - 5$

First, $f_x = 12x + 6y + 36$ and
$f_y = 12y + 6x$.

We must solve the system

$$12x + 6y + 36 = 0$$
$$12y + 6x = 0.$$

Multiply both sides of the first
equation by -2 and add.

$$-24x - 12y - 72 = 0$$
$$\underline{6x + 12y \qquad = 0}$$
$$-18x \qquad - 72 = 0$$
$$x = -4$$

Substitute into either equation to get $y = 2$.

31. $f(x, y) = 9xy - x^3 - y^3 - 6$

First, $f_x = 9y - 3x^2$ and $f_y = 9x - 3y^2$.

We must solve the system

$$9y - 3x^2 = 0$$
$$9x - 3y^2 = 0.$$

From the first equation, $y = \frac{1}{3}x^2$.

Substitute into the second equation to get

$$9x - 3\left(\frac{1}{3}x^2\right)^2 = 0$$

$$9x - 3\left(\frac{1}{9}x^4\right) = 0$$

$$9x - \frac{1}{3}x^4 = 0.$$

Multiply by 3 to get

$$27x - x^4 = 0.$$

Now factor:

$$x(27 - x^3) = 0.$$

Set each factor equal to 0.

$$x = 0 \quad \text{or} \quad 27 - x^3 = 0$$
$$x = 3$$

Substitute into $y = \frac{x^2}{3}$.

$$y = 0 \quad \text{or} \quad y = 3$$

The solutions are $x = 0$, $y = 0$ and $x = 3$, $y = 3$.

33. $f(x, y, z) = x^2 + yz + z^4$

$f_x = 2x$

$f_y = z$

$f_z = y + 4z^3$

$f_{yz} = 1$

35. $f(x, y, z) = \dfrac{6x - 5y}{4z + 5}$

$f_x = \dfrac{6}{4z + 5}$

$f_y = \dfrac{-5}{4z + 5}$

$f_z = \dfrac{-4(6x - 5y)}{(4z + 5)^2}$

$f_{yz} = \dfrac{20}{(4z + 5)^2}$

37. $f(x, y, z) = \ln \left| x^2 - 5xz^2 + y^4 \right|$

$f_x = \dfrac{2x - 5z^2}{x^2 - 5xz^2 + y^4}$

$f_y = \dfrac{4y^3}{x^2 - 5xz^2 + y^4}$

$f_z = \dfrac{-10xz}{x^2 - 5xz^2 + y^4}$

$f_{yz} = \dfrac{4y^3(10zx)}{(x^2 - 5xz^2 + y^4)^2}$

$\phantom{f_{yz}} = \dfrac{40xy^3z}{(x^2 - 5xz^2 + y^4)^2}$

39. $M(x, y) = 40x^2 + 30y^2 - 10xy + 30$

(a) $\quad M_y = 60y - 10x$

$\quad M_y(4, 2) = 120 - 40 = 80$

(b) $\quad M_x = 80x - 10y$

$\quad M_x(3, 6) = 240 - 60 = 180$

(c) $\dfrac{\partial M}{\partial x}(2, 5) = 80(2) - 10(5)$

$\qquad = 110$

(d) $\dfrac{\partial M}{\partial y}(6, 7) = 60(7) - 10(6)$

$\qquad = 360$

41. $f(p, i) = 132p - 2pi - .01p^2$

(a) $f(9400, 8)$

$$= 132(9400) - 2(9400)(8)$$
$$- .01(9400)^2$$
$$= \$206,800$$

The weekly sales are $206,800.

(b) $f_p = 132 - 2i - .02p$, which represents the rate of change in weekly sales revenue per unit change in price when the interest rate remains constant.

$f_i = -2p$, which represents the rate of change in weekly sales revenue per unit change in interest rate when the list price remains constant.

(c) $p = 9400$ remains constant and i changes by 1 unit from 8 to 9.

$$f_i(p, i) = f_i(9400, 8)$$
$$= -2(9400)$$
$$= -18,800$$

Therefore, sales revenue declines by $18,800.

43. $f(x, y) = \left(\frac{1}{3}x^{-1/3} + \frac{2}{3}y^{-1/3}\right)^{-3}$

(a) $f(27, 64)$

$$= \left[\frac{1}{3}(27)^{-1/3} + \frac{2}{3}(64)^{-1/3}\right]^{-3}$$
$$= \left[\frac{1}{3}\left(\frac{1}{3}\right) + \frac{2}{3}\left(\frac{1}{4}\right)\right]^{-3}$$
$$= \left(\frac{1}{9} + \frac{1}{6}\right)^{-3}$$
$$= \left(\frac{5}{18}\right)^{-3}$$
$$= \left(\frac{18}{5}\right)^{3}$$
$$= 46.656$$

The production is 46.656 hundred units.

(b) $f_x = -3\left(\frac{1}{3}x^{-1/3} + \frac{2}{3}y^{-1/3}\right)\left(-\frac{1}{9}x^{-4/3}\right)$

$f_x(27, 64)$

$$= -3\left[\frac{1}{3}(27)^{-1/3} + \frac{2}{3}(64)^{-1/3}\right]^{-4}$$
$$\cdot \left(-\frac{1}{9}\right)(27)^{-4/3}$$
$$= -3\left(\frac{5}{18}\right)^{-4}\left(-\frac{1}{9}\right)\left(\frac{1}{81}\right)$$
$$= \frac{432}{625}$$

$f_x(27, 64) = .6912$ hundred units, which represents the rate at which production is changing when labor changes by 1 unit from 27 to 28 and capital remains constant.

$f_y = -3\left(\frac{1}{3}x^{-1/3} + \frac{2}{3}y^{-1/3}\right)^{-4}\left(-\frac{2}{9}y^{-4/3}\right)$

$f_y(27, 64)$

$$= -3\left[\frac{1}{3}(27)^{-1/3} + \frac{2}{3}(64)^{-1/3}\right]^{-4}$$
$$\cdot \left(-\frac{2}{9}\right)(64)^{-4/3}$$
$$= -3\left(\frac{5}{18}\right)^{-4}\left(-\frac{2}{9}\right)\left(\frac{1}{256}\right)$$
$$= \frac{2187}{5000}$$
$$= .4374 \text{ hundred units,}$$

which represents the rate at which production is changing when capital changes by 1 unit from 64 to 65 and labor remains constant.

(c) If labor increases by 1 unit then production would increase at the rate of

$f_x(x, y)$

$= \frac{1}{3}x^{-4/3}\left(\frac{1}{3}x^{-1/3} + \frac{2}{3}y^{-1/3}\right)^{-4}.$

(See part (b) of this solution.)

45. $z = x^{.4}y^{.6}$

The marginal productivity of labor is

$\frac{\partial z}{\partial x} = .4x^{-.6}y^{.6} + x^{.4}\cdot 0$

$= .4x^{-.6}y^{.6}.$

The marginal productivity of capital is

$\frac{\partial z}{\partial y} = x^{.4}(.6y^{-.4}) + y^{.6}\cdot 0$

$= .6x^{.4}y^{-.4}.$

47. $M(x, y) = 2xy + 10xy^2 + 30y^2 + 20$

(a) $\frac{\partial M}{\partial x} = 2y + 10y^2$

$\frac{\partial M}{\partial x}(20, 4) = 168$

(b) $\frac{\partial M}{\partial y} = 2x + 20yx + 60y$

$\frac{\partial M}{\partial y}(24, 10) = 5448$

(c) An increase in days since rain causes more of an increase in matings.

49. $A(W, H) = .202W^{.425}H^{.725}$

(a) $\frac{\partial A}{\partial W}(72, 1.8)$

$= .08585(72)^{-.575}(1.8)^{.725}$

$= .0112$

(b) $\frac{\partial A}{\partial H}(70, 1.6)$

$= .14645(70)^{.425}(1.6)^{-.275}$

$\approx .783$

51. $f(n, c) = \frac{1}{8}n^2 - \frac{1}{5}c + \frac{1937}{8}$

(a) $f(4, 1200)$

$= \frac{1}{8}(4)^2 - \frac{1}{5}(1200) + \frac{1937}{8}$

$= 2 - 240 + \frac{1937}{8}$

$= 4.125$ lb

(b) $\frac{\partial f}{\partial n} = \frac{1}{8}(2n) - \frac{1}{5}(0) + 0$

$= \frac{1}{4}n,$

which represents the rate of change of weight loss per unit change in number of workouts.

(c) $f_n(3, 1100) = \frac{1}{4}(3)$

$= \frac{3}{4}$ lb

represents an additional weight loss by adding the fourth workout.

53. $p = f(s, n, a) = .003a + .1(sn)^{1/2}$

(a) $f(8, 6, 450)$

$= .003(450) + .1[(8)(6)]^{1/2}$

$= 1.35 + .1(48)^{1/2}$

$= 2.0428$

$p \approx 2.04\%$

(b) $f(3, 3, 320)$

$= .003(320) + .1[(3)(3)]^{1/2}$

$= .96 + .1(9)^{1/2}$

$= 1.26$

$p = 1.26\%$

(c) $f_n = .003(0) + .1\left[\frac{1}{2}(sn)^{-1/2}(s)\right]$

$f_n(3, 3, 320) = .1\left(\frac{1}{2}\right)[(3)(3)]^{-1/2}(3)$

$\qquad = (.1)\left(\frac{1}{2}\right)\left(\frac{1}{3}\right)(3)$

$\qquad = .05$

$\qquad p_n = .05\%$

$f_a = .003$ for all ordered triples (s, n, a). Therefore, $p_a = .003\%$.

$p_n = .05\%$ is the rate of change of the probability for an additional semester of high school math.

$p_a = .003\%$ is the rate of change of the probability per unit of change in an SAT score.

Section 8.3

1. $f(x, y) = xy + x - y$

$f_x = y + 1, \; f_y = x - 1$

If $f_x = 0$, $y = -1$.

If $f_y = 0$, $x = 1$.

Therefore, $(1, -1)$ is the critical point.

$\qquad f_{xx} = 0,$
$\qquad f_{yy} = 0,$
$\qquad f_{xy} = 1.$

For $(1, -1)$,

$\qquad D = 0 \cdot 0 - 1^2 = -1 < 0.$

A saddle point is at $(1, -1)$.

3. $f(x, y) = x^2 - 2xy + 2y^2 + x - 5$

$f_x = 2x - 2y + 1, \; f_y = -2x + 4y$

Solve the system $f_x = 0$, $f_y = 0$.

$\qquad 2x - 2y + 1 = 0$
$\qquad \underline{-2x + 4y \qquad = 0}$
$\qquad \qquad 2y + 1 = 0$

$\qquad \qquad y = -\frac{1}{2}$

$\qquad -2x + 4\left(-\frac{1}{2}\right) = 0$

$\qquad \qquad -2x = 2$

$\qquad \qquad x = -1$

Therefore, $(-1, -1/2)$ is the critical point.

$\qquad f_{xx} = 2,$
$\qquad f_{yy} = 4,$
$\qquad f_{xy} = -2.$

For $\left(-1, -\frac{1}{2}\right)$,

$\qquad D = 2 \cdot 4 - (-2)^2 = 4 > 0.$

Since $f_{xx} = 2 > 0$, then a relative minimum is at $(-1, -1/2)$.

5. $f(x, y) = x^2 - xy + y^2 + 2x + 2y + 6$

$f_x = 2x - y + 2, \; f_y = -x + 2y + 2$

Solve the system $f_x = 0$, $f_y = 0$.

$\qquad 2x - y + 2 = 0$
$\qquad \underline{-x + 2y + 2 = 0}$

$\qquad 2x - y + 2 = 0$
$\qquad \underline{-2x + 4y + 4 = 0}$
$\qquad \qquad 3y + 6 = 0$

$\qquad \qquad y = -2$

$\qquad -x + 2(-2) + 2 = 0$

$\qquad \qquad x = -2$

$(-2, -2)$ is the critical point.

$$f_{xx} = 2,$$
$$f_{yy} = 2,$$
$$f_{xy} = -1$$

For $(-2, -2)$

$$D = (2)(2) - (-1)^2 = 3 > 0.$$

Since $f_{xx} > 0$, a relative minimum is at $(-2, -2)$.

7. $f(x, y) = x^2 + 3xy + 3y^2 - 6x + 3y$
    $f_x = 2x + 3y - 6, f_y = 3x + 6y + 3$

Solve the system $f_x = 0$, $f_y = 0$.

$$2x + 3y - 6 = 0$$
$$3x + 6y + 3 = 0$$

$$\begin{array}{r} -4x - 6y + 12 = 0 \\ \underline{3x + 6y + 3 = 0} \\ -x + 15 = 0 \\ x = 15 \end{array}$$

$$3(15) + 6y + 3 = 0$$
$$6y = -48$$
$$y = -8$$

$(15, -8)$ is the critical point.

$$f_{xx} = 2,$$
$$f_{yy} = 6,$$
$$f_{xy} = 3$$

For $(15, -8)$,

$$D = 2 \cdot 6 - 9 = 3 > 0.$$

Since $f_{xx} > 0$, a relative minimum is at $(15, -8)$.

9. $f(x, y) = 4xy - 10x^2 - 4y^2 + 8x + 8y + 9$
    $f_x = 4y - 20x + 8, f_y = 4x - 8y + 8$

$$4y - 20x + 8 = 0$$
$$4x - 8y + 8 = 0$$

$$\begin{array}{r} 4y - 20x + 8 = 0 \\ \underline{-4y + 2x + 4 = 0} \\ -18x + 12 = 0 \\ x = \frac{2}{3} \end{array}$$

$$4y - 20\left(\frac{2}{3}\right) + 8 = 0$$

The critical point is $\left(\frac{2}{3}, \frac{4}{3}\right)$.

$$f_{xx} = -20,$$
$$f_{yy} = -8,$$
$$f_{xy} = 4$$

For $\left(\frac{2}{3}, \frac{4}{3}\right)$,

$$D = (-20)(-8) - 16 = 144 > 0.$$

Since $f_{xx} < 0$, a relative maximum is at $\left(\frac{2}{3}, \frac{4}{3}\right)$.

11. $f(x, y) = x^2 + xy - 2x - 2y + 2$
     $f_x = 2x + y - 2, f_y = x - 2$

$$2x + y - 2 = 0$$
$$x \qquad - 2 = 0$$
$$x = 2$$

$$2(2) + y - 2 = 0$$
$$y = -2$$

The critical point is $(2, -2)$.

$$f_{xx} = 2$$
$$f_{yy} = 0$$
$$f_{xy} = 1$$

For $(2, -2)$,

$$D = 2 \cdot 0 - 1^2 = -1 < 0.$$

A saddle point is at $(2, -2)$.

13. $f(x, y) = 2x^3 + 3y^2 - 12xy + 4$

$f_x = 6x^2 - 12y$, $f_y = 6y - 12x$

$$6x^2 - 12y = 0$$
$$6y - 12x = 0$$

If $6y - 12x = 0$, $y = 2x$.
Substitute for y in the first
equation.

$$6x^2 - 12(2x) = 0$$
$$6x(x - 4) = 0$$
$$x = 0 \quad \text{or} \quad x = 4$$

Then, $y = 0$ or $y = 8$.

The critical points are $(0, 0)$ and
$(4, 8)$.

$$f_{xx} = 12x,$$
$$f_{yy} = 6,$$
$$f_{xy} = -12$$

For $(0, 0)$,

$$D = 12(0)6 - (-12)^2$$
$$= -144 > 0.$$

A saddle point is at $(0, 0)$.
For $(4, 8)$,

$$D = 12(4)6 - (-12)^2$$
$$= 144 > 0.$$

Since $f_{xx} = 12(4) = 48 > 0$, a
relative minimum is at $(4, 8)$.

15. $f(x, y) = x^2 + 4y^3 - 6xy - 1$

$f_x = 2x - 6y$, $f_y = 12y^2 - 6x$

Solve $f_x = 0$ for x.

$$2x + 6y = 0$$
$$x = 3y$$

Substitute for x in $12y^2 - 6x = 0$.

$$12y^2 - 6(3y) = 0$$
$$6y(2y - 3) = 0$$
$$y = 0 \quad \text{or} \quad y = \frac{3}{2}$$

Then $x = 0$ or $x = \frac{9}{2}$.

The critical points are $(0, 0)$ and
$(9/2, 3/2)$.

$$f_{xx} = 2,$$
$$f_{yy} = 24y,$$
$$f_{xy} = -6$$

For $(0, 0)$,

$$D = 2 \cdot 24(0) - (-6)^2$$
$$= -36 < 0.$$

A saddle point is at $(0, 0)$.
For $(9/2, 3/2)$,

$$D = 2 \cdot 24\left(\frac{3}{2}\right) - (-6)^2$$
$$= 36 > 0.$$

Since $f_{xx} > 0$, a relative minimum is
at $(9/2, 3/2)$.

17. $f(x, y) = e^{xy}$

$$f_x = ye^{xy}$$
$$f_y = xe^{xy}$$
$$ye^{xy} = 0$$
$$xe^{xy} = 0$$
$$x = y = 0$$

The critical point is $(0, 0)$.

$$f_{xx} = y^2e^{xy},$$
$$f_{yy} = x^2e^{xy},$$
$$f_{xy} = e^{xy} + xye^{xy}$$

For $(0, 0)$,

$$D = 0 \cdot 0 - (e^0)^2 = -1 < 0.$$

A saddle point is at $(0, 0)$.

21. $z = -3xy + x^3 - y^3 + \dfrac{1}{8}$

$f_x = -3y + 3x^2$, $f_y = -3x - 3y^2$

Solve the system $f_x = 0$, $f_y = 0$.

$$-3y + 3x^2 = 0$$
$$-3x - 3y^2 = 0$$
$$-y + x^2 = 0$$
$$\underline{-x - y^2 = 0}$$

Solve the first equation for y, substitute into the second, and solve for x.

$$y = x^2$$
$$-x - x^4 = 0$$
$$x(1 + x^3) = 0$$
$$x = 0 \quad \text{or} \quad x = -1$$

Then $y = 0$ or $y = 1$.

The critical points are $(0, 0)$ and $(-1, 1)$.

$$f_{xx} = 6x$$
$$f_{yy} = -6y$$
$$f_{xy} = -3$$

For $(0, 0)$,

$$D = 0 \cdot 0 - (-3)^2 = -9 < 0.$$

A saddle point is at $(0, 0)$.
For $(-1, 1)$,

$$D = -6(-6) - (-3)^2 = 27 > 0.$$

$f_{xx} = 6(-1) = -6 < 0.$

$f(-1, 1)$

$\quad = -3(-1)(1) + (-1)^3 - 1^2 + \dfrac{1}{8}$

$\quad = 1\dfrac{1}{8}$

A relative maximum of $1\dfrac{1}{8}$ is at $(-1, 1)$.

The equation matches the graph in (a).

23. $z = y^4 - 2y^2 + x^2 - \dfrac{17}{16}$

$f_x = 2x$, $f_y = 4y^3 - 4y$

Solve the system $f_x = 0$, $f_y = 0$.

$$2x = 0 \quad (1)$$
$$4y^3 - 4y = 0 \quad (2)$$
$$4y(y^2 - 1) = 0$$
$$4y(y + 1)(y - 1) = 0$$

Equation 1 gives $x = 0$ and equation 2 gives $y = 0$, $y = -1$, or $y = 1$.
The critical points are $(0, 0)$, $(0, -1)$, and $(0, 1)$.

$$f_{xx} = 2,$$
$$f_{yy} = 12y^2 - 4,$$
$$f_{xy} = 0$$

For $(0, 0)$,

$$D = 2(12 \cdot 0^2 - 4) = -8 < 0.$$

A saddle point is at $(0, 0)$.
For $(0, -1)$,

$$D = 2[12(-1)^2 - 4] = 16 > 0.$$

$$f_{xx} = 2 > 0$$

$f(0, -1) = (-1)^4 - 2(-1)^2 + 0^2 - \dfrac{17}{16}$

$$= -2\dfrac{1}{16}$$

A relative minimum of $-2\dfrac{1}{16}$ is at $(0, -1)$.

For (0, 1),

$$D = 2(12 \cdot 1^2 - 4) = 16 > 0$$

$$f_{xx} = 2 > 0$$

$$f(0, 1) = 1^4 - 2 \cdot 1^2 + 0^2 - \frac{17}{16}$$

$$= -2\frac{1}{16}$$

A relative minimum of $-2\frac{1}{16}$ is at (0, 1).

The equation matches the graph of (b).

25. $z = -x^4 + y^4 + 2x^2 - 2y^2 + \frac{1}{16}$

$f_x = -4x^3 + 4x, \ f_y = 4y^3 - 4y$

Solve $f_x = 0, \ f_y = 0$.

$$-4x^3 + 4x = 0 \quad (1)$$
$$4y^3 - 4y = 0 \quad (2)$$

$$-4x(x^2 - 1) = 0 \quad (1)$$
$$-4x(x + 1)(x - 1) = 0$$

$$4y(y^2 - 1) = 0 \quad (2)$$
$$4y(y + 1)(y - 1) = 0$$

Equation 1 gives x = 0, −1, or 1.
Equation 2 gives y = 0, −1, or 1.

Critical points are (0, 0), (0, −1), (0, 1), (−1, 0), (−1, −1), (−1, 1), (1, 0), (1, −1), (1, 1).

$$f_{xx} = -12x^2 + 4,$$
$$f_{yy} = 12y^2 - 4$$
$$f_{xy} = 0$$

For (0, 0),

$$D = 4(-4) - 0 = 16 < 0.$$

For (0, −1),

$$D = 4(8) - 0 = 32 > 0,$$

and $f_{xx} = 4 > 0$.

$$f(0, -1) = -\frac{15}{16}$$

For (0, 1),

$$D = 4(8) - 0 = 32 > 0,$$

and $f_{xx} = 4 > 0$.

$$f(0, 1) = -\frac{15}{16}$$

For (−1, 0),

$$D = -8(-4) - 0 = 32 > 0,$$

and $f_{xx} = -8 < 0$.

$$f(-1, 0) = 1\frac{1}{16}$$

For (−1, −1),

$$D = -8(8) - 0 = -64 < 0.$$

For (−1, 1),

$$D = -8(8) - 0 = -64 < 0.$$

For (1, 0),

$$D = -8(-4) = 32 > 0.$$

and $f_{xx} = -8 < 0$.

$$f(1, 0) = 1\frac{1}{16}$$

For (1, −1),

$$D = -8(8) - 0 = -64 < 0.$$

For (1, 1),

$$D = -8(8) - 0 = -64 < 0.$$

Saddle points are at (0, 0), (−1, −1), (−1, 1), (1, −1), and (1, 1).

Relative maximum of $1\frac{1}{16}$ is at

$(-1, 0)$ and $(1, 0)$.

Relative minimum of $-\frac{15}{16}$ is at

$(0, -1)$ and $(0, 1)$.

27. $f(x, y) = 1 - x^4 - y^4$

$f_x = -4x^3$, $f_y = -4y^3$

The system

$f_x = -4x^3 = 0$, $f_y = -4y^3 = 0$

gives the critical point $(0, 0)$.

$$f_{xx} = -12x^2,$$
$$f_{yy} = -12y^3,$$
$$f_{xy} = 0$$

For $(0, 0)$,

$D = 0 \cdot 0 - 0^2 = 0.$

Therefore, the test gives no information. Examine a graph of
the function drawn by using level
curves.

If $f(x, y) = 1$, then $x^4 + y^4 = 0$.
The level curve is the point
$(0, 0, 1)$.

If $f(x, y) = 0$, then $x^4 + y^4 = 1$.
The level curve is the circle with
center $(0, 0, 0)$ and radius 1.

If $(x, y) = -15$, then $x^4 + y^4 = 16$.
The level curve is the curve with
center $(0, 0, -15)$ and radius 2.
The xz-trace is

$z = 1 - x^4.$

This curve has a maximum at
$(0, 0, 1)$ and opens downward.

The yz-trace is

$z = 1 - y^4.$

This curve also has a maximum at
$(0, 0, 1)$ and opens downward.

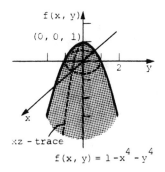

$f(x, y) = 1 - x^4 - y^4$

If $f(x, y) > 1$, then $x^4 + y^4 < 0$,
which is impossible, so the function does not exist. Thus, the
function has a relative maximum of
1 at $(0, 0)$.

29. $P(x, y) = 1000 + 24x - x^2 + 80y - y^2$

$$P_x = 24 - 2x$$
$$P_y = 80 - 2y$$
$$24 - 2x = 0$$
$$80 - 2y = 0$$
$$12 = x$$
$$40 = y$$

A critical point is $(12, 40)$.

$P_{xx} = -2$, $P_{yy} = -2$, $P_{xy} = 0$

For $(12, 40)$,

$D = -2(-2) - 0^2 = 4 > 0.$

Since $P_{xx} < 0$,

$$P(12, 40) = 1000 + 24(12) - (12)^2$$
$$+ 80(40) - (40)^2$$
$$= 2744$$

or $274,400 is the maximum profit.

31. $C(x, y) = 2x^2 + 3y^2 - 2xy$
$$+ 2x - 126y + 3800$$

$$C_x = 4x - 2y + 2$$
$$C_y = 6y - 2x - 126$$

$$0 = 4x - 2y + 2$$
$$0 = 6y - 2x - 126$$

$$0 = 2x - y + 1$$
$$\underline{0 = -2x + 6y - 126 \quad Add}$$
$$0 = \qquad 5y - 125$$
$$y = 25$$
$$0 = 2x - 25 + 1$$

If $y = 25$, $x = 12$.

(12, 25) is a critical point.

$$C_{xx} = 4$$
$$C_{yy} = 6$$
$$C_{xy} = -2$$

For (12, 25).

$$D = 4 \cdot 6 - 4 = 20 > 0.$$

Since $C_{xx} > 0$, 12 units of electrical tape and 25 units of packing tape should be produced to yield a minimum cost.

C(12, 25)
$$= 2(12)^2 + 3(25)^2 - 2(12 \cdot 25)$$
$$+ 2(12) - 126(25) + 3800$$
$$= 2237$$

The minimum cost is $2237.

Section 8.4

1. Maximize $f(x, y) = 2xy$
subject to $x + y = 12$.

1. Constraint $g(x, y) = x + y - 12$

2. Let $F(x, y, \lambda)$
$$= 2xy + \lambda(x + y - 12)$$

3. $F_x = 2y + \lambda$
$F_y = 2x + \lambda$
$F_\lambda = x + y - 12$

4. $2y + \lambda = 0$
$2x + \lambda = 0$
$x + y = 12$

5. $\lambda = -2y$ and $\lambda = -2x$
$$-2y = -2x$$
$$y = x$$

Substituting into the third equation gives

$$x + x = 12$$
$$x = 6.$$
Thus, $\qquad y = 6.$

Maximum is $f(6, 6) = 2 \cdot 6 \cdot 6 = 72.$

3. Maximize $f(x, y) = x^2y$,
subject to $2x + y = 4$.

1. $g(x, y) = 2x + y - 4$

2. $F(x, y, \lambda)$
$$= x^2y + \lambda(2x + y - 4)$$

3. $F_x = 2xy + 2\lambda$
$F_y = x^2 + \lambda$
$F_\lambda = 2x + y - 4$

4. $2xy + 2\lambda = 0$
$x^2 + \lambda = 0$
$2x + y = 4$

5. $xy = -\lambda$

$x^2 = -\lambda$

$$xy = x^2$$
$$x(y - x) = 0$$
$$y = 0 \quad \text{or} \quad y = x$$

Substituting $x = 0$ into the third equation gives

$$2(0) + y = 4$$
$$y = 4.$$

Substituting $y = x$ into the third equation gives

$$2x + x = 4$$
$$x = \frac{4}{3}.$$

Thus, $y = \frac{4}{3}.$

$$f(0, 4) = 0^2(4) = 0$$
$$f\left(\frac{4}{3}, \frac{4}{3}\right) = \frac{16}{9} \cdot \frac{4}{3} = \frac{64}{27} > 0$$

Therefore, $f\left(\frac{4}{3}, \frac{4}{3}\right) = \frac{64}{27} \approx 2.4$ is a maximum.

5. Minimize $f(x, y) = x^2 + 2y^2 - xy$, subject to $x + y = 8$.

1. $g(x, y) = x + y - 8$

2. $F(x, y, \lambda)$
   $= x^2 + 2y^2 - xy + \lambda(x + y - 8)$

3. $F_x = 2x - y + \lambda$
   $F_y = 4y - x + \lambda$
   $F_\lambda = x + y - 8$

4. $2x - y + \lambda = 0$
   $4y - x + \lambda = 0$
   $x + y - 8 = 0$

5. Subtracting the second equation from the first equation to eliminate λ gives the new system of equations

$$x + y = 8$$
$$3x - 5y = 0$$

$$5x + 5y = 40$$
$$\underline{3x - 5y = 0}$$
$$8x = 40$$
$$x = 5$$

But $x + y = 8$, so $y = 3$.

$$f(5, 3) = 25 + 18 - 15 = 28.$$

$f(5, 3) = 28$ is a minimum.

7. Maximize $f(x, y) = x^2 - 10y^2$, subject to $x - y = 18$.

1. $g(x, y) = x - y - 18$

2. $F(x, y, \lambda)$
   $= x^2 - 10y^2 + \lambda(x - y - 18)$

3. $F_x = 2x + \lambda$
   $F_y = -20y - \lambda$
   $F_\lambda = x - y - 18$

4. $2x + \lambda = 0$
   $-20y - \lambda = 0$
   $x - y - 18 = 0$

5. Adding the first two equations to elimimate λ gives

$$2x - 20y = 0$$
$$x = 10y.$$

Substituting $x = 10y$ in the third equation gives

$$10y - y = 18$$
$$y = 2$$

Thus, $x = 20.$

$$f(20, 2) = 20^2 - 10(2)^2$$
$$= 400 - 40 = 360.$$

$f(20, 2) = 360$ is a relative maximum.

9. Maximize $f(x, y, z) = xyz^2$, subject to $x + y + z = 6$.

1. $g(x, y, z) = x + y + z - 6$

2. $F(x, y, \lambda)$
$$= xyz^2 + \lambda(x + y + z - 6)$$

3. $F_x = yz^2 + \lambda$
$F_y = xz^2 + \lambda$
$F_z = 2zxy + \lambda$
$F_\lambda = x + y + z - 6$

4. Setting F_x, F_y, F_z and F_λ equal to zero yields

$$yz^2 + \lambda = 0$$
$$xz^2 + \lambda = 0$$
$$2xyz + \lambda = 0$$
$$x + y + z - 6 = 0.$$

5. $\lambda = -yz^2$, $\lambda = -xz^2$, and $\lambda = -2xyz$

$$-yz^2 = -xz^2$$
$$z^2(x - y) = 0$$
$$x = y \quad \text{or} \quad z = 0$$
$$-yz^2 = -2xyz$$
$$2xyz - yz^2 = 0$$
$$yz(2x - z) = 0$$
$$y = 0, \ z = 0, \ \text{or} \ z = 2x$$

If $y = x$ and $z = 2x$, substituting into the fourth equation gives

$$x + x + 2x - 6 = 0$$
$$x = \frac{3}{2}.$$

Thus, $y = \frac{3}{2}$ and $z = 3$.

If $z = 0$, then $x + y = 6$ from the fourth equation.

If $y = 0$, then $x = 6$.

If $z = 2x$, then $x = 0$ or $y = 6$.

If $y = 2x$, then $x \pm y = 3$.

Possible extrema:

$(\frac{3}{2}, \frac{3}{2}, 3)$, $(6, 0, 0)$, $(0, 6, 0)$,

$(3, 3, 0)$

But $f(6, 0, 0) = f(0, 6, 0)$
$$= f(3, 3, 0) = 0$$
$$f\left(\frac{3}{2}, \frac{3}{2}, 3\right) = \frac{3}{2} \cdot \frac{3}{2} \cdot 9$$
$$= \frac{81}{4} > 0.$$

So $f\left(\frac{3}{2}, \frac{3}{2}, 3\right) = \frac{81}{4} = 20.25$

is a maximum.

11. Let $f(x, y) = xy^2$.

1. $g(x, y) = x + y - 18$

2. $F(x, y, \lambda)$
$$= xy^2 + \lambda(x + y - 18)$$

3. $F_x = y^2 + \lambda$
$F_y = 2yx + \lambda$
$F_\lambda = x + y - 18$

4.
$$y^2 + \lambda = 0$$
$$2yx + \lambda = 0$$
$$x + y - 18 = 0$$

5. $\lambda = -y^2$ and $\lambda = -2yx$

$$-y^2 = -2yx$$
$$y^2 - 2yx = 0$$
$$y(y - 2x) = 0.$$
$$y = 0 \quad \text{or} \quad y = 2x$$

If y = 0, from the third equation, x = 18.

If y = 2x, from the third equation,

$$x + 2x - 18 = 0$$
$$3x - 18 = 0$$
$$x = 6 \text{ and } y = 12.$$

f(18, 0) = 18 · 0 = 0

f(6, 12) = 6(144) = 864 > 0

So x = 6, y = 12 will maximize

$f(x, y) = xy^2$.

13. Let x, y, and z be three numbers such that

$$x + y + z = 90$$
and $f(x, y, z) = xyz.$

1. g(x, y, z) = x + y + z - 90

2. F(x, y, z)

   $= xyz + \lambda(x + y + z - 90)$

3. $F_x = yz + \lambda$
   $F_y = xz + \lambda$
   $F_\lambda = xy + \lambda$

   $F_\lambda = x + y + z - 90$

4. $yz + \lambda = 0$ (1)
          $xz + \lambda = 0$ (2)
          $xy + \lambda = 0$ (3)
   $x + y + z - 90 = 0$ (4)

5. $\lambda = -yz$, $\lambda = -xz$, and $\lambda = -xy$

          $-yz = -xz$
          $yz - xz = 0$
          $(y - x)z = 0$
          $y - x = 0 \text{ or } z = 0$
   $xz - xy = 0$
   $x(z - y) = 0$
          $x = 0 \text{ or } z - y = 0$

Since x = 0 or z = 0 would not maximize f(x, y, z) = xyz, then y − x = 0 and z − y = 0 imply that y = x = z.

Substituting into the fourth equation gives

$$x + x + x - 90 = 0$$
$$x = 30.$$

x = y = z = 30 will maximize

f(x, y, z) = xyz.

The numbers are 30, 30, and 30.

17. Let x be the width and y be the length of a field such that the cost in dollars to enclose the field is

$$6x + 6y + 4x + 4y = 1200$$
$$10x + 10y = 1200.$$

The area is

f(x, y) = xy.

1. g(x, y) = 10x + 10y - 1200

2. F(x, y) = xy + λ(10x + 10y − 1200)

3. $F_x = y + 10\lambda$
   $F_y = x + 10\lambda$
   $F_\lambda = 10x + 10y - 1200$

4. $y + 10\lambda = 0$
          $x + 10\lambda = 0$
   $10x + 10y - 1200 = 0$

5. 10λ = −y and 10λ = −x
          $-y = -x$
          $y = x$

Substituting into the third equation gives

$$10x + 10x - 1200 = 0$$
$$20x - 1200 = 0$$
$$x = 60$$
$$y = 60.$$

The dimensions, 60 ft by 60 ft, will maximize the area.

19. $C(x, y) = 2x^2 + 6y^2 + 4xy + 10$, subject to $x + y = 10$

1. $g(x, y) = x + y - 10$

2. $F(x, y)$
$$= 2x^2 + 6y^2 + 4xy + 10$$
$$+ \lambda(x + y - 10)$$

3. $F_x = 4x + 4y + \lambda$
$F_y = 12y + 4x + \lambda$
$F_\lambda = x + y - 10$

4. $4x + 4y + \lambda = 0$
$12y + 4x + \lambda = 0$
$x + y - 10 = 0$

5. $\lambda = -4x + 4y$ and $\lambda = -12y - 4x.$

$$-4x + 4y = -12y - 4x$$
$$8y = 0$$
$$y = 0$$

Since $x + y - 10$, $x = 10$.
10 large kits and no small kits will maximize the cost.

21. $f(x, y) = 3x^{1/3} y^{2/3}$, subject to
$80x + 150y = 40,000$

1. $g(x, y) = 80x + 150y - 40,000$

2. $F(x, y)$
$$= 3x^{1/3} y^{2/3} + \lambda(80x + 150y - 40,000)$$

3. and 4.

$F_x = x^{-2/3} + 80\lambda = 0$
$F_y = 2x^{1/3} y^{-1/3} + 150\lambda = 0$
$F_\lambda = 80x + 150y - 40,000 = 0$

5. $\dfrac{x^{-2/3} y^{2/3}}{80} = \dfrac{2x^{1/3} y^{-1/3}}{150}$

$$\dfrac{15y}{16} = x$$

Substitute into the third equation.

$$80\left(\dfrac{15y}{16}\right) + 150y - 40,000 = 0$$

$$y = 178 \text{ (rounded)}$$
$$= \dfrac{15(178)}{16}$$
$$x = 167$$

Use 167 units of labor and 178 units of capital to maximize production.

23. Let x and y be the dimensions of the field such that $2x + 2y = 200$, and the area is $f(x, y) = xy$.

1. $g(x, y) = 2x + 2y - 200$

2. $F(x, y)$
$$= xy + \lambda(2x + 2y - 200)$$

3. $F_x = y + 2\lambda$

$F_y = x + 2\lambda$
$F_\lambda = 2x + 2y - 200$

4.
$$y + 2\lambda = 0$$
$$x + 2\lambda = 0$$
$$2x + 2y - 200 = 0$$

5. $2\lambda = -y$ and $2\lambda = -x$ so $x = y$.

$$2x + 2x - 200 = 0$$
$$4x - 200 = 0$$
$$x = 50$$

Thus, $y = 50$.

Dimensions of 50 m by 50 m will maximize the area.

25. Let x be the radius r of the circular base and y the height h of the can, such that the volume is

$$\pi x^2 y = 250\pi.$$

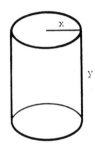

The surface is

$$f(x, y) = 2\pi xy + 2\pi x^2.$$

1. $g(x, y) = \pi x^2 y - 250\pi$

2. $F(x, y)$
$$= 2\pi xy + 2\pi x^2 + \lambda(\pi x^2 y - 250\pi)$$

3. $F_x = 2\pi y + 4\pi x + \lambda(2\pi xy)$
$F_y = 2\pi x + \lambda(\pi x^2)$
$F_\lambda = \pi x^2 y - 250\pi$

4. $2\pi y + 4\pi x + \lambda(2\pi xy) = 0$
$$2\pi x + \lambda\pi x^2 = 0$$
$$\pi x^2 y - 250\pi = 0$$

Simplifying these equations gives

$$y + 2x + \lambda xy = 0$$
$$2x + \lambda x^2 = 0$$
$$x^2 y - 250 = 0.$$

5. From the second equation

$$x(2 + \lambda x) = 0$$
$$x = 0 \quad \text{or} \quad \lambda = -\frac{2}{x}.$$

If $x = 0$, the volume will be 0, which is not possible.

Substituting $x = -\frac{2}{\lambda}$ into the first equation gives

$$y + 2\left(-\frac{2}{\lambda}\right) + \lambda\left(-\frac{2}{\lambda}\right)y = 0$$
$$y - \frac{4}{\lambda} - 2y = 0$$
$$-\frac{4}{\lambda} = y$$
$$\lambda = -\frac{4}{y}.$$

Since $\lambda = -2/x$, $y = 2x$.

Substituting into the third equation gives

$$x^2(2x) - 250 = 0$$
$$2x^3 - 250 = 0$$
$$x = 5$$
$$y = 10.$$

Since $g(1, 250) = 0$, and $f(1, 250) = 502\pi > f(5, 10) = 300\pi$, a can with radius of 5 in and height of 10 in will have minimum surface area.

27. Let x, y, and z be the dimensions of the box such that the surface area is

$$xy + 2yz + 2xz = 500$$

and the volume is

$$f(x, y, z) = xyz.$$

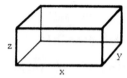

1. $g(x, y, z) - 500 = 0$

2. $F(x, y, z)$
$$= xyz + \lambda(xy + 2yz + 2xz - 500)$$

3. and **4.**
$$F_x = yz + \lambda(y + 2z) = 0 \qquad (1)$$
$$F_y = xz + \lambda(x + 2z) = 0 \qquad (2)$$
$$F_z = xy + \lambda(2y + 2x) = 0 \qquad (3)$$
$$F_\lambda = xy + 2xz + 2yz - 500 = 0 \quad (4)$$

Multiplying Equation (1) by x, Equation (2) by y, Equation (3) by z gives

$$xyz + \lambda x(y + 2z) = 0$$
$$xyz + \lambda y(x + 2z) = 0$$
$$xyz + \lambda z(2y + 2x) = 0.$$

5. Subtracting the second equation from the first equation gives

$$\lambda x(y + 2z) - \lambda y(x + 2z) = 0$$
$$2\lambda xz - 2\lambda yz = 0$$
$$\lambda z(x - y) = 0,$$

so $\qquad x = y.$

Subtracting the second equation from the third equation gives

$$\lambda z(2y + 2x) - \lambda y(x + 2z) = 0$$
$$2\lambda xz - \lambda xy = 0$$
$$\lambda x(2z - y) = 0$$

so $\qquad z = \dfrac{y}{2}.$

Substituting into the fourth equation gives

$$y^2 + 2y\left(\frac{y}{2}\right) + 2y\left(\frac{y}{2}\right) - 500 = 0$$
$$3y^2 = 500$$
$$y = \sqrt{\frac{500}{3}}$$
$$\approx 12.9099$$
$$x \approx 12.9099$$
$$z \approx \frac{12.9099}{2}$$
$$\approx 6.4649$$

The dimensions are 12.91 m by 12.91 m by 6.45 m.

29. Let x, y, and z be the dimensions of the box.
The surface area is

$$2xy + 2xz + 2yz.$$

We must minimize

$$f(x, y, z) = 2xy + 2xz + 2yz$$

subject to $xyz = 27$.

1. $g(x, y, z) = xyz - 27$

2. $F(x, y, z)$
$$= 2xy + 2xz + 2yz + \lambda(xyz - 27)$$

3. $F_x = 2y + 2z + \lambda yz$
$$F_y = 2x + 2z + \lambda xz$$
$$F_z = 2x + 2y + \lambda xy$$
$$F_\lambda = xyz - 27$$

4. $2y + 2z + \lambda yz = 0$

$2x + 2z + \lambda xz = 0$

$2x + 2y + \lambda xy = 0$

$xyz - 27 = 0$

5. The first and second equation gives

$\dfrac{2y + 2z}{yz} = -\lambda$ and $\dfrac{2x + 2z}{xz} = -\lambda.$

Thus,

$$\frac{2y + 2z}{yz} = \frac{2x + 2z}{xz}$$

$$2xyz + 2xz^2 = 2xyz + 2yz^2$$

$$2xz^2 - 2yz^2 = 0$$

$2z^2 = 0$ or $x - y = 0$

$z = 0$ (impossible) or $x = y$

The second and third equations give

$\dfrac{2x + 2z}{xz} = -\lambda$ and $\dfrac{2x + 2y}{xy} = -\lambda.$

Thus,

$$\frac{2x + 2z}{xz} = \frac{2x + 2y}{xy}$$

$$2x^2y + 2xyz = 2x^2z + 2xyz$$

$$2x^2y - 2x^2z = 0$$

$$2x^2(y - z) = 0$$

$2x^2 = 0$ or $y - z = 0$

$x = 0$ (impossible) or $y = z$

Therefore, $x = y = z$. Substituting into the fourth equation gives

$$x^3 - 27 = 0$$

$$x^3 = 27$$

$$x = 3$$

Thus, $y = 3$ and $z = 3$.

The dimensions that will minimize the surface area are 3 m by 3 m by 3 m.

Section 8.5

3. See the answer graph at the back of the textbook.

x	y	x^2	xy
4	3	16	12
5	7	25	35
8	17	64	136
12	28	144	336
14	35	196	490
Totals 43	90	445	1009

$n = 5$

$$m = \frac{(43)90 - 5(1009)}{(43)^2 - 5(445)}$$

$$= \frac{-1175}{-376}$$

$$= 3.125$$

$$b = \frac{90 - 3.125(43)}{5}$$

$$= -8.875$$

Therefore, the least squares equation is

$$y' = 3.125x - 8.875.$$

When $y' = 12$,

$$12 = 3.125x - 8.875$$

$$\frac{20.875}{3.125} = x$$

$$6.68 = x.$$

When $y' = 32$,

$$\frac{32 + 8.875}{3.125} = x$$

$$13.08 = x.$$

5. **(a)** See the answer graph at the back of the textbook.

(b)

x	y	x^2	xy
250	9.3	62,500	2325
300	10.8	90,000	3240
375	14.8	140,625	5550
425	14.0	180,625	5950
450	14.2	202,500	6390
475	15.3	225,625	7267.5
500	15.9	250,000	7950
575	19.1	330,625	10,982.5
600	19.2	360,000	11,520
650	21.0	422,500	13,650
Totals 4600	153.6	2,265,000	74,825

$$m = \frac{4600(153.6) - 10(74,825)}{(4600)^2 - 10(2,265,000)}$$

$$\approx .02798$$

$$\approx .03$$

$$b = \frac{153.6 - m(4600)}{10} \approx 2.49.$$

The least squares equation is

$$y' = .03x + 2.49$$

See graph.

(c) If x = 700, then $y' \approx 23.5$;

if x = 750, then $y' \approx 25.0$

7. **(a)**

x	y	x^2	xy
369	299	136,161	110,331
379	280	143,641	106,120
482	393	232,324	189,426
493	388	243,049	191,284
496	385	246,016	190,960
567	423	321,489	239,841
521	374	277,441	194,854
482	357	232,324	172,074
391	106	152,881	41,446
417	332	173,889	138,444
Totals 4597	3337	2,153,215	1,574,780

$$m = \frac{4597(337) - 10(1,574,780)}{(4597)^2 - 10(2,153,215)} \approx 1.02$$

$$b = \frac{3337 - m(4597)}{10} \approx -135$$

Least squares equation is

$$y' = 1.02x - 135.$$

(b) If x = 500, then y' = 375, so we predict $375,000 in repair costs.

(c) There appears to be an approximately linear relationship.

9. **(a)** All answers are rounded to four significant digits.

$$y' = 49.2755 - 1.1924(42)$$
$$+ .1631(170.0)$$
$$\approx \$26,920$$

$$y' = 49.2755 - 1.1924(45)$$
$$+ .1631(170.0)$$
$$\approx \$23,340$$

$$y' = 49.2755 - 1.1924(48)$$
$$+ .1631(170.0)$$
$$\approx \$19,770$$

(b) y = 49.2755 - 1.1924(42)

  + .1631(185.0)

  ≈ \$29,370

y = 49.2755 - 1.1924(45)

  + .1631(185.0)

  ≈ \$24,790

y = 49.2755 - 1.1924(48)

  + .1631(185.0)

  ≈ \$22,210

(c) \$42, which is obvious, since y is largest when the term being subtracted is smallest.

11. (a) and (b) See the answer graphs at the back of the textbook.

(c)

x	log, y	x^2	xy
0	3.00	0	0.00
1	3.21	1	3.21
2	3.43	4	6.86
3	3.65	9	10.95
4	3.87	16	15.48
5	4.09	25	20.45
Totals 15	21.25	55	56.95

$$m = \frac{(15)(21.25) - 6(56.95)}{(15)^2 - 6(55)}$$

$$= \frac{-22.95}{-105}$$

  ≈ .22

$$b = \frac{21.25 - m(15)}{6}$$

  ≈ 3.00

  y' = .22x + 3.00

(d) If x = 7 hr,

  y' = .22(7) + 3.00 = 4.54.

The antilogarithm of 4.54 is the predicted number, or $10^{4.54}$ ≈ 35,000 bacteria.

13. Σx = 175,878; Σy = 9989; Σxy = 46,209,266; $Σx^2$ = 872,066,218; $Σy^2$ = 2,629,701; and n = 38

(a)

$$m = \frac{(175,878)(9989) - 38(46,209,266)}{(175,878)^2 - 38(872,066,218)}$$

$$= \frac{893,234}{-2,205,445,400}$$

  ≈ -.00041

$$b = \frac{9989 - m(175,878)}{38}$$

  ≈ 265

  y' = -.00041x + 265

(b) If x = 2838, then

  y' = -.00041(2838) + 265 ≈ 264.

For Idaho, the average mathematics proficiency is 264.

If x = 7850, then

  y' = -.00041(7850) + 265 ≈ 262.

For Washington, D.C., the average mathematics proficiency is 262.

(c) The negative sign indicates that as expenditure per pupil increases, the mathematics proficiency decreases.

15.

	x	y	x^2	xy
	540	20	291,600	10,800
	510	16	260,100	8160
	490	10	240,100	4900
	560	8	313,600	4480
	470	12	220,900	5640
	600	11	360,000	6600
	540	10	291,600	5400
	580	8	336,400	4640
	680	15	462,400	10,200
	560	8	313,600	4480
	560	13	313,600	7280
	500	14	250,000	7000
	470	10	220,900	4700
	440	10	193,600	4400
	520	11	270,400	5720
	620	11	384,400	6820
	680	8	462,400	5440
	550	8	302,500	4400
	620	7	384,400	4340
Totals	10,490	210	5,872,500	115,400

(a) $m = \dfrac{(10,490)(210) - 19(115,400)}{(10,490)^2 - 19(5,872,500)}$

$= \dfrac{10,300}{-1,537,400}$

$\approx -.0067$

$b = \dfrac{210 - m(10,490)}{19}$

≈ 14.75

$y' = -.0067x + 14.75$

(b) If $x = 420$, then

$y' = -.0067(420) + 14.75 \approx 12.$

The predicted mathematics placement test score for a math SAT score of 420 is 12.

(c) If x = 620, then

$$y' = -.0067(620) + 14.75 \approx 11.$$

The predicted mathematics placement test score for a math SAT score of 620 is 11.

17.

	x	y	x^2	xy
	150	5000	22,500	750,000
	175	5500	30,625	962,500
	215	6000	46,225	1,290,000
	250	6500	62,500	1,625,000
	280	7000	78,400	1,960,000
	310	7500	96,100	2,325,000
	350	8000	122,500	2,800,000
	370	8500	136,900	3,145,000
	420	9000	176,400	3,780,000
	450	9500	202,500	4,275,000
Totals	2970	72,500	974,650	22,912,500

$$m = \frac{(2970)(72,500) - 10(22,912,500)}{(2970)^2 - 10(974,650)}$$

$$= \frac{-13,800,000}{-925,600}$$

$$\approx 14.9$$

$$b = \frac{72,500 - m(2970)}{10}$$

$$\approx 2820$$

$$y' = 14.9x + 2820$$

(a) If x = 150, y' = 14.9(150) + 2820 ≈ 5060.

If x = 280, y' = 14.9(280) + 2820 ≈ 6990.

If x = 420, y' = 14.9(420) + 2820 ≈ 9080.

The following table compares the actual data with the predicted values.

Square feet	Actual BTUs	Predicted BTUs
150	5000	5060
280	7000	6990
420	9000	9080

The largest discrepancy is 80 BTUs for 420 ft². This is a good agreement.

(b) If x = 230, y′ = 14.9(230) + 2820 ≈ 6250.

Since air conditioners are available only with the given BTU choices, 6500 BTUs must be used.

Section 8.6

1. $z = 9x^4 - 5y^3$

   $dz = 36x^3\ dx - 15y^2\ dy$

3. $z = \dfrac{x + y}{x - y}$

   $f_x = \dfrac{(x - y)(1) - (x + y)(1)}{(x - y)^2}$

   $\quad = \dfrac{-2y}{(x - y)^2}$

   $f_y = \dfrac{(x - y)(1) - (x + y)(-1)}{(x - y)^2}$

   $\quad = \dfrac{2x}{(x - y)^2}$

   $dz = \dfrac{-2y}{(x - y)^2}\ dx + \dfrac{2x}{(x - y)^2}\ dy$

5. $z = 2\sqrt{xy} - \sqrt{x + y}$

   $\quad = 2(xy)^{1/2} - (x + y)^{1/2}$

   $f_x = 2y^{1/2}\left(\dfrac{1}{2}\right)(x^{-1/2}) - \dfrac{1}{2}(x + y)^{-1/2}\ (1)$

   $\quad = y^{1/2}(x^{-1/2}) - \dfrac{1}{2}(x + y)^{-1/2}$

   $f_y = 2x^{1/2}\left(\dfrac{1}{2}\right)(y^{-1/2})$

   $\quad\quad - \dfrac{1}{2}(x + y)^{-1/2}\ (1)$

   $\quad = x^{1/2}(y^{-1/2}) - \dfrac{1}{2}(x + y)^{-1/2}$

   $dz = \left[\dfrac{y^{1/2}}{x^{1/2}} - \dfrac{1}{2(x + y)^{1/2}}\right] dx$

   $\quad\quad + \left[\dfrac{x^{1/2}}{y^{1/2}} - \dfrac{1}{2(x + y)^{1/2}}\right] dy$

7. $z = (3x + 2)\sqrt{1 - 2y}$

   $\quad = (3x + 2)(1 - 2y)^{1/2}$

   $f_x = (1 - 2y)^{1/2}\ (3) + (3x + 2)(0)$

   $\quad = 3(1 - 2y)^{1/2}$

   $f_y = (3x + 2)\left(\dfrac{1}{2}\right)(1 - 2y)^{-1/2}\ (-2)$

   $\quad\quad + (1 - 2y)^{1/2}\ (0)$

   $\quad = -(3x + 2)(1 - 2y)^{-1/2}$

   $dz = 3(1 - 2y)^{1/2}\ dx - \dfrac{3x + 2}{(1 - 2y)^{1/2}}\ dy$

9. $z = \ln(x^2 + 2y^4)$

   $f_x = \dfrac{2x}{x^2 + 2y^4}$

   $f_y = \dfrac{8y^3}{x^2 + 2y^4}$

   $dz = \dfrac{2x}{x^2 + 2y^4}\ dx + \dfrac{8y^3}{x^2 + 2y^4}\ dy$

11. $z = xy^2e^{x+y}$

   $f_x = y^2(xe^{x+y} + e^{x+y})$

   $\quad = y^2e^{x+y}(x + 1)$

   $f_y = x[y^2e^{x+y} + e^{x+y}(2y)]$

   $\quad = xye^{x+y}(y + 2)$

   $dz = y^2e^{x+y}(x + 1)dx$

   $\quad\quad + xye^{x+y}(y + 2)dy$

13. $z = x^2 - y \ln x$

   $f_x = 2x - \dfrac{y}{x}$

   $f_y = -\ln x$

   $dz = \left(2x - \dfrac{y}{x}\right)dx + (-\ln x\ dy)$

15. $w = x^4yz^3$

   $dw = 4x^3yz^3\ dx + x^4z^3\ dy$

   $\quad\quad + 3x^4yz^2\ dz$

17. $z = x^2 + 3xy + y^2$

$x = 4$, $y = -2$, $dx = .02$, $dy = -.03$

$f_x = 2x + 3y$

$f_y = 3x + 2y$

$dz = (2x + 3y)dx + (3x + 2y)dy$

Substitute the given values to get

$dz = [2(4) + 3(-2)](.02)$

$\quad + [3(4) + 2(-2)](-.03)$

$\quad = .04 + (-.24) = -.2.$

19. $z = \dfrac{x - 4y}{x + 2y}$

$x = 0$, $y = 5$, $dx = -.03$,

$dy = .05$

$f_x = \dfrac{(x + 2y)(1) - (x - 4y)(1)}{(x + 2y)^2}$

$\quad = \dfrac{6y}{(x + 2y)^2}$

$f_y = \dfrac{(x + 2y)(-4) - (x - 4y)(2)}{(x + 2y)^2}$

$\quad = \dfrac{-6x}{(x + 2y)^2}$

$dz = \dfrac{6y}{(x + 2y)^2}\, dx - \dfrac{6x}{(x + 2y)^2}\, dy$

Substitute the given information.

$dz = \dfrac{30}{10^2}(-.03) - \dfrac{0}{10^2}(.05)$

$\quad = \dfrac{3}{10}(-.03) = -.009$

21. $z = \ln(x^2 + y^2)$

$x = 2$, $y = 3$, $dx = .02$,

$dy = -.03$

$dz = \dfrac{2x\ dx}{x^2 + y^2} + \dfrac{2y\ dy}{x^2 + y^2}$

Substitute the given information.

$dz = \dfrac{2(2)(.02)}{2^2 + 3^2} + \dfrac{2(3)(-.03)}{2^2 + 3^2}$

$\quad = -.00769$

23. $w = \dfrac{5x^2 + y^2}{z + 1}$

$x = -2$, $y = 1$, $z = 1$,

$dx = .02$, $dy = -.03$, $dz = .02$

$f_x = \dfrac{(z + 1)10x - (5x^2 + y^2)(0)}{(z + 1)^2}$

$\quad = \dfrac{10x}{z + 1}$

$f_y = \dfrac{(z + 1)(2y) - (5x^2 + y^2)(0)}{(z + 1)^2}$

$\quad = \dfrac{2y}{z + 1}$

$f_z = \dfrac{(z + 1)(0) - (5x^2 + y^2)(1)}{(z + 1)^2}$

$\quad = \dfrac{-5x^2 - y^2}{(z + 1)^2}$

$dz = \dfrac{10x}{z + 1}\, dx + \dfrac{2y}{z + 1}\, dy$

$\quad + \dfrac{-5x^2 - y^2}{(z + 1)^2}\, dz$

Substitute the given information.

$= \dfrac{-20}{2}(.02) + \dfrac{2}{2}(-.03)$

$\quad + \dfrac{[-5(4) - 1](.02)}{(2)^2}$

$= -.2 - .03 - \dfrac{21}{4}(.02) = -.355$

25. The volume of the can is

$$V = \pi r^2 h$$

with $r = 2.5$ cm, $h = 14$ cm,

$dr = .08$ cm, $dh = .16$ cm.

$dV = 2\pi rh\ dr + \pi r^2\ dh$

$dV = 2\pi(2.5)(14)(.08) + \pi(2.5)^2(.16)$

$dV \approx 20.73$

Approximately 20.73 cm³ are needed.

27. The volume of the box is

$$V = LWH$$

with L = 10 in, W = 9 in, H = 14 in.
Since .2 in is applied to each side
and each dimension has a side at
each end,

$$dL = dW = dH = 2(.2) = .4 \text{ in.}$$
$$dV = WH\,dL + LH\,dW + LW\,dH$$

Substitute.

$$dV = (9)(14)(.4) + (10)(14)(.4)$$
$$+ (10)(9)(.4)$$
$$dV = 142.4$$

Approximately 142.4 in³ are needed.

29. $z = x^{.65}y^{.35}$

$x = 50$, $y = 29$, $dx = 52 - 50 = 2$,
$dy = 27 - 29 = -2$

$$f_x = y^{.35}(.65)(x^{-.35})$$
$$= .65\left(\frac{y}{x}\right)^{.35}$$

$$f_y = (.65)(.35)(y^{-.65})$$
$$= .35\left(\frac{x}{y}\right)^{.65}$$

$$dz = .65\left(\frac{y}{x}\right)^{.35}dx + .35\left(\frac{x}{y}\right)^{.65}dy$$

Substitute.

$$dz = .65\left(\frac{29}{50}\right)^{.35}(2) + .35\left(\frac{50}{29}\right)^{.65}(-2)$$

$$dz = .0769 \text{ units}$$

31.   The volume of the bone is

$$V = \pi r^2 h,$$

with h = 7 cm, r = 1.4 cm,

$dr = .09$ cm, $dh = 2(.09) = 18$ cm
$dV = 2\pi rh\,dr + \pi r^2\,dh$
$dV = 2\pi(1.4)(7)(.09) + \pi(1.4)^2(.18)$
$dV = 6.65$

6.65 cm³ are used.

33. $C = \dfrac{b}{a - v} = b(a - v)^{-1}$, $a = 160$,

$b = 200$, $v = 125$, $da = 145 - 160$
$= -15$, $db = 190 - 200 = -10$,
$v = 125$, $da = 145 - 160$

$dC = -b(a - v)^{-2}da$

$\qquad + \dfrac{1}{a - v}\,db + b(a - v)^{-2}\,dv$

$\qquad = \dfrac{-b}{(a - v)^2}\,da + \dfrac{1}{a - v}\,db$

$\qquad + \dfrac{b}{(a - v)^2}\,dv$

$\qquad = \dfrac{-200}{(160 - 125)^2}(-15) + \dfrac{1}{160 - 125}(-10)$

$\qquad + \dfrac{200}{(160 - 125)^2}(5)$

$\qquad \approx 2.98 \text{ liters}$

35.   The area is

$$A = \frac{1}{2}bh$$

with h = 42.6 cm, b = 23.4 cm,
dh = 1.2 cm, db = .9 cm.

$$dA = \frac{1}{2}h\,db + \frac{1}{2}b\,dh$$

Substitute.

$$dA = \frac{1}{2}(42.6)(.9) + \frac{1}{2}(23.4)(1.2)$$
$$= 33.2 \text{ cm}^2$$

Section 8.7

1. $\int_0^3 (x^3y + y)dx$

$= (\frac{x^4}{4}y + yx)\Big|_0^3$

$= \frac{81y}{4} + 3y$

$= \frac{93y}{4}$

3. $\int_4^8 \sqrt{6x + y}\ dx = \int_4^8 (6x + y)^{1/2}\ dx$

Let $u = 6x + y$.

Then $du = 6\ dx$.

When $x = 8$, $u = 48 + y$.

When $x = 4$, $u = 24 + y$.

$\int_{y+24}^{y+48} u^{1/2} \cdot \frac{1}{6}\ du$

$= \frac{1}{6} \cdot \frac{2}{3}u^{3/2}\Big|_{y+24}^{y+48}$

$= \frac{1}{9}u^{3/2}\Big|_{y+24}^{y+48}$

$= \frac{1}{9}[(48 + y)^{3/2} - (24 + y)^{3/2}]$

5. $\int_4^5 x\sqrt{x^2 + 3y}\ dy$

$= \int_4^5 x(x^2 + 3y)^{1/2}\ dy$

$= \frac{2x}{9}(x^2 + 3y)^{3/2}\Big|_4^5$

$= \frac{2x}{9}[(x^2 + 15)^{3/2} - (x^2 + 12)^{3/2}]$

7. $\int_4^9 \frac{3 + 5y}{\sqrt{x}}\ dx$

$= (3 + 5y)\int_4^9 x^{-1/2}\ dx$

$= (3 + 5y)2x^{1/2}\Big|_4^9$

$= (3 + 5y)2[\sqrt{9} - \sqrt{4}]$

$= 6 + 10y$

9. $\int_{-1}^1 e^{x+4y}dy$

Let $u = x + 4y$.

Then $du = 4\ dy$.

When $y = 1$, $u = x + 4$.

When $y = -1$, $u = x - 4$.

$\frac{1}{4}\int_{x-4}^{x+4} e^u\ du = \frac{1}{4}e^u\Big|_{x-4}^{x+4}$

$= \frac{1}{4}(e^{x+4} - e^{x-4})$

$= \frac{1}{4}e^{x+4} - \frac{1}{4}e^{x-4}$

11. $\int_0^5 xe^{x^2+9y}\ dx$

Let $u = x^2 + 9y$.

Then $du = 2x\ dx$.

When $x = 5$, $u = 25 + 9y$.

When $x = 0$, $u = 9y$.

$\frac{1}{2}\int_{9y}^{25+9y} e^u\ du = \frac{1}{2}(e^{25+9y} - e^{9y})$

$= \frac{1}{2}e^{25+9y} - \frac{1}{2}e^{9y}$

13. $\int_{1}^{2}\left[\int_{0}^{3}(x^3y + y)dx\right]dy$

From Exercise 1,

$$\int_{0}^{3}(x^3y + y)dx = \frac{93y}{4}.$$

$$\int_{1}^{2}\left[\int_{0}^{3}(x^3y + y)dx\right]dy$$

$$= \int_{1}^{2}\frac{93y}{4}\,dy$$

$$= \frac{93}{8}y^2\bigg|_{1}^{2}$$

$$= \frac{93}{8}(4 - 1)$$

$$= \frac{279}{8}$$

15. $\int_{0}^{1}\left[\int_{3}^{6}x\sqrt{x^2 + 3y}\ dx\right]dy$

From Exercise 6,

$$\int_{3}^{6}x\sqrt{x^2 + 3y}\ dx$$

$$= \frac{1}{3}[(36 + 3y)^{3/2} - (9 + 3y)^3]$$

$$\int_{0}^{1}\left[\int_{3}^{6}x\sqrt{x^2 + 3y}\ dx\right]dy$$

$$= \int_{0}^{1}\frac{1}{3}[(36 + 3y)^{3/2} - (9 + 3y)^{3/2}]dy$$

Let $u = 36 + 3y$.

Then $du = 3\ dy$.

When $x = 0$, $u = 36$.

When $x = 1$, $w = 39$.

Let $z = 9 + 3y$.

Then $dz = 3\ dy$.

When $y = 0$, $z = 9$.

When $y = 1$, $z = 12$.

$$\frac{1}{9}\left[\int_{36}^{39}u^{3/2}\ du - \int_{9}^{12}z^{3/2}\ dz\right]$$

$$= \frac{1}{9}\cdot\frac{2}{5}[(39)^{5/2} - (36)^{5/2}$$

$$\qquad - (12)^{9/2} + (9)^{5/2})]$$

$$= \frac{2}{45}[(39)^{5/2} - (12)^{5/2} - 6^5 + 3^5]$$

$$= \frac{2}{45}(39^{5/2} - 12^{5/2} - 7533)$$

17. $\int_{1}^{2}\left[\int_{4}^{9}\frac{3 + 5y}{\sqrt{x}}\ dx\right]dy$

From Exercise 7,

$$\int_{4}^{9}\frac{3 + 5y}{\sqrt{x}}\ dx = 6 + 10y$$

$$\int_{1}^{2}\left[\int_{4}^{9}\frac{3 + 5y}{\sqrt{x}}\ dx\right]dy$$

$$= \int_{1}^{2}(6 + 10y)dy$$

$$= 6y\bigg|_{1}^{2} + 5y^2\bigg|_{1}^{2}$$

$$= 6(2 - 1) + 5(4 - 1)$$

$$= 6 + 15 = 21$$

19. $\int_{1}^{2}\int_{1}^{2}\frac{dx\ dy}{xy}$

$$= \int_{1}^{2}\left(\frac{\ln|x|}{y}\right)\bigg|_{1}^{2}\ dy$$

$$= \int_{1}^{2}\frac{1}{y}(\ln 2 - \ln 1)dy$$

$$= \ln 2 \int_{1}^{2} \frac{1}{y} \, dy$$

$$= \ln 2 (\ln |y|) \Big|_{1}^{2}$$

$$= \ln 2 (\ln 2 - \ln 1)$$

$$= (\ln 2)^2$$

$$= \int_{1}^{5} \left(\frac{4}{2} + 6y^2 - 0 \right) dy$$

$$= (2y + 2y^3) \Big|_{1}^{5}$$

$$= 2 \cdot 5 + 2 \cdot 125 - (2 \cdot 1 + 2 \cdot 1)$$

$$= 256$$

21. $\displaystyle\int_{2}^{4}\int_{3}^{5} \left(\frac{x}{y} + \frac{y}{3} \right) dx \, dy$

$$= \int_{2}^{4} \left(\frac{x^2}{2y} + \frac{yx}{3} \right) \Big|_{3}^{5} dy$$

$$= \int_{2}^{4} \left[\frac{25}{2y} + \frac{5y}{3} - \left(\frac{9}{2y} + \frac{3y}{3} \right) \right] dy$$

$$= \int_{2}^{4} \left(\frac{16}{2y} + \frac{2y}{3} \right) dy$$

$$= \left(8 \ln |y| + \frac{y^2}{3} \right) \Big|_{2}^{4}$$

$$= 8(\ln 4 - \ln 2) + \frac{16}{3} - \frac{4}{3}$$

$$= 8 \ln \frac{4}{2} + \frac{12}{3}$$

$$= 8 \ln 2 + 4$$

23. $\displaystyle\int_{R}\int (x + 3y^2) dx \, dy; \ 0 \le x \le 2,$

$1 \le y \le 5$

$$\int_{R}\int (x + 3y^2) dx \, dy$$

$$= \int_{1}^{5}\int_{0}^{2} (x + 3y^2) dx \, dy$$

$$= \int_{1}^{5} \left(\frac{x^2}{2} + 3y^2 x \right) \Big|_{0}^{2} dy$$

25. $\displaystyle\int_{R}\int \sqrt{x + y} \, dy \, dx; \ 1 \le x \le 3,$

$0 \le y \le 1$

$$\int_{R}\int \sqrt{x + y} \, dy \, dx$$

$$= \int_{1}^{3}\int_{0}^{1} (x + y)^{1/2} dy \, dx$$

$$= \int_{1}^{3} \left[\frac{2}{3}(x + y)^{3/2} \right] \Big|_{0}^{1} dx$$

$$= \int_{1}^{3} \frac{2}{3} [(x + 1)^{3/2} - x^{3/2}] dx$$

$$= \frac{2}{3} \cdot \frac{2}{5} [(x + 1)^{5/2} - x^{5/2}] \Big|_{1}^{3}$$

$$= \frac{4}{15}(4^{5/2} - 3^{5/2} - 2^{5/2} + 1^{5/2})$$

$$= \frac{4}{15}(32 - 3^{5/2} - 2^{5/2} + 1)$$

$$= \frac{4}{15}(33 - 3^{5/2} - 2^{5/2})$$

27. $\iint\limits_{R} \dfrac{2}{(x + y)^2}\, dy\, dx;\ 2 \le x \le 3,$

$1 \le y \le 5$

$\iint\limits_{R} \dfrac{2}{(x + y)^2}\, dy\, dx$

$= \displaystyle\int_{2}^{3}\int_{1}^{5} 2(x + y)^{-2} dy\, dx$

$= \displaystyle\int_{2}^{3} -2(x + y)^{-1}\Big|_{1}^{5}\, dy$

$= -2\displaystyle\int_{2}^{3}\left[\dfrac{1}{(5 + y)} - \dfrac{1}{(1 + y)}\right]dy$

$= -2(\ln\,|5 + y| - \ln\,|1 + y|)\Big|_{2}^{3}$

$= -2\left(\ln\,\left|\dfrac{5 + y}{1 + y}\right|\right)\Big|_{2}^{3}$

$= -2\left(\ln\, 2 - \ln\,\dfrac{7}{3}\right)$

$= -2\,\ln\,\dfrac{6}{7}\quad$ or $\quad 2\,\ln\,\dfrac{7}{6}$

29. $\iint\limits_{R} ye^{(x+y^2)}\, dx\, dy;\ 2 \le x \le 3,$

$0 \le y \le 2$

$\iint\limits_{R} ye^{(x+y^2)}\, dx\, dy$

$= \displaystyle\int_{0}^{2}\int_{2}^{3} ye^{x+y^2} dx\, dy$

$= \displaystyle\int_{0}^{2} ye^{x+y^2}\Big|_{2}^{3}\, dy$

$= \displaystyle\int_{0}^{2} (ye^{3+y^2} - ye^{2+y^2})\, dy$

$= e^3\displaystyle\int_{0}^{2} ye^{y^2} dy - e^2\displaystyle\int_{0}^{2} ye^{y^2} dy$

$= \dfrac{e^3}{2}(e^{y^2})\Big|_{0}^{2} - \dfrac{e^2}{2}(e^{y^2})\Big|_{0}^{2}$

$= \dfrac{e^3}{2}(e^4 - e^0) - \dfrac{e^2}{2}(e^4 - e^0)$

$= \dfrac{1}{2}(e^7 - e^6 - e^3 + e^2)$

31. $z = 6x + 2y + 5;\ -1 \le x \le 1,$

$0 \le y \le 3$

$V = \displaystyle\int_{-1}^{1}\int_{0}^{3} (6x + 2y + 5)\, dy\, dx$

$= \displaystyle\int_{-1}^{1} (6xy + 5y + y^2)\Big|_{0}^{3}\, dx$

$= \displaystyle\int_{-1}^{1} (18x + 15 + 9)\, dx$

$= (9x^2 + 24x)\Big|_{-1}^{1}$

$= 9 + 24 - (9 - 24) = 48$

33. $z = x^2;\ 0 \le x \le 1,\ 0 \le y \le 4$

$V = \displaystyle\int_{0}^{1}\int_{0}^{4} x^2 dy\, dx$

$= \displaystyle\int_{0}^{1} (x^2 y)\Big|_{0}^{4}\, dx$

$= \displaystyle\int_{0}^{1} 4x^2 dx$

$= \dfrac{4}{3}x^3\Big|_{0}^{1} = \dfrac{4}{3}$

35. $z = x\sqrt{x^2 + y}; \; 0 \le x \le 1, \; 0 \le y \le 1$

$$V = \int_0^1 \int_0^1 x\sqrt{x^2 + y} \; dx \; dy$$

Let $u = x^2 + y$.

Then $du = 2x \; dx$.

When $x = 0$, $u = y$.

When $x = 1$, $u = 1 + y$.

$$= \int_0^1 \left[\int_y^{1+y} u^{1/2} \; du \right] dy$$

$$= \int_0^1 \frac{1}{2}\left(\frac{2}{3}u^{3/2}\right) \Big|_y^{1+y} \; dy$$

$$= \int_0^1 \frac{1}{3}[(1 + y)^{3/2} - y^{3/2}] \, dy$$

$$= \frac{1}{3} \cdot \frac{2}{5}[(1 + y)^{5/2} - y^{5/2}] \Big|_0^1$$

$$= \frac{2}{15}(2^{5/2} - 1 - 1)$$

$$= \frac{2}{15}(2^{5/2} - 2)$$

37. $z = \dfrac{xy}{(x^2 + y^2)^2}; \; 1 \le x \le 2, \; 1 \le y \le 4$

$$V = \int_1^2 \int_1^4 \frac{xy}{(x^2 + y^2)^2} \; dy \; dx$$

$$= \int_1^2 \left[\int_1^4 xy(x^2 + y^2)^{-2} dy \right] dx$$

$$= \int_1^2 \left[\int_1^4 \frac{1}{2}x(x^2 + y^2)^{-2}(2y) \, dy \right] dx$$

$$= \int_1^2 \left[-\frac{1}{2}x(x^2 + y^2)^{-1} \right] \Big|_1^4 \; dx$$

$$= \int_1^2 \left[-\frac{1}{2}x(x^2 + 16)^{-1} + \frac{1}{2}x(x^2 + 1)^{-1} \right] dx$$

$$= -\frac{1}{2} \int_1^2 \frac{1}{2}(x^2 + 16)^{-1}(2x)(dx$$

$$+ \frac{1}{2} \int_1^2 \frac{1}{2}(x^2 + 1)^{-1}(2x) \, dx$$

$$= -\frac{1}{2} \cdot \frac{1}{2} \ln |x^2 + 16| \Big|_1^2$$

$$+ \frac{1}{2} \cdot \frac{1}{2} \ln |x^2 + 1| \Big|_1^2$$

$$= -\frac{1}{4} \cdot \ln 20 + \frac{1}{4} \ln 17$$

$$+ \frac{1}{4} \ln 5 - \frac{1}{4} \ln 2$$

$$= \frac{1}{4}(-\ln 20 + \ln 17 + \ln 5 - \ln 2)$$

$$= \frac{1}{4} \ln \frac{(17)(5)}{(20)(2)}$$

$$= \frac{1}{4} \ln \frac{17}{8}$$

39. $\displaystyle\iint_R xe^{xy} \; dx \; dy; \; 0 \le x \le 2; \; 0 \le y \le 1$

$$\iint_R xe^{xy} \; dx \; dy$$

$$= \int_0^2 \int_0^1 xe^{xy} \; dy \; dx$$

$$= \int_0^2 \frac{x}{x}e^{xy} \Big|_0^1 \; dx$$

$$= \int_0^2 (e^x - e^0) \, dx$$

$$= (e^x - x) \Big|_0^2$$

$$= e^2 - 2 - e^0 + 0$$

$$= e^2 - 3$$

41. $\displaystyle\int_2^4\int_2^{x^2} (x^2 + y^2)\,dy\,dx$

$= \displaystyle\int_2^4 \left.(x^2 y + \frac{y^3}{3})\right|_2^{x^2} dx$

$= \displaystyle\int_2^4 \left(x^4 + \frac{x^6}{3} - 2x^2 - \frac{8}{3}\right) dx$

$= \left.\left(\frac{x^5}{5} + \frac{x^7}{21} - \frac{2}{3}x^3 - \frac{8}{3}x\right)\right|_2^4$

$= \dfrac{1024}{5} + \dfrac{16,384}{21} - \dfrac{2}{3}(64) - \dfrac{8}{3}(4)$

$\quad - \left(\dfrac{32}{5} + \dfrac{128}{21} - \dfrac{16}{3} - \dfrac{16}{3}\right)$

$= \dfrac{1024}{5} - \dfrac{32}{5} + \dfrac{16,384 - 128}{21}$

$\quad - \dfrac{128}{3} - \dfrac{32}{3} - \left(\dfrac{-32}{3}\right)$

$= \dfrac{992}{5} + \dfrac{16,256}{21} - \dfrac{128}{3}$

$= \dfrac{20,832}{105} + \dfrac{81,280}{105} - \dfrac{4480}{105}$

$= \dfrac{97,632}{105} \approx 929.83$

43. $\displaystyle\int_0^4\int_0^x \sqrt{xy}\,dy\,dx$

$= \displaystyle\int_0^4 \left.\left[\frac{2(xy)^{3/2}}{3x}\right]\right|_0^x dx$

$= \dfrac{2}{3}\displaystyle\int_0^4 \left[\frac{(\sqrt{x^2})^3}{x} - \frac{0}{x}\right] dx$

$= \dfrac{2}{3}\displaystyle\int_0^4 x^2\,dx = \frac{2}{3}\cdot\left.\frac{x^3}{3}\right|_0^4 = \frac{2}{9}(64)$

$= \dfrac{128}{9}$

45. $\displaystyle\int_1^2\int_y^{3y} \frac{1}{x}\,dx\,dy$

$= \displaystyle\int_1^2 \left.(\ln|x|)\right|_y^{3y} dy$

$= \displaystyle\int_1^2 (\ln|3y| - \ln|y|)\,dy$

$= \displaystyle\int_1^2 \ln\left|\frac{3y}{y}\right| dy$

$= \left.(\ln 3)(y)\right|_1^2 = \ln 3$

47. $\displaystyle\int_0^4\int_1^{e^x} \frac{x}{y}\,dy\,dx$

$= \displaystyle\int_0^4 \left.(x\ln|y|)\right|_1^{e^x} dx$

$= \displaystyle\int_0^4 (x\ln e^x - x\ln 1)\,dx$

$= \displaystyle\int_0^4 x^2\,dx = \left.\frac{x^3}{3}\right|_0^4 = \frac{64}{3}$

49. $\displaystyle\iint_R (4x + 7y)\,dy\,dx;\ 1 \le x \le 3;$

$0 \le y \le x + 1$

$\displaystyle\iint_R (4x + 7y)\,dy\,dx$

$= \displaystyle\int_1^3\int_0^{x+1} (4x + 7y)\,dy\,dx$

$= \displaystyle\int_1^3 \left.\left(4xy + \frac{7y^2}{2}\right)\right|_0^{x+1} dx$

$$= \int_1^3 [4x^2 + 4x + \frac{7}{2}(x^2 + 2x + 1)]\,dx$$

$$= \int_1^3 (\frac{15}{2}x^2 + 11x + \frac{7}{2})\,dx$$

$$= (\frac{15}{2} \cdot \frac{x^3}{3} + \frac{11x^2}{2} + \frac{7x}{2})\Big|_1^3$$

$$= \frac{5}{2} \cdot 27 + \frac{11(9)}{2} + \frac{7(3)}{2}$$

$$\quad - (\frac{5}{2} + \frac{11}{2} + \frac{7}{2})$$

$$= \frac{130}{2} + \frac{88}{2} + \frac{14}{2}$$

$$= \frac{232}{2} = 116$$

51. $\displaystyle\iint_R (4 - 4x^2)\,dy\ dx;\ 0 \le x \le 1,$

$0 \le y \le 2 - 2x$

$$\iint_R (4 - 4x^2)\,dy\ dx$$

$$= \int_0^2 \int_0^{2-2x} 4(1 - x^2)\,dy\ dx$$

$$= \int_0^1 [4(1 - x^2)y]\Big|_0^{2(1-x)}\,dx$$

$$= \int_0^1 4(1 - x^2)(2)(1 - x)\,dx$$

$$= 8\int_0^1 (1 - x - x^2 + x^3)\,dx$$

$$= 8(x - \frac{x^2}{2} - \frac{x^3}{3} + \frac{x^4}{4})\Big|_0^1$$

$$= 8(1 - \frac{1}{2} - \frac{1}{3} + \frac{1}{4})$$

$$= 8(\frac{1}{2} - \frac{1}{12})$$

$$= 8 \cdot \frac{5}{12} = \frac{10}{3}$$

53. $\displaystyle\iint_R e^{x/y^2}\,dx\ dy;\ 1 \le y \le 2;$

$0 \le x \le y^2$

$$\iint_R e^{x/y^2}\,dx\ dy$$

$$= \int_1^2 \int_0^{y^2} e^{x/y^2}\,dx\ dy$$

$$= \int_1^2 [y^2 e^{x/y^2}]\Big|_0^{y^2}\ dy$$

$$= \int_1^2 (y^2 e^{y^2/y^2} - y^2 e^0)\,dy$$

$$= \int_1^2 (ey^2 - y^2)\,dy$$

$$= (e - 1)\frac{y^3}{3}\Big|_1^2$$

$$= (e - 1)(\frac{8}{3} - \frac{1}{3})$$

$$= \frac{7(e - 1)}{3}$$

55. $\displaystyle\iint_R x^3 y\,dx\ dy;\ R \text{ bounded by } y = x^2,$

$y = 2x$

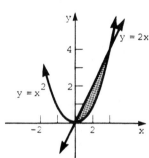

The points of intersection can be determined by solving the following system for x.

$$y = x^2$$
$$\underline{y = 2x}$$

$$x^2 = 2x$$
$$x(x - 2) = 0$$
$$x = 0 \quad \text{or} \quad x = 2$$

Therefore,

$$\iint_R x^3 y \, dx \, dy$$

$$= \int_0^2 \int_{x^2}^{2x} x^3 y \, dy \, dx$$

$$= \int_0^2 \left(x^3 \frac{y^2}{2} \right) \Big|_{x^2}^{2x} dx$$

$$= \int_0^2 \left[x^3 \frac{(4x^2)}{2} - x^3 \frac{(x^4)}{2} \right] dx$$

$$= \int_0^2 \left(2x^5 - \frac{x^7}{2} \right) dx$$

$$= \left(\frac{1}{3}x^6 - \frac{1}{16}x^8 \right) \Big|_0^2 = \frac{1}{3} \cdot 2^6 - \frac{1}{16} \cdot 2^8$$

$$= \frac{64}{3} - 16 = \frac{16}{3}.$$

57. $\displaystyle\iint_R \frac{dy \, dx}{y}$; R bounded by y = x,

$y = \dfrac{1}{x}$, x = 2

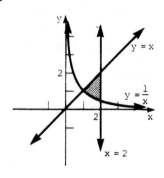

$y = x$ and $y = \dfrac{1}{x}$

intersect at (1, 1).

$$\int_1^2 \int_{1/x}^x \frac{dy}{y} \, dx$$

$$= \int_1^2 \ln y \Big|_{1/x}^x dx$$

$$= \int_1^2 \left(\ln x - \ln \frac{1}{x} \right) dx$$

$$= \int_1^2 2 \ln x \, dx$$

$$= 2(x \ln x - x) \Big|_1^2$$

$$= 2[(2 \ln 2 - 2) - (\ln 1 - 1)]$$
$$= 4 \ln 2 - 2$$

59. $f(x, y) = x^2 + y^2$; $0 \le x \le 2$,
$0 \le y \le 3$

The area of region R is

$$A = (2 - 0)(3 - 0) = 6.$$

The average value of

$$f(x, y) = x^2 + y^2 \text{ is}$$

$$\frac{1}{A} \iint_R (x^2 + y^2) dy \, dx$$

$$= \frac{1}{6} \int_0^2 \int_0^3 (x^2 + y^2) dy \, dx$$

$$= \frac{1}{6} \int_0^2 \left(x^2 y + \frac{y^3}{3} \right) \Big|_0^3 dx$$

$$= \frac{1}{6} \int_0^2 (3x^2 + 9) dx$$

$$= \frac{1}{6} (x^3 + 9x) \Big|_0^2$$

$= \frac{1}{6}(8 + 18 - 0)$

$= \frac{1}{6} \cdot 26$

$= \frac{13}{3}.$

61. $f(x, y)^{2x+y}; \ 1 \le x \le 2, \ 2 \le y \le 3$

The area of region R is

    $A = (2 - 1)(3 - 2) = 1.$

The average value is

$\frac{1}{A} \iint\limits_{R} e^{2x+y} \, dy \, dx$

$= \iint\limits_{R} e^{2x+y} \, dy \, dx$

$= \int_{1}^{2} \int_{2}^{3} e^{2x+y} \, dy \, dx$

$= \int_{1}^{2} (e^{2x+y}) \Big|_{2}^{3} \, dx$

$= \int_{1}^{2} (e^{2x+3} - e^{2x+2}) \, dx$

$= \frac{1}{2}(e^{2x+3} - e^{2x+2}) \Big|_{1}^{2}$

$= \frac{1}{2}(e^{4+3} - e^{2+3} - e^{4+2} + e^{4})$

$= \frac{e^7 - e^6 - e^5 + e^4}{2}.$

63. $C(x, y) = \frac{1}{9}x^2 + 2x + y^2 + 5y + 100$

    Area $= (75 - 48)(60 - 20)$

        $= 27(40)$

        $= 1080$

The average cost is

$\frac{1}{1080} \iint\limits_{R} (\frac{1}{9}x^2 + 2x + y^2 + 5y + 100) \, dx \, dy$

$= \frac{1}{1080} \int_{48}^{75} \int_{20}^{60} \Big[\frac{1}{9}x^2 + 2x + 100$

$\quad + y^2 + 5y \Big] dy \, dx$

$= \frac{1}{1080} \int_{48}^{75} \Big[(\frac{1}{9}x^2 + 2x + 100)y + \frac{y^3}{3}$

$\quad + \frac{5}{2}y^2 \Big] \Big|_{20}^{60} dx$

$= \frac{1}{1080} \int_{48}^{75} \Big[(\frac{1}{9}x^2 + 2x + 100)(60 - 20)$

$\quad + \frac{1}{3}(60^3 - 20^3) + \frac{5}{2}(60^2 - 20^2) \Big] dx$

$= \frac{40}{1080}(\frac{1}{27}x^3 + x^2 + 100x) \Big|_{48}^{75}$

$\quad + \frac{1}{1080}\Big[\frac{208,000}{3} + \frac{5}{2}(3200)\Big]x \Big|_{48}^{75}$

$= \frac{40}{1080}$

$\quad \cdot \Big[\frac{1}{27}(75^3 - 48^3) + 75^2 - 48^2 + 100(75 - 48)\Big]$

$\quad + \frac{1}{1080}(\frac{208,000}{3} + 8000)(75 - 48)$

$= \frac{40}{1080}(\frac{311,283}{27} + 3321 + 2700)$

$\quad + \frac{1}{1080}(\frac{232,000}{3})27$

$= \frac{40}{1080}(11,529 + 6021) + \frac{23,200}{12}$

$\approx \$2583$

65. $P(x, y)$

$$= -(x - 100)^2 - (y - 50)^2 + 2000$$

Area $= (150 - 100)(80 - 40)$

$$= (50)(40)$$

$$= 2000$$

The average weekly profit is

$$\frac{1}{2000} \int \int_R [-(x - 100)^2 - (y - 50)^2 + 2000]dy\,dx$$

$$= \frac{1}{2000} \int_{100}^{150} \int_{40}^{80} [-(x - 100)^2 - (y - 50)^2$$

$$+ 2000]dy\,dx$$

$$= \frac{1}{2000} \int_{100}^{150} \Big[-(x - 100)^2 y - \frac{(y - 50)^3}{3}$$

$$+ 2000y\Big]\Big|_{40}^{80} dx\Big|$$

$$= \frac{1}{2000} \int_{100}^{150} \Big[-(x - 100)^2(80 - 40)$$

$$- \frac{(80 - 50)^3}{3} + \frac{(40 - 50)^3}{3}$$

$$+ 2000(80 - 40)\Big]dx$$

$$= \frac{1}{2000} \int_{100}^{150} \Big[-40(x - 100)^2 - \frac{30^3}{3}$$

$$+ \frac{(-10)^3}{3} + 2000(40)\Big]dx$$

$$= \frac{1}{2000} \int_{100}^{150} \Big[-40(x - 100)^2 - \frac{28,000}{3}$$

$$+ 80,000\Big]dx$$

$$= \frac{1}{2000}$$

$$\cdot \Big[\frac{-40(x - 100)^3}{3} - \frac{28,000}{3}x + 80,000x\Big]\Big|_{100}^{150}$$

$$= \frac{1}{2000}\Big[-\frac{40}{3}(150 - 100)^3 + \frac{40}{3}(100 - 100)^3\Big]$$

$$- \frac{28,000}{3}(150 - 100) + 80,000(150 - 1000)\Big]$$

$$= \frac{1}{2000}\Big[-\frac{40}{3}(50)^3 + \frac{40}{3}\cdot 0$$

$$- \frac{28,000}{3}(50) + 80,000(50)\Big]$$

$$= \frac{1}{2000}\Big(\frac{-5,000,000 - 1,400,000 + 12,000,000}{3}\Big)$$

$$= \frac{1}{2000}\Big(\frac{5,600,000}{3}\Big) = \$933.33$$

Chapter 8   Review Exercises

1. $f(x, y) = -4x^2 + 6xy - 3$

$$f(-1, 2) = -4(-1)^2 + 6(-1)(2) - 3$$

$$= -19$$

$$f(6, -3) = -4(6)^2 + 6(6)(-3) - 3$$

$$= -4(36) + (-108) - 3$$

$$= -255$$

3. $f(x, y) = \dfrac{x - 3y}{x + 4y}$

$$f(-1, 2) = \frac{-1 - 6}{-1 + 8} = \frac{-7}{7} = -1$$

$$f(6, -3) = \frac{6 - 3(-3)}{6 + 4(-3)} = \frac{6 + 9}{6 - 12}$$

$$= \frac{15}{-6} = -\frac{5}{2}$$

For Exercises 5–9, see the answer graphs at the back of the textbook.

5.   The plane $x + y + z = 4$ intersects the axes at $(4, 0, 0)$, $(0, 4, 0)$, and $(0, 0, 4)$.

7. The plane $5x + 2y = 10$ intersects the x- and y-axes at $(2, 0, 0)$ and $(0, 5, 0)$. Note that there is no z-intercept since $x = y = 0$ is not a solution of the equation of the plane.

9. $x = 3$

   The plane is parallel to the yz-plane. It intersects the x-axis at $(3, 0, 0)$.

11. Let $z = f(x, y) = -5x^2 + 7xy - y^2$

   (a) $\frac{\partial z}{\partial x} = -10x + 7y$

   (b) $\frac{\partial z}{\partial y} = 7x - 2y$

   $\left(\frac{\partial z}{\partial y}\right)(-1, 4) = 7(-1) - 2(4)$
   $= -15$

   (c) $f_{xy}(x, y) = 7$
   $f_{xy}(2, -1) = 7$

13. $f(x, y) = 9x^3y^2 - 5x$
   $f_x = 27x^2y^2 - 5$
   $f_y = 18x^3y$

15. $f(x, y) = \sqrt{4x^2 + y^2}$
   $f_x = \frac{1}{2}(4x^2 + y^2)^{-1/2}(8x)$

   $= \frac{4x}{(4x^2 + y^2)^{1/2}}$

   $f_y = \frac{1}{2}(4x^2 + y^2)^{-1/2}(2y)$

   $= \frac{y}{(4x^2 + y^2)^{1/2}}$

17. $f(x, y) = x^2 \cdot e^{2y}$
   $f_x = 2xe^{2y}$
   $f_y = 2x^2e^{2y}$

19. $f(x, y) = \ln |2x^2 + y^2|$
   $f_x = \frac{1}{2x^2 + y^2} \cdot 4x$

   $= \frac{4x}{2x^2 + y^2}$

   $f_y = \frac{1}{2x^2 + y^2} \cdot 2y$

   $= \frac{2y}{2x^2 + y^2}$

21. $f(x, y) = 4x^3y^2 - 8xy$
   $f_x = 12x^2y^2 - 8y$
   $f_{xx} = 24xy^2$
   $f_{xy} = 24x^2y - 8$

23. $f(x, y) = \frac{2x}{x - 2y}$

   $f_x = \frac{2(x - 2y) - (2x)}{(x - 2y)^2}$

   $= \frac{-4y}{(x - 2y)^2}$

   $= -4y(x - 2y)^{-2}$

   $f_{xx} = -4y(-2(x - 2y)^{-3})$

   $= 8y(x - 2y)^{-3}$

   $= \frac{8y}{(x - 2y)^3}$

   Use the quotient rule on

   $f_x = \frac{-4y}{(x - 2y)^2}$

to get

$$f_{xy} = \frac{-4(x-2y)^2 + 4y[2(x-2y)(-2)]}{(x-2y)^4}$$

$$= \frac{-4(x-2y)[(x-2y)+4y]}{(x-2y)^4}$$

$$= \frac{-4(x+2y)}{(x-2y)^3}$$

$$= \frac{-4x-8y}{(x-2y)^3}.$$

25. $f(x, y) = x^2 e^y$

$$f_x = 2xe^y$$

$$f_{xx} = 2e^y$$

$$f_{xy} = 2xe^y$$

27. $f(x, y) = \ln |2 - x^2y|$

$$f_x = \frac{1}{2 - x^2y} \cdot -2xy$$

$$= \frac{2xy}{x^2y - 2}$$

$$f_{xx} = \frac{(x^2y - 2)2y - 2xy(2xy)}{(x^2y - 2)^2}$$

$$= \frac{2y[(x^2y - 2) - 2x^2y]}{(x^2y - 2)^2}$$

$$= \frac{2y(-x^2y - 2)}{(x^2y - 2)^2}$$

$$= \frac{-2x^2y^2 - 4y}{(2 - x^2y)^2}$$

$$f_{xy} = \frac{2x(x^2y - 2) - x^2(2xy)}{(x^2y - 2)^2}$$

$$= \frac{2x[(x^2y - 2) - x^2y]}{(x^2y - 2)^2}$$

$$= \frac{2x(-2)}{(x^2y - 2)^2}$$

$$= \frac{-4x}{(2 - x^2y)^2}$$

29. $z = x^2 + 2y^2 - 4y$

$$z_x = 2x$$

$$z_y = 4y - 4$$

Setting z_x and z_y equal to zero simultaneously implies $x = 0$ and $y = 1$.

$$z_{xx} = 2, \ z_{yy} = 4, \ z_{xy} = 0$$

For $(0, 1)$,

$$D = 2 \cdot 4 - 0 = 8 > 0.$$

Since

$$z_{xx}(0, 1) = 2 > 0,$$

then z has a relative minimum at $(0, 1)$.

31. $f(x, y) = x^2 + 5xy - 10x + 3y^2 - 12y$

$$f_x = 2x + 5y - 10$$

$$f_y = 5x + 6y - 12$$

Setting f_x and f_y equal to zero and solving yields

$$2x + 5y = 10$$

$$5x + 6y = 12$$

$$-10x - 25y = -50$$

$$\underline{10x + 12y = \ \ \ 24}$$

$$-13y = -26$$

$$y = 2$$

$$2x + 10 = 10$$

$$x = 0$$

$$f_{xx} = 2, \ f_{yy} = 6, \ f_{xy} = 5,$$

For $(0, 2)$,

$$D = 2 \cdot 6 - 5^2 = -13 < 0.$$

Therefore f has a saddle point at $(0, 2)$.

33. $z = \frac{1}{2}x^2 + \frac{1}{2}y^2 + 2xy - 5x - 7y + 10$

$z_x = x + 2y - 5$

$z_y = y + 2x - 7$

Setting $z_x = z_y = 0$ and solving yields

$$x + 2y = 5$$
$$2x + y = 7.$$
$$-2x - 4y = -10$$
$$\underline{2x + y = \quad 7}$$
$$-3y = \; -3$$
$$y = 1, \; x = 3.$$

$z_{xx} = 1, \; z_{yy} = 1, \; z_{xy} = 2$

For $(3, 1)$,

$$D = 1 \cdot 1 - 4 = -3 < 0.$$

Therefore z has a saddle point at $(3, 1)$.

35. $z = x^3 + y^2 + 2xy - 4x - 3y - 2$

$z_x = 3x^2 + 2y - 4$

$z_y = 2y + 2x - 3$

Setting $z_x = z_y = 0$ yields

$$3x^2 + 2y - 4 = 0 \quad (1)$$
$$2y + 2x - 3 = 0. \quad (2)$$

Solving for 2y in Equation (2) gives $2y = -2x + 3$. Substitute into Equation (1).

$$3x^2 + (-2x) + 3 - 4 = 0$$
$$3x^2 - 2x - 1 = 0$$
$$(3x + 1)(x - 1) = 0$$

$$x = \frac{-1}{3} \quad \text{or} \quad x = 1$$

$$y = \frac{11}{6} \quad \text{or} \quad y = \frac{1}{2}$$

$z_{xx} = 6x, \; z_{yy} = 2, \; z_{xy} = 2$

For $(-1/3, 11/6)$,

$$D = 6\left(\frac{-1}{3}\right)(2) - 4$$
$$= -4 - 4 = -8 < 0, \text{ so}$$

so z has a saddle point at $(-1/3, 11/6)$.

For $(1, 1/2)$,

$$D = 6(1)(2) - 4 = 8 > 0.$$

$z_{xx}\left(1, \frac{1}{2}\right) = 6 > 0$, so

z has a relative minimum at $(1, 1/2)$.

37. $f(x, y) = x^2y; \; x + y = 4$

1. $g(x) = x + y - 4$

2. $F(x, y) = x^2y + \lambda(x + y - 4)$

3. $F_x = 2xy + \lambda$

   $F_y = x^2 + \lambda$

   $F_\lambda = x + y - 4$

4. $2xy + \lambda = 0$

   $x^2 + \lambda = 0$

   $x + y - 4 = 0$

5. $\lambda = -2xy$

   $\lambda = -x^2$

   $-2xy = -x^2$

$$2xy - x^2 = 0$$
$$x(2y - x) = 0$$
$$x = 0 \quad \text{or} \quad 2y = x$$

Substituting into the third equation gives

$$y = 4 \quad \text{or} \quad y = \frac{4}{3}.$$

If $y = \frac{4}{3}$, $x = \frac{8}{3}$.

The critical points are (0, 4) and

$\left(\frac{8}{3}, \frac{4}{3}\right)$.

$$f(0, 4) = 0$$

$$f\left(\frac{8}{3}, \frac{4}{3}\right) = \frac{64}{9} \cdot \frac{4}{3}$$

$$= \frac{256}{27}$$

Therefore, f has a minimum of 0 at (0, 4) and a maximum of 256/27 at (8/3, 4/3).

39. Let x and y be the numbers such that x + y = 80 and $f(x, y) = x^2y$.

1. $g(x) = x + y - 80$

2. $F(x, y) = x^2y + \lambda(x + y - 80)$

3. $F_x = 2xy + \lambda$
   $F_y = x^2 + \lambda$
   $F_\lambda = x + y - 80$

4. $2xy + \lambda = 0$
       $x^2 + \lambda = 0$
      $x + y - 80 = 0$

5. $\lambda = -2xy$
      $\lambda = -x^2$
     $-2xy = -x^2$

   $$2xy - x^2 = 0$$
   $$x(2y - x) = 0$$
   $$x = 0 \quad \text{or} \quad x = 2y$$

Substituting into the third equation gives

$y = 80$ or $2y + y - 80 = 0$
                    $3y = 80$
                    $y = \frac{80}{3}$
          and $x = \frac{160}{3}$.

$$f(0, 80) = 0 \cdot 80^2 = 0$$

$$f\left(\frac{160}{3}, \frac{80}{3}\right) = \frac{(160)^2}{9} \frac{(80)}{3}$$

$$= \frac{2,048,000}{27} > f(0, 80)$$

f has maximum at (160/3, 80/3). Therefore, if x = 160/3 and y = 80/3, then x^2y is maximized.

41. Let x be the length of each of the square faces of the box and y be the length of the box.

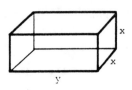

Since the volume must be 125, the constraint is $125 = x^2y$.
$f(x, y) = 2x^2 + 4xy$ is the surface area of the box.

1. $g(x) = x^2y - 125$

2. $F(x, y) = 2x^2 + 4xy + \lambda(x^2y - 125)$

3. $F_x = 4x + 4y + 2xy\lambda$
   $F_y = 4x + x^2\lambda$
   $F_\lambda = x^2y - 125$

4. $4x + 4y + 2xy\lambda = 0$ (1)
        $4x + x^2\lambda = 0$ (2)
         $x^2y - 125 = 0$ (3)

5. Factoring Equation (2) gives

   $$x(4 + x\lambda) = 0$$
   $$x = 0 \quad \text{or} \quad 4 + x\lambda = 0.$$

Since x = 0 is not a solution of Equation (3), then

$$4 + x\lambda = 0$$

$$\lambda = \frac{-4}{x}.$$

Substituting into Equation (1) gives

$$4x + 4y + 2xy\left(\frac{-4}{x}\right) = 0$$

or $\qquad 4x + 4y - 8y = 0$

$$x = y.$$

Substituting $x = y$ into Equation (3) gives

$$x^2y - 125 = 0$$
$$y^3 = 125$$
$$y = 5.$$

Therefore $x = y = 5$. The dimensions are 5 in by 5 in by 5 in.

43. $\quad z = 7x^3y - 4y^3$

$\quad dz = 21x^2y\ dx + (7x^3 - 12y^2)dy$

45. $\quad z = x^2ye^{x-y}$

$\quad f_x = x^2ye^{x-y} + e^{x-y} \cdot y(2x)$

$\qquad = xye^{x-y}(x + 2)$

$\quad f_y = x^2ye^{x-y}(-1) + x^2e^{x-y}$

$\qquad = x^2e^{x-y}(1 - y)$

$\quad dz = xye^{x-y}(2 + x)dx + x^2e^{x-y}(1 - y)dy$

47. $\quad w = x^5 + y^4 - z^3$

$\quad dw = 5x^4dx + 4y^3dy - 3z^2dz$

49. $\quad z = 2x^2 - 4y^2 + 6xy;\ x = 2,\ y = -3,$

$\quad dx = .01,\ dy = .05$

$\quad f_x = 4x + 6y$

$\quad f_y = -8y + 6x$

$\quad dz = (4x + 6y)dx + (-8y + 6x)dy$

Substitute.

$\quad dz = (4(2) + 6(-3))(.01)$

$\qquad + (-8(-3) + 6(2))(.05)$

$\quad = -.1 + 1.8 = 1.7$

51. $\displaystyle\int_0^4 (x^2y^2 + 5x)dx$

$$= \left(\frac{x^3}{3}y^2 + \frac{5}{2}x^2\right)\Bigg|_0^4$$

$$= \frac{64}{3}y^2 + \frac{5(4)^2}{2}$$

$$= \frac{64y^2}{3} + 40$$

53. $\displaystyle\int_2^5 \sqrt{6x + 3y}\ dx$

Let $u = 6x + 3y$

$\qquad du = 6\ dx$

When $x = 2$, $u = 12 + 3y$.

When $x = 5$, $u = 30 + 3y$.

$$= \frac{1}{6}\int_{12+3y}^{30+3y} u^{1/2}\ du$$

$$= \frac{1}{6}\left(\frac{2}{3}u^{3/2}\right)\Bigg|_{12+3y}^{30+3y}$$

$$= \frac{1}{9}[(30 + 3y)^{3/2} - (12 + 3y)^{3/2}]$$

55. $\displaystyle\int_4^9 \frac{6y - 8}{\sqrt{x}}\ dx$

$$= (6y - 8)(2\sqrt{x})\Bigg|_4^9$$

$$= 2(3y - 4)(2)(\sqrt{9} - \sqrt{4})$$

$$= 4(3y - 4)$$

$$= 12y - 16$$

57. $\displaystyle\int_0^5 \frac{6x}{\sqrt{4x^2 + 2y^2}}\ dx$

Let $u = 4x^2 + 2y^2$

$\qquad du = 8x\ dx$

When x = 0, u = $2y^2$.

When x = 5, u = $100 + 2y^2$.

$$= \frac{3}{4}\int_{2y^2}^{100+2y^2} u^{-1/2}\,du$$

$$= \frac{3}{4}(2u^{1/2})\Big|_{2y^2}^{100+2y^2}$$

$$= \frac{3}{4}\cdot 2[(100 + 2y^2)^{1/2} - (2y^2)^{1/2}]$$

$$= \frac{3}{2}[(100 + 2y^2)^{1/2} - (2y^2)^{1/2}]$$

59. $\displaystyle\int_0^2\left[\int_0^4 (x^2y^2 + 5x)\,dx\right]dy$

See Exercise 51.

$$= \int_0^2\left(\frac{64}{3}y^2 + 40\right)dy$$

$$= \left(\frac{64}{3}y^3 + 40y\right)\Big|_0^2$$

$$= \frac{64}{9}(8) + 40(2)$$

$$= \frac{512}{9} + \frac{720}{9}$$

$$= \frac{1232}{9}$$

61. $\displaystyle\int_3^4\left[\int_2^5 \sqrt{6x + 3y}\ dx\right]dy$

See Exercise 53.

$$= \int_3^4 \frac{1}{9}\left[(30 + 3y)^{3/2} - (12 + 3y)^{3/2}\right]dy$$

$$= \frac{1}{3}\cdot\frac{1}{9}\cdot\frac{2}{5}$$

$$\quad \cdot[(30 + 3y)^{5/2} - (12 + 3y)^{5/2}]\Big|_3^4$$

$$= \frac{2}{135}[(42)^{5/2} - (24)^{5/2} - (39)^{5/2}$$

$$+ (21)^{5/2}]$$

63. $\displaystyle\int_2^4\int_2^4 \frac{dx\ dy}{y}$

$$= \int_2^4 \left(\frac{1}{y}x\right)\Big|_2^4 dy$$

$$= \int_2^4 \left[\frac{1}{y}(4 - 2)\right]dy$$

$$= 2\ln |y|\Big|_2^4$$

$$= 2\ln\left|\frac{4}{2}\right| = 2\ln 2\ \text{or}\ \ln 4$$

65. $\displaystyle\iint_R (x^2 + y^2)\,dx\,dy;\ 0 \le x \le 2,$

$$0 \le y \le 3$$

$$\iint_R (x^2 + y^2)\,dx\,dy$$

$$= \int_0^3\int_0^2 (x^2 + y^2)\,dx\,dy$$

$$= \int_0^3 \left(\frac{x^3}{3} + y^2x\right)\Big|_0^2 dy$$

$$= \int_0^3 \left(\frac{8}{3} + 2y^2\right)dy$$

$$= \left(\frac{8}{3}y + \frac{2}{3}y^3\right)\Big|_0^3$$

$$= 8 + 18 = 26$$

67. $\iint\limits_{R} \sqrt{y + x} \; dx \; dy; \; 0 \le x \le 7,$

$1 \le y \le 9$

$\iint\limits_{R} \sqrt{y + x} \; dx \; dy$

$= \int_{0}^{7}\int_{1}^{9} \sqrt{y + x} \; dy \; dx$

$= \int_{0}^{7} \left[\frac{2}{3}(y + x)^{3/2} \right] \Big|_{1}^{9} \; dx$

$= \int_{0}^{7} \frac{2}{3}[(9 + x)^{3/2} - (1 + x)^{3/2}] dx$

$= \frac{2}{3} \cdot \frac{2}{5}[(9 + x)^{5/2} - (1 + x)^{5/2}] \Big|_{0}^{7}$

$= \frac{4}{15}[(16)^{5/2} - (8)^{5/2} - (9)^{5/2}$

$\quad + (1)^{5/2}]$

$= \frac{4}{15}[4^5 - (2\sqrt{2})^5 - 3^5 + 1]$

$= \frac{4}{15}(1024 - 32(4\sqrt{2}) - 243 + 1)$

$= \frac{4}{15}(782 - 128\sqrt{2})$

$= = \frac{4}{15}(782 - 8^{5/2})$

69. $z = x + 9y + 8; \; 1 \le x \le 6,$

$0 \le y \le 8$

$V = \iint\limits_{R} (x + 9y + 8) dx \; dy$

$= \int_{0}^{8}\int_{1}^{6} (x + 9y + 8) dx \; dy$

$= \int_{0}^{8} \left[\frac{x^2}{2} + (9y + 8)x \right] \Big|_{1}^{6} \; dy$

$= \int_{0}^{8} \left[\frac{36 - 1}{2} + (9y + 8)(6 - 1) \right] dy$

$= \left[\frac{35}{2}y + 5\left(\frac{9}{2}y^2 + 8y\right) \right] \Big|_{0}^{8}$

$= \frac{35}{2}(8) + 5\left[\left(\frac{9}{2}\right)(64) + 8 \cdot 8) \right]$

$= \frac{35(8)}{2} + 5\left(\frac{9(64)}{2} + 8 \cdot 8\right)$

$= 140 + 5(352)$

$= 1900$

71. $\int_{0}^{1}\int_{0}^{2x} xy \; dy \; dx$

$= \int_{0}^{1} \left(\frac{xy^2}{2}\right) \Big|_{0}^{2x} \; dx$

$= \int_{0}^{1} \frac{x}{2}(4x^2 - 0) dx$

$= \int_{0}^{1} 2x^3 \; dx$

$= \left(\frac{1}{2}x^4\right) \Big|_{0}^{1} = \frac{1}{2}$

73. $\int_{0}^{1}\int_{x^2}^{x} x^3y \; dy \; dx$

$\int_{0}^{1} \left(\frac{x^3}{2}y^2\right) \Big|_{x^2}^{x} \; dx$

$= \int_{0}^{1} \frac{x^3}{2}(x^2 - x^4) dx$

$= \frac{1}{2}\int_{0}^{1} (x^5 - x^7) dx$

$= \frac{1}{2}\left(\frac{x^6}{6} - \frac{x^8}{8}\right) \Big|_{0}^{1}$

$$= \frac{1}{2}\left(\frac{1}{6} - \frac{1}{8}\right)$$

$$= \frac{1}{2} \cdot \frac{1}{24}$$

$$= \frac{1}{48}$$

75. $\displaystyle\int\int_R (2x + 3y)dx\ dy;\ 0 \le y \le 1,$

$y \le x \le 2 - y$

$\displaystyle\int_0^1\int_y^{2-y} (2x + 3y)dx\ dy$

$$= \int_0^1 (x^2 + 3xy)\Big|_y^{2-y}\ dy$$

$$= \int_0^1 [(2-y)^2 - y^2 + 3y(2-y-y)]dy$$

$$= \int_0^1 (4 - 4y + y^2 - y^2 + 6y - 6y^2)dy$$

$$= \int_0^1 (4 + 2y - 6y^2)dy$$

$$= (4y + y^2 - 2y^3)\Big|_0^1$$

$$= 4 + 1 - 2$$

$$= 3$$

77. $C(x, y) = 2x^2 + 4y^2 - 3xy + \sqrt{x}$

(a) $C(10, 5)$

$\quad = 2(10)^2 + 4(5)^2 - 3(10)(5) + \sqrt{10}$

$\quad = 200 + 100 - 150 + \sqrt{10}$

$\quad = \$(150 + \sqrt{10})$

(b) $C(15, 10)$

$\quad = 2(15)^2 + 4(10)^2 - 3(15)(10)$

$\quad\quad + \sqrt{15}$

$\quad = 450 + 400 - 450 + \sqrt{15}$

$\quad = \$(400 + \sqrt{15})$

(c) $C(20, 20)$

$\quad = 2(20)^2 + 4(20)^2 - 3(20)(20)$

$\quad\quad + \sqrt{20}$

$\quad = 800 + 1600 - 1200 + \sqrt{20}$

$\quad = \$(1200 + 2\sqrt{5})$

79. $z = x^{.6}y^{.4}$

(a) The marginal productivity of labor is

$$\frac{\partial z}{\partial x} = .6x^{-.4}y^{.4}$$

$$= \frac{.6y^{.4}}{x^{.4}}$$

(b) The marginal productivity of capital is

$$\frac{\partial z}{\partial y} = .4x^{.6}y^{-.6}$$

$$= \frac{.4x^{.6}}{y^{.6}}$$

81. (a)

x	y	x^2	xy
3	4	9	12
5	11	25	55
7	20	49	140
8	23	64	184
Totals 23	58	147	391

$$m = \frac{23(58) - 4(391)}{(23)^2 - 4(147)} \approx 3.90$$

$$b = \frac{58 - m(23)}{4} \approx -7.92$$

Least squares equation is

$$y' = 3.90x - 7.92.$$

(b) If $x = 6$,

$$y' = 3.90(6) - 7.92 = 15.48$$

(in ten-thousands).

The predicted earnings are $154,800.

83. $P(x, y) = \dfrac{x}{x + 5y} + \dfrac{y + x}{y}$

$$= x(x + 5y)^{-1} + \dfrac{y + x}{y}$$

$x = 75$, $y = 50$, $dx = -3$, $dy = 2$

$dP = \left[\dfrac{(x + 5y) \cdot 1 - x \cdot 1}{(x + 5y)^2} + \dfrac{1}{y}\right] dx$

$\quad + \left[-5x(x + 5y)^{-2} + \dfrac{y \cdot 1 - (y + x)}{y^2}\right] dy$

$= \left[\dfrac{5y}{(x + 5y)^2} + \dfrac{1}{y}\right] dx$

$\quad + \left[\dfrac{-5x}{(x + 5y)^2} - \dfrac{x}{y^2}\right] dy$

$= \left[\dfrac{5 \cdot 50}{(75 + 5 \cdot 50)^2} + \dfrac{1}{50}\right](-3)$

$\quad + \left[\dfrac{-5(75)}{(75 + 5 \cdot 50)^2} - \dfrac{75}{50^2}\right](2)$

$\approx -.1342$

Expect profits to decrease by $13.42.

85. $V = \dfrac{4}{3}\pi r^3$, $r = 2$ ft,

$dr = 1$ in $= \dfrac{1}{12}$ ft

$dV = 4\pi r^2 dr = 4\pi(2)^2\left(\dfrac{1}{12}\right) \approx 4.19$ ft^3

87. **(a)** $m = \dfrac{(\Sigma x)(\Sigma y) - n(\Sigma xy)}{(\Sigma x)^2 - n(\Sigma x^2)}$

$\quad = \dfrac{(1394)(1607) - 8(291,990)}{(1394)^2 - 8(255,214)}$

$\quad = \dfrac{-95,762}{-98,476} \approx .97$

$b = \dfrac{\Sigma y - m\Sigma x}{n}$

$\quad = \dfrac{1607 - .97(1394)}{8}$

$\quad \approx 31.5$

$y' = .97x + 31.5$

(b) When $x = 190$,

$$y' = .97(190) + 31.5$$

$$\approx 216.$$

(c) If $x = 130$,

$$y' = .97(130) + 31.5$$

$$= 157.6 \approx 158.$$

If $x = 142$,

$$y' = .97(142) + 31.5$$

$$= 169.24 \approx 169.$$

If $x = 200$,

$$y' = .97(200) + 31.5$$

$$= 225.5 \approx 226.$$

Blood Sugar Level	Predicted Cholesterol Level	Actual Cholesterol Level
130	158	170
142	169	173
200	226	192

The predicted values are in the vicinity of the actual values, but not "close."

89. Assume blood vessel is cylindrical.

$V = \pi r^2 h$, $r = .7$, $h = 2.7$,
$dr = dh = \pm.1$

$dV = 2\pi rh\,dr + \pi r^2\,dh$
$= 2\pi(.7)(2.7)(\pm.1) + \pi(.7)^2(\pm.1)$
$\approx \pm 1.3$

The possible error is 1.3 cm³.

91. $P(x, y)$

$= .01(-x^2 + 3xy + 160x - 5y^2$
$+ 200y + 2600)$

with $x + y = 280$.

(a) $y = 280 - x$

$P(x) = .01[-x^2 + 3x(280 - x) + 160x$
$- 5(280 - x)^2 + 200(280 - x)$
$+ 2600]$
$= .01(-x^2 + 840x - 3x^2 + 160x$
$- 392{,}000 + 2800x - 5x^2$
$+ 56{,}000 - 200x + 2600)$

$P(x) = .01(-9x^2 + 3600x - 333{,}400)$

$P'(x) = .01(-18x + 3600)$
$.01(-18x + 3600) = 0$
$-18x = -3600$
$x = 200$

If $x < 200$, $P'(x) > 0$, and if
$x > 200$, $P'(x) < 0$. Therefore,
P is maximum when $x = 200$. If
$x = 200$, $y = 80$.

$P(200, 80)$
$= .01[-200^2 + 3(200)(80) + 160(200)$
$- 5(80)^2 + 200(80) + 2600]$
$= .01(26{,}600)$
$= 266$

Thus, \$200 spent on fertilizer and
\$80 spent on seed will produce a
maximum profit of \$266 per acre.

(b) $P(x, y)$
$= .01(-x^2 + 3xy + 160x - 5y^2$
$+ 200y + 2600)$

$P_x = .01(-2x + 3y + 160)$
$P_y = .01(3x - 10y + 200)$
$.01(-2x + 3y + 160) = 0$
$.01(3x - 10y + 200) = 0$

These equations simplify to

$-2x + 3y = -160$
$3x - 10y = -200.$

Solve the system.

$-6x + 9y = -480$
$\underline{6x - 20y = -400}$
$-11y = -880$
$y = 80$

If $y = 80$,

$3x - 10(80) = -200$
$3x = 600$
$x = 200.$

$P_{xx} = .01(-2) = -.02$
$P_{yy} = .01(-10) = -.1$
$P_{xy} = 0$

For (200, 80), $D = (-.02)(-.1) - 0^2$
$= .002 > 0$ since $P_{xx} < 0$, there is a
relative maximum at (200, 80).

$P(200, 80) = 266$ as in part (a).
Thus, \$200 spent on fertilizer and
\$80 spent on seed will produce a
maximum profit of \$266 per acre.

(c) Maximize P(x, y)

$= .01(-x^2 + 3xy + 160x - 5y^2$
$+ 200y + 2600)$

subject to $x + y = 280$.

1. $g(x, y) = x + y - 280$

2. $F(x, y, \lambda)$

$= .01(-x^2 + 3xy + 160x - 5y^2$
$+ 200y + 2600)$
$+ \lambda(x + y - 280)$

3. $F_x = .01(-2x + 3y + 160) + \lambda$
$F_y = .01(3x - 10y + 200) + \lambda$
$F_\lambda = x + y - 280$

4. $.01(-2x + 3y + 160) + \lambda = 0$ (1)
$.01(3x - 10y + 200) + \lambda = 0$ (2)
$x + y - 280 = 0$ (3)

5. Equations (1) and (2) give

$.01(-2x + 3y + 160) = .01(3x - 10y + 200)$
$-2x + 3y + 160 = 3x - 10y + 200$
$-5x + 13y = 40$

Multiplying equation (3) by 5 gives

$5x + 5y - 1400 = 0$

$-5x + 13y = \quad 40$
$\underline{5x + \quad 5y = 1400}$
$18y = 1440$
$y = 80$

If $y = 80$,

$5x + 5(80) = 1400$
$5x = 1000$
$x = 200.$

Thus, P(200, 80) is a maximum. As before, P(200, 80) = 266.

Thus, $200 spent on fertilizer and $80 spent on seed will produce a maximum profit of $266 per acre.

Extended Application

1. $\dfrac{\partial F}{\partial x_1} = 2(.4 - .7x_1 - .4x_2)(-.7)$
$+ 2(.5)(.15 - .4x_1 - .1x_2)(-.4)$
$+ 2(.4)(.25 - .3x_1 - .3x_2)(.3) + \lambda$
$= -1.4(.4 - .7x_1 - .4x_2)$
$- .4(.15 - .4x_1 - .1x_2)$
$- .24(.25 - .3x_1 - .3x_2) + \lambda$
$= -.56 + .98x_1 + .56x_2 - .06 + .16x_1$
$+ .04x_2 - .06 + .072x_1 + .072x_2 + \lambda$
$= -.68 + 1.212x_1 + .672x_2 + \lambda$

$\dfrac{\partial F}{\partial x_2} = 2(.4 - .7x_1 - .4x_2)(-.4)$
$+ 2(.5)(.15 - .4x_1 - .1x_2)(-.1)$
$+ 2(.4)(.25 - .3x_1 - .3x_2)(-.3)$
$+ \lambda$
$= -.8(.4 - .7x_1 - .4x_2)$
$- .1(.15 - .4x_1 - .1x_2)$
$- .24(.25 - .3x_1 - .3x_2) + \lambda$
$= -.32 + .56x_1 + .32x_2 - .015$
$+ .04x_1 + .01x_2 - .06 + .072x_1$
$+ .072x_2 + \lambda$
$= -.395 + .672x_1 + .402x_2 + \lambda$

2. $-.68 + 1.212x_1 + .672x_2 + \lambda = 0$
$-.395 + .672x_1 + .402x_2 + \lambda = 0$
$-.06 + .06x_3 + \lambda = 0$
$x_1 + x_2 + x_3 - 1 = 0$

We will solve this system of linear equations using Cramer's Rule.

$1.212x_1 + .672x_2 + 0x_3 + 1\lambda = .68$
$.672x_1 + .402x_2 + 0x_3 + 1\lambda = .395$
$0x_1 + 0x_2 + .06x_3 + 1\lambda = .06$
$1x_1 + 1x_2 + 1x_3 + 0\lambda = 1$

$$x_1 = \frac{\begin{vmatrix} .68 & .672 & 0 & 1 \\ .395 & .402 & 0 & 1 \\ .06 & 0 & .06 & 1 \\ 1 & 1 & 1 & 0 \end{vmatrix}}{\begin{vmatrix} 1.212 & .672 & 0 & 1 \\ .672 & .402 & 0 & 1 \\ 0 & 0 & .06 & 1 \\ 1 & 1 & 1 & 0 \end{vmatrix}}$$

$$= \frac{-.02502}{-.05184} \approx .48$$

$$x_2 = \frac{\begin{vmatrix} 1.212 & .68 & 0 & 1 \\ .672 & .395 & 0 & 1 \\ 0 & .06 & .06 & 1 \\ 1 & 1 & 1 & 0 \end{vmatrix}}{-.05184}$$

$$= \frac{-.00468}{-.05184} \approx .09$$

$$x_3 = \frac{\begin{vmatrix} 1.212 & .672 & .68 & 1 \\ .672 & .402 & .395 & 1 \\ 0 & 0 & .06 & 1 \\ 1 & 1 & 1 & 0 \end{vmatrix}}{-.05184}$$

$$= \frac{-.02214}{-.05184} \approx .43$$

3. Suppose $w_1 = .4$ and $w_2 = .5$ and all other values remain the same.   In this case,

$$F = [.4 - .7x_1 - .4x_2]^2$$
$$+ .4[.15 - .4x_1 - .1x_2]^2$$
$$+ .5[.25 - .3x_1 - .3x_2]^2$$
$$+ .03(1 - x_3)^2 + \lambda(x_1 + x_2 + x_3 - 1).$$

Then,

$$\frac{\partial F}{\partial x_1} = 2(.4 - .7x_1 - .4x_2)(-.7)$$
$$+ 2(.4)(.15 - .4x_1 - .1x_2)(-.4)$$
$$+ 2(.5)(.25 - .3x_1 - .3x_2)(-.3) + \lambda$$
$$= -1.4(.4 - .7x_1 - .4x_2)$$
$$- .32(.15 - .4x_1 - .1x_2)$$
$$- .3(.25 - .3x_1 - .3x_2) + \lambda$$

$$= -.56 + .98x_1 + .56x_2 - .048$$
$$+ .128x_1 + .032x_2 - .075$$
$$+ .09x_1 + .09x_2 + \lambda$$
$$= -.683 + 1.198x_1 + .682x_2 + \lambda.$$

$$\frac{\partial F}{\partial x_2} = 2(.4 - .7x_1 - .4x_2)(-.4)$$
$$+ 2(.4)(.15 - .4x_1 - .1x_2)(-.1)$$
$$+ 2(.5)(.25 - .3x_1 - .3x_2)(-.3) + \lambda$$
$$= -.8(.4 - .7x_1 - .4x_2)$$
$$- .08(.15 - .4x_1 - .1x_2)$$
$$- .3(.25 - .3x_1 - .3x_2) + \lambda$$
$$= -.32 + .56x_1 + .32x_2 - .012$$
$$+ .032x_1 + .008x_2 - .075$$
$$+ .09x_1 + .09x_2 + \lambda$$
$$= -.407 + .682x_1 + .418x_2 + \lambda$$

$$\frac{\partial F}{\partial x_3} = 2(.03)(1 - x_3)(-1) + \lambda$$
$$= -.06 + .06x_3 + \lambda$$

$$\frac{\partial F}{\partial \lambda} = x_1 + x_2 + x_3 - 1$$

We set these equal to zero.

$$-.683 + 1.198x_1 + .682x_2 + \lambda = 0$$
$$-.407 + .682x_1 + .418x_2 + \lambda = 0$$
$$-.06 + .06x_3 + \lambda = 0$$
$$x_1 + x_2 + x_3 - 1 = 0$$

We will solve this system of linear equations using Cramer's Rule.

$$1.198x_1 + .682x_2 + 0x_3 + 1\lambda = .683$$
$$.682x_1 + .418x_2 + 0x_3 + 1\lambda = .407$$
$$0x_1 + 0x_2 + .06x_3 + 1\lambda = .06$$
$$1x_1 + 1x_2 + 1x_3 + 0\lambda = 1$$

$$x_1 = \frac{\begin{vmatrix} .683 & .682 & 0 & 1 \\ .407 & .418 & 0 & 1 \\ .06 & 0 & .06 & 1 \\ 1 & 1 & 1 & 0 \end{vmatrix}}{\begin{vmatrix} 1.198 & .682 & 0 & 1 \\ .682 & .418 & 0 & 1 \\ 0 & 0 & .06 & 1 \\ 1 & 1 & 1 & 0 \end{vmatrix}}$$

$$= \frac{-.02448}{-.05076} \approx .48$$

$$x_2 = \frac{\begin{vmatrix} 1.198 & .683 & 0 & 1 \\ .682 & .407 & 0 & 1 \\ 0 & .06 & .06 & 1 \\ 1 & 1 & 1 & 0 \end{vmatrix}}{-.05076}$$

$$= \frac{-.00522}{-.05076} \approx .10$$

$$x_3 = \frac{\begin{vmatrix} 1.198 & .682 & .683 & 1 \\ .682 & .418 & .407 & 1 \\ 0 & 0 & .06 & 1 \\ 1 & 1 & 1 & 0 \end{vmatrix}}{-.05076}$$

$$= \frac{-.02106}{-.05076} \approx .41$$

The predator should spend .48 of its
time feed in location 1, .10 of its
time in location 2, and .41 of its
time on nonfeeding activities.

CHAPTER 8   TEST

[8.1]   1. Find $f(-2, 1)$ for $f(x, y) = 2x^2 - 4xy + 7y$.

[8.1]   2. Find $g(-1, 3)$ for $g(x, y) = \sqrt{2x^2 - xy^2}$.

[8.1]   3. Complete the ordered triples $(0, 0,\ \ )$, $(0,\ \ , 0)$ and $(\ \ , 0, 0)$ for the plane $2x - 4y + 8z = 8$.

[8.1]   4. Graph the first octant portion of the plane $4x + 2y + 8z = 8$.

[8.1]   5. Let $f(x, y) = x^2 + 2y^2$. Find $\dfrac{f(x + h, y) - f(x, y)}{h}$.

[8.2]   6. Let $z = f(x, y) = 3x^3 - 5x^2y + 4y^2$. Find

        (a) $\dfrac{\partial z}{\partial x}$ **(b)** $\dfrac{\partial z}{\partial y}(1, 1)$ **(c)** f_{xx}.

[8.2]   7. Let $f(x, y) = \dfrac{x}{2x + y^2}$. Find

        (a) $f_x(1, -1)$ **(b)** $f_y(2, 1)$.

[8.2]   8. Let $f(x, y) = \sqrt{2x^2 - y^2}$. Find

        (a) f_x **(b)** f_y **(c)** f_{xy}.

[8.2]   9. Let $f(x, y) = xe^{y^2}$. Find

        (a) f_x **(b)** f_y **(c)** f_{yx}.

[8.2]   10. Let $f(x, y) = \ln (2x^2y^2 + 1)$. Find

        (a) f_{xx} **(b)** f_{xy} **(c)** f_{yy}.

[8.2] 11. The production function for a certain country is

$$z = 2x^5y^4$$

where x represents the amount of labor and y the amount of capital. Find the marginal productivity of

(a) labor   (b) capital.

[8.3] **Find all points where the functions defined below have any relative extrema. Find any saddle points.**

12. $z = 4x^2 + 2y^2 - 8x$ 13. $z = 2x^2 - 4xy + y^3$

14. $z = 1 - 2y - x^2 - 2xy - 2y^2$

[8.3] 15. A company manufactures two calculator models. The total revenue from x thousand solar calculators and y thousand battery–operated calculators is

$$R(x, y) = 4000 - 5y^2 - 8x^2 - 2xy + 42y + 102x.$$

Find x and y so that revenue is maximized.

[8.4] 16. Use Lagrange multipliers to find extrema of $f(x, y) = x^2 - 6xy$ subject to the constraint $x + y = 7$.

[8.4] 17. Find two numbers whose sum is 30 such that x^2y is maximized.

[8.4] 18. A closed box with square ends must have a volume of 64 cubic inches. Find the dimensions of such a box that has minimum surface area.

[8.5] 19. Find an equation for the least squares line for the following data.

x	1	2	3	5	9
y	18	26	36	50	82

Estimate y when x = 6.

[8.6] 20. (a) Find dz for $z = e^{x+y} \ln xy$.

(b) Evaluate dz when x = 1, y = 1, dx = .02 and dy = .01.

[8.6]   21. A sphere of radius 3 ft is to receive an insulating coating
              1/2 inch thick. Use differentials to approximate the volume
              of the coating needed.

[8.7]   22. Evaluate $\displaystyle\int_0^4\int_1^4 \sqrt{x + 3y}\ dx\ dy$.

[8.7]   23. Find the volume under the surface z = 2xy and above the
              rectangle $0 \le x \le 1,\ 0 \le y \le 4$.

[8.7]   24. Evaluate $\displaystyle\int_0^2\int_{y/2}^3 (x + y)dx\ dy$.

[8.7]   25. Use the region R with boundaries $0 \le y \le 2$ and $0 \le x \le y$ to
              evaluate

$$\iint_R (6 - x - y)dx\ dy.$$

CHAPTER 8   TEST ANSWERS

1. 23   **2.** 11   **3.** (0, 0, 1), (0, -2, 0), (4, 0, 0)

4.   **5.** $2x + h$ **6.** **(a)** $9x^2 - 10xy$ **(b)** 3

(c) $18x - 10y$

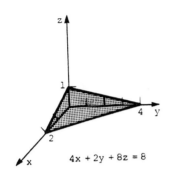

$4x + 2y + 8z = 8$

7. **(a)** $\frac{1}{9}$ **(b)** $\frac{-4}{25}$ **8.** **(a)** $\frac{2x}{\sqrt{2x^2 - y^2}}$ **(b)** $\frac{-y}{\sqrt{2x^2 - y^2}}$ **(c)** $\frac{2xy}{(2x^2 - y^2)^{3/2}}$

9. **(a)** e^{y^2} **(b)** $2xye^{y^2}$ **(c)** $2ye^{y^2}$

10. **(a)** $\frac{-8x^2y^4 + 4y^2}{(2x^2y^2 + 1)^2}$ **(b)** $\frac{8xy}{(2x^2y^2 + 1)^2}$ **(c)** $\frac{4x^2 - 8x^4y^2}{(2x^2y^2 + 1)^2}$

11. **(a)** $10x^4y^4$ **(b)** $8x^5y^3$

12. Relative minimum of -4 at $(1, 0)$

13. Relative minimum of $-\frac{32}{27}$ at $\left(\frac{4}{3}, \frac{4}{3}\right)$, saddle point at $(0, 0)$

14. Relative maximum of 2 at $(1, -1)$

15. 6 thousand solar calculators and 3 thousand battery-operated calculators

16. Minimum of -63 at $(3, 4)$ **17.** $x = 20, y = 10$

18. 4 inches by 4 inches by 4 inches   **19.** $y' = 7.96x + 10.56$; 58.32

20.   **(a)** $\left(e^{x+y} \ln (xy) + \dfrac{e^{x+y}}{x}\right)dx + \left(e^{x+y} \ln xy + \dfrac{e^{x+y}}{y}\right)dy$ **(b)** $.03e^2 \approx .222$

21. 4.71 cu ft   22. $\dfrac{4}{45}(993 - 13^{5/2})$ 23. 8   24. $\dfrac{40}{3}$ 25. 8

CHAPTER 9 DIFFERENTIAL EQUATIONS

Section 9.1

1. $\dfrac{dy}{dx} = -2x + 3x^2$

$$y = \int (-2x + 3x^2)\,dx$$

$$= \dfrac{-2x^2}{2} + \dfrac{3x^3}{3} + C$$

$$= -x^2 + x^3 + C$$

3. $3x^2 - 2\dfrac{dy}{dx} = 0$

Solve for $\dfrac{dy}{dx}$.

$$\dfrac{dy}{dx} = \dfrac{3x^3}{2}$$

$$y = \dfrac{3}{2} \int x^3\,dx$$

$$= \dfrac{3}{2}\left(\dfrac{x^4}{4}\right) + C$$

$$= \dfrac{3x^4}{8} + C$$

5. $y\dfrac{dy}{dx} = x$

Separate the variables and take antiderivatives.

$$\int y\,dy = \int x\,dx$$

$$\dfrac{y^2}{2} = \dfrac{x^2}{2} + K$$

$$y^2 = \dfrac{2}{2}x^2 + 2K$$

$$y^2 = x^2 + C$$

7. $\dfrac{dy}{dx} = 2xy$

$$\int \dfrac{dy}{y} = \int 2x\,dx$$

$$\ln|y| = \dfrac{2x^2}{2} + C$$

$$\ln|y| = x^2 + C$$

$$y = \pm e^{x^2} + C$$

$$e^{\ln|y|} = e^{x^2+C}$$

$$y = \pm e^{x^2} \cdot e^C$$

$$y = ke^{x^2}$$

9. $\dfrac{dy}{dx} = 3x^2y - 2xy$

$$\dfrac{dy}{dx} = y(3x^2 - 2x)$$

$$\int \dfrac{dy}{y} = \int (3x^2 - 2x)\,dx$$

$$\ln|y| = \dfrac{3x^3}{3} - \dfrac{2x^2}{2} + C$$

$$e^{\ln|y|} = e^{x^3-x^2+C}$$

$$y = \pm(e^{x^3-x^2})e^C$$

$$y = ke^{(x^3-x^2)}$$

11. $\dfrac{dy}{dx} = \dfrac{y}{x}, \ x > 0$

$$\int \dfrac{dy}{y} = \int \dfrac{dx}{x}$$

$$\ln|y| = \ln x + C$$

$$e^{\ln|y|} = e^{\ln x+C}$$

$$y = \pm e^{\ln x} \cdot e^C$$

$$y = Me^{\ln x}$$

$$y = Mx$$

13. $\dfrac{dy}{dx} = y - 5$

$$\int \dfrac{dy}{y - 5} = \int dx$$

$$\ln|y - 5| = x + C$$

$$e^{\ln|y-5|} = e^{x+C}$$

$$y - 5 = \pm e^x \cdot e^C$$
$$y - 5 = Me^x$$
$$y = Me^x + 5$$

Thus,

$$y = -2xe^{-x} - 2e^{-x} + 44.$$

15. $\quad \dfrac{dy}{dx} = y^2 e^x$

$$\int y^{-2}\, dy = \int e^x\, dx$$
$$-y^{-1} = e^x + C$$
$$-\frac{1}{y} = e^x + C$$
$$y = \frac{-1}{e^x + C}$$

17. $\dfrac{dy}{dx} + 2x = 3x^2$

$$\frac{dy}{dx} = 3x^2 - 2x$$
$$y = \frac{3x^3}{3} - \frac{2x^2}{2} + C$$
$$y = x^3 - x^2 + C$$

Since $y = 2$ when $x = 0$,

$$2 = 0 - 0 + C$$
$$C = 2.$$

Thus,

$$y = x^3 - x^2 + 2.$$

19. $2\,\dfrac{dy}{dx} = 4xe^{-x}$

$$\frac{dy}{dx} = 2xe^{-x}$$

Use tables for antiderivatives or integrate by parts.

$$y = 2(-x - 1)e^{-x} + C$$

Since $y = 42$ when $x = 0$,

$$42 = 2(0 - 1)(1) + C$$
$$42 = -2 + C$$
$$C = 44.$$

21. $\quad \dfrac{dy}{dx} = \dfrac{x^2}{y};\ y = 3$ when $x = 0$.

$$\int y\, dy = \int x^2\, dx$$
$$\frac{y^2}{2} = \frac{x^3}{3} + C$$
$$y^2 = \frac{2}{3}x^3 + 2C$$
$$y^2 = \frac{2}{3}x^3 + k$$

Since $y = 3$ when $x = 0$,

$$9 = 0 + k$$
$$k = 0.$$

So $y^2 = \dfrac{2}{3}x^3 + 9$.

23. $\quad (2x + 3)y = \dfrac{dy}{dx};\ y = 1$ when $x = 0$.

$$\int (2x + 3)\, dx = \int \frac{dy}{y}$$
$$\frac{2x^2}{2} + 3x + C = \ln|y|$$
$$e^{x^2 + 3x + C} = e^{\ln|y|}$$
$$y = (e^{x^2 + 3x})(\pm e^C)$$
$$y = ke^{x^2 + 3x}$$

Since $y = 1$ when $x = 0$,

$$1 = ke^{0 + 0}$$
$$k = 1.$$

So $y = e^{x^2 + 3x}$.

25. $\dfrac{dy}{dx} = \dfrac{y^2}{x}$; $y = 5$ when $x = e$.

$$\int y^{-2}\, dy = \int \frac{dx}{x}$$

$$-y^{-1} = \ln|x| + C$$

$$-\frac{1}{y} = \ln|x| + C$$

$$y = \frac{-1}{\ln|x| + C}$$

Since $y = 5$ when $x = e$,

$$5 = \frac{-1}{\ln e + C}$$

$$5 = \frac{-1}{1 + C}$$

$$5 + 5C = -1$$

$$5C = -6$$

$$C = -\frac{6}{5}.$$

So $y = \dfrac{-1}{\ln|x| - \dfrac{6}{5}}$

$$= \frac{-5}{5\ln|x| - 6}$$

27. $\dfrac{dy}{dx} = \dfrac{2x + 1}{y - 3}$; $y = 4$ when $x = 0$.

$$\int (y - 3)\,dy = \int (2x + 1)\,dx$$

$$\frac{y^2}{2} - 3y = \frac{2x^2}{2} + x + C$$

Since $y = 4$ when $x = 0$,

$$\frac{16}{2} - 12 = 0 + 0 + C$$

$$C = -4.$$

So $\dfrac{y^2}{2} - 3y = x^2 + x - 4.$

29. $\dfrac{dy}{dx} = (y - 1)^2 e^x$; $y = 2$ when $x = 0$.

$$\int (y - 1)^{-2}\, dy = \int e^x\, dx$$

$$-(y - 1)^{-1} = e^x + C$$

$$-\frac{1}{y - 1} = e^x + C$$

$$y - 1 = \frac{-1}{e^x + C}$$

$$y = 1 - \frac{1}{e^x + C}$$

Since $y = 2$ when $x = 0$,

$$2 = 1 - \frac{1}{1 + C}$$

$$1 = -\frac{1}{1 + C}$$

$$1 + C = -1$$

$$C = -2.$$

So $y = 1 - \dfrac{1}{e^x - 2}$

$$= \frac{e^x - 2 - 1}{e^x - 1}$$

$$= \frac{e^x - 3}{e^x - 2}.$$

31. $\dfrac{dy}{dx} = \dfrac{k}{N}(N - y)y$

(a) $\dfrac{N\, dy}{(N - y)y} = k\, dx$

Since $\dfrac{1}{y} + \dfrac{1}{N - y} = \dfrac{N}{(N - y)y}$,

$$\int \frac{dy}{y} + \int \frac{dy}{N - y} = kx$$

$$\ln\left|\frac{y}{N - y}\right| = kx + C$$

$$\frac{y}{N - y} = Ce^{kx}.$$

For $0 < y < N$, $Ce^{kx} > 0$.

For $0 < N < y$, $Ce^{kx} < 0$.

Solve for y.

$$y = \frac{Ce^{kx}N}{1 + Ce^{kx}} = \frac{N}{1 + C^{-1}e^{-kx}}$$

Let $b = C^{-1}$ for $0 < y < N$.

$$y = \frac{N}{1 + be^{-kx}}$$

Let $-b = C^{-1}$ for $0 < N < y$.

$$y = \frac{N}{1 + be^{-kx}}$$

(b) For $0 < y < N$; $t = 0$, $y = y_0$.

$$y_0 = \frac{N}{1 + be^0} = \frac{N}{1 + b}$$

Solve for b.

$$b = \frac{N - y_0}{y_0}$$

(c) For $0 < N < y$; $t = 0$, $y = y_0$.

$$y_0 = \frac{N}{1 - be^0} = \frac{N}{1 - b}$$

Solve for b.

$$b = \frac{y_0 - N}{y_0}$$

33.   **(a)** $0 < y_0 < N$ implies that $y_0 > 0$, $N > 0$, and $N - y_0 > 0$.

Therefore, $b = \dfrac{N - y_0}{y_0} > 0$.

Also $e^{-kx} > 0$ for all x which implies that $1 + be^{-kx} > 1$.

(1) $y(x) = \dfrac{N}{1 + be^{-kx}} < N$ since $1 + be^{-kx} > 1$.

(2) $y(x) = \dfrac{N}{1 + be^{-kx}} > 0$ since $N > 0$ and $1 + be^{-kx} > 0$.

Combining statements (1) and (2), we have

$$0 < \frac{N}{1 + be^{-kx}} = y(x)$$

$$= \frac{N}{1 + be^{-kx}} < N$$

or $0 < y(x) < N$ for all x.

(b) $\lim\limits_{x \to \infty} \dfrac{N}{1 + be^{-kx}} = \dfrac{N}{1 + b(0)} = N$

$\lim\limits_{x \to -\infty} \dfrac{N}{1 + be^{-kx}} = 0$

Note that as $x \to -\infty$, $1 + be^{-kx}$ becomes infinitely large.

Therefore, the horizontal asymptotes are $y = N$ and $y = 0$.

(c) $y'(x) = \dfrac{(1 + be^{-kx})(0) - N(-kbe^{-kx})}{(1 + be^{-kx})^2}$

$= \dfrac{Nkbe^{-kx}}{(1 + be^{-kx})^2} > 0$ for all x.

Therefore, $y(x)$ is an increasing function.

(d) $y''(x) = \dfrac{(1 + be^{-kx})^2(-Nk^2be^{-kx}) - Nkbe^{-kx}[-2kbe^{-kx}(1 + be^{-kx})]}{(1 + be^{-kx})^4}$

$= \dfrac{-Nkbe^{-kx}(1 + be^{-kx})[k(1 + be^{-kx}) - 2kbe^{-kx}]}{(1 + be^{-kx})^4}$

$= \dfrac{-Nkbe^{-kx}(k - kbe^{-kx})}{(1 + be^{-kx})^3}$

$y''(x) = 0$ when

$k - kbe^{-kx} = 0$

$be^{-kx} = 1$

$e^{-kx} = \dfrac{1}{b}$

$-kx = \ln\left(\dfrac{1}{b}\right)$

$x = -\dfrac{\ln\left(\dfrac{1}{b}\right)}{k}$

$= \dfrac{\ln\left(\dfrac{1}{b}\right)^{-1}}{k}$

$= \dfrac{\ln b}{k}.$

When $x = \dfrac{\ln b}{k}$,

$$y = \frac{N}{1 + be^{-k\left(\frac{\ln b}{k}\right)}}$$

$$= \frac{N}{1 + be^{(-\ln b)}}$$

$$= \frac{N}{1 + be^{\ln(1/b)}}$$

$$= \frac{N}{1 + b\left(\frac{1}{b}\right)}$$

$$= \frac{N}{2}.$$

Therefore $\left(\dfrac{\ln b}{k}, \dfrac{N}{2}\right)$ is a point of inflection.

(e) To locate the maximum of $\dfrac{dy}{dx}$ we must consider, from part (d),

$$\frac{d}{dx}\left(\frac{dy}{dx}\right) = \frac{-Nkbe^{-kx}(k - kbe^{-kx})}{(1 + be^{-kx})^3}.$$

Since $y''(x) > 0$ for $x < \dfrac{\ln b}{k}$ and

$$y''(x) < 0 \text{ for } x > \frac{\ln b}{k},$$

we know that $x = \dfrac{\ln b}{k}$ locates a relative maximum of $\dfrac{dy}{dx}$.

35. $\dfrac{dy}{dx} = \dfrac{100}{32 - 4x}$

$$y = 100\left(-\frac{1}{4}\right) \ln |32 - 4x| + C$$

$$y = -25 \ln |32 - 4x| + C$$

Now, $y = 1000$ when $x = 0$.

$$1000 = -25 \ln |32| + C$$

$$C = 1000 + 25 \ln 32$$

$$C \approx 1086.64$$

Thus,

$$y = -25 \ln |32 - 4x| + 1086.64$$

(a) Let $x = 3$.

$$y = -25 \ln |32 - 12| + 1086.64$$

$$\approx \$1011.75$$

(b) Let $x = 5$.

$$y = -25 \ln |32 - 20| + 1086.64$$

$$\approx \$1024.52$$

(c) Advertising expenditures can never reach \$8000. If $x = 8$, the denominator becomes zero.

37. $\dfrac{dy}{dt} = -.06y$

See Example 4.

$$\int \frac{dy}{y} = \int -.06 \, dt$$

$$\ln |y| = -.06t + C$$

$$e^{\ln|y|} = e^{-.06t+C}$$

$$e^{\ln|y|} = e^{-.06t} \cdot e^C$$

$$|y| = e^{-.06t} \cdot e^C$$

$$y = Me^{-.06t}$$

Let $y = 1$ when $t = 0$.
Solve for M:

$$1 = Me^0$$

$$M = 1.$$

So $y = e^{-.06t}$.
If $y = .50$,

$$.50 = e^{-.06t}$$

$$\ln .5 = -.06t \ln e$$

$$t = \frac{-\ln .5}{.06}$$

$$\approx 11.6 \text{ yr.}$$

39. Let y = the number of firms that will be bankrupt.

$$\frac{dy}{dt} = .06(1500 - y)$$

See Example 5.

$$\int \frac{dy}{1500 - y} = .06 \int dt$$

$$-\ln|1500 - y| = .06t + C$$

$$\ln|1500 - y| = -.06t - C$$

$$|1500 - y| = e^{-.06t - C}$$

$$1500 - y = ke^{-.06t}$$

$$y = 1500 - ke^{-.06t}$$

Since $t = 0$, when $y = 100$.

$$100 = 1500 - k(1)$$

$$k = 1400.$$

$$y = 1500 - 1400e^{-.06t}$$

When $t = 2$ yr,

$$y = 1500 - 1400e^{-.06(2)}$$

$$y = 258.31.$$

About 260 firms will be bankrupt in 2 yr.

41. $E = -\frac{p}{q} \cdot \frac{dq}{dp}$

If $E = \frac{4p^2}{q^2}$,

$$\frac{4p^2}{q^2} = -\frac{p}{q} \cdot \frac{dq}{dp}$$

$$4p\,dp = -q\,dq.$$

$$\int 4p\,dp = -\int q\,dq$$

$$2p^2 = -\frac{1}{2}q^2 + C_1$$

Multiplying by 2, we have

$$4p^2 = -q^2 + 2C_2$$

$$q^2 = -4p^2 + 2C_2.$$

Let $C = 2C_2$. Then

$$q^2 = -4p^2 + C$$

$$q = \pm\sqrt{-4p^2 + C}.$$

Since q cannot be negative,

$$q = \sqrt{-4p^2 + C}.$$

43. $$\frac{dy}{dt} = -.03y$$

$$\int \frac{dy}{y} = -.03 \int dt$$

$$\ln|y| = -.03t + C$$

$$e^{\ln|y|} = e^{-.03t + C}$$

$$y = Me^{-.03t}$$

Since $y = 6$ when $t = 0$,

$$6 = Me^0$$

$$M = 6.$$

So $y = 6e^{-.03t}$.

If $t = 10$,

$$y = 6e^{-.03(10)}$$

$$\approx 4.4 \text{ cc.}$$

45. $$\frac{dy}{dx} = .01(5000 - y)$$

$$\int \frac{dy}{5000 - y} = \int .01\,dx$$

$$-\ln|5000 - y| = .01x + C$$

$$\ln|5000 - y| = -.01x + C$$

$$|5000 - y| = e^{-.01x - C}$$

$$|5000 - y| = e^{-.01x} \cdot e^C$$

$$5000 - y = Me^{-.01x}$$

$$y = 5000 - Me^{-.01x}$$

Since $y = 150$ when $x = 0$,

$$150 = 5000 - Me^0$$
$$M = 4850$$

So $y = 5000 - 4850e^{-.01x}$.

If $x = 5$,

$$y = 5000 - 4850e^{-.01(5)}$$
$$\approx 387.$$

47. **(a)** $\dfrac{dy}{dt} = ky$

$$\int \frac{dy}{y} = \int k \, dt$$
$$\ln |y| = kt + C$$
$$e^{\ln|y|} = e^{kt+C}$$
$$y = \pm(e^{kt})(e^C)$$
$$y = Me^{kt}$$

If $y = 1$ when $t = 0$ and $y = 5$ when $t = 2$, we have the system of equations:

$$1 = Me^{k(0)}$$
$$\frac{5 = Me^{2k}}{1 = M(1)}$$
$$M = 1$$

Substitute.

$$5 = (1)e^{2k}$$
$$e^{2k} = 5$$
$$2k \ln e = \ln 5$$
$$k = \frac{\ln 5}{2}$$
$$\approx .8$$

(b) If $k = .8$ and $M = 1$,
$$y = e^{.8t}.$$

When $t = 3$,

$$y = e^{.8(3)}$$
$$= e^{2.4}$$
$$\approx 11.$$

(c) When $t = 5$,

$$y = e^{.8(5)}$$
$$= e^4$$
$$\approx 55.$$

(d) When $t = 10$,

$$y = e^{.8(10)}$$
$$= e^8$$
$$\approx 2981.$$

49. **(a)** $\dfrac{dy}{dt} = -.05y$

(b) $\displaystyle\int \frac{dy}{y} = -\int .05 \, dt$

$$\ln |y| = -.05t + C$$
$$e^{\ln|y|} = e^{-.05+C}$$
$$y = \pm e^{-.05t} \cdot e^C$$
$$y = Me^{-.05t}$$

(c) Since $y = 90$ when $t = 0$.

$$90 = Me^0$$
$$M = 90.$$

So $y = 90e^{-.05t}$.

(d) At $t = 10$,

$$y = 90e^{-.05(10)}$$
$$\approx 55 \text{ g.}$$

Section 9.2

1. $y' + 2y = 5$

$$I(x) = e^{2\int dx} = e^{2x}$$

Multiply each term by e^{2x}.

$$e^{2x}y' + 2e^{2x}y = 5e^{2x}$$
$$D_x(e^{2x}y) = 5e^{2x}$$

Integrate both sides.

$$e^{2x}y = \int 5e^{2x}\ dx$$

$$e^{2x}y = \frac{5}{2}e^{2x} + C$$

$$y = \frac{5}{2} + Ce^{-2x}$$

3. $y' + xy = 3x$

$$I(x) = e^{\int x\ dx} = e^{x^2/2}$$

$$e^{x^2/2}y' + xe^{x^2/2}y = 3xe^{x^2/2}$$

$$D_x(e^{x^2/2}y) = 3xe^{x^2/2}$$

$$e^{x^2/2}y = \int 3xe^{x^2/2}\ dx$$

$$e^{x^2/2}y = 3e^{x^2/2} + C$$

$$y = 3 + Ce^{-x^2/2}$$

5. $xy' - y - x = 0;\ x > 0$

$$y' - \frac{1}{x}y = 1$$

$$I(x) = e^{-\int 1/x\ dx}$$

$$= e^{-\ln x} = \frac{1}{x}$$

$$\frac{1}{x}y' - \frac{1}{x^2}y = \frac{1}{x}$$

$$D_x\left(\frac{1}{x}y\right) = \frac{1}{x}$$

$$\frac{y}{x} = \int \frac{1}{x}\ dx$$

$$\frac{y}{x} = \ln x + C$$

$$y = x\ln x + Cx$$

7. $2y' - 2xy - x = 0$

$$y' - xy = \frac{x}{2}$$

$$I(x) = e^{-\int x\ dx}$$

$$= e^{-x^2/2}$$

$$e^{-x^2/2}y' - xe^{-x^2/2}y = \frac{x}{2}e^{-x^2/2}$$

$$D_x(e^{-x^2/2}y) = \frac{x}{2}e^{-x^2/2}$$

$$e^{-x^2/2}y = \int \frac{x}{2}e^{-x^2/2}\ dx$$

$$e^{-x^2/2}y = \frac{-1}{2}e^{-x^2/2} + C$$

$$y = -\frac{1}{2} + Ce^{x^2/2}$$

9. $xy' + 2y = x^2 + 3x;\ x > 0$

$$y' + \frac{2}{x}y = x + 3$$

$$I(x) = e^{\int 2/x\ dx}$$

$$= e^{2\ln x} = x^2$$

$$x^2y' + 2xy = x^3 + 3x^2$$

$$D_x(x^2y) = x^3 + 3x^2$$

$$x^2y = \int(x^3 + 3x^2)\,dx$$

$$x^2y = \frac{x^4}{4} + x^3 + C$$

$$y = \frac{x^2}{4} + x + \frac{C}{x^2}$$

11. $y - x\frac{dy}{dx} = x^3;\ x > 0$

$$y - xy' = x^3$$

$$y' - \frac{y}{x} = -x^2$$

$$I(x) = e^{-\int 1/x\ dx}$$

$$= e^{-\ln x} = x^{-1}$$

$$\frac{1}{x}y' - \frac{y}{x^2} = -x$$

$$D_x\left(\frac{1}{x}y\right) = -x$$

$$\frac{y}{x} = \int -x\ dx$$

$$\frac{y}{x} = \frac{-x^2}{2} + C$$

$$y = \frac{-x^3}{2} + Cx$$

13. $y' + y = 2e^x$; $y = 100$ when $x = 0$.

$$I(x) = e^{\int dx} = e^x$$

$$e^x y' + ye^x = 2e^{2x}$$

$$D_x(e^x y) = 2e^{2x}$$

$$e^x y = \int 2e^{2x}\ dx$$

$$e^x y = e^{2x} + C$$

$$y = e^x + Ce^{-x}$$

Since $y = 100$ when $x = 0$,

$$100 = e^0 + Ce^0$$

$$100 = 1 + C$$

$$C = 99.$$

Therefore,

$$y = e^x + 99e^{-x}.$$

15. $y' - xy - x = 0$; $y = 10$ when $x = 1$.

$$y' - xy = x$$

$$I(x) = e^{-\int x\ dx} = e^{-x^2/2}$$

$$e^{-x^2/2}\,y' - xe^{-x^2/2}\,y = xe^{-x^2/2}$$

$$D_x\,(e^{-x^2/2}\,y) = xe^{-x^2/2}$$

$$e^{-x^2/2}\,y = \int xe^{-x^2/2}\ dx$$

$$e^{-x^2/2}\,y = -e^{-x^2/2} + C$$

$$y = -1 + Ce^{x^2/2}$$

Since $y = 10$ when $x = 1$,

$$10 = -1 + Ce^{1/2}$$

$$11 = Ce^{1/2}$$

$$C = \frac{11}{e^{1/2}}.$$

Therefore,

$$y = -1 + \frac{11}{e^{1/2}}(e^{x^2/2})$$

$$= -1 + 11e^{x^2/2\ -\ 1/2}$$

$$= -1 + 11e^{(x^2-1)/2}.$$

17. $x\dfrac{dy}{dx} + 5y = x^2$; $y = 12$ when $x = 2$.

$$xy' + 5y = x^2$$

$$y' + \frac{5}{x}y = x$$

$$I(x) = e^{\int 5/x\ dx}$$

$$= e^{5\ \ln x} = x^5$$

$$x^5 y' + 5x^4 y = x^6$$

$$D_x\,(x^5 y) = x^6$$

$$x^5 y = \int x^6\ dx$$

$$x^5 y = \frac{x^7}{7} + C$$

$$y = \frac{x^2}{7} + \frac{C}{x^5}$$

Since $y = 12$, when $x = 2$,

$$12 = \frac{4}{7} + \frac{C}{32}$$

$$\frac{80}{7} = \frac{C}{32}$$

$$C = \frac{2560}{7}.$$

Therefore,

$$y = \frac{x^2}{7} + \frac{2560}{7x^5}.$$

19. $xy' + (1 + x)y = 3$; $y = 50$ when $x = 4$

$$y' + \left(\frac{1 + x}{x}\right)y = \frac{3}{x}$$

$$I(x) = e^{\int (1+x)\,dx/x}$$

$$= e^{\int (1/x)\ dx + dx}$$

$$= e^{(\ln x) + x}$$

$$= e^{\ln x} \cdot e^x$$

$$= xe^x$$

$$xe^x y' + (1 + x)e^x y = 3e^x$$

$$D_x (xe^x y) = 3e^x$$

$$xe^x y = \int 3e^x \, dx$$

$$xe^x y = 3e^x + C$$

$$y = \frac{3}{x} + \frac{C}{xe^x}$$

Since $y = 50$ when $x = 4$,

$$50 = \frac{3}{4} + \frac{C}{4e^4}$$

$$\frac{197}{4} = \frac{C}{4e^4}$$

$$C = 197e^4.$$

Therefore,

$$y = \frac{3}{x} + \frac{197e^4}{xe^x}$$

$$= \frac{3}{x} + \frac{197}{x}e^{4-x}$$

$$= \frac{3 + 197e^{4-x}}{x}.$$

21. $\dfrac{dy}{dx} = cy - py^2$

(a) Let $y = \dfrac{1}{z}$ and $y' = -\dfrac{z'}{z^2}$.

$$-\frac{z'}{z^2} = c\left(\frac{1}{z}\right) - p\left(\frac{1}{z^2}\right)$$

$$z' = -cz + p$$

$$z' + cz = p$$

$$I(x) = e^{\int c \, dx} = e^{cx}$$

$$D_x (e^{cx} \cdot z) = \int pe^{cx} \, dx$$

$$e^{cx} \cdot z = \frac{p}{c}e^{cx} + k$$

$$z = \frac{p}{c} + ke^{-cx}$$

$$= \frac{p + kce^{-cx}}{c}$$

Therefore,

$$y = \frac{c}{p + kce^{-cx}}.$$

(b) Let $z(0) = \dfrac{1}{y_0}$.

$$\frac{1}{y_0} = \frac{p + kce^0}{c} = \frac{p + kc}{c}$$

$$\frac{c}{y_0} = p + kc$$

$$kc = \frac{c}{y_0} - p = \frac{c - py_0}{y_0}$$

$$k = \frac{c - py_0}{cy_0}$$

$$y = \frac{c}{p + \left(\dfrac{c - py_0}{cy_0}\right)ce^{-cx}}$$

$$\hspace{3cm} \textit{From part (a)}$$

$$= \frac{cy_0}{py_0 + (c - py_0)e^{-cx}}$$

(c) $\displaystyle\lim_{x \to \infty} y$

$$= \lim_{x \to \infty} \left(\frac{cy_0}{py_0 + (c - py_0)e^{-cx}}\right)$$

$$= \frac{cy_0}{py_0 - 0}$$

$$= \frac{c}{p}$$

23. $\dfrac{dy}{dt} = .02y + e^t$; $y = 10{,}000$ when $t = 0$.

$$\frac{dy}{dt} - .02y = e^t$$

$$I(t) = e^{\int -.02 \, dt} = e^{-.02t}$$

$$e^{-.02t}\frac{dy}{dt} - .02e^{-.02t}y$$

$$= e^{-.02t} \cdot e^t$$

$$D_t (e^{-.02t}y) = e^{.98t}$$

$$e^{-.02t}y = \int e^{.98t} \, dt$$

$$= \frac{e^{.98t}}{.98} + C$$

$$y = \frac{e^t}{.98} + Ce^{.02t}$$

$$10,000 = \frac{1}{.98} + C$$

$$C \approx 9999$$

$$y \approx \frac{e^t}{.98} + 9999e^{.02t}$$

$$= 1.02e^t + 9999e^{.02t}$$

25. $\frac{dy}{dt} = .02y - t$; $y = 10,000$ when $t = 0$.

$$\frac{dy}{dt} - .02y = -t$$

$$I(t) = e^{\int -.02} = e^{-.02t}$$

$$e^{-.02t}\frac{dy}{dt} - .02e^{-.02t}y$$

$$= -te^{-.02t}$$

$$D_t(e^{-.02t}y) = -te^{-.02t}$$

$$e^{-.02t}y = \int -te^{-.02t}\ dt$$

Integration by parts:

Let $u = -t$ $dv = e^{-.02t}\ dt$

$\quad du = -dt$ $v = \dfrac{e^{-.02t}}{-.02}$

$$e^{-.02t} = \frac{te^{-.02t}}{.02} - \int \frac{e^{-.02t}}{.02}\ dt$$

$$e^{-.02t}y = \frac{te^{-.02t}}{.02} + \frac{e^{-.02t}}{.0004} + C$$

$$y = 50t + 2500 + Ce^{.02t}$$

$$10,000 = 2500 + C$$

$$C = 7500$$

$$y = 50t + 2500 + 7500e^{.02t}$$

27. $\quad \dfrac{dT}{dt} = -k(T - T_F)$

$$\frac{dT}{dt} = -kT + kT_F$$

$$\frac{dT}{dt} + kT = kT_F$$

$$I(t) = e^{\int k\ dt} = e^{kt}$$

Multiply both sides by e^{kt}.

$$e^{kt}\frac{dT}{dt} + ke^{kt}T = kT_Fe^{kt}$$

$$D_t(Te^{kt}) = kT_Fe^{kt}$$

$$Te^{kt} = \int kT_Fe^{kt}\ dt$$

$$Te^{kt} = T_Fe^{kt} + c$$

$$T = T_F + ce^{-kt}$$

$$T = ce^{-kt} + T_F$$

29. If $T = ce^{-kt} + T_F$,

$$\lim_{t \to \infty} T = \lim_{t \to \infty}(ce^{-kt} + T_F)$$

$$= c(0) + T_F \text{ since } k > 0$$

$$= T_F$$

Thus, the temperature approaches T_F according to Newton's law of cooling. We would expect the temperature of the object to approach the temperature of the surrounding medium.

31. At $t = 0$, $T = 50°$; $T_F = 22°$; at $t = 1$, $T = 35°$

(a)
$$T = ce^{-kt} + 22$$

$$50 = ce^{-k(0)} + 22$$

$$50 = c + 22$$

$$c = 28$$

$$T = 28e^{-kt} + 22$$

$$35 = 28e^{-k} + 22$$

$$13 = 28e^{-k}$$

$$\frac{13}{28} = e^{-k}$$

$$\ln\left(\frac{13}{28}\right) = -k$$

$$-\ln\left(\frac{13}{28}\right) = k$$

$$T = 28e^{t \ln(13/28)} + 22$$

$$T = 28e^{-.767t} + 22$$

(b) At $t = 2$,

$$T = 28e^{-.767(2)} + 22$$
$$\approx 28.04°C$$

Section 9.3

Answers may vary when values are rounded at intermediary steps.

1. $y' = x^2 + y^2$; $f(0) = 1$, $h = .1$;
find $f(.5)$.

$g(x, y) = x^2 + y^2$
$x_0 = 0$; $y_0 = 1$
$g(x_0, y_0) = 0 + 1 = 1$

$x_1 = .1$; $y_1 = 1 + 1(.1) = 1.1$
  $g(x_1, y_1) = (.1)^2 + (1.1)^2 = 1.22$

$x_2 = .2$; $y_2 = 1.1 + 1.22(.1) = 1.222$
  $g(x_2, y_2) = (.2)^2 + (1.222)^2$
             $= 1.533$

$x_3 = .3$; $y_3 = 1.222 + 1.533(.1)$
             $= 1.375$
  $g(x_3, y_3) = (.3)^2 + (1.375)^2$
             $= 1.981$

$x_4 = .4$; $y_4 = 1.375 + (1.981)(.1)$
             $= 1.573$

$g(x_4, y_4) = (.4)^2 + (1.573)^2$
           $= 2.635$

$x_5 = .5$; $y_5 = 1.573 + (2.635)(.1)$
           $= 1.837$

These results are tabulated as follows.

x_i	y_i
0	1
.1	1.1
.2	1.222
.3	1.375
.4	1.573
.5	1.837

$f(.5) = 1.837$

3. $y' = 1 + y = g(x, y)$; $f(0) = 2$,
$h = .1$; find $f(.6)$.

$x_0 = 0$; $y_0 = 2$
 $g(x_0, y_0) = 1 + 2 = 3$

$x_1 = .1$; $y_1 = 2 + 3(.1) = 2.3$
  $g(x_1, y_1) = 1 + 2.3 = 3.3$

$x_2 = .2$; $y_2 = 2.3 + 3.3(.1) = 2.63$
  $g(x_2, y_2) = 1 + 2.63 = 3.63$

$x_3 = .3$; $y_3 = 2.63 + 3.63(.1)$
            $= 2.993$
  $g(x_3, y_3) = 1 + 2.993 = 3.993$

$x_4 = .4$; $y_4 = 2.993 + 3.993(.1)$
            $= 3.3923$
  $g(x_4, y_4) = 1 + 3.3923 = 4.3923$

$x_5 = .5$; $y_5 = 3.3923 + 4.392(.1)$
            $= 3.8315$
  $g(x_5, y_5) = 1 + 3.8315 = 4.8315$

$x_6 = .6$; $y_6 = 3.8315 + 4.8315(.1)$
            $= 4.31465$

x_i	y_i
0	2
.1	2.3
.2	2.63
.3	2.993
.4	3.3923
.5	3.8315
.6	4.31465

$f(.6) \approx 4.315$

5. $y' = x + \sqrt{y} = g(x, y); \; f(0) = 1,$
$h = .1;$ find $f(.4)$.

$x_0 = 0; \; y_0 = 1$
$\quad g(x_0, y_0) = 0 + 1 = 1$

$x_1 = .1; \; y_1 = 1 + 1(.1) = 1.1$
$\quad g(x_1, y_1) = .1 + \sqrt{1.1} = 1.149$

$x_2 = .2; \; y_2 = 1.1 + 1.149(.1)$
$\qquad\qquad = 1.215$
$\quad g(x_2, y_2) = .2 + \sqrt{1.215} = 1.302$

$x_3 = .3; \; y_3 = 1.215 + 1.302(.1)$
$\qquad\qquad = 1.345$
$\quad g(x_3, y_3) = .3 + \sqrt{1.345} = 1.460$

$x_4 = .4; \; y_4 = 1.345 + 1.460(.1)$
$\qquad\qquad = 1.491$

x_i	y_i
0	1
.1	1.1
.2	1.215
.3	1.345
.4	1.491

$f(.4) \approx 1.491$

7. $y' = 1 - e^{-x} = g(x, y); \; f(0) = 0,$
$h = .1;$ find $f(.5)$.

$x_0 = 0; \; f_0 = 0$
$\qquad\qquad = 1 - e^0 = 0$

$x_1 = .1; \; y_1 = 0 + 0(.1) = 0$
$\quad g(x_1, y_1) = 1 - e^{-.1} = .095$

$x_2 = .2; \; y_2 = 0 + .095 + (.1)$
$\qquad\qquad = .0095$
$\quad g(x_2, y_2) = 1 - e^{-.2} = .181$

$x_3 = .3; \; y_3 = .0095 + .181(.1)$
$\qquad\qquad = .0276$
$\quad g(x_3, y_3) = 1 - e^{-.3} = .259$

$x_4 = .4; \; = .0276 + .259(.1)$
$\qquad\qquad = .0535$
$\quad g(x_4, y_4) = 1 - e^{-.4} = .330$

$x_5 = .5; \; y_5 = .0535 + .330(.1)$
$\qquad\qquad = .0865$

x_i	y_i
0	0
.1	0
.2	.0095
.3	.0276
.4	.0535
.5	.0865

$f(.5) \approx .087$°

9. $y' = -4 + x = g(x, y); \; f(0) = 1,$
$h = .1,$ find $f(.4)$.

$x_0 = 0; \; y_0 = 1$
$\quad g(x_0, y_0) = -4 + 0 = -4$

$x_1 = .1; \; y_1 = 1 + (-4)(.1) = .6$
$\quad g(x_1, y_1) = -4 + .1 = -3.9$

$x_2 = .2; \; y_2 = .6 + (-3.9)(.1) = .21$
$\quad g(x_2, y_2) = -4 + .2 = -3.8$

$x_3 = .3; \; y_3 = .21 + (-3.8)(.1)$
$\qquad\qquad = -.17$
$\quad g(x_3, y_3) = -4 + .3 = -3.7$

$x_4 = .4; \; y_4 = -.17 + (-3.7)(.1)$
$\qquad\qquad = -.540$

x_i	y_i
0	1
.1	.6
.2	.21
.3	−.17
.4	−.540

$f(.4) \approx -.540$

Actual solution:

$$\frac{dy}{dx} = -4 + x$$

$$y = -4x + \frac{x^2}{2} + C$$

At $f(0) = 1$,

$$1 = -4(0) + \frac{0}{2} + C$$

$$C = 1.$$

Therefore,

$$y = -4x + \frac{x^2}{2} + 1$$

$$f(.4) = -4(.4) + \frac{(.4)^2}{2} + 1$$

$$= -.520.$$

11. $y' = x^2$; $f(0) = 2$, $h = .1$, find $f(.5)$.

$x_0 = 0$; $y_0 = 2$
$\quad = 0^2 = 0$

$x_1 = .1$; $y_1 = 2 + 0(.1) = 2$
$\quad g(x_1, y_1) = (.1)^2 = .01$

$x_2 = .2$; $y_2 = 2 + .01(.1) = 2.001$
$\quad g(x_2, y_2) = (.2)^2 = .04$

$x_3 = .3$; $y_3 = 2.001 + .04(.1)$
$\quad\quad = 2.005$
$\quad g(x_3, y_3) = (.3)^2 = .09$

$x_4 = .4$; $y_4 = 2.005 + .09(.1)$
$\quad\quad = 2.014$
$\quad g(x_4, y_4) = (.4)^2 = .16$

$x_5 = .5$; $y_5 = 2.014 + .16(.1)$
$\quad\quad = 2.030$

x_i	y_i
0	2
.1	2
.2	2.001
.3	2.005
.4	2.014
.5	2.030

$f(.5) \approx 2.030$

Actual solution:

$$\frac{dy}{dx} = x^2$$

$$y = \frac{x^3}{3} + C$$

At $f(0) = 2$,

$$2 = \frac{0}{3} + C$$

$$C = 2.$$

Therefore,

$$y = \frac{x^3}{3} + 2.$$

$$f(.5) = \frac{(.5)^3}{3} + 2 = 2.042.$$

13. $y' = 2xy$; $f(1) = 1$, $h = .1$, find $f(1.6)$.

$$g(x, y) = 2xy$$

$x_0 = 1$; $y_0 = 1$
$\quad g(x_0, y_0) = 2(1)(1) = 2$

$x_1 = 1.1$; $y_1 = 1 + 2(.1) = 1.2$
$\quad g(x_1, y_1) = 2(1.1)(1.2) = 2.64$

$x_2 = 1.2;\ y_2 = 1.2 + 2.64(.1)$
$$= 1.464$$
$g(x_2,\ y_2) = 2(1.2)(1.464)$
$$\approx 3.514$$

$x_3 = 1.3;\ y_3 = 1.464 + 3.514(.1)$
$$= 1.815$$
$g(x_3,\ y_3) = 2(1.3)(1.815)$
$$= 4.720$$

$x_4 = 1.4;\ y_4 = 1.815 + 4.720(.1)$
$$= 2.287$$
$g(x_4,\ y_4) = 2(1.4)(2.287)$
$$= 6.404$$

$x_5 = 1.5;\ y_5 = 2.287 + 6.404(.1)$
$$= 2.927$$
$g(x_5,\ y_5) = 2(1.5)(2.927)$
$$= 8.782$$

$x_6 = 1.6;\ y_6 = 2.927 + 8.782(.1)$
$$= 3.805$$

x_i	y_i
1	1
1.1	1.2
1.2	1.464
1.3	1.815
1.4	2.287
1.5	2.927
1.6	3.805

$f(1.6) \approx 3.805$

Actual solution:

$$\int \frac{dy}{y} = \int 2x\ dx$$
$$\ln |y| = x^2 + C$$
$$|y| = e^{x^2+C}$$
$$y = ke^{x^2}$$

At $f(1) = 1$,

$$1 = ke^1 = ke$$
$$k = \frac{1}{e}.$$

Therefore,

$$y = \frac{1}{e}(e^{x^2}) = e^{x^2-1}.$$
$$f(1.6) = e^{(1.6)^2-1} = 4.759$$

15. $y' = x^2 - x;\ f(0) = 0,\ h = .1;$ find $f(.3)$.

$x_0 = 0,\ y_0 = 0$

$x_1 = .1;\ y_1 = 0 + (.1)0 = 0$

$x_2 = .2;\ y_2 = 0 + .1(.1^2 - .1)$
$$= -.009$$

$x_3 = .3;\ y_3 = -.009 + .1(.2^2 - .2)$
$$= -.025$$

x_i	y_i
0	0
.1	0
.2	−.009
.3	−.025

$f(.3) \approx -.025$

Actual solution:

$$\frac{dy}{dx} = x^2 - x$$
$$y = \frac{x^3}{3} - \frac{x^2}{2} + C$$

At $f(0) = 0$,

$$0 = C.$$

Therefore,

$$y = \frac{x^3}{3} - \frac{x^2}{2}.$$
$$f(.3) = \frac{(.3)^3}{3} - \frac{(.3)^2}{2} = -.036.$$

17. $y' = \sqrt[3]{x};\ f(0) = 0;\ h = .2,$
$$g(x,\ y) = \sqrt[3]{x}$$

$x_0 = 0,\ y_0 = 0$
$g(x_0,\ y_0) = \sqrt[3]{0} = 0$

$x_1 = .2$; $y_1 = 0 + 0(.2) = 0$

$g(x_1, y_1) = \sqrt[3]{.2} = .58480360$

$x_2 = .4$; $y_2 = 0 + .5848036(.2)$
$$= .11696071$$

$g(x_2, y_2) = \sqrt[3]{.4} = .73680630$

$x_3 = .6$; $y_3 = .11696071$
$$+ .7368063(.2)$$
$$= .26432197$$

$g(x_3, y_3) = \sqrt[3]{.6} = .84343270$

$x_4 = .8$; $y_4 = .26432197$
$$+ .8434327(.2)$$
$$= .43300850$$

$g(x_4, y_4) = \sqrt[3]{.8} = .92831780$

$x_5 = 1.0$; $y_5 = .43300850$
$$+ .9283178(.2)$$
$$= .61867206$$

Solving the differential equation gives

$$\frac{dy}{dx} = x^{1/3}$$

$$y = \frac{3}{4}x^{4/3} + C$$

$$0 = \frac{3}{4}(0) + C$$

$$C = 0$$

$$f(x) = \frac{3}{4}x^{4/3}.$$

$$f(x_0) = f(0) = \frac{3}{4}(0) = 0$$

$$y_0 - f(x_0) = 0$$

$$f(x_1) = f(.2)$$
$$= \frac{3}{4}(.2)^{4/3} = .08772053$$

$$y_1 - f(x_1) = 0 - .08772053$$
$$= -.08772053$$

$$f(x_2) = f(.4)$$
$$= \frac{3}{4}(.4)^{4/3} = .22104189$$

$$y_2 - f(x_2) = .11696071 - .22104189$$
$$= -.10408118$$

$$f(x_3) = f(.6)$$
$$= \frac{3}{4}(.6)^{4/3} = .37954470$$

$$y_3 - f(x_3) = .26432197 - .37954470$$
$$= -.11522273$$

$$f(x_4) = f(.8)$$
$$= \frac{3}{4}(.8)^{4/3} = .55699066$$

$$y_4 - f(x_4) = .43300850 - .55699066$$
$$= -.12398216$$

$$f(x_5) = f(1)$$
$$= \frac{3}{4}(1)^{4/3} = .75000000$$

$$y_5 - f(x_5) = .61867206 - .75000000$$
$$= -.13132794$$

See the table in the answer section in the back of your book.

19. $y' = 1 - y$; $f(0) = 0$; $h = .2$,
$$g(x, y) = 1 - y$$

$x_0 = 0$; $y_0 = 0$
$g(x_0, y_0) = 1 - 0 = 1$

$x_1 = .2$; $y_1 = 0 + 1(.2) = .2$
$g(x_1, y_1) = 1 - .2 = .8$

$x_2 = .4$; $y_2 = .2 + .8(.2) = .36$
$g(x_2, y_2) = 1 - .36 = .64$

$x_3 = .6$; $y_3 = .36 = .64(.2) = .488$
$g(x_3, y_3) = 1 - .488 = .512$

$x_4 = .8$; $y_4 = .488 + .512(.2)$
$$= .5904$$

$g(x_4, y_4) = 1 - .5904 = .4096$

$x_5 = 1.0$; $y_5 = .5904 + .4096(.2)$
$$= .67232$$

Solving the differential equation gives

$$\frac{dy}{dx} = 1 - y$$

$$\frac{dy}{dx} + y = 1$$

$$I(x) = e^x$$

$$D_x (e^x \cdot y) = e^x$$

$$e^x \cdot y = e^x + C$$

$$y = 1 + Ce^{-x}.$$

If $x = 0$ and $y = 0$, $C = -1$.

$$y = 1 - e^{-x}$$

$f(x_0) = f(0) = 1 - e^0 = 1 - 1 = 0$

$y_0 - f(x_0) = 0$

$f(x_1) = f(.2)$
$= 1 - e^{-.2} = .1812692$

$y_1 - f(x_1) = .0187308$

$f(x_2) = f(.4)$
$= 1 - e^{-.4} = .32967995$

$y_2 - f(x_2) = .03032005$

$f(x_3) = f(.6)$
$= 1 - e^{-.6} = .45118836$

$y_3 - f(x_3) = .03681164$

$f(x_4) = f(.8)$
$= 1 - e^{-.8} = .55067104$

$y_4 - f(x_4) = .03972896$

$f(x_5) = f(1)$
$= 1 - e^{-1} = .63212056$

$y_5 - f(x_5) = .04019944$

See the table in the answer section in the back of your book.

21. $y' = \sqrt[3]{x}$; $f(0) = 0$

See the solution in Exercise 17 for plotting points for approximation of this equation.
See the graph in the answer section in the back of your book.

23. $y' = 1 - y$; $f(0) = 0$

See the solution in Exercise 19 for plotting points for approximations of this equation.
See the graph in the answer section in the back of your book.

25. $y' = y^2$; $f(0) = 1$

(a)

x_i	y_i
0	1
.2	1.2
.4	1.488
.6	1.9308288
.8	2.676448771
1.0	4.109124376

Thus, $f(1.0) \approx 4.109$

(b) $y' = y^2$; $y = 1$ when $x = 0$

$$\frac{dy}{dx} = y^2$$

$$\frac{1}{y^2} dy = dx$$

$$\int \frac{1}{y^2} dy = \int dx$$

$$-\frac{1}{y} = x + C$$

When x = 0, y = 1.

$$-\frac{1}{1} = 0 + C$$

$$C = -1$$

$$-\frac{1}{y} = x - 1$$

$$-1 = (x - 1)y$$

$$y = \frac{-1}{x - 1}$$

$$y = \frac{1}{1 - x}$$

As x approaches 1 from the left, y
approaches ∞.

27. Let y = the number of algae (in
thousands) at time t = f(t).

$$y \le 100; \quad f(0) = 3$$

(a) $\frac{dy}{dt} = .01y(100 - y)$

$$= y - .01y^2$$

(b) Find f(2); h = .5.

x_i	y_i
0	3
.5	4.455
1	6.583265
1.5	9.6582
2	14.0209

Therefore, f(2) ≈ 14 so about 14,000
algae are present when t = 2.

29. $\frac{dy}{dt} = .05y - .1y^{1/2}$; f(0) = 60,

h = 1; find f(6).

x_i	y_i
0	60
1	62.22541
2	64.54785
3	66.97182
4	69.50205
5	72.14347
6	74.90127

Therefore, f(6) ≈ 75. So about 75
insects are present after 6 weeks.

31. $\frac{dN}{dt} = .04(500 - N)N^{1/2}$; f(0) = 2,

h = .5; find f(3).

x_i	y_i
0	2
.5	16.08557
1	54.9021
1.5	120.862
2	204.2248
2.5	288.7616
3	360.553

Therefore, f(3) ≈ 360, so about 360
people have heard the rumor.

Exercises 33 and 35 are to be completed
using a computer. The answers may vary
depending on the computer and software
used.

33. y′ = e^x - e^{-y}; f(1) = .8
When x = 1.5, y = 2.334.

35. y′ = $\frac{x^2 + y}{y}$; f(0) = 2.3
When x = 1.5, y = 4.091.

Section 9.4

1. $\dfrac{dA}{dt} = rA + D$; $r = .06$, $D = \$2000$

$$\dfrac{dA}{dt} = .06A + 2000$$

$$\int \dfrac{dA}{.06A + 2000} = \int dt$$

$$\ln |.06A + 2000| = .06t + C$$

$$|.06A + 2000| = e^{.06t+C}$$

$$.06A + 2000 = Me^{.06t}$$

$$A = \dfrac{M}{.06}e^{.06t} - \dfrac{2000}{.06}$$

$$A = 16.66667Me^{.06t}$$
$$- 33,333.33$$

$t = 0$, $A = 2000$

$2000 = 16.66667Me^{0} - 33,333.33$

$M = 2120$

$A = 16.66667(2120)e^{.06t} - 33,333.33$

$A = 35,333.33e^{.06t} - 33,333.33$

When $t = 10$ yr,

$A = 35,333.33e^{.06(10)} - 33,333.33$

$= \$31,048.20$.

3. $\dfrac{dA}{dt} = rA + D$; $r = .07$; $D = \$10,000$;

$t = 0$, $A = 0$

$$\dfrac{dA}{dt} = .07A + 10,000$$

$$\dfrac{dA}{.07A + 10,000} = dt$$

$$\dfrac{1}{.07} \ln |.07A + 10,000| = t + C$$

$$\ln |.07A + 10,000| = .07t + C$$

$$|.07A + 10,000| = e^{.07t+C}$$

$$.07A + 10,000 = Me^{.07t}$$

$$A = \dfrac{M}{.07}e^{.07t} - \dfrac{10,000}{.07}$$

When $t = 0$, $A = 0$.

$$0 = \dfrac{M}{.07}(1) - \dfrac{10,000}{.07}$$

$M = 10,000$

$$A = \dfrac{10,000}{.07}e^{.07t} - \dfrac{10,000}{.07}$$

$$= 142,857.14(e^{.07t} - 1)$$

Find t when $A = \$70,000$.

$$70,000 = 142,857.14(e^{.07t} - 1)$$

$$e^{.07t} = \dfrac{70,000}{142,857.14} + 1 = 1.49$$

$$t = \dfrac{1}{.07} \ln 1.49 \approx 5.7 \text{ yr}$$

It will take about 5.7 yr to accumulate $70,000.

5. (a) $\dfrac{dy}{dt} = 3y - 2xy = y(3 - 2x)$

$$\dfrac{dy}{dt} = -2x + 3xy$$

$$= x(-2 + 3y)$$

$$\dfrac{dy}{dx} = \dfrac{\dfrac{dy}{dt}}{\dfrac{dx}{dt}} = \dfrac{y(3 - 2x)}{x(-2 + 3y)}$$

$$\int \left(\dfrac{3y - 2}{y}\right) dy = \int \left(\dfrac{3 - 2x}{x}\right) dx$$

$$\int \left(3 - \dfrac{2}{y}\right) dy = \int \left(\dfrac{3}{x} - 2\right) dx$$

$$3y - 2 \ln y = 3 \ln x - 2x + C$$

$$3 \ln x - 2x - 3y + 2 \ln y = C$$

When $x = 1$ and $y = 2$,

$$3 \ln 1 - 2 - 6 + 2 \ln 2 = C$$

$$-8 + \ln 4 = C.$$

Therefore,

$$3 \ln x - 2x - 3y + 2 \ln y$$
$$= -8 + \ln 4.$$

(b) $0 = 3y - 2xy = -y(-3 + 2x)$

$$0 = -2x + 3xy = x(-2 + 3y)$$

$0 = -y(-3 + 2x)$ and $0 = x(-2 + 3y)$

$y = 0$	and $x = 0$ or
$-3 + 2x = 0$	$-2 + 3y = 0$
$x = \dfrac{3}{2}$	and $y = \dfrac{2}{3}$

7. **(a)** Let $y =$ the number of individuals infected. The differential equation is

$$\frac{dy}{dt} = a(N - y)y.$$

The solution is Equation 12 in Example 4, which is

$$y = \frac{N}{1 + (N - 1)e^{-aNt}}.$$

The number of individuals uninfected at time t is

$$y = N - \frac{N}{1 + (N - 1)e^{-aNt}}$$

$$= \frac{N + N(N - 1)e^{-aNt} - N}{1 + (N - 1)e^{-aNt}}$$

$$= \frac{N(N - 1)}{N - 1 + e^{aNt}}.$$

Now substitute $N = 5000$ and $a = .00005$.

$$y = \frac{5000(5000 - 1)}{5000 - 1 + e^{(.00005)(5000)t}}$$

$$= \frac{24,995,000}{4999 + e^{.25t}}$$

(b) $t = 30$

$$y = \frac{24,995,000}{4999 + e^{.25(30)}} = 3672$$

(c) $t = 50$

$$\frac{24,995,000}{4999 + e^{.25(50)}} = 91$$

(d) From Example 4,

$$t_m = \frac{\ln (N - 1)}{aN}$$

$$= \frac{\ln (5000 - 1)}{(.00005)(5000)}$$

$$= 34.$$

The maximum infection rate will occur on the 34th day.

9. **(a)** The differential equation is

$$\frac{dy}{dt} = a(N - y)y.$$

$y_0 = 50$; $y = 300$ when $t = 10$, $N = 10,000$.

The solution is Equation 11 in Example 4, which is

$$y = \frac{N}{1 + be^{-kt}},$$

where $b = \dfrac{N - y_0}{y_0}$ and $k = aN$.

Since $y_0 = 50$ and $N = 10,000$,

$$b = \frac{10,000 - 50}{50} = 199; \quad k = 10,000a.$$

Therefore,

$$y = \frac{10,000}{1 + 199e^{-10,000at}}.$$

$y = 300$ when $t = 10$.

$$300 = \frac{10,000}{1 + 99e^{-10,000(10)a}}$$

$$300 + 300(199)e^{-100,000a}$$

$$= 10,000$$

$$e^{-100,000a} = \frac{9700}{300(199)} = .1624791$$

$$a = \frac{\ln{(.1624791)}}{-100,000} = .000018$$

$k = 10,000a = 10,000(.000018) = .18$

Therefore,

$$y = \frac{10,000}{1 + 199e^{-.18t}} \quad \text{or} \quad \frac{10,000e^{.18t}}{e^{.18t} + 199}$$

(b) One half the community is
y = 5000.   Find t for y = 5000.

$$5000 = \frac{10,000}{1 + 199e^{-.18t}}$$

$$5000 + 5000(199)e^{-.18t} = 10,000$$

$$e^{-.18t} = \frac{5000}{5000(199)} = .005$$

$$t = \frac{\ln{(.005)}}{-.18}$$

$$= 29.44$$

Half the community will be infected
in about 29 days.

11.   (a) $\frac{dy}{dt} = -ay + b(f - y)Y$

a = 1, b = 1, f = .5, Y = .01; and
y = .02 when t = 0

$$\frac{dy}{dt} = -y + 1(.5 - y)(.01)$$

$$= -1.010y + .005$$

$$\int \frac{dy}{-1.010y + .005} = \int dt$$

$$\frac{1}{-1.010} \ln{|-1.010y + .005|}$$

$$= t + C_2$$

$$\ln{|-1.010y + .005|} = -1.010t + C_1$$

$$|-1.010y + .005| = e^{-1.010t+C_1}$$

$$= e^{C_1}e^{-1.010t}$$

$$-1.010y + .005 = Ce^{-1.010t}$$

$$y = .005 - .990Ce^{-1.010t}$$

Since y = .02 when t = 0,

$$.02 = .005 - .990Ce^{0}$$

$$-.990C = .015.$$

Therefore,

$$y = .005 + .015e^{-1.010t}.$$

(b) $\frac{dY}{dt} = -AY + B(F - Y)y$

A = 1, B = 1, y = .1, F = .03; and
Y = .01 when t = 0

$$\frac{dY}{dt} = -Y + 1(.03 - Y)(.1)$$

$$= -1.1Y + .003$$

$$\frac{dY}{-1.1Y - .003} = dt$$

$$-\frac{1}{1.1} \ln{|-1.1Y + .003|} = t + C_2$$

$$\ln{|-1.1Y + .003|} = -1.1t + C_1$$

$$|-1.1Y + .003| = e^{-1.1t+C_1}$$

$$= e^{C_1}e^{-1.1t}$$

$$-1.1Y + .003 = Ce^{-1.1t}$$

$$Y = \frac{C}{-1.1}e^{-1.1t} - \frac{.003}{-1.1}$$

$$= .909Ce^{-1.1t}$$

$$+ .00273$$

Since Y = .01 when t = 0,

$$.01 = -.909Ce^{0} + .00273$$

$$-.909C = .00727.$$

Therefore,

$$Y = .00727e^{-1.1t}$$

$$+ .00273.$$

13. (a) $\frac{dy}{dt} = a(N - y)y$

$y_0 = 5$; $y = 15$ when $t = 3$; $N = 50$

The solution to this differential equation is Equation 11 in Example 4, which is

$y = \frac{N}{1 + be^{-kt}}$ where $b = \frac{N - y_0}{y_0}$ and $k = aN$.

Since $y = 5$ and $N = 50$,

$b = \frac{50 - 5}{5} = 9$; $k = 50a$.

$y = \frac{50}{1 + 9e^{-50at}}$

$y = 15$ when $t = 3$.

$$15 = \frac{50}{1 + 9e^{-50a}}$$

$15 + 135e^{-150a} = 50$

$e^{-150a} = \frac{35}{135} = \frac{7}{27}$

$-150a = \ln \frac{7}{27} = -1.350$

$-50a = -.45$

Therefore,

$y = \frac{50}{1 + 9e^{-.45t}}.$

(b) When $y = 30$,

$$30 = \frac{50}{1 + 9e^{-.45t}}$$

$30 + 270e^{-.45t} = 50$

$e^{-.45t} = \frac{20}{270} = \frac{2}{27}$

$t = -\frac{1}{.45} \ln \frac{2}{27} = 5.78.$

In about 6 days, 30 employees have heard the rumor.

15. (a) $\frac{dy}{dt} = kye^{-at}$; $a = .1$; $y = 5$ when $t = 0$; $y = 15$ when $t = 3$.

$$\int \frac{dy}{y} = k \int e^{-.1t}\, dt$$

$\ln |y| = -10ke^{-.1t} + C_1$

$|y| = e^{-10ke^{-.1t}+C_1}$

$\quad = e^{C_1}e^{-10ke^{-.1t}}$

$y = Ce^{-10ke^{-.1t}}$

Since $y = 5$ when $t = 0$,

$5 = Ce^{-10k}$

$C = 5e^{10k}.$

Since $y = 15$ when $t = 3$,

$15 = Ce^{-10ke^{-.3}} = Ce^{-7.41k}$

$C = 15e^{7.41k}$

Solve the system

$C = 5e^{10k}$

$C = 15e^{7.41k}.$

$5e^{10k} = 15e^{7.41k}$

$e^{10k} = 3e^{7.41k}$

$10k \ln e = \ln 3 + 7.41k \ln e$

$2.59k = \ln 3$

$k = \frac{1}{2.59} \ln 3 = .424$

$C = 5e^{10(.424)} = 347$

Therefore,

$y = 347e^{-10(.424)e^{-.1t}}$

$\quad = 347e^{-4.24e^{-.1t}}$

(b) If $y = 30$ employees,

$$30 = 347e^{-4.24e^{-.1t}}$$

$$e^{-4.24e^{-.1t}} = \frac{30}{347} = .0865$$

$$-4.24e^{-.1t} \ln e = \ln .0865$$

$$e^{-.1t} = -\frac{1}{4.24} \ln .0865$$

$$= .5773$$

$$-.1t \ln e = \ln .5773$$

$$t = -10 \ln .5773$$

$$= 5.493$$

30 employees have heard the rumor in about 5.5 days.

17. Let y = the amount of salt present at time t.

(a)

$$\frac{dy}{dt} = (\text{Rate of Salt In})$$

$$- (\text{Rate of Salt Out})$$

Rate of Salt In

$$= (3 \text{ gal/min})(2 \text{ lb/gal})$$

$$= 6 \text{ lb/min}$$

Rate of Salt Out

$$= \left(\frac{y}{V} \text{ lb/gal}\right)(2 \text{ gal/min}) = \frac{2y}{V} \text{ lb/min}$$

$$\frac{dy}{dt} = 6 - \frac{2y}{V}; \ y(0) = 20 \text{ lb}$$

$$\frac{dV}{dt} = (\text{Rate of Liquid in})$$

$$- (\text{Rate of Liquid out})$$

$$= 3 \text{ gal/min} - 2 \text{ gal/min}$$

$$= 1 \text{ gal/min}$$

$$\frac{dV}{dt} = 1$$

$$V = t + C_1$$

When $t = 0$, $V = 100$. Thus, $C_1 = 100$.

$$V = t + 100$$

Therefore,

$$\frac{dy}{dt} = 6 - \frac{2y}{t + 100}$$

$$\frac{dy}{dt} + \frac{2}{t + 100}y = 6.$$

$$I(t) = e^{\int 2dt/(t+100)}$$

$$= e^{2 \ln |t+100|}$$

$$= (t + 100)^2$$

$$y'(t + 100)^2 + 2y(t + 100) = 6(t + 100)^2$$

$$D_t \ [y(t + 100)^2] = 6(t + 100)^2$$

$$y(t + 100)^2 = 6\int (t + 100)^2 \ dt$$

$$y(t + 100)^2 = 2(t + 100)^3 + C$$

$$y = 2(t + 100)$$

$$+ \frac{C}{(t + 100)^2}$$

Since $t = 0$ when $y = 20$,

$$20 = 2(100) + \frac{C}{100^2}$$

$$C = -1,800,000.$$

$$y = 2(t + 100) - \frac{1,800,000}{(t + 100)^2}$$

$$= \frac{2(t + 100)^3 - 1,800,000}{(t + 100)^2}.$$

(b) $t = 1 \text{ hr} = 60 \text{ min}$

$$y = \frac{2(160)^3 - 1,800,000}{(160)^2} = 249.69$$

After 1 hr, about 250 lb of salt are present.

(c) As time increases, salt concentration continues to increase.

19. Let y = the amount of salt present at time t minutes.

(a) $\frac{dy}{dt}$ = (Rate of Salt In)

   − (Rate of Salt Out)

Rate of Salt In = 0

Rate of Salt Out

$= \left(\frac{y}{V}\text{ lb/gal}\right)(2\text{ gal/min})$

$= \frac{2y}{V}$ lb/min

$\frac{dy}{dt} = -\frac{2y}{V}$; y(0) = 20

$\frac{dV}{dt}$ = (Rate of Liquid In)

   − (Rate of Liquid Out)

$= 2\text{ gal/min} - 2\text{ gal/min} = 0$

$\frac{dV}{dt} = 0$

$V = C_1$

When t = 0, V = 100, so C_1 = 100. Therefore,

$\frac{dy}{dt} = -\frac{2y}{100} = -.02y$

$\frac{dy}{y} = -.02\, dt$

$\ln|y| = -.02t + C_1$

$|y| = e^{-.02t+C_1} = e^{C_1}e^{-.02t}$

$y = Ce^{-.02t}$

Since t = 0 when y = 20,

$20 = Ce^0$

$C = 20.$

$y = 20e^{-.02t}$

(b) t = 1 hr = 60 min

$y = 20e^{-.02(60)} = 6.024$

After 1 hr, about 6 lb of salt are present.

(c) As time increases, salt concentration continues to decrease.

21. Let y = amount of the chemical at time t.

(a) $\frac{dy}{dt}$ = (Rate of Chemical In)

   − (Rate of Chemical Out)

Rate of Chemical In

$= (2\text{ liters/min})(.1\text{ g/liter})$

$= .2$ g/min

Rate of Chemical Out

$= \left(\frac{y}{V}\text{ g/liter}\right)(1\text{ liter/min})$

$= \frac{y}{V}$ g/liter

$\frac{dy}{dt} = .2 - \frac{y}{V}$; y(0) = 5

$\frac{dV}{dt}$ = (Rate of Liquid In)

   − (Rate of Liquid Out)

$= 2\text{ liter/min} - 1\text{ liter/min}$

$= 1$ liter/min

$\frac{dV}{dt} = 1$

$V = t + C_1$

When t = 0, V = 100, so C_1 = 100.

$V = t + 100$

Therefore,

$\frac{dy}{dt} = .2 - \frac{y}{t + 100}$

$\frac{dy}{dt} + \frac{1}{t + 100} \cdot y = .2.$

$$I(t) = e^{\int dt/(t+100)} = e^{\ln |t+100|}$$

$$= t + 100$$

$$y'(t + 100) + y = .2(t + 100)$$

$$D_x (t + 100)y = .2(t + 100)$$

$$(t + 100)y = \int .2(t + 100)dt$$

$$(t + 100)y = .1(t + 100)^2 + C$$

$$y = .1(t + 100)$$

$$+ \frac{C}{t + 100}$$

$$t = 0, \ y = 5$$

$$5 = .1(100) + \frac{C}{100}$$

$$500 = 1000 + C$$

$$C = -500$$

Therefore,

$$y = .1(t + 100) + \frac{-500}{t + 100}$$

$$= \frac{.1(t + 100)^2 - 500}{t + 100}.$$

(b) When $t = 30$ min,

$$y = \frac{.1(130)^2 - 500}{130} = 9.154.$$

After 30 min, about 9.2 g of chemical are present.

Chapter 9   Review Exercises

5. $\dfrac{dy}{dx} = 2x^3 + 6x$

$$dy = (2x^3 + 6x)dx$$

$$y = \frac{x^4}{2} + 3x^2 + C$$

7. $\dfrac{dy}{dx} = 4e^x$

$$dy = 4e^x \ dx$$

$$y = 4e^x + C$$

9. $\dfrac{dy}{dx} = \dfrac{3x + 1}{y}$

$$y \ dy = (3x + 1)dx$$

$$\frac{y^2}{2} = \frac{3x^2}{2} + x + C_1$$

$$y^2 = 3x^2 + 2x + C$$

11.
$$\frac{dy}{dx} = \frac{2y + 1}{x}$$

$$\frac{dy}{2y + 1} = \frac{dx}{x}$$

$$\frac{1}{2}\left(\frac{2 \ dy}{2y + 1}\right) = \frac{dx}{x}$$

$$\frac{1}{2} \ln |2y + 1| = \ln |x| + C_1$$

$$\ln |2y + 1|^{1/2} = \ln |x| + \ln k$$

$$\text{\textit{Let} } \ln k = C_1$$

$$\ln |2y + 1|^{1/2} = \ln k|x|$$

$$|2y + 1|^{1/2} = k \ |x|$$

$$2y + 1 = k^2 x^2$$

$$2y + 1 = Cx^2$$

$$2y = Cx^2 - 1$$

$$y = \frac{Cx^2 - 1}{2}$$

13. $\dfrac{dy}{dx} + 5y = 12$

$$I(x) = e^{\int 5 \ dx} = e^{5x}$$

$$e^{5x}y' + 5e^{5x}y = 12e^{5x}$$

$$D_x (e^{5x}y) = 12e^{5x}$$

$$e^{5x}y = \frac{12}{5}e^{5x} + C$$

$$y = \frac{12}{5} + Ce^{-5x}$$

15. $3 \frac{dy}{dz} + xy - x = 0$

$$y' + \frac{x}{3}y = \frac{x}{3}$$

$$I(x) = e^{\int x/3 \, dx}$$

$$= e^{x^2/6}$$

$$e^{x^2/6} y' + \frac{x}{3}e^{x^2/6} y = \frac{x}{3}e^{x^2/6}$$

$$D_x (e^{x^2/6} y) = \frac{x}{3}e^{x^2/6}$$

$$e^{x^2/6} y = \int \frac{x}{3}e^{x^2/6} \, dx$$

$$= e^{x^2/6} + C$$

$$y = 1 + Ce^{-x^2/6}$$

17. $\frac{dy}{dx} = x^2 - 5x; \ y = 1$ when $x = 0$

$$dy = (x^2 - 5x)dx$$

$$y = \frac{x^3}{3} - \frac{5x^2}{2} + C$$

When $x = 0, \ y = 1.$

$$1 = 0 - 0 + C$$

$$C = 1$$

$$y = \frac{x^3}{3} - \frac{5x^2}{2} + 1$$

19. $\frac{dy}{dx} = 5(e^{-x} - 1); \ y = 17$ when $x = 0.$

$$dy = 5(e^{-x} - 1)dx$$

$$y = 5(-e^{-x} - x) + C$$

$$= -5e^{-x} - 5x + C$$

$$17 = -5e^0 - 0 + C$$

$$17 = -5 + C$$

$$22 = C$$

$$y = -5e^{-x} - 5x + 22$$

21. $(5 - 2x)y = \frac{dy}{dx}; \ y = 2$ when $x = 0.$

$$(5 - 2x)dx = \frac{dy}{y}$$

$$5x - x^2 + C = \ln |y|$$

$$e^{5x-x^2+C} = y$$

$$e^{5x-x^2} \cdot e^C = y$$

$$Me^{5x-x^2} = y$$

$$Me^0 = 2$$

$$M = 2$$

$$y = 2e^{5x-x^2}$$

23. $\frac{dy}{dx} = \frac{1 - 2x}{y + 3}; \ y = 16$ when $x = 0.$

$$(y + 3)dy = (1 - 2x)dx$$

$$\frac{y^2}{2} + 3y = x - x^2 + C$$

$$\frac{16^2}{2} + 3(16) = 0 + C$$

$$176 = C$$

$$\frac{y^2}{2} + 3y = x - x^2 + 176$$

$$y^2 + 6y = 2x - 2x^2 + 352$$

25. $y' + x^2y = x^2; \ y = 8$ when $x = 0.$

$$I(x) = e^{\int x^2 \, dx} = e^{x^3/3}$$

$$e^{x^3/3} y' + x^2 e^{x^3/3} y = x^2 e^{x^3/3}$$

$$D_x (e^{x^3/3} y) = x^2 e^{x^3/3}$$

$$e^{x^3/3} y = \int x^2 e^{x^3/3} \, dx$$

$$= e^{x^3/3} + C$$

$$y = 1 + Ce^{-x^3/3}$$

Since $x = 0$ when $y = 8,$

$$8 = 1 + Ce^0$$

$$C = 7.$$

Therefore,

$$y = 1 + 7e^{-x^3/3}$$

27. $x \dfrac{dy}{dx} - 2x^2y + 3x^2 = 0$

$xy' - 2x^2y + 3x^2 = 0;$

$y = 15$ when $x = 0$

$$y' - 2xy = -3x$$

$$I(x) = e^{\int -2x\, dx} = e^{-x^2}$$

$$e^{-x^2}y' - 2xe^{-x^2}y = -3xe^{-x^2}$$

$$D_x\,(e^{-x^2}y) = -3xe^{-x^2}$$

$$e^{-x^2}y = \int -3xe^{-x^2}\, dx$$

$$= \frac{3}{2}e^{-x^2} + C$$

$$y = \frac{3}{2} + Ce^{x^2}$$

Since $x = 0$ when $y = 15$,

$$15 = \frac{3}{2} + Ce^0$$

$$C = \frac{27}{2}.$$

Therefore,

$$y = \frac{3}{2} + \frac{27e^{x^2}}{2}.$$

31. $y' = e^x + 1$, $f(0) = 1$, $h = .2$,

find $f(.6)$.

x_i	y_i
0	1
.2	1.4
.4	1.844281
.6	2.342646

Therefore, $f(.6) \approx 2.343$.

33. $y' = 3 + \sqrt{y}$, $f(0) = 0$, $h = .2$,

find $f(1)$.

x_i	y_i
0	0
.2	.6
.4	1.354919
.6	2.187722
.8	3.083541
1	4.034741

Therefore, $f(1) \approx 4.035$.

See the answer graph in the back of the textbook.

35. **(a)** $\dfrac{dy}{dx} = 5e^{.2x}$

$$dy = 5e^{.2x}\, dx$$

$$y = \frac{5}{.2}e^{.2x} + C$$

$$y = 25e^{.2x} + C$$

When $x = 0$, $y = 0$.

$$0 = 25e^0 + C$$

$$C = -25 \qquad e^0 = 1$$

$$y = 25e^{.2x} - 25$$

When $x = 6$,

$$y = 25e^{1.2} - 25$$

$$\approx 58.$$

Sales are $5800.

(b) When $x = 12$,

$$y = 25e^{2.4} - 25$$

$$\approx 251.$$

Sales are $25,100.

37. $\dfrac{dA}{dt} = rA - D$; $t = 0$, $A = \$10,000$;

$r = .05$; $D = \$1000$

$$\dfrac{dA}{dt} = .05A - 1000$$

$$\int \dfrac{dA}{.05A - 1000} = \int dt$$

$\dfrac{1}{.05} \ln |.05A - 1000| = t + C_2$

$\ln |.05A - 1000| = .05t + C_1$

$.05A - 1000 = Ce^{.05t}$

$A = 20Ce^{.05t} + 20,000$

Since $t = 0$ when $A = 10,000$,

$10,000 = 20C(1) + 20,000$

$C = -500.$

$A = 20(-500)e^{.05t} + 20,000$

$A = 20,000 - 10,000e^{.05t}$

Find t for $A = 0$.

$0 = 20,000 - 10,000e^{.05t}$

$e^{.05t} = \dfrac{20,000}{10,000} = 2$

$\ln e^{.05t} = \ln 2$

$t = 20 \ln 2 = 13.86$

It will take about 13.9 yr.

39. $\dfrac{dy}{dt} = ky$; $k = .1$, $t = 0$, $y = 120$

$\dfrac{dy}{y} = k\, dt$

$\ln |y| = kt + C_1$

$|y| = e^{kt+C_1}$

$y = Me^{kt}$

$y = Me^{.1t}$

$120 = Me^0$

$M = 120$

$y = 120e^{.1t}$

Let $t = 6$ and find y.

$y = 120e^{.6}$

≈ 219

After 6 wk, about 219 are present.

41. $\dfrac{dy}{dt} =$ Rate of Smoke In

$-$ Rate of Smoke Out

Rate of Smoke In $= 0$

Rate of Smoke Out

$= \left(\dfrac{y}{V}\right)1200 = \dfrac{1200y}{V}$

$\dfrac{dy}{dt} = \dfrac{-1200y}{V}$

$\dfrac{dV}{dt} = 1200 - 1200 = 0$

$V(t) = C_1$ at $t = 0$, $V = 15,000$;

$C_1 = 15,000$

$V(t) = 15,000$

$\dfrac{dy}{dt} = \dfrac{-1200y}{15,000} = -.08y$

$\dfrac{dy}{y} = -.08\, dt$

$\ln |y| = -.08t + C$

$y = ke^{-.08t}$

At $t = 0$, $y = 20$.

$20 = ke^0$

$k = 20$

Therefore,

$y = 20e^{-.08t}.$

At $y = 5$,

$5 = 20e^{-.08t}$

$.25 = e^{-.08t}$

$t = \dfrac{-\ln (.25)}{.08}$

$= 17.3$ min.

43. **(a)** The differential equation for y, the number of individuals infected, is

$$\frac{dy}{dt} = a(N - y)y.$$

The solution is Equation 12 in Example 4 of Section 9.4 which is:

$$y = \frac{N}{1 + (N - 1)e^{-aNt}}; \quad t \text{ is in weeks.}$$

The number of individuals uninfected at time t is

$$y = N - \frac{N}{1 + (N - 1)e^{-aNt}}$$

$$= \frac{N(N - 1)}{N - 1 + e^{aNt}}.$$

Substitute N = 700.

$$y = \frac{700(699)}{699 + e^{700at}}$$

$$= \frac{489,300}{699 + e^{700at}}$$

Substitute t = 6 wk, y = 300.

$$300 = \frac{489,300}{699 + e^{700(6)a}}$$

$$= \frac{489,300}{699 + e^{4200a}}$$

$$209,700 + 300e^{4200a}$$

$$= 489,300$$

$$e^{4200a} = 932$$

$$a = \frac{\ln 932}{4200} = .00163$$

$$700a = 700(.00163) = 1.140$$

Therefore,

$$y = \frac{489,300}{699 + e^{1.140t}}.$$

(b) At t = 7 wk,

$$y = \frac{489,300}{699 + e^{1.140(7)}}$$

$$\approx 135 \text{ people.}$$

(c) From Example 4,

$$t_m = \frac{\ln (N - 1)}{aN}$$

$$= \frac{\ln (700 - 1)}{(.00163)(700)}$$

$$t_m = 5.7 \text{ wk}$$

45. **(a)** $\frac{dy}{dx} = a(N - y)y$; N = 100; $y_0 = 4$; y = 15 when x = 3 da

The solution to the differential equation is Equation 11 in Example 4, which is

$$y = \frac{N}{1 + be^{-kx}},$$

where $b = \frac{N - y_0}{y_0}$ and k = aN.

$$b = \frac{100 - 4}{4} = 24; \quad k = 100a$$

$$y = \frac{100}{1 + 24e^{-100ax}}$$

Since y = 15 when x = 3,

$$15 = \frac{100}{1 + 24e^{-100a(3)}}$$

$$15 + 360e^{-300a} = 100$$

$$e^{-300a} = \frac{85}{360}$$

$$-300a = \ln \left(\frac{85}{360}\right)$$

$$-100a = \frac{1}{3} \ln \left(\frac{85}{360}\right) = -.481.$$

Therefore,

$$y = \frac{100}{1 + 24e^{-.481x}}.$$

(b) For x = 5 days,

$$y = \frac{100}{1 + 24e^{-.481(5)}}$$

$$y = 31.58$$

In 5 days about 32 people have heard the rumor.

47. $N = \dfrac{\dfrac{1}{y_1} + \dfrac{1}{y_3} - \dfrac{2}{y_2}}{\dfrac{1}{y_1 y_3} - \dfrac{1}{y_2{}^2}}$

(a) For 1800, $y_1 = 5$.

For 1850, $y_2 = 23$.

For 1900, $y_3 = 76$.

$$N = \frac{\dfrac{1}{5} + \dfrac{1}{76} - \dfrac{2}{23}}{\dfrac{1}{5(76)} - \dfrac{1}{(23)^2}} \approx 170 \text{ million}$$

(b) For 1850, $y_1 = 23$.

For 1900, $y_2 = 76$.

For 1950, $y_3 = 151$.

$$N = \frac{\dfrac{1}{23} + \dfrac{1}{151} - \dfrac{2}{76}}{\dfrac{1}{23(151)} - \dfrac{1}{(76)^2}} \approx 207 \text{ million}$$

(c) For 1870, $y_1 = 40$.

For 1920, $y_2 = 106$.

For 1970, $y_3 = 204$.

$$N = \frac{\dfrac{1}{40} + \dfrac{1}{204} - \dfrac{2}{106}}{\dfrac{1}{40(204)} - \dfrac{1}{(106)^2}} \approx 329 \text{ million}$$

49.

$$\frac{dT}{dt} = -k(T - T_F)$$

$$\frac{dT}{dt} = -k(T - 300°)$$

$$T(0) = 40°$$

$$T(1) = 150°$$

$$\frac{dT}{T - 300} = -k \, dt$$

$$\ln |T - 300| = -kt + C_1$$

$$|T - 300| = e^{-kt+C_1}$$

$$T - 300 = Ce^{-kt}$$

$$T = Ce^{-kt} + 300$$

Since $T(0) = 40°$,

$$40 = Ce^0 + 300$$

$$C = -260.$$

Therefore,

$$T = -260e^{-kt} + 300.$$

At $T(1) = 150°$,

$$150 = -260e^{-k} + 300$$

$$-150 = -260e^{-k}$$

$$\frac{15}{26} = e^{-k}$$

$$\ln\left(\frac{15}{26}\right) = -k \ln e$$

$$k = -\ln\left(\frac{15}{26}\right)$$

$$k = .55.$$

Therefore,

$$T = -260e^{-.55t} + 300.$$

At $t = 2$,

$$T = -260e^{-.55(2)} + 300$$

$$= 213°.$$

51. (a)

$$\frac{dv}{dt} = B^2 - K^2v^2$$

$$dv = (B^2 - K^2v^2)dt$$

$$\frac{1}{B^2 - K^2v^2}\, dv = dt$$

$$\int \frac{1}{B^2 - K^2v^2}\, dv = \int dt$$

$$\frac{1}{K^2}\int \frac{1}{\left(\frac{B}{K}\right)^2 - v^2}\, dv = \int dt$$

Use entry 7 from the table of integrals.

$$\frac{1}{K^2}\cdot\frac{1}{2\frac{B}{K}}\ln\left|\frac{\frac{B}{K} + v}{\frac{B}{K} - v}\right| = t + C_1$$

$$\frac{1}{2BK}\ln\left|\frac{B + Kv}{B - Kv}\right| = t + C_1$$

$$\ln\left|\frac{B + Kv}{B - Kv}\right| = 2BKt + C_2$$

Since $v < \frac{B}{K}$, $Kv < B$ and $\frac{B + Kv}{B - Kv}$ is positive.

$$\ln\left(\frac{B + Kv}{B - Kv}\right) = 2BKt + C_2$$

When $t = 0$, $v = 0$, so

$$\ln\left(\frac{B + 0}{B - 0}\right) = 2BK(0) + C_2$$

$$\ln 1 = C_2$$

$$C_2 = 0.$$

Thus,

$$\ln\left(\frac{B + Kv}{B - Kv}\right) = 2BKt$$

$$\frac{B + Kv}{B - Kv} = e^{2BKt}$$

$$B + Kv = Be^{2BKt} - Kve^{2BKt}$$

$$kv + kve^{2BKt} = Be^{2BKt} - B$$

$$vK(e^{2BKt} + 1) = B(e^{2BKt} - 1)$$

$$v = \frac{B(e^{2BKt} - 1)}{K(e^{2BKt} + 1)}$$

$$v = \frac{B}{K}\cdot\frac{e^{2BKt} - 1}{e^{2BKt} + 1}$$

(b) $\lim\limits_{t\to\infty} v = \lim\limits_{t\to\infty}\left(\frac{B}{K}\cdot\frac{e^{2BKt} - 1}{e^{2BKt} - 1}\right)$

$$\lim\limits_{t\to\infty} v = \frac{B}{K}\lim\limits_{t\to\infty}\frac{e^{2BKt} - 1}{e^{2BKt} + 1}$$

$$\lim\limits_{t\to\infty} v = \frac{B}{K}\cdot 1 = \frac{B}{K}$$

A falling object in the presence of air resistance has a limiting velocity, $\frac{B}{K}$.

Extended Application

1. $t = \frac{1}{k}\ln\left(\frac{P_L(0)}{P_L(t)}\right)$

(a) $\frac{1}{k} = 2.6$; $\frac{P_L(t)}{P_L(0)} = .4$

$$t = 2.6\ln\left(\frac{1}{.4}\right) = 2.4\text{ yr}$$

(b) $\frac{1}{k} = 2.6$; $\frac{P_L(t)}{P_L(0)} = .3$

$$t = 2.6\ln\left(\frac{1}{.3}\right) = 3.1\text{ yr}$$

2. $t = \frac{1}{k}\ln\left(\frac{P_L(0)}{P_L(t)}\right)$

(a) $\frac{1}{k} = 30.8$; $\frac{P_L(t)}{P_L(0)} = .4$

$$t = 30.8\ln\left(\frac{1}{.4}\right) = 28.2\text{ yr}$$

(b) $\frac{1}{k} = 30.8$; $\frac{P_L(t)}{P_L(0)} = .3$

$t = 30.8$; $\left(\frac{1}{.3}\right) = 37.1$ yr

3. $t = \frac{1}{k} \ln \left(\frac{P_L(0)}{P_L(t)}\right)$

(a) $\frac{1}{k} = 189$; $\frac{P_L(t)}{P_L(0)} = .4$

$t = 189 \ln \left(\frac{1}{.4}\right) = 173$ yr

(b) $\frac{1}{k} = 189$; $\frac{P_L(t)}{P_L(0)} = .3$

$t = 189 \ln \left(\frac{1}{.3}\right) = 228$ yr

CHAPTER 9   TEST

[9.1]   Find general solutions for the following differential equations.

1. $\dfrac{dy}{dx} = 3x^2 + 4x - 5$ 2. $\dfrac{dy}{dx} = 5e^{2x}$

3. $\dfrac{dy}{dx} = \dfrac{2x}{x^2 + 5}$

[9.1]   Find particular solutions of the following differential equations.

4. $\dfrac{dy}{dx} = 4x^3 + 2x^2 + 1$; $y = 4$ when $x = 1$

5. $\dfrac{dy}{dx} = \dfrac{1}{3x + 1}$; $y = 4$ when $x = 3$

6. $\dfrac{dy}{dx} = 4(e^{-x} - 1)$; $y = 5$ when $x = 0$

7. After use of an insecticide, the rate of decline of an insect population is

$$\frac{dy}{dt} = \frac{-8}{1 + 4t}$$

where t is the number of hours after the insecticide is applied. If there were 40 insects initially, how many are left after 20 hours?

[9.1]   Find general solutions for the following differential equations. (Some solutions may give y implicitly.)

8. $\dfrac{dy}{dx} = \dfrac{2x + 1}{y - 1}$ 9. $\dfrac{dy}{dx} = \dfrac{3y}{x + 1}$ 10. $\dfrac{dy}{dx} = \dfrac{e^x - x}{y + 1}$

[9.1]   Find particular solutions of the following differential equations.

11. $\sqrt{y}\,\dfrac{dy}{dx} = xy$; $y = 4$ when $x = 2$

12. $\dfrac{dy}{dx} = e^y \cdot x^3$; $y = 0$ when $x = 0$

[9.2]   **Find the general solution of each differential equation.**

13. $y' + 4y = 12$

14. $y' - xy = 2x$

[9.2]   **Find the particular solution of each differential equation.**

15. $y' + 3y = 10$; $y = 5$ when $x = 0$

16. $x^2 \dfrac{dy}{dx} + 2x^3y + 2x^3 = 0$; $y = 4$ when $x = 0$

17. $e^xy' + e^xy = x + 1$; $y = \dfrac{2}{e}$ when $x = 1$

[9.3]   **Use Euler's method to approximate the indicated function value for**
$y = f(x)$ to three decimal places using $h = .1$.

18. $y' = 2x - y^{-1}$; $f(0) = 1$; find $(.3)$.

19. $y' = xy$; $f(0) = 1$; find $f(.4)$.

20. Let $y = f(x)$ and $y' = \dfrac{x^2}{2} + 3$ with $f(0) = 0$. Use Euler's method
with $h = .1$ to approximate $f(.3)$ to three decimal places. Then
solve the differential equation and find $f(.3)$ to three decimal
places. Also, find $y_3 - f(x_3)$.

[9.4]   **21.** At her birth, Sally's grandparents deposit $5000 into a savings
account earning 8% interest compounded continuously. If they
make continuous deposits at a rate of $2000 per year through
her twenty-first birthday, how much money will accumulate in the
account?

[9.4]   **22.** A deposit of $20,000 is made into a savings account earning 5%
interest compounded continuously. If continuous withdrawals are
made at a rate of $2000 per year, how much money will be left in
the account after 2 years? Approximately how long will it take
to deplete the account?

[9.4]   **23.** Find an equation relating x to y given the following equations,
which describe the interaction of two competing species and their
growth rates.

$$\frac{dx}{dt} = 6x - 4xy$$

$$\frac{dy}{dt} = -8y + 12xy$$

Find the values of x and y for which both growth rates are 0.

[9.4]   24. In a population of size N, the rate at which an influenza epidemic spreads is given by

$$\frac{dy}{dt} = a(N - y) \cdot y,$$

where y is the number of people infected and a is a constant.

(a) If 40 people in a community of 10,000 people are infected at the beginning of an epidemic, and 100 people are infected 5 days later, write an equation for the number of people infected after t days.

(b) How many people are infected after 10 days?

[9.4]   25. A tank presently contains 100 gal of a solution of dissolved salt and water, which is kept uniform by stirring. While pure water is allowed to flow into the tank at a rate of 5 gal/min, the mixture flows out of the tank at a rate of 3 gal/min. How much salt will remain in the tank after t min if 10 lb of salt are in the mixture originally?

[9.4]   26. Explain why differential equations are useful in problem solving.

[9.3]   27. In Euler's method, large errors may occur. What can be done to reduce the error?

[9.4]   28. A tank filled with a salt solution has solution flowing in and out. The number of pounds y of salt in the tank is given by

$$y = 25(1 - e^{-.02t})$$

where t is the time in hours. Will there ever be 25 lbs of salt in in the solution? Explain.

CHAPTER 9 TEST ANSWERS

1. $y = x^3 + 2x^2 - 5x + C$ **2.** $y = \frac{5}{2}e^{2x} + C$ **3.** $y = \ln(x^2 + 5) + C$

4. $y = x^4 + \frac{2}{3}x^3 + x + \frac{4}{3}$ **5.** $y = \frac{1}{3}\ln|3x + 1| + 4 - \frac{1}{3}\ln 10$

6. $y = -4e^{-x} - 4x + 9$ **7.** 31   **8.** $\frac{y^2}{2} - y = x^2 + x + C$ **9.** $y = C(x + 1)^3$

10. $\frac{y^2}{2} + y = e^x - \frac{x^2}{2} + C$ **11.** $y^{1/2} = \frac{x^2}{4} + 1$

12. $e^{-y} = 1 - \frac{x^4}{4}$ or $y = -\ln\left|1 - \frac{x^4}{4}\right|$ **13.** $y = 3 + Ce^{-4x}$ **14.** $y = Ce^{x^2/2} - 2$

15. $y = \frac{10}{3} + \frac{5}{3}e^{-3x}$ **16.** $y = 5e^{-x^2} - 1$ **17.** $y = \frac{x^2}{2}e^{-x} + xe^{-x} + \frac{1}{2}e^{-x}$

18. .794   **19.** 1.104   **20.** .907; $y = \frac{1}{6}x^3 + 3x$; .905; .002   **21.** \$135,966.68

22. \$17,896.58; 13.9 yr   **23.** $3\ln y - 2y = -4\ln x + 6x + C$; $x = \frac{2}{3}$, $y = \frac{3}{2}$

24. **(a)** $y = \dfrac{10,000}{1 + 249e^{-.184t}}$ **(b)** 247   **25.** $y = 10 \cdot \left(\dfrac{50}{t + 50}\right)^{3/2}$

26. Many problems involve rates of change which in turn lead to differential equations.

27. The error can be reduced by using more subintervals of smaller width.

28. As $t \to \infty$, $e^{-.02t} \to 0$, so $\lim\limits_{t \to \infty} y = \lim\limits_{t \to \infty} 25(1 - e^{-.02t}) = 25$.

The limiting value is 25. Theoretically, there will never be 25 lb of salt in the solution. However, as time increases the amount of salt will approach 25 lb.

CHAPTER 10   PROBABILITY AND CALCULUS

Section 10.1

1. $f(x) = \frac{1}{9}x - \frac{1}{18}$; [2, 5]

Show that condition 1 holds.

$$\int_2^5 \left(\frac{1}{9}x - \frac{1}{18}\right)dx = \frac{1}{9}\int_2^5\left(x - \frac{1}{2}\right)dx$$

$$= \frac{1}{9}\left(\frac{x^2}{2} - \frac{1}{2}x\right)\Big|_2^5$$

$$= \frac{1}{9}\left(\frac{25}{2} - \frac{5}{2} - \frac{4}{2} + 1\right)$$

$$= \frac{1}{9}(8 + 1)$$

$$= 1$$

Show that condition 2 holds.

Since $2 \le x \le 5$,

$$\frac{2}{9} \le \frac{1}{9}x \le \frac{5}{9}$$

$$\frac{1}{6} \le \frac{1}{9}x - \frac{1}{18} \le \frac{1}{2}.$$

Hence, $f(x) \ge 0$ on [2, 5].
Yes, $f(x)$ is a probability density function.

3. $f(x) = \frac{1}{21}x^2$; [1, 4]

$$\frac{1}{21}\int_1^4 x^2\, dx = \frac{1}{21}\left(\frac{x^3}{3}\right)\Big|_1^4$$

$$= \frac{1}{21}\left(\frac{64}{3} - \frac{1}{3}\right)$$

$$= 1$$

Since $x^2 \ge 0$, $f(x) \ge 0$ on [1, 4].
Yes, $f(x)$ is a probability density function.

5. $f(x) = 4x^3$; [0, 3].

$$4\int_0^3 x^3\, dx = 4\left(\frac{x^4}{4}\right)\Big|_0^3$$

$$= 4\left(\frac{81}{4} - 0\right)$$

$$= 81 \ne 1$$

No, $f(x)$ is not a probability density function.

7. $f(x) = \frac{x^2}{16}$; [-2, 2].

$$\frac{1}{16}\int_{-2}^2 x^2\, dx = \frac{1}{16}\left(\frac{x^3}{3}\right)\Big|_{-2}^2$$

$$= \frac{1}{16}\left(\frac{8}{3} + \frac{8}{3}\right)$$

$$= \frac{1}{3} \ne 1$$

No, $f(x)$ is not a probability density function.

9. $f(x) = kx^{1/2}$; [1, 4]

$$\int_1^4 kx^{1/2}\, dx = \frac{2}{3}kx^{3/2}\Big|_1^4$$

$$= \frac{2}{3}k(8 - 1)$$

$$= \frac{14}{3}k$$

If $\frac{14}{3}k = 1$,

$$k = \frac{3}{14}.$$

Notice that $f(x) = \frac{3}{4}x^{1/2} \ge 0$ for all x in [1, 4].

11. $f(x) = kx^2$; $[0, 5]$

$$\int_0^5 kx^2 \, dx = k\frac{x^3}{3}\Big|_0^5$$

$$= k\left(\frac{125}{3} - 0\right)$$

$$= k\left(\frac{125}{3}\right)$$

If $k\left(\frac{125}{3}\right) = 1$,

$$k = \frac{3}{125}.$$

Notice that $f(x) = \frac{3}{125}x^2 \geq 0$ for all x in $[0, 5]$.

13. $f(x) = kx$; $[0, 3]$

$$\int_0^3 kx \, dx = k\frac{x^2}{2}\Big|_0^3$$

$$= k\left(\frac{9}{2} - 0\right)$$

$$= \frac{9}{2}k$$

If $\frac{9}{2}k = 1$,

$$k = \frac{2}{9}.$$

Notice that $f(x) = \frac{2}{9}x \geq 0$ for all x in $[0, 3]$.

15. $f(x) = kx$; $[1, 5]$

$$\int_1^5 kx \, dx = k\frac{x^2}{2}\Big|_1^5$$

$$= k\left(\frac{25}{2} - \frac{1}{2}\right)$$

$$= 12k$$

If $12k = 1$,

$$k = \frac{1}{12}.$$

17. The total area under the graph of a probability density function always equals 1.

21. $f(x) = \frac{1}{2}(1 + x)^{-3/2}$; $[0, \infty)$

$$\frac{1}{2}\int_0^\infty (1 + x)^{-3/2} \, dx$$

$$= \lim_{a \to \infty} \frac{1}{2}\int_0^a (1 + x)^{-3/2} \, dx$$

$$= \lim_{a \to \infty} \frac{1}{2}(1 + x)^{-1/2}\left(\frac{-2}{1}\right)\Big|_0^a$$

$$= \lim_{a \to \infty} [-(1 + a)^{-1/2} + 1]$$

$$= \lim_{a \to \infty} \left(\frac{-1}{\sqrt{1 + a}} + 1\right)$$

$$= 0 + 1 = 1$$

Since $x \geq 0$, $f(x) \geq 0$.
$f(x)$ is a probability density function.

(a) $P(0 \leq x \leq 2)$

$$= \frac{1}{2}\int_0^2 (1 + x)^{-3/2} \, dx$$

$$= -(1 + x)^{-1/2}\Big|_0^2$$

$$= -3^{-1/2} + 1$$

$$\approx .4226$$

(b) $P(1 \leq x \leq 3)$

$$= \frac{1}{2}\int_1^3 (1 + x)^{-3/2} \, dx$$

$$= -(1 + x)^{-1/2}\Big|_1^3$$

$$= -4^{-1/2} + 2^{-1/2}$$

$$\approx .2071$$

(c) $P(x \leq 5)$

$$= \frac{1}{2} \int_{5}^{\infty} (1 + x)^{-3/2} \, dx$$

$$= \lim_{a \to \infty} \frac{1}{2} \int_{5}^{a} (1 + x)^{-3/2} \, dx$$

$$= \lim_{a \to \infty} [-(1 + x)^{-1/2}] \Big|_{5}^{a}$$

$$= \lim_{a \to \infty} [-(1 + a)^{-1/2} + 6^{-1/2}]$$

$$= \lim_{a \to \infty} \left(\frac{-1}{\sqrt{1 + a}} + 6^{-1/2} \right)$$

$$\approx 0 + .4082$$

$$\approx .4082$$

23. $f(x) = \frac{1}{2} e^{-x/2} ; \; [0, \infty),$

$$\frac{1}{2} \int_{0}^{\infty} e^{-x/2} \, dx$$

$$= \lim_{a \to \infty} \frac{1}{2} \int_{0}^{a} e^{-x/2} \, dx$$

$$= \lim_{a \to \infty} \frac{1}{2} (\frac{-2}{1} e^{-x/2}) \Big|_{0}^{a}$$

$$= \lim_{a \to \infty} -e^{-x/2} \Big|_{0}^{a}$$

$$= \lim_{a \to \infty} \left(\frac{-1}{e^{a/2}} + 1 \right)$$

$$= 0 + 1$$

$$= 1$$

$f(x) > 0$ for all x.

$f(x)$ is a probability density function.

(a) $P(0 \leq x \leq 1) = \frac{1}{2} \int_{0}^{1} e^{-x/2} \, dx$

$$= -e^{-x/2} \Big|_{0}^{1}$$

$$= \frac{-1}{e^{1/2}} + 1$$

$$\approx .3935$$

(b) $P(1 \leq x \leq 3) = \frac{1}{2} \int_{1}^{3} e^{-x/2} \, dx$

$$= -e^{-x/2} \Big|_{1}^{3}$$

$$= \frac{-1}{e^{3/2}} + \frac{1}{e^{1/2}}$$

$$\approx .3834$$

(c) $P(x \geq 2) = \frac{1}{2} \int_{2}^{\infty} e^{-x/2} \, dx$

$$= \lim_{a \to \infty} \frac{1}{2} \int_{2}^{a} e^{-x/2} \, dx$$

$$= \lim_{a \to \infty} (-e^{-x/2}) \Big|_{2}^{a}$$

$$= \lim_{a \to \infty} \left(\frac{-1}{e^{a/2}} + \frac{1}{e} \right)$$

$$\approx .3679$$

25. $f(x) = \frac{1}{2} e^{-x/2} ; \; [0, \infty)$

(a) $P(0 \leq x \leq 12) = \frac{1}{2} \int_{0}^{12} e^{-x/2} \, dx$

$$= -e^{-x/2} \Big|_{0}^{12}$$

$$= \frac{-1}{e^{6}} + 1$$

$$\approx .9975$$

(b) $P(12 \leq x \leq 20) = \dfrac{1}{2} \displaystyle\int_{12}^{20} e^{-x/2} \, dx$

$\qquad = -e^{-x/2} \Big|_{12}^{20}$

$\qquad = \dfrac{-1}{e^{10}} + \dfrac{1}{e^{6}}$

$\qquad \approx .0024$

27. $f(x) = \dfrac{1}{2\sqrt{x}};\ [1,\ 4]$

(a) $P(3 \leq x \leq 4) = \displaystyle\int_{3}^{4} \left(\dfrac{1}{2\sqrt{x}} \right) dx$

$\qquad = \dfrac{1}{2} \displaystyle\int_{3}^{4} x^{-1/2} \, dx$

$\qquad = \dfrac{1}{2}(2)x^{1/2} \Big|_{3}^{4}$

$\qquad = (2 - 3^{1/2})$

$\qquad = .2679$

(b) $P(1 \leq x \leq 2) = \displaystyle\int_{1}^{2} \left(\dfrac{1}{2\sqrt{x}} \right) dx$

$\qquad = \dfrac{1}{2}(2)x^{1/2} \Big|_{1}^{2}$

$\qquad = 2^{1/2} - 1$

$\qquad = .4142$

(c) $P(2 \leq x \leq 3) = \displaystyle\int_{2}^{3} \left(\dfrac{1}{2\sqrt{x}} \right) dx$

$\qquad = \dfrac{1}{2}(2)x^{1/2} \Big|_{2}^{3}$

$\qquad = 3^{1/2} - 2^{1/2}$

$\qquad = .3178$

29. $f(x) = \dfrac{8}{7(x-2)^2}$ for $[3,\ 10]$

(a) $P(3 \leq x \leq 4) = \dfrac{8}{7} \displaystyle\int_{3}^{4} (x-2)^{-2} dx$

$\qquad = -\dfrac{8}{7}(x-2)^{-1} \Big|_{3}^{4}$

$\qquad = -\dfrac{8}{7}\left(\dfrac{1}{2} - 1 \right)$

$\qquad = -\dfrac{8}{7}\left(-\dfrac{1}{2} \right)$

$\qquad = \dfrac{4}{7}$

$\qquad \approx .5714$

(b) $P(5 \leq x \leq 10) = \dfrac{8}{7} \displaystyle\int_{5}^{10} (x-2)^{-2} dx$

$\qquad = -\dfrac{8}{7}(x-2)^{-1} \Big|_{5}^{10}$

$\qquad = \dfrac{8}{7}\left(\dfrac{1}{8} - \dfrac{1}{3} \right)$

$\qquad = -\dfrac{8}{7}\left(-\dfrac{5}{24} \right)$

$\qquad = \dfrac{5}{21}$

$\qquad \approx .2381$

31. $f(x) = \dfrac{105}{4x^2};\ [15,\ 35]$

$\displaystyle\int_{a}^{b} \dfrac{105}{4x^2} \, dx = \dfrac{105}{4} \displaystyle\int_{a}^{b} x^{-2} dx$

$\qquad = \dfrac{105}{4}\left(-\dfrac{1}{x} \right) \Big|_{a}^{b}$

$\qquad = \dfrac{105}{4}\left(-\dfrac{1}{b} + \dfrac{1}{a} \right)$

$\qquad = \dfrac{105(b-a)}{4ab}$

(a) $P(x \leq 25) = \displaystyle\int_{15}^{25} \frac{105}{4x^2}\, dx$

$\qquad = \dfrac{105(25 - 15)}{4(15)(25)}$

$\qquad = .7$

(b) $P(x \leq 21) = \displaystyle\int_{15}^{21} \frac{105}{4x^2}\, dx$

$\qquad = \dfrac{105(21 - 15)}{4(15)(21)}$

$\qquad = .5$

(c) $P(21 \leq x \leq 25) =$

$\qquad = P(x \leq 25) - P(x \leq 21)$

$\qquad = .7 - .5$

$\qquad = .2$

Section 10.2

1. $f(x) = \dfrac{1}{4}$; [3, 7]

$E(x) = \mu = \displaystyle\int_{3}^{7} \frac{1}{4}x\, dx = \frac{1}{4}\Big(\frac{x^2}{2}\Big)\Big|_{3}^{7}$

$\qquad = \dfrac{49}{8} - \dfrac{9}{8}$

$\qquad = 5$

$Var(x) = \displaystyle\int_{3}^{7} (x - 5)^2 \Big(\frac{1}{4}\Big)\, dx$

$\qquad = \dfrac{1}{4} \cdot \dfrac{(x - 5)^3}{3}\Big|_{3}^{7}$

$\qquad = \dfrac{8}{12} + \dfrac{8}{12}$

$\qquad = \dfrac{4}{3} \approx 1.33$

$\sigma \approx \sqrt{Var(x)}\ \sqrt{4/3} \approx 1.15$

3. $f(x) = \dfrac{x}{8} - \dfrac{1}{4}$; [2, 6]

$\mu = \displaystyle\int_{2}^{6} x\Big(\frac{x}{8} - \frac{1}{4}\Big)\, dx$

$\qquad = \displaystyle\int_{2}^{6} \Big(\frac{x^2}{8} - \frac{x}{4}\Big)\, dx$

$\qquad = \Big(\dfrac{x^3}{24} - \dfrac{x^2}{8}\Big)\Big|_{2}^{6}$

$\qquad = \Big(\dfrac{216}{24} - \dfrac{36}{8}\Big) - \Big(\dfrac{8}{24} - \dfrac{4}{8}\Big)$

$\qquad = \dfrac{208}{24} - 4$

$\qquad = \dfrac{26}{3} - 4$

$\qquad = \dfrac{14}{3} \approx 4.67$

Use the alternative formula to find

$Var(x) = \displaystyle\int_{2}^{6} x^2\Big(\frac{x}{8} - \frac{1}{4}\Big)\, dx - \Big(\frac{14}{3}\Big)^2$

$\qquad = \displaystyle\int_{2}^{6} \Big(\frac{x^3}{8} - \frac{x^2}{4}\Big)\, dx - \dfrac{196}{9}$

$\qquad = \Big(\dfrac{x^4}{32} - \dfrac{x^3}{12}\Big)\Big|_{2}^{6} - \dfrac{196}{9}$

$\qquad = \Big(\dfrac{1296}{32} - \dfrac{216}{12}\Big) - \Big(\dfrac{16}{32} - \dfrac{8}{12}\Big)$

$\qquad\quad - \dfrac{196}{9}$

$\qquad \approx .89.$

$\sigma = \sqrt{Var(x)} \approx \sqrt{.89} \approx .94$

5. $f(x) = 1 - \dfrac{1}{\sqrt{x}}$; [1, 4]

$\mu = \displaystyle\int_{1}^{4} x(1 - x^{-1/2})\, dx$

$\qquad = \displaystyle\int_{1}^{4} (x - x^{1/2})\, dx$

$= \left(\dfrac{x^2}{2} - \dfrac{2x^{3/2}}{3} \right) \Big|_1^4$

$= \dfrac{16}{2} - \dfrac{16}{3} - \dfrac{1}{2} + \dfrac{2}{3}$

$= \dfrac{17}{6} \approx 2.83$

$\mathrm{Var}(x) = \displaystyle\int_1^4 x^2 (1 - x^{-1/2})\, dx$

$\qquad - \left(\dfrac{17}{6} \right)^2$

$= \displaystyle\int_1^4 (x^2 - x^{3/2})\, dx - \dfrac{289}{36}$

$= \left(\dfrac{x^3}{3} - \dfrac{2x^{5/2}}{5} \right) \Big|_1^4 - \dfrac{289}{36}$

$= \dfrac{64}{3} - \dfrac{64}{5} - \dfrac{1}{3} + \dfrac{2}{5} - \dfrac{289}{36}$

$\approx .57$

$\sigma \approx \sqrt{\mathrm{Var}(x)} \approx .76$

7. $f(x) = 4x^{-5}$; for $[1, \infty)$

$\mu = \displaystyle\int_1^\infty x(4x^{-5})\, dx$

$= \lim_{a \to \infty} \displaystyle\int_1^a 4x^{-4}\, dx$

$= \lim_{a \to \infty} \left(\dfrac{4x^{-3}}{-3} \right) \Big|_1^a$

$= \lim_{a \to \infty} \left(\dfrac{-4}{3a^3} + \dfrac{4}{3} \right)$

$= \dfrac{4}{3} \approx 1.33$

$\mathrm{Var}(x) = \displaystyle\int_1^\infty x^2 (4x^{-5})\, dx - \left(\dfrac{4}{3} \right)^2$

$= \lim_{a \to \infty} \displaystyle\int_1^a 4x^{-3}\, dx - \dfrac{16}{9}$

$= \lim_{a \to \infty} \left(\dfrac{4x^{-2}}{-2} \right) \Big|_1^a - \dfrac{16}{9}$

$= \lim_{a \to \infty} \left(\dfrac{-2}{a^2} + 2 \right) - \dfrac{16}{9}$

$= 2 - \dfrac{16}{9} = \dfrac{2}{9} \approx .22$

$\sigma = \sqrt{\mathrm{Var}(x)} = \sqrt{\dfrac{2}{9}} \approx .47$

11. $f(x) = \dfrac{\sqrt{x}}{18}$; $[0, 9]$

(a) $E(x) = \mu = \displaystyle\int_0^9 \dfrac{x\sqrt{x}}{18}\, dx$

$= \displaystyle\int_0^9 \dfrac{x^{3/2}}{18}\, dx$

$= \dfrac{2x^{5/2}}{90} \Big|_0^9 = \dfrac{x^{5/2}}{45} \Big|_0^9$

$= \dfrac{243}{45} = \dfrac{27}{5} = 5.40$

(b) $\mathrm{Var}(x) = \displaystyle\int_0^9 \dfrac{x^2\sqrt{x}}{18}\, dx - \left(\dfrac{27}{5} \right)^2$

$= \displaystyle\int_0^9 \dfrac{x^{5/2}}{18}\, dx - \left(\dfrac{27}{5} \right)^2$

$= \dfrac{x^{7/2}}{63} \Big|_0^9 - \left(\dfrac{27}{5} \right)^2$

$= \dfrac{2187}{63} - \left(\dfrac{27}{5} \right)^2 \approx 5.55$

(c) $\sigma = \sqrt{\mathrm{Var}(x)} \approx 2.36$

(d) $P(5.40 < x \leq 9)$

$$= \int_{5.4}^{9} \frac{x^{1/2}}{18} \, dx$$

$$= \frac{x^{3/2}}{27} \Big|_{5.4}^{9}$$

$$= \frac{27}{27} - \frac{(5.4)^{1.5}}{27} \approx .54$$

(e) $P(5.40 - 2.36 \leq x \leq 5.40 + 2.36)$

$$= \int_{3.04}^{7.76} \frac{x^{1/2}}{18} \, dx$$

$$= \frac{x^{3/2}}{27} \Big|_{3.04}^{7.76}$$

$$= \frac{7.76^{3/2}}{27} - \frac{3.04^{3/2}}{27}$$

$$\approx .60$$

13. $f(x) = \frac{1}{2}x;$ $[0, 2]$

(a) $E(x) = \mu = \int_{0}^{2} \frac{1}{2}x^2 \, dx = \frac{x^3}{6} \Big|_{0}^{2}$

$$= \frac{8}{6} = \frac{4}{3} \approx 1.33$$

(b) $\text{Var}(x) = \int_{0}^{1} \frac{1}{2}x^3 \, dx - \frac{16}{9}$

$$= \frac{x^4}{8} \Big|_{0}^{2} - \frac{16}{9}$$

$$= 2 - \frac{16}{9} = \frac{2}{9} \approx .22$$

(c) $\sigma = \sqrt{\text{Var}(x)} = \sqrt{\frac{2}{9}} \approx .47$

(d) $P(\frac{4}{3} < x \leq 2) = \int_{4/3}^{2} \frac{x}{2} \, dx$

$$= \frac{x^2}{4} \Big|_{4/3}^{2}$$

$$= 1 - \frac{16}{36} \approx .56$$

(e) $P(\frac{4}{3} - .47 \leq x \leq \frac{4}{3} + .47)$

$$= \int_{.86}^{1.8} \frac{x}{2} \, dx = \frac{x^2}{4} \Big|_{.86}^{1.8}$$

$$= \frac{1.8^2}{4} - \frac{.86^2}{4} \approx .63$$

15. $f(x) = \frac{1}{4};$ $[3, 7]$

(a) $m = \text{median:}$ $\int_{3}^{m} \frac{1}{4} \, dx = \frac{1}{2}$

$$\frac{1}{4}x \Big|_{3}^{m} = \frac{1}{2}$$

$$\frac{m}{4} - \frac{3}{4} = \frac{1}{2}$$

$$m - 3 = 2$$

$$m = 5$$

(b) $E(x) = \mu = 5$ (from Exercise 1)

$$P(x = 5) = \int_{5}^{5} \frac{1}{4} \, dx = 0$$

17. $f(x) = \frac{x}{8} - \frac{1}{4}$; [2, 6]

(a) m = median:

$$\int_2^m \left(\frac{x}{8} - \frac{1}{4}\right) dx = \frac{1}{2}$$

$$\left(\frac{x^2}{16} - \frac{x}{4}\right)\Big|_2^m = \frac{1}{2}$$

$$\frac{m^2}{16} - \frac{m}{4} - \frac{1}{4} + \frac{1}{2} = \frac{1}{2}$$

$$m^2 - 4m - 4 + 8 = 8$$

$$m^2 - 4m - 4 = 0$$

$$m = \frac{4 \pm \sqrt{16 + 16(1)}}{2}$$

Reject $\frac{4 - \sqrt{32}}{2}$ since it is not in

[2, 6].

$$m = \frac{4 + \sqrt{32}}{2} \approx 4.83$$

(b) $E(x) = \mu = \frac{14}{3}$ (from Exercise 3)

$$P\left(\frac{14}{3} \le x \le 4.83\right) = \int_{4.67}^{4.83} \left(\frac{x}{8} - \frac{1}{4}\right) dx$$

$$= \left(\frac{x^2}{16} - \frac{x}{4}\right)\Big|_{4.67}^{4.83}$$

$$= \frac{4.83^2}{16} - \frac{4.83}{4}$$

$$- \frac{4.67^2}{16} + \frac{4.67}{4}$$

$$\approx .055$$

19. $f(x) = 4x^{-5}$; [1, ∞)

(a) m = median:

$$\int_1^m 4x^{-5} \, dx = \frac{1}{2}$$

$$\frac{4x^{-4}}{-4}\Big|_1^m = \frac{1}{2}$$

$$-m^{-4} + 1 = \frac{1}{2}$$

$$1 - \frac{1}{m^4} = \frac{1}{2}$$

$$2m^4 - 2 = m^4$$

$$m^4 = 2$$

$$m = \sqrt[4]{2} \approx 1.19$$

(b) $E(x) = \mu = \frac{4}{3}$ (From Exercise 7)

$$P\left(1.19 \le x \le \frac{4}{3}\right) \approx \int_{1.19}^{1.33} 4x^{-5} dx$$

$$\approx -x^{-4}\Big|_{1.19}^{1.33}$$

$$\approx -\frac{1}{1.33^4} + \frac{1}{1.19^4}$$

$$\approx .18$$

21. $f(x) = \frac{1}{58\sqrt{x}}$; [1, 900]

(a) $E(x) = \int_1^{900} x \cdot \frac{1}{58\sqrt{x}} \, dx$

$$= \frac{1}{58} \int_1^{900} x^{1/2} \, dx$$

$$= \frac{1}{87} x^{3/2} \Big|_1^{900}$$

$$= \frac{1}{87}(900x^{3/2} - 1^{3/2})$$

$$\approx 310.3 \text{ hr}$$

(b) Var(x)

$$= \int_{1}^{900} x^2 \frac{1}{58\sqrt{x}} dx - (310.3)^2$$

$$= \frac{1}{58} \int_{1}^{900} x^{3/2} dx - (310.3)^2$$

$$= \frac{1}{145} x^{5/2} \Big|_{1}^{900} - (310.3)^2$$

$$= \frac{1}{145}(900^{5/2} - 1^{5/2}) - (310.3)^2$$

$$= 71,300.11$$

$$\sigma = \sqrt{71,300.11} \approx 267 \text{ hr}$$

(c) P(310.3 + 267 < x ≤ 900)

$$= \int_{577.3}^{9.00} \frac{1}{58\sqrt{x}} dx$$

$$= \frac{2}{58} x^{1/2} \Big|_{577.3}^{900}$$

$$= \frac{2}{58}(900^{1/2} - 577.3^{1/2})$$

$$\approx .206$$

23. $f(x) = \frac{1}{2}e^{-x/2}$; [0, ∞]

(a) E(x) = μ

$$= \int_{0}^{\infty} x \cdot \frac{1}{2}e^{-x/2} dx$$

$$= \frac{1}{2} \lim_{a \to \infty} \int_{0}^{a} xe^{-x/2} dx$$

Use integration by parts.

$$= \frac{1}{2} \lim_{a \to \infty} (-2xe^{-x/2}) \Big|_{0}^{a}$$

$$- \frac{1}{2} \lim_{a \to \infty} \int_{0}^{a} -2e^{-x/2} dx$$

$$= \frac{1}{2} \lim_{a \to \infty} (-2xe^{-x/2}) \Big|_{0}^{a}$$

$$+ \lim_{a \to \infty} \int_{0}^{a} e^{-x/2} dx$$

$$= -\lim_{a \to \infty} (xe^{-x/2}) \Big|_{0}^{a}$$

$$- 2 \lim_{a \to \infty} (e^{-x/2}) \Big|_{0}^{a}$$

$$= -\lim_{a \to \infty} \left(\frac{a}{e^{a/2}} - \frac{0}{e^0}\right)$$

$$- 2 \lim_{a \to \infty} \left(\frac{1}{e^{a/2}} - \frac{1}{e^0}\right)$$

$$= (0 - 0) - 2(0 - 1)$$

$$= -(0) + 2$$

$$E(x) = 2 \text{ mo}$$

(b) Var(x) $= \frac{1}{2} \int_{0}^{\infty} x^2 e^{-x/2} dx - 2^2$

Use integration by parts twice.

$$\frac{1}{2} \int x^2 e^{-x/2} dx$$

$$= \frac{1}{2}(-2x^2 e^{-x/2} + 4 \int xe^{-x/2} dx)$$

$$= \frac{1}{2}[-2x^2 e^{-x/2} + 4(-2xe^{-x/2}$$

$$+ 2 \int e^{-x/2} dx)]$$

$$= \frac{1}{2}(-2x^2 e^{-x/2} - 8xe^{-x/2} - 16e^{-x/2})$$

$$= -x^2 e^{-x/2} - 4xe^{-x/2} - 8e^{-x/2}$$

Var(x)

$$= \lim_{a \to \infty} (-x^2 e^{-x/2} - 4xe^{-x/2} - 8e^{-x/2}) \Big|_0^a - 4$$

$$= \lim_{a \to \infty} (-a^2 e^{-a/2} - 4ae^{-a/2} - 8e^{-a/2} + 0$$

$$+ 0 + 8e^0) - 4$$

$$= -0 - 0 - 0 + 0 + 0 + 8 - 4$$

Var(x) = 4 mo

$$\sigma = \sqrt{4} = 2 \text{ mo}$$

(c) $P(0 < x < 2) = \dfrac{1}{2} \displaystyle\int_0^2 e^{-x/2} \, dx$

$$= -e^{-x/2} \Big|_0^2$$

$$= -e^{-1} + 1$$

$$\approx .632$$

25. $f(x) = \dfrac{3}{32}(4x - x^2); \ [0, 4]$

(a) $E(x) = \displaystyle\int_0^4 \dfrac{3x}{32}(4x - x^2) \, dx$

$$= \int_0^4 \left(\dfrac{3}{8}x^2 - \dfrac{3x^3}{32}\right) dx$$

$$= \left(\dfrac{x^3}{8} - \dfrac{3x^4}{128}\right)\Big|_0^4$$

$$= 8 - \dfrac{768}{128}$$

$$= 2$$

(b) $Var(x) = \displaystyle\int_0^4 \dfrac{3x^2}{32}(4x - x^2) \, dx - 4$

$$= \int_0^4 \left(\dfrac{3x^3}{8} - \dfrac{3x^4}{32}\right) dx - 4$$

$$= \left(\dfrac{3x^4}{32} - \dfrac{3x^5}{160}\right)\Big|_0^4$$

$$= \dfrac{3(4^4)}{32} - \dfrac{3(4^5)}{160} - 4$$

$$= 24 - 19.2 - 4$$

$$= .8$$

$$\sigma = \sqrt{Var(x)} \approx .89$$

(c) $P(2 - .89 \le x \le 2 + .89)$

$$\approx \int_{1.11}^{2.89} \dfrac{3}{32}(4x - x^2) \, dx$$

$$= \left(\dfrac{3}{16}x^2 - \dfrac{x^3}{32}\right)\Big|_{1.11}^{2.89}$$

$$= \dfrac{3(2.89)^2}{16} - \dfrac{2.89^3}{32} - \dfrac{3(1.11)^2}{16}$$

$$+ \dfrac{1.11^3}{32}$$

$$\approx .62$$

27. $f(x) = \dfrac{5.5 - x}{15}$ for $[0, 5]$

(a) $\mu = \displaystyle\int_0^5 x\left(\dfrac{5.5 - x}{15}\right) dx$

$$= \int_0^5 \left(\dfrac{5.5}{15}x - \dfrac{1}{15}x^2\right) dx$$

$$= \left(\dfrac{5.5}{30}x^2 - \dfrac{1}{45}x^3\right)\Big|_0^5$$

$$= \left(\dfrac{5.5}{30} \cdot 25 - \dfrac{1}{45} \cdot 125\right) - 0$$

$$\approx 1.806$$

(b) Var(x)

$$= \int_0^5 x^2\left(\frac{5.5 - x}{15}\right)dx - \mu^2$$

$$= \int_0^5 \left(\frac{5.5}{15}x^2 - \frac{1}{15}x^3\right)dx - \mu^2$$

$$= \left(\frac{5.5}{45}x^3 - \frac{1}{60}x^4\right)\Big|_0^5 - \mu^2$$

$$= \frac{5.5}{45}\cdot 125 - \frac{1}{60}\cdot 625 - 0 - \mu^2$$

$$\approx 1.60108$$

$$\sigma = \sqrt{\text{Var}(x)} \approx 1.265$$

(c) $P(x \le \mu - \sigma)$

$$= P(x \le 1.806 - 1.265)$$

$$= P(x \le .541)$$

$$= \int_0^{.541} \frac{5.5 - x}{15}\, dx$$

$$= \left(\frac{5.5}{15}x - \frac{1}{30}x^2\right)\Big|_0^{.541}$$

$$= \left(\frac{5.5}{15}(.541) - \frac{1}{30}(.541)^2 - 0\right)$$

$$\approx .1886$$

Section 10.3

1. $f(x) = \frac{5}{4}$ for $[4, 4.8]$

   This is a uniform distribution:
   $a = 4$, $b = 4.8$.

   (a) $\mu = \frac{1}{2}(4.8 + 4) = \frac{1}{2}(8.8)$

   $$= 4.4 \text{ cm}$$

(b) $\sigma = \frac{1}{\sqrt{12}}(4.8 - 4)$

$$= \frac{1}{\sqrt{12}}(.8)$$

$$\approx .23 \text{ cm}$$

(c) $P(4.4 < x < 4.4 + .23)$

$$= \int_{4.4}^{4.63} \frac{5}{4}\, dx$$

$$= \frac{5}{4}x \Big|_{4.4}^{4.63} = \frac{5}{4}(4.63 - 4.4)$$

$$\approx .29$$

3. $f(t) = .03e^{-.03t}$ for $[0, \infty)$

   This is an exponential distribution:
   $a = .03$.

   (a) $\mu = \frac{1}{.03} \approx 33.33$ yr

   (b) $\sigma = \frac{1}{.03} \approx 33.33$ yr

   (c) $P(33.33 < x < 33.33 + 33.33)$

   $$= \int_{33.33}^{66.66} .03e^{-.03t}\, dt$$

   $$= -e^{-.03t}\Big|_{33.33}^{66.66}$$

   $$= \frac{-1}{e^{.03(66.66)}} + \frac{-1}{e^{.03(33.33)}}$$

   $$\approx .23$$

5. $f(x) = e^{-t}$ for $[0, \infty)$

   This is an exponential distribution:
   $a = 1$.

   (a) $\mu = \frac{1}{1} = 1$ day

   (b) $\sigma = \frac{1}{1} = 1$ day

(c) $P(1 < x < 1 + 1)$

$$= \int_{1}^{2} e^{-t} dt$$

$$= -e^{-t} \Big|_{1}^{2}$$

$$= -\frac{1}{e^2} + \frac{1}{e}$$

$$\approx .23$$

In Exercises 7–13, use the table in the appendix for areas under the normal curve.

7. $z = 3.50$

Area to the left of $z = 3.50$ is .9998. Given mean $\mu = z = 0$, so area to left of μ is .5. Area between μ and z is

$$.9998 - .5 = .4998.$$

Therefore, this area represents 49.98% of total area under normal curve.

9. Between $z = 1.28$ and $z = 2.05$
Area to left of $z = 2.05$ is .9798 and area to left of $z = 1.28$ is .8997.

$$.9798 - .8997 = .0801$$

Percent of total area = 8.01%.

11. Since 10% = .10, the z-score that corresponds to the area of .10 to the left of z is −1.28.

13. 18% of the total area to the right of z means $1 - .18$ of the total area is to the left of z.

$$1 - .18 = .82$$

The closest z-score that corresponds to the area of .82 is .92.

19. Let m be the median of the exponential distribution $f(x) = ae^{-ax}$ for $[0, \infty)$.

$$\int_{0}^{m} ae^{-ax} dx = .5$$

$$-e^{-ax} \Big|_{0}^{m} = .5$$

$$-e^{-am} + 1 = .5$$

$$.5 = e^{-am}$$

$$-am = \ln .5$$

$$m = -\frac{\ln .5}{a}$$

or
$$-am = \ln \frac{1}{2}$$

$$-am = -\ln 2$$

$$m = \frac{\ln 2}{a}$$

21. For a uniform distribution,

$$f(x) = \frac{1}{b - a} \text{ for } [a, b].$$

Thus, we have

$$f(x) = \frac{1}{85 - 10} = \frac{1}{75} \text{ for } [10, 85].$$

(a) $\mu = \frac{1}{2}(10 + 85) = \frac{1}{2}(95)$

$$= 47.5 \text{ thousands}$$

Therefore, the agent sells $47,500 in insurance.

(b) $P(50 < x < 85) = \displaystyle\int_{50}^{85} \frac{1}{75} \, dx$

$$= \frac{x}{75}\Big|_{50}^{85}$$

$$= \frac{87}{75} - \frac{50}{75}$$

$$= \frac{35}{75}$$

$$= .47$$

23. **(a)** Since we have an exponential distribution with $\mu = 4.25$,

$$\mu = \frac{1}{a} = 4.25.$$

$$a = .235.$$

Therefore,

$$f(x) = .235e^{-.235x} \text{ on } [0, \infty).$$

(b) $P(x > 10)$

$$= \int_{10}^{\infty} .235e^{-.235x}$$

$$= \lim_{a \to \infty} \int_{10}^{a} .235e^{-.235x}$$

$$= \lim_{a \to \infty} (-e^{-.235x})\Big|_{10}^{a}$$

$$= \lim_{a \to \infty} \left(-\frac{1}{e^{.235a}} + \frac{1}{e^{2.35}}\right)$$

$$= \frac{1}{e^{2.35}} = .095$$

25. **(a)** $\mu = 2.5$, $\sigma = .2$, $x = 2.7$

$$z = \frac{2.7 - 2.5}{.2} = 1$$

Area to the right of $z = 1$ is

$$1 - .8413 = .1587.$$

Probability $= .1587$

(b) Within 1.2 standard deviations of the mean is the area between $z = -1.2$ and $z = 1.2$.

Area to left of $1.2 = .8849$

Area to the left of $-1.2 = .1151$

$$.8849 - .1151 = .7698$$

Probability $= .7698$

27. For a uniform distribution,

$$f(x) = \frac{1}{b - a} \text{ for } [a, b].$$

$$f(x) = \frac{1}{36 - 20} = \frac{1}{16} \text{ for } [20, 36]$$

(a) $\mu = \frac{1}{2}(20 + 36) = \frac{1}{2}(56)$

$$= 28 \text{ days}$$

(b) $P(30 < x \le 36)$

$$= \int_{30}^{36} \frac{1}{16} \, dx = \frac{1}{16}x \Big|_{30}^{36}$$

$$= \frac{1}{16}(36 - 30)$$

$$= .375$$

29. We have an exponential distribution, with $a = 1$.

$$f(t) = e^{-t}, \ [0, \infty)$$

(a) $\mu = \frac{1}{1} = 1 \text{ hr}$

(b) $P(t < 30 \text{ min})$

$\quad = P(t < .5 \text{ hr})$

$\quad = \displaystyle\int_0^{.5} e^{-t} \, dt$

$\quad = -e^{-t} \Big|_0^{.5}$

$\quad = 1 - e^{-.5} \approx .39$

31. $f(x) = ae^{-ax}$ for $[0, \infty)$

Since $\mu = 25$ and $\mu = \dfrac{1}{a}$,

$\quad a = \dfrac{1}{25} = .04.$

Thus, $f(x) = .04e^{-.04x}$.

(a) We must find t such that $P(x \leq t) = .90$.

$\quad \displaystyle\int_0^t .04e^{-.04x} dx = .90$

$\quad\quad -e^{-.04x} \Big|_0^t = .90$

$\quad\quad -e^{-.04t} + 1 = .90$

$\quad\quad\quad .10 = -e^{-.04t}$

$\quad\quad\quad -.04t = \ln .10$

$\quad\quad\quad\quad t = \dfrac{\ln .10}{-.04}$

$\quad\quad\quad\quad t \approx 57.56$

The longest time within which the predator will be 90% certain of finding a prey is approximately 58 min.

(b) $P(x \geq 60)$

$\quad = \displaystyle\int_{60}^{\infty} .04e^{-.04x} \, dx$

$\quad = \lim_{b \to \infty} \displaystyle\int_{60}^{b} .04e^{-.04x} \, dx$

$\quad = \lim_{b \to \infty} (-e^{-.04x}) \Big|_{60}^{b}$

$\quad = \lim_{b \to \infty} [-e^{-.04b} + e^{-(.04)(60)}]$

$\quad = 0 + e^{-2.4}$

$\quad \approx .0907$

The probability that the predator will have to spend more than one hour looking for a prey is approximately .09.

Exercises 33–37 are to be solved using a computer. The answers may vary depending on the computer and the software that is used.

33. $\displaystyle\int_0^{50} .5e^{-.5x} \, dx \approx 1.00002$

35. $\displaystyle\int_0^{50} .5x^2 e^{-.5x} \, dx = 8.000506$

37. $\displaystyle\int_{-\infty}^{\infty} \dfrac{1}{\sqrt{2\pi}} e^{-x^2/2} \, dx$

(a) $\mu = 2.7416 \times 10^8 \approx 0$

(b) $\sigma = .999433 \approx 1$

Chapter 10   Review Exercises

1. In a probability function, the y-values (or function values) represent probabilities.

3. A probability density function f for [a, b] must satisfy the following two conditions:

    (1) $\int_a^b f(x)\ dx = 1$

    (2) $f(x) \geq 0$ for all x in the interval [a, b].

5. $f(x) = \sqrt{x} \geq 0$; [4, 9]

$$\int_4^9 x^{1/2}\ dx = \frac{2}{3}x^{3/2}\ \Big|_4^9$$

$$= \frac{2}{3}(27 - 8)$$

$$= \frac{38}{3} \neq 1$$

f(x) is not a probability density function.

7. $f(x) = e^{-x}$; [0, ∞]

$$\int_0^\infty e^{-x}\ dx = \lim_{b \to \infty} \int_0^b e^{-x}\ dx$$

$$= \lim_{b \to \infty} -e^{-x}\ \Big|_0^b$$

$$= \lim_{b \to \infty} \left(1 - \frac{1}{e^b}\right) = 1$$

f(x) > 0 for all x.
f(x) is a probability density function.

9. $f(x) = kx^2$; [0, 3]

$$\int_0^3 kx^2\ dx = \frac{kx^3}{3}\ \Big|_0^3$$

$$= 9k$$

Since f(x) is a probability density function,

$$9k = 1$$

$$k = \frac{1}{9}.$$

11. $f(x) = \frac{1}{10}$ for [10, 20]

    (a) $P(x \leq 12)$

    $$= \int_{10}^{12} \frac{1}{10}\ dx$$

    $$= \frac{x}{10}\Big|_{10}^{12}$$

    $$= \frac{1}{5} = .2$$

    (b) $P\left(x \geq \frac{31}{2}\right)$

    $$= \int_{31/2}^{20} \frac{1}{10}\ dx$$

    $$= \frac{x}{20}\Big|_{31/2}^{20}$$

    $$= \frac{9}{20} = .45$$

    (c) $P(10.8 \leq x \leq 16.2)$

    $$= \int_{10.8}^{16.2} \frac{1}{10}\ dx$$

    $$= \frac{x}{10}\Big|_{10.8}^{16.2} = .54$$

13. The distribution that is tallest or most peaked has the smallest standard deviation. This is the distribution pictured in graph (b).

15. $f(x) = \frac{2}{9}(x - 2)$; [2, 5]

$$\mu = \int_2^5 \frac{2x}{9}(x - 2)\,dx$$

$$= \int_2^5 \frac{2}{9}(x^2 - 2x)\,dx$$

$$= \frac{2}{9}\left(\frac{x^3}{3} - x^2\right)\Big|_2^5$$

$$= \frac{2}{9}\left(\frac{125}{3} - 25 - \frac{8}{3} + 4\right) = 4$$

$$\text{Var}(x) = \int_2^5 \frac{2x^2}{9}(x - 2)\,dx - (4)^2$$

$$= \int_2^5 \frac{2}{9}(x^3 - 2x^2)\,dx - 16$$

$$= \frac{2}{9}\left(\frac{x^4}{4} - \frac{2x^3}{3}\right)\Big|_2^5 - 16$$

$$= \frac{2}{9}\left(\frac{625}{4} - \frac{250}{3} - 4 + \frac{8}{3}\right) - 16$$

$$= .5$$

$$\sigma = \sqrt{.5} \approx .71$$

17. $f(x) = 5x^{-6}$; [1, ∞)

$$\mu = \int_1^\infty x \cdot 5x^{-6}\,dx$$

$$= \int_1^\infty 5x^{-5}\,dx$$

$$= \lim_{b \to \infty} \int_1^b 5x^{-5}\,dx$$

$$= \lim_{b \to \infty} \frac{5x^{-4}}{-4}\Big|_1^b$$

$$= \lim_{b \to \infty} \frac{5}{4}\left(1 - \frac{1}{b^4}\right)$$

$$= \frac{5}{4}$$

$$\text{Var}(x) = \int_1^\infty x^2 \cdot 5x^{-6}\,dx - \left(\frac{5}{4}\right)^2$$

$$= \lim_{b \to \infty} \int_1^b 5x^{-4}\,dx - \frac{25}{16}$$

$$= \lim_{b \to \infty} \frac{5x^{-3}}{-3}\Big|_1^b - \frac{25}{16}$$

$$= \lim_{b \to \infty} \frac{5}{3}\left(1 - \frac{1}{b^3}\right) - \frac{25}{16}$$

$$= \frac{5}{3} - \frac{25}{16} = \frac{5}{48} \approx .10$$

$$\sigma \approx \sqrt{\text{Var}(x)} \approx .32$$

19. $f(x) = 4x - 3x^2$; [0, 1]

$$\mu = \int_0^1 x(4x - 3x^2)\,dx$$

$$= \int_0^1 (4x^2 - 3x^3)\,dx$$

$$= \left(\frac{4x^3}{3} - \frac{3x^4}{4}\right)\Big|_0^1$$

$$= \frac{4}{3} - \frac{3}{4} = \frac{7}{12} \approx .58$$

Find m such that

$$\int_0^m (4x - 3x^2)\,dx = \frac{1}{2}.$$

$$\int_0^m (4x - 3x^2)\,dx = (2x^2 - x^3)\Big|_0^m$$

$$= 2m^2 - m^3 = \frac{1}{2}$$

Therefore,

$$2m^3 - 4m^2 + 1 = 0.$$

This equation has no rational roots, but trial and error used with synthetic division reveals that $m \approx -.44$, .60, and 1.87. The only one of these in [0, 1] is .60.

$$P\left(\frac{7}{12} < x < .60\right)$$

$$= \int_{7/12}^{.60} (4x - 3x^2)\,dx$$

$$= 2x^2 - x^3 \Big|_{7/12}^{.60}$$

$$= 2(.60)^2 - (.60)^3 - 2\left(\frac{7}{12}\right)^2$$

$$+ \left(\frac{7}{12}\right)^3 \approx .02$$

21. $f(x) = .01e^{-.01x}$ for $[0, \infty)$ is an exponential distribution.

(a) $\mu = \dfrac{1}{.01} = 100$

(b) $\sigma = \dfrac{1}{.01} = 100$

(c) $P(100 - 100 < x < 100 + 100)$

$$= P(0 < x < 200)$$

$$= \int_0^{200} .01e^{-.01x}\,dx$$

$$= -e^{-.01x}\Big|_0^{200}$$

$$= 1 - e^{-2} \approx .86$$

For Exercises 23–29, use the table in the Appendix for area under the normal curve.

23. Area to the left of z = -.49 is .3121.
Percent of area is 31.21%.

25. Area between z = -.98 and z = -.15 is

$$.4404 - .1635 = .2769.$$

Percent of area is 27.69%.

27. Region up to 1.2 standard deviations below the mean is region to the left of z = -1.2. Area is .1151 so percent of area is 11.51%.

29. 52% of area is to the right implies that 48% is to the left.

$$P(z < a) = .48 \text{ for } a = -.05$$

Thus, 52% of the area lies to the right of z = -.05.

31. $f(t) = \frac{5}{112}(1 - t^{-3/2})$; [, 25]

P(No repairs in years 1–3)

  = P(First repair needed in
      years 4–25)

  $= \int_{4}^{25} \frac{5}{112}(1 - t^{-3/2})dt$

  $= \frac{5}{112}(t + 2t^{-1/2})\Big|_{4}^{25}$

  $= \frac{5}{112}\Big[25 + \frac{2}{5} - 4 - 1\Big]$

  $= \frac{51}{56} \approx .911$

33. $f(x) = \frac{1}{6}e^{-x/6}$ for [0, ∞) is an

exponential distribution.

(a) $\mu = \frac{1}{1/6} = 6$

(b) $\sigma = \frac{1}{1/6} = 6$

(c) $P(x > 6) = \int_{6}^{\infty} \frac{1}{6}e^{-x/6} dx$

  $= \lim_{b \to \infty} \int_{6}^{b} \frac{1}{6}e^{-x/6} dx$

  $= \lim_{b \to \infty} -e^{-x/6}\Big|_{6}^{b}$

  $= \lim_{b \to \infty} \left(e^{-1} - \frac{1}{e^{b/6}}\right)$

  $= \frac{1}{e} \approx .37$

35. $f(x) = .01e^{-.01x}$ for [0, ∞) is an

exponential distribution.

P(0 ≤ x ≤ 100)

  $= \int_{0}^{100} .01e^{-.01x} dx$

  $= -e^{-.01x}\Big|_{0}^{100}$

  $= 1 - \frac{1}{e} \approx .63$

37. $f(x) = \frac{6}{15,925}(x^2 + x)$ for [20, 25]

(a)

  $\mu = \int_{20}^{25} \frac{6}{15,925}(x^2 + x)dx$

  $= \frac{6}{15,925}\int_{20}^{25} (x^3 + x^2)dx$

  $= \frac{6}{15,925}\Big[\frac{x^4}{4} + \frac{x^3}{3}\Big]\Big|_{20}^{25}$

  $= \frac{6}{15,925}$

  $\cdot \Big[\frac{(25)^4}{4} + \frac{(25)^3}{3} - \frac{(20)^4}{4} - \frac{(20)^3}{3}\Big]$

  ≈ 22.68°C

(b)

P(x < μ)

  $= \int_{20}^{22.68} \frac{6}{15,925}(x^2 + x)dx$

  $= \frac{6}{15,925}\Big[\frac{x^3}{3} + \frac{x^2}{2}\Big]\Big|_{20}^{22.68}$

  $= \frac{6}{15,925}$

  $\cdot \Big[\frac{(22.68)^3}{3} + \frac{(22.68)^2}{2} - \frac{(20)^3}{3} - \frac{(20)^2}{2}\Big]$

  ≈ .48

39. Normal distribution, $\mu = 2.4$ g,

$\sigma = .4$ g, x = tension

$$P(x < 1.9) = P\left(\frac{x - 2.4}{.4} < \frac{1.9 - 2.4}{.4}\right)$$

$$= P(z < -1.25)$$

$$= .1056$$

41. Normal distribution, $\mu = 40$,

$\sigma = 13$, x = "take"

$$P(x > 50) = P\left(\frac{x - 40}{13} > \frac{50 - 40}{13}\right)$$

$$= P(z > .77)$$

$$= 1 - P(z \le .77)$$

$$= 1 - .7794$$

$$= .2206$$

43. $f(x) = e^{-x}$ for $[0, \infty)$

$f(x) = 1e^{-1x}$

(a) This is an exponential distribution with a = 1.

(b) The domain of f is $[0, \infty)$.

The range of f is $(0, 1]$.

(c) See the graph in the answer section of your textbook.

(d) For an exponential distribution, $\mu = \frac{1}{a}$ and $\sigma = \frac{1}{a}$.

Thus,

$$\mu = \frac{1}{1} = 1 \quad \text{and} \quad \sigma = \frac{1}{1} = 1.$$

(e) $P(\mu - \sigma \le x \le \mu + \sigma) = P(1 - 1 \le x \le 1 + 1)$

$$= P(0 \le x \le 2)$$

$$= \int_0^2 e^{-x} dx$$

$$= -e^{-x} \Big|_0^2$$

$$= -e^{-2} + 1$$

$$\approx .86$$

Extended Application

1. Expected profit

$$= (S - Cp)\int_a^T Df(D)\,dD + (S - C_p)(T)\int_T^b f(D)\,dD - C_C \int_a^T (T - D)f(D)\,dD - C_A A$$

$$= (10,000 - 5000)\int_{500}^T D\frac{1}{1000}\,dD + (10,000 - 5000)T\int_T^{1500} \frac{1}{1000}\,dD$$

$$- 3000\int_{500}^T (T - D)\frac{1}{1000}\,dD - 100A$$

$$= 5\cdot\frac{D^2}{2}\Big|_{500}^T + 5T\cdot D\Big|_T^{1500} - 3\left(TD - \frac{D^2}{2}\right)\Big|_{500}^T - 100A$$

$$= 2.5T^2 - 625,000 + 7500T - 5T^2 - 3\left(T^2 - \frac{T^2}{2}\right) + 3(500T - 125,000) - 100A$$

$$= 2.5T^2 - 625,000 + 7500T - 5T^2 - 1.5T^2 + 1500T - 375,000 - 100A$$

Expected profit $= -4.0T^2 + 9000T - 1,000,000 - 100A$

Since $T = .1A + 200$,

Expected profit $= -4.0(.1A + 200)^2 + 9000(.1A + 200) - 1,000,000 - 100A$

$$= -4(.01A^2 + 40A + 40,000) + 900A + 1,800,000 - 1,000,000 - 100A$$

$$= -.04A^2 - 160A - 160,000 + 900A + 1,800,000 - 1,000,000 - 100A$$

$$= -.04A^2 + 640A + 640,000.$$

2. Let $P(A) = -.04A^2 + 640A + 640,000$.   We want to maximize this profit function.

$$P'(A) = -.08A + 640$$

If $-.08A + 640 = 0$, then $A = 8000$.

The graph of this function is a parabola opening downward.   It has a maximum value when $A = 8000$.   Thus, planting 8000 acres will maximize profit.

CHAPTER 10   TEST

[10.1]   In Exercises 1–4, which of the functions are probability density functions on the given intervals? If a function is not a probability density function, tell why.

1. $f(x) = 4$; $[3, 7]$ 2. $f(x) = \frac{1}{5}$; $[5, 10]$

3. $f(x) = \frac{1}{2}(x - 4)$; $[4, 6]$ 4. $f(x) = e^{-3x}$; $[0, \infty)$

[10.1]   5.   Find a value of k that will make $f(x) = kx^{1/3}$ on the interval $[1, 8]$ a probability density function.

[10.1]   6.   The probability density function for a random variable x is defined by

$$f(x) = .2 \quad \text{for } [12, 17].$$

Find the following probabilities.

(a) $P(x \le 14)$ **(b)** $P(13 < x < 16)$ **(c)** $P(x < 15)$

[10.1]   7.   The probability density function of a random variable x is defined by

$$f(x) = 3x^2 \quad \text{for } [0, 1].$$

Find the following probabilities.

(a) $P(x \le 1)$ **(b)** $P(.2 \le x \le .7)$ **(c)** $P(x \ge .3)$

[10.1]   8.   The time in years until a particular radioactive particle decays is a random variable with probability density function

$$f(t) = .04e^{-.04t} \quad \text{for } [0, \infty).$$

Find the probability that a certain such particle decays in less than 60 yr.

[10.2]   **Find the expected value and the standard deviation for each probability density function defined as follows.**

9. $f(x) = \frac{1}{4}$; $[1, 5]$ 10. $f(x) = 2x$; $[0, 1]$

11. $f(x) = \frac{3}{2}(x - 1)^2$; $[0, 2]$ 12. $f(x) = \frac{1}{4\sqrt{x}}$; $[1, 9]$

[10.2]   13. Suppose that the time spent waiting in the waiting room of
Dr. Jones' office is a random variable with probability
density function defined by

$$f(x) = \frac{x(30 - x)}{4500} \quad \text{for } [0, 30].$$

Find the average waiting time. Find the standard deviation.

[10.2]   **For the probability density functions defined in Exercises 14 and 15
find (a) the probability that the value of the random variable will
be less than the mean and (b) the probability that the value of the
random variable will be within one standard deviation of the mean.**

14. $f(x) = 2(1 - x)$ for $[0, 1]$ 15. $f(x) = 1 - x^{-1/2}$ for $[1, 4]$

[10.2]   16. For the probability density function defined by

$$f(x) = \frac{1}{8}x \quad \text{for } [0, 4].$$

find the probability that the value of the random variable will be
between the median and the mean.

[10.3]   17. The maximum daily temperature in a certain city is uniformly distributed over the interval $[10, 40]$.

(a) What is the expected maximum temperature?

(b) What is the probability that the temperature
will be less than 25°?

[10.3]   18. The number of repairs required by a new product each month is
exponentially distributed with an average of 6.

(a) What is the probability density function for this distribution?

(b) What is the probability that the number of repairs is between
3 and 5?

(c) What is the standard deviation?

[10.3]   **Find the percent of area under a normal curve for each of the following.**

19. The region to the left of $z = -.57$

20. The region to the right of $z = 1.49$

21. The region between $z = 1.78$ and $z = -1.30$

[10.3]   Find a z–score satisfying the condition for a normal curve in each of the following.

22.   19% of the area to the left of z.

23.   91% of the area is to the right of z.

[10.3]   24.   The heights of the students in a calculus class are normally distributed with mean 65 inches and standard deviation 10 inches.

(a)   Find the probability that a student's height is between 60 inches and 70 inches.

(b)   Find the probability that a student's height is more than 72 inches.

[10.3]   25.   On a certain day, the amount spent for lunch in a particular restaurant was $3.75 with a standard deviation of $.25. What percent of the customers spent between $3.60 and $4.10? Assume that the amount spent is normally distributed.

[10.1]   26.   Why is the term "random" used in naming a random variable?

[10.1]   27. $\int_0^1 (3 - 4x)\,dx = 1$ but $f(x) = 3 - 4x$ for $[0, 1]$ is not a probability density function. Explain.

[10.2]   28.   What does the variance of a probability distribution tell about the distribution?

[10.3]   29.   Explain why a normal distribution is sometimes called a "bell–shaped" distribution.

[10.3]   30.   If the table for the area under a normal curve is not given, how can probabilities for a normal distribution be found?

CHAPTER 10 TEST ANSWERS

1. No; $\int_{3}^{7} 4 \, dx \neq 1$ 2. Yes   3. Yes   4. No; $\int_{0}^{\infty} e^{-3x} \, dx \neq 1$

5. 4/45   6. (a) .4 (b) .6 (c) .6   7. (a) 1 (b) .335 (c) .973

8. .909   9. 3; 1.155   10. .667; .236   11. 1; .775

12. 4.333; 2.329   13. 15; 6.708   14. (a) .556 (b) .630

15. (a) .467 (b) .606   16. .055   17. (a) 25° (b) .5

18. (a) $f(x) = .167e^{-.167x}$ (b) .172 (c) 6   19. 28.43%   20. 6.81%

21. 86.57%   22. −.88   23. −1.34   24. (a) .383 (b) .242   25. 64.49%

26. The variable's value is determined by chance due to a random event.

27. On [0, 1] $f(x) = 3 - 4x$ takes on negative values. For example, $f(.9) = -.6$.

28. The variance of a probability distribution gives an indication of how closely the values of the distribution cluster about the mean. A small value of the variance tells that the values are closely grouped about the mean.

29. The graph of a probability distribution which is normal has a shape that looks like a bell.

30. Use Simpson's rule to approximate an integral to find probabilities for a normal distribution.

CHAPTER 11   THE TRIGONOMETRIC FUNCTIONS

Section 11.1

1. $60° = 60\left(\frac{\pi}{180}\right) = \frac{\pi}{3}$

3. $150° = 150\left(\frac{\pi}{180}\right) = \frac{5\pi}{6}$

5. $210° = 210\left(\frac{\pi}{180}\right) = \frac{7\pi}{6}$

7. $390° = 390\left(\frac{\pi}{180}\right) = \frac{13\pi}{6}$

9. $\frac{7\pi}{4} = \frac{7\pi}{4}\left(\frac{180°}{\pi}\right) = 315°$

11. $\frac{11\pi}{6} = \frac{11\pi}{6}\left(\frac{180°}{\pi}\right) = 330°$

13. $\frac{8\pi}{5} = \frac{8\pi}{5}\left(\frac{180°}{\pi}\right) = 288°$

15. $\frac{4\pi}{15} = \frac{4\pi}{15}\left(\frac{180°}{\pi}\right) = 48°$

17. Let α = the angle with terminal side through $(-3, 4)$. Then $x = -3$, $y = 4$, and
$$r = \sqrt{x^2 + y^2} = \sqrt{(-3)^2 + (4)^2}$$
$$= \sqrt{25} = 5.$$

$\sin \alpha = \dfrac{y}{r} = \dfrac{4}{5}$ $\qquad$ $\cot \alpha = \dfrac{x}{y} = -\dfrac{3}{4}$

$\cos \alpha = \dfrac{x}{r} = -\dfrac{3}{5}$ $\qquad$ $\sec \alpha = \dfrac{r}{x} = -\dfrac{5}{3}$

$\tan \alpha = \dfrac{y}{x} = -\dfrac{4}{3}$ $\qquad$ $\csc \alpha = \dfrac{r}{y} = \dfrac{5}{4}$

19. Let α = the angle with terminal side through $(6, 8)$. Then $x = 6$, $y = 8$, and
$$r = \sqrt{x^2 + y^2} = \sqrt{36 + 64}$$
$$= \sqrt{100} = 10.$$

$\sin \alpha = \dfrac{y}{r} = \dfrac{4}{5}$ $\qquad$ $\cot \alpha = \dfrac{x}{y} = \dfrac{3}{4}$

$\cos \alpha = \dfrac{x}{r} = \dfrac{3}{5}$ $\qquad$ $\sec \alpha = \dfrac{r}{x} = \dfrac{5}{3}$

$\tan \alpha = \dfrac{y}{x} = \dfrac{4}{3}$ $\qquad$ $\csc \alpha = \dfrac{r}{y} = \dfrac{5}{4}$

21. In quadrant I, all six trigonometric functions are positive, so their sign is +.

23. In quadrant III, $x < 0$ and $y < 0$. Furthermore, $r > 0$.

$\sin \theta = \dfrac{y}{r} < 0$, so the sign is $-$.

$\cos \theta = \dfrac{x}{r} < 0$, so the sign is $-$.

$\tan \theta = \dfrac{y}{x} > 0$, so the sign is $+$.

$\cot \theta = \dfrac{x}{y} > 0$, so the sign is $+$.

$\sec \theta = \dfrac{r}{x} < 0$, so the sign is $-$.

$\csc \theta = \dfrac{r}{y} < 0$, so the sign is $-$.

25. When an angle θ of 30° is drawn in standard position, one choice of a point on its terminal side is $(x, y) = (\sqrt{3}, 1)$. Then
$$r = \sqrt{x^2 + y^2} = \sqrt{3 + 1} = 2.$$

$$\tan \theta = \frac{y}{x} = \frac{1}{\sqrt{3}} = \frac{\sqrt{3}}{3}$$

$$\cot \theta = \frac{x}{y} = \sqrt{3}$$

$$\csc \theta = \frac{r}{y} = 2$$

27. When an angle θ of 60° is drawn in standard position, one choice of a point on its terminal side is $(x, y) = (1, \sqrt{3})$. Then

$$r = \sqrt{x^2 + y^2} = \sqrt{1 + 3} = 2.$$

$$\sin \theta = \frac{y}{r} = \frac{\sqrt{3}}{2}$$

$$\cot \theta = \frac{x}{y} = \frac{1}{\sqrt{3}} = \frac{\sqrt{3}}{3}$$

$$\csc \theta = \frac{r}{y} = \frac{1}{\sqrt{3}} = \frac{2\sqrt{3}}{3}$$

29. When an angle θ of 135° is drawn in standard position, one choice of a point on its terminal side is $(x, y) = (-1, 1)$. Then

$$r = \sqrt{x^2 + y^2} = \sqrt{1 + 1} = \sqrt{2}.$$

$$\tan \theta = \frac{y}{x} = -1$$

$$\cot \theta = \frac{x}{y} = -1$$

31. When an angle θ of 210° is drawn in standard position, one choice of a point on its terminal side is $(x, y) = (-\sqrt{3}, -1)$. Then

$$r = \sqrt{x^2 + y^2} = \sqrt{3 + 1} = 2.$$

$$\cos \theta = \frac{x}{r} = -\frac{\sqrt{3}}{2}$$

$$\sec \theta = \frac{r}{x} = \frac{2}{-\sqrt{3}} = -\frac{2\sqrt{3}}{3}$$

33. When an angle of $\frac{\pi}{3}$ is drawn in standard position, one choice of a point on its terminal side is $(x, y) = (1, \sqrt{3})$. Then

$$r = \sqrt{x^2 + y^2} = \sqrt{1 + 3} = 2.$$

$$\sin \frac{\pi}{3} = \frac{y}{r} = \frac{\sqrt{3}}{2}$$

35. When an angle of $\frac{\pi}{4}$ is drawn in standard position, one choice of a point on its terminal side is $(x, y) = (1, 1)$.

$$\tan \frac{\pi}{4} = \frac{y}{x} = 1$$

37. When an angle of $\frac{\pi}{6}$ is drawn in standard position, one choice of a point on its terminal side. is $(x, y) = (\sqrt{3}, 1)$. Then

$$r = \sqrt{x^2 + y^2} = \sqrt{3 + 1} = 2.$$

$$\sec \frac{\pi}{6} = \frac{r}{x} = \frac{2}{\sqrt{3}} = \frac{2\sqrt{3}}{3}$$

39. When an angle of 3π is drawn in standard position, one choice of a point on its terminal side is $(x, y) = (-1, 0)$. Then

$$r = \sqrt{x^2 + y^2} = \sqrt{1} = 1.$$

$$\cos 3\pi = \frac{x}{r} = -1$$

41. When an angle of $\frac{4\pi}{3}$ is drawn in standard position, one choice of a point on it terminal side is $(x, y) = (-1, -\sqrt{3})$. Then

$$r = \sqrt{x^2 + y^2} = \sqrt{1 + 3} = 2.$$

$$\sin \frac{4\pi}{3} = \frac{y}{r} = -\frac{\sqrt{3}}{2}$$

43. When an angle of $\frac{5\pi}{4}$ is drawn in standard position, one choice of a point on its terminal side is $(x, y) = (-1, -1)$. Then

$$r = \sqrt{x^2 + y^2} = \sqrt{1 + 1} = \sqrt{2}.$$

$$\csc \frac{5\pi}{4} = \frac{r}{y} = -\sqrt{2}$$

45. When an angle of $-\frac{\pi}{3}$ is drawn in standard position, one choice of a point on its terminal side is $(x, y) = (1, -\sqrt{3})$. Then

$$\tan -\frac{\pi}{3} = \frac{y}{x} = -\sqrt{3}$$

47. When an angle of $-\frac{7\pi}{6}$ is drawn in standard position, one choice of a point on its terminal side is $(x, y) = (-\sqrt{3}, 1)$. Then

$$r = \sqrt{x^2 + y^2} = \sqrt{3 + 1} = 2.$$

$$\sin -\frac{7\pi}{6} = \frac{y}{r} = \frac{1}{2}$$

49. $\sin 39° = .6293$

51. $\tan 82° = 7.1154$

53. $\sin .4014 = .3907$

55. $\cos 1.4137 = .1564$

For Exercises 57–65, see the answer graphs in the back of your textbook.

57. The graph of $y = 2 \sin x$ is similar to the graph of $y = \sin x$ except that it has twice the amplitude. (That is, its height is twice as great.)

59. The graph of $y = \cos 2x$ is similar to the graph of $y = \cos x$ except that it oscillates twice as fast. Therefore, its period is π.

61. The graph of $y = -\sin x$ is similar to the graph of $y = \sin x$ except that it is reflected about the y–axis.

63. The graph of $y = \frac{1}{2} \tan x$ is similar to the graph of $y = \tan x$ except that the y–values of points on the graph are one–half the y–values of points on the graph of $y = \tan x$.

65. $S(t) = 500 + 500 \cos \frac{\pi}{6}t$

(a) November corresponds to $t = 0$. Therefore,

$$S(0) = 500 + 500 \cos \left(\frac{\pi}{6}\right)(0)$$

$$= 500 + 500 \cos 0$$

$$= 1000.$$

(b) January corresponds to $t = 2$.
Therefore,

$$S(2) = 500 + 500 \cos \left(\frac{\pi}{6}\right)(2)$$

$$= 500 + 500 \cos \frac{\pi}{3}$$

$$= 500 + 500\left(\frac{1}{2}\right)$$

$$= 750.$$

(c) February corresponds to $t = 3$.
Therefore,

$$S(3) = 500 + 500 \cos \left(\frac{\pi}{6}\right)(3)$$

$$= 500 + 500 \cos \frac{\pi}{2}$$

$$= 500 + 500(0)$$

$$= 500.$$

(d) May corresponds to $t = 6$.
Therefore,

$$S(6) = 500 + 500 \cos \left(\frac{\pi}{6}\right)(6)$$

$$= 500 + 500 \cos \pi$$

$$= 500 + 500(-1)$$

$$= 0.$$

(e) August corresponds to $t = 9$.
Therfore,

$$S(8) = 500 + 500 \cos \left(\frac{\pi}{6}\right)(9)$$

$$= 500 + 500 \cos \frac{3\pi}{2}$$

$$= 500 + 500(0)$$

$$= 500.$$

(f) Use the ordered pairs obtained in parts (a) – (e) to plot the graph.

67. Solving $\dfrac{c_1}{c_2} = \dfrac{\sin \theta_1}{\sin \theta_2}$ for c_2 gives

$$c_2 = \frac{c_1 \sin \theta_2}{\sin \theta_1}.$$

$c_1 = 3 \cdot 10^8$, $\theta_1 = 46°$, and $\theta_2 = 31°$,
so

$$c_2 = \frac{3 \cdot 10^8 (\sin 31°)}{\sin 46°}$$

$$= 214,796,150$$

$$\approx 2 \times 10^8 \text{ m/sec.}$$

69. On the horizontal scale, one whole period clearly spans four squares, so $4 \cdot 30° = 120°$ is the period.

71. $T(x) = 60 - 30 \cos (x/2)$

(a) $x = 0$ represents January, so the maximum afternoon temperature in January is

$$T(0) = 60 - 30 \cos 0 = 30°\text{F.}$$

(b) $x = 2$ represents March, so the maximum afternoon temperature in March is

$$T(2) = 60 - 30 \cos 1 \approx 44°\text{F.}$$

(c) $x = 9$ represents October, so the maximum afternoon temperature in October is

$$T(9) = 60 - 30 \cos \frac{9}{2} \approx 66°\text{F.}$$

(d) $x = 5$ represents June, so the maximum afternoon temperature in June is

$$T(5) = 60 - 30 \cos \frac{5}{2} \approx 84°\text{F.}$$

(e) $x = 7$ represents August, so the maximum afternoon temperature in August is

$$T(7) = 60 - 30 \cos \frac{7}{2} \approx 88°F.$$

73. $w_1 = \dfrac{u^2(\tan \theta)}{L + u(\tan \theta)}$ and

$w_2 = \dfrac{u^2(\tan \theta)}{L - u(\tan \theta)}$

If $\theta = \dfrac{1}{30}°$, $L = .00625$, and $u = 6$, then

$$w_1 = \frac{6^2\left(\tan \dfrac{1}{30}°\right)}{.00625 + 6\left(\tan \dfrac{1}{30}°\right)} \approx 2.2 \text{ and}$$

$$w_2 = \frac{6^2\left(\tan \dfrac{1}{30}°\right)}{.00625 + 6\left(\tan \dfrac{1}{30}°\right)} \approx 7.6.$$

The near and far limits of the depth of field when the object being photographed is 6 m from the camera, are 2.2 m and 7.6 m, respectively.

75. Let x be the distance to the opposite side of the canyon. Then

$$\tan 24° = \frac{80}{x}$$

$$x \tan 24° = 80$$

$$x = \frac{80}{\tan 24°}$$

$$x \approx 179.68.$$

The distance to the opposite side of the canyon is approximately 180 ft.

Section 11.2

1. $y = 2 \sin 6x$

$y' = 2(\cos 6x) \cdot D_x (6x)$

$\quad = 2(\cos 6x) \cdot 6$

$\quad = 12 \cos 6x$

3. $y = 12 \tan (9x + 1)$

$y' = [12 \sec^2(9x + 1)] \cdot D_x (9x + 1)$

$\quad = [12 \sec^2(9x + 1)] \cdot 9$

$\quad = 108 \sec^2(9x + 1)$

5. $y = \cos^4 x$

$y' = [4(\cos x)^3]D_x(\cos x)$

$\quad = (4 \cos^3 x)(-\sin x)$

$\quad = -4 \sin x \cos^3 x$

7. $y = \tan^5 x$

$y' = 5(\tan x)^4 \cdot D_x(\tan x)$

$\quad = 5 \tan^4 x \sec^2 x$

9. $y = -5x \cdot \sin 4x$

$y' = -5x \cdot D_x(\sin 4x)$

$\qquad + (\sin 4x) \cdot D_x (-5x)$

$\quad = -5x(\cos 4x) \cdot D_x (4x)$

$\qquad + (\sin 4x)(-5)$

$\quad = -5x(\cos 4x) \cdot 4 - 5 \sin 4x$

$\quad = -5(4x \cos 4x + \sin 4x)$

11. $y = \dfrac{\sin x}{x}$

$y' = \dfrac{x \cdot D_x(\sin x) - (\sin x) \cdot D_x x}{x^2}$

$\quad = \dfrac{x \cos x - \sin x}{x^2}$

13. $y = \sin e^{5x}$

$y' = (\cos e^{5x}) \cdot D_x(e^{5x})$

$= (\cos e^{5x}) \cdot e^{5x} \cdot D_x(5x)$

$= (\cos e^{5x}) \cdot e^{5x} \cdot 5$

$= 5e^{5x} \cos e^{5x}$

15. $y = e^{\sin x}$

$y' = (e^{\sin x}) \cdot D_x (\sin x)$

$= (\cos x)e^{\sin x}$

17. $y = \sin (\ln 4x^2)$

$y' = [\cos (\ln 4x^2)] \cdot D_x(\ln 4x^2)$

$= \cos (\ln 4x^2) \cdot \dfrac{D_x(4x^2)}{4x^2}$

$= \cos (\ln 4x^2)\dfrac{8x}{4x^2}$

$= \cos (\ln 4x_2) \cdot \dfrac{2}{x}$

$= \dfrac{2}{x} \cos (\ln 4x^2)$

19. $y = \ln \left| \sin x^2 \right|$

$y' = \dfrac{D_x(\sin x^2)}{\sin x^2}$

$= \dfrac{(\cos x^2) \cdot D_x(x^2)}{\sin x^2}$

$= \dfrac{(\cos x^2) \cdot 2x}{\sin x^2}$

$= \dfrac{2x \cos x^2}{\sin x^2}$

or $2x \cot x^2$

21. $y = \dfrac{2 \sin x}{3 - 2 \sin x}$

$y' = \dfrac{(3 - 2 \sin x)D_x (2 \sin x) - (2 \sin x) \cdot D_x (3 - 2 \sin x)}{(3 - 2 \sin x)^2}$

$= \dfrac{(3 - 2 \sin x) \cdot 2D_x(\sin x) - (2 \sin x) \cdot [-2D_x(\sin x)]}{(3 - 2 \sin x)^2}$

$= \dfrac{6 \cos x - 4 \sin x \cos x + 4 \sin x \cos x}{(3 - 2 \sin x)^2}$

$= \dfrac{6 \cos x}{(3 - 2 \sin x)^2}$

23. $y = \sqrt{\dfrac{\sin x}{\sin 3x}} = \left(\dfrac{\sin x}{\sin 3x}\right)^{1/2}$

$y' = \dfrac{1}{2}\left(\dfrac{\sin x}{\sin 3x}\right)^{-1/2} \cdot D_x\left(\dfrac{\sin x}{\sin 3x}\right)$

$= \dfrac{1}{2}\left(\dfrac{\sin 3x}{\sin x}\right)^{-1/2} \cdot \left[\dfrac{(\sin 3x) \cdot D_x(\sin x) - (\sin x) \cdot D_x(\sin 3x)}{(\sin 3x)^2}\right]$

$= \dfrac{1}{2}\left(\dfrac{\sin 3x}{\sin x}\right)^{1/2} \cdot \left(\dfrac{(\sin 3x)(\cos x) - (\sin x)(\cos 3x) \cdot D_x(3x)}{\sin^2 3x}\right)$

$= \dfrac{1(\sin 3x)^{1/2}}{2(\sin x)^{1/2}} \cdot \dfrac{\sin 3x \cos x - 3 \sin x \cos 3x}{\sin^2 3x}$

$= \dfrac{(\sin 3x)^{1/2}(\sin 3x \cos x - 3 \sin x \cos 3x)}{2(\sin x)^{1/2}(\sin^2 3x)}$

$= \dfrac{\sqrt{\sin 3x}[\sin 3x \cos x - 3 \sin x \cos 3x]}{2\sqrt{\sin x}(\sin^2 3x)}$

25.   Since $\cot x = \dfrac{\cos x}{\sin x}$, then by using the quotient rule,

$D_x (\cot x) = D_x\left(\dfrac{\cos x}{\sin x}\right)$

$= \dfrac{(\sin x)(-\sin x) - (\cos x)(\cos x)}{\sin^2 x}$

$= \dfrac{-\sin^2 x - \cos^2 x}{\sin^2 x}$

$= -\dfrac{\sin^2 x + \cos^2 x}{\sin^2 x}$

$= -\dfrac{1}{\sin^2 x}$

$= -\csc^2 x$

27.   Since $\csc x = \dfrac{1}{\sin x} = (\sin x)^{-1}$, then

$D_x (\csc x) = D_x\left(\dfrac{1}{\sin x}\right) = D_x (\sin x)^{-1}$

$= -1(\sin x)^{-2} \cos x$

$= -\dfrac{\cos x}{\sin^2 x}$

$= -\dfrac{1}{\sin x} \cdot \dfrac{\cos x}{\sin x}$

$= -\csc x \cot x.$

29. $s(t) = \sin t + 2 \cos t$

$v(t) = s'(t) = D_t (\sin t + 2 \cos t)$

$\qquad = \cos t - 2 \sin t$

(a) $v(0) = 1 - 2(0) = 1$

(b) $v\left(\frac{\pi}{2}\right) = 0 - 2(1) = -2$

(c) $v(\pi) = -1 - 2(0) = -1$

$a(t) = v'(t) = s''(t)$

$\qquad = D_t (\cos t - 2 \sin t)$

$\qquad = -\sin t - 2 \cos t$

(d) $a(0) = 0 - 2(1) = -2$

(e) $a\left(\frac{\pi}{2}\right) = -1 - 2(0) = -1$

(f) $a(\pi) = 0 - 2(-1) = 2$

31. **(a)** The following is a table of values for

$$y = \frac{1}{5} \sin \pi(t - 1).$$

t	0	.5	1.0	1.5	2.0	2.5	3.0
y	0	-.2	0	.2	0	-.2	0

See the answer graph in the back of your textbook.

(b)

$v(t) = \dfrac{dy}{dt}$

$\qquad = \left[\frac{1}{5} \cos \pi(t - 1)\right] \cdot D_x[\pi(t - 1)]$

$\qquad = \frac{\pi}{5} \cos \pi(t - 1)$

$a(t) = v'(t) = \dfrac{d^2y}{dt^2}$

$\qquad = \left[-\frac{\pi}{5} \sin \pi(t-1)\right] \cdot D_x[\pi(t-1)]$

$\qquad = -\frac{\pi^2}{5} \sin \pi(t - 1)$

(c) $\dfrac{d^2y}{dt^2} + \pi^2 y$

$\qquad = -\frac{\pi^2}{5} \sin [\pi(t - 1)]$

$\qquad\quad + (\pi^2)\frac{1}{5} \sin \pi(t - 1)$

$\qquad = 0$

(d) Since the constant of proportionality is positive, the force and acceleration are in the same direction. At $t = 1.5$ sec,

$$a(1.5) = -\frac{\pi^2}{5} < 0$$

so the force is in a clockwise direction. At $t = 2.5$ sec,

$$a(2.5) = -\frac{\pi^2}{5} > 0$$

so the force is in a counterclockwise direction. At $t = 3.5$ sec,

$$a(3.5) = -\frac{\pi^2}{5} < 0$$

so the force is in a clockwise direction.

33. From the figure, we see that

$$\tan \theta = \frac{x}{60}$$

$$60 \tan \theta = x.$$

Differentiating with respect to time, t, gives

$$60 \sec^2 \theta \cdot \frac{d\theta}{dt} = \frac{dx}{dt}$$

$$\frac{d\theta}{dt} = \frac{\frac{dx}{dt}}{60 \sec^2 \theta}.$$

Since $\dfrac{dx}{dt} = 600,$

$$\dfrac{d\theta}{dt} = \dfrac{600}{60 \sec^2 \theta}$$

$$= \dfrac{10}{\sec^2 \theta}.$$

(a) When the car is at the point on the road closest to the camera, $\theta = 0$. Thus,

$$\dfrac{d\theta}{dt} = \dfrac{10}{(\sec 0)^2}$$

$$= \dfrac{10}{1^2} = 10 \text{ radians/min}$$

$$\dfrac{d\theta}{dt} = \dfrac{10 \text{ radians}}{\text{min}} \cdot \dfrac{1 \text{ rev}}{2\pi \text{ radians}}$$

$$= \dfrac{5}{\pi} \text{ rev/min}.$$

(b) Six seconds later is $\dfrac{1}{10}$ of a minute later, and the car has traveled 60 ft. Thus,

$\tan \theta = \dfrac{60}{60}$ and $\theta = \dfrac{\pi}{4}$.

$$\dfrac{d\theta}{dt} = \dfrac{10}{\left(\sec \dfrac{\pi}{4}\right)^2} = \dfrac{10}{(\sqrt{2})^2}$$

$$= 5 \text{ radians/min}$$

This is one-half the previous value, so

$$\dfrac{d\theta}{dt} = \dfrac{1}{2} \cdot \dfrac{5}{\pi} \text{ rev/min}$$

$$= \dfrac{5}{2\pi} \text{ rev/min}.$$

35. Let x represent the length of the ladder.

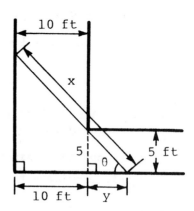

$\tan \theta = \dfrac{5}{y}$ and $\cos \theta = \dfrac{10 + y}{x}$

$y = 5 \cot \theta$ and $x = \dfrac{10 + y}{\cos \theta}$

Thus,

$$x = \dfrac{10 + 5 \cot \theta}{\cos \theta}$$

$\dfrac{dx}{d\theta}$

$$= \dfrac{\cos \theta(-5 \csc^2 \theta) - (10 + 5 \cot \theta)(-\sin \theta)}{\cos^2 \theta}$$

$$= \dfrac{-5 \cos \theta \csc^2 \theta + 10 \sin \theta + 5 \cos \theta}{\cos^2 \theta}$$

$$= \dfrac{-5(\cos \theta \csc^2 \theta - 2 \sin \theta + \cos \theta)}{\cos^2 \theta}$$

If $\dfrac{dx}{d\theta} = 0$, then

$$\cos \theta \csc^2 \theta - 2 \sin \theta - \cos \theta = 0$$

$$\dfrac{\cos \theta - 2 \sin^3 \theta - \cos \theta \sin^2 \theta}{\sin^2 \theta} = 0$$

$$\cos \theta - 2 \sin^3 \theta - \cos \theta \sin^2 \theta = 0$$

$$\cos \theta(1 - \sin^2 \theta) - 2 \sin^3 \theta = 0$$

$$\cos \theta \cos^2 \theta - 2 \sin^3 \theta = 0$$

$$\cos^3 \theta = 2 \sin^3 \theta$$

$$\dfrac{1}{2} = \tan^3 \theta$$

$$\tan \theta = \sqrt[3]{\dfrac{1}{2}}$$

$$\theta \approx .6709 \text{ radian}.$$

If $\theta < .6709$ radian, $\frac{dx}{d\theta} < 0$.

If $\theta > .6709$ radian, $\frac{dx}{d\theta} > 0$.

Thus, x will be minimum when
$\theta = .6709$.

If $\theta = .6709$, then

$$x = \frac{10 + 5 \cot .6709}{\cos .6709} \approx 20.81.$$

The length of the longest possible
ladder is approximately 20.81 ft.

Section 11.3

1. $\int \cos 5x \, dx$

Let $u = 5x$.
Then $du = 5 \, dx$

$\frac{1}{5} du = dx$.

$$\int \cos 5x \, dx = \int \cos u \cdot \frac{1}{5} \, du$$

$$= \frac{1}{5} \int \cos u \, du$$

$$= \frac{1}{5} \sin u + C$$

$$= \frac{1}{5} \sin 5x + C$$

3. $\int (5 \cos x + 2 \sin x) dx$

$$= \int 5 \cos x \, dx + \int 2 \sin x \, dx$$

$$= 5 \int \cos x \, dx + 2 \int \sin x \, dx$$

$$= 5 \sin x - 2 \cos x + C$$

5. $\int x \sin x^2 \, dx$

Let $u = x^2$.
Then $du = 2x \, dx$

$\frac{1}{2} du = x \, dx$.

$$\int x \sin x^2 \, dx = \int \sin u \cdot \frac{1}{2} \, du$$

$$= \frac{1}{2} \int \sin u \, du$$

$$= -\frac{1}{2} \cos u + C$$

$$= -\frac{1}{2} \cos x^2 + C$$

7. $-\int 6 \sec^2 2x \, dx$

Let $u = 2x$.
Then $du = 2 \, dx$

$\frac{1}{2} du = dx$.

$$-\int 6 \sec^2 2x \, dx = -\int 6 \sec^2 u \cdot \frac{1}{2} \, du$$

$$= -3 \int \sec^2 u \, du$$

$$= -3 \tan u + C$$

$$= -3 \tan 2x + C$$

9. $\int \sin^7 x \cos x \, dx$

Let $u = \sin x$.
Then $du = \cos x \, dx$.

$$\int \sin^7 x (\cos x) dx = \int u^7 \, du$$

$$= \frac{1}{8} u^8 + C$$

$$= \frac{1}{8} \sin^8 x + C$$

11. $\displaystyle\int \sqrt{\sin x}(\cos x)\,dx$

Let $u = \sin x$.

Then $du = \cos x\,dx$.

$\displaystyle\int \sqrt{\sin x}(\cos x)\,dx$

$\displaystyle = \int \sqrt{u}\,du$

$\displaystyle = \int u^{1/2}\,du$

$\displaystyle = \frac{2}{3}u^{3/2} + C$

$\displaystyle = \frac{2}{3}(\sin x)^{3/2} + C$

or $\displaystyle\frac{2}{3}\sin x\,\sqrt{\sin x} + C$

13. $\displaystyle\int \frac{\sin x}{1 + \cos x}\,dx$

Let $u = 1 + \cos x$.

Then $du = -\sin x$

$-du = \sin x\,dx$.

$\displaystyle\int \frac{\sin x}{1 + \cos x}\,dx$

$\displaystyle = \int \frac{1}{u}(-du)$

$\displaystyle = -\int \frac{1}{u}\,du$

$= -\ln |u| + C$

$= -\ln |1 + \cos x| + C$

15. $\displaystyle\int x^5 \cos x^6\,dx$

Let $u = x^6$.

Then $du = 6x^5\,dx$

$\displaystyle\frac{1}{6}\,du = x^5\,dx$.

$\displaystyle\int x^5 \cos x^6\,dx = \int (\cos u)\cdot\frac{1}{6}\,du$

$\displaystyle = \frac{1}{6}\int \cos u\,du$

$\displaystyle = \frac{1}{6}\sin u + C$

$\displaystyle = \frac{1}{6}\sin x^6 + C$

17. $\displaystyle\int \tan \frac{1}{4}x\,dx$

Let $u = \frac{1}{4}x$.

Then $4\,du = dx$.

$\displaystyle\int \tan \frac{1}{4}x\,dx = \int (\tan u)(4\,du)$

$\displaystyle = 4\int \tan u\,du$

$= -4\ln |\cos u| + C$

$\displaystyle = -4\ln \left|\cos \frac{1}{4}x\right| + C$

19. $\displaystyle\int x^2 \cot x^3\,dx$

Let $u = x^3$.

Then $\frac{1}{3}\,du = x^2\,dx$.

$\displaystyle\int x^2 \cot x^3\,dx = \int (\cot u)\left(\frac{1}{3}\,du\right)$

$\displaystyle = \frac{1}{3}\int \cot u\,du$

$\displaystyle = \frac{1}{3}\ln |\sin u| + C$

$\displaystyle = \frac{1}{3}\ln \left|\sin x^3\right| + C$

21. $\displaystyle\int e^x \sin e^x \, dx$

Let $u = e^x$.

Then $du = e^x \, dx$.

$\displaystyle\int e^x \sin e^x \, dx = \int \sin u \, du$

$\qquad\qquad\qquad\quad = -\cos u + C$

$\qquad\qquad\qquad\quad = -\cos e^x + C$

23. $\displaystyle\int -6x \cos 5x \, dx$

Let $u = -6x$ and $dv = \cos 5x \, dx$.

Then $du = -6 \, dx$ and $v = \displaystyle\int \cos 5x \, dx$

$\qquad\qquad\qquad\qquad\qquad = \dfrac{1}{5} \sin 5x.$

$\displaystyle\int -6x \cos 5x \, dx$

$= (-6x)\left(\dfrac{1}{5} \sin 5x\right) - \displaystyle\int\left(\dfrac{1}{5} \sin 5x\right)(-6 \, dx)$

$= -\dfrac{6}{5}x \sin 5x + \dfrac{6}{5}\displaystyle\int \sin 5x \, dx$

$= -\dfrac{6}{5}x \sin 5x + \dfrac{6}{5}\cdot\dfrac{1}{5}(-\cos 5x) + C$

$= -\dfrac{6}{5}x \sin 5x - \dfrac{6}{25}\cos 5x + C$

25. $\displaystyle\int 8x \sin x \, dx$

Let $u = 8x$ and $\dfrac{dv}{dx} = \sin x$.

Then $du = 8 \, dx$ and $v = \displaystyle\int \sin x \, dx$

$\qquad\qquad\qquad\qquad\qquad = -\cos x.$

$\displaystyle\int 8x \sin x \, dx$

$= (8x)(-\cos x) - \displaystyle\int(-\cos x)(8 \, dx)$

$= -8x \cos x + 8 \displaystyle\int \cos x \, dx$

$= -8x \cos x + 8 \sin x + C$

27. $\displaystyle\int -6x^2 \cos 8x \, dx$

Let $u = -6x^2$ and $dv = \cos 8x \, dx$.

Then $du = -12x \, dx$ and $v = \dfrac{1}{8} \sin 8x$.

$\displaystyle\int -6x^2 \cos 8x \, dx$

$= (-6x^2)\left(\dfrac{1}{8} \sin 8x\right) - \displaystyle\int\left(\dfrac{1}{8} \sin 8x\right)(-12x \, dx)$

$= -\dfrac{3}{4}x^2 \sin 8x + \dfrac{3}{2}\displaystyle\int x \sin 8x \, dx$

In $\displaystyle\int x \sin 8x \, dx$, let

$\qquad u = x$ and $dv = \sin 8x \, dx$.

Then $du = dx$ and $v = -\dfrac{1}{8} \cos 8x$.

$\displaystyle\int -6x^2 \cos 8x \, dx$

$= -\dfrac{3}{4}x^2 \sin 8x + \dfrac{3}{2}\left[-\dfrac{1}{8}x \cos 8x - \displaystyle\int\left(-\dfrac{1}{8} \cos 8x\right) dx\right]$

$= -\dfrac{3}{4}x^2 \sin 8x - \dfrac{3}{16}x \cos 8x + \dfrac{3}{16}\displaystyle\int \cos 8x \, dx$

$= -\dfrac{3}{4}x^2 \sin 8x - \dfrac{3}{16}x \cos 8x + \dfrac{3}{16}\cdot\dfrac{1}{8} \sin 8x + C$

$= -\dfrac{3}{4}x^2 \sin 8x - \dfrac{3}{16}x \cos 8x + \dfrac{3}{128} \sin 8x + C$

29. $\displaystyle\int_0^{\pi/4} \sin x \, dx = -\cos x \Big|_0^{\pi/4}$

$\qquad\qquad\qquad\qquad = -\cos \dfrac{\pi}{4} - (-\cos 0)$

$\qquad\qquad\qquad\qquad = -\dfrac{\sqrt{2}}{2} + 1$

$\qquad\qquad\qquad\qquad = 1 - \dfrac{\sqrt{2}}{2}$

31. $\displaystyle\int_0^{\pi/3} \tan x \, dx$

$\qquad = -\ln |\cos x| \Big|_0^{\pi/3}$

$\qquad = -\ln \left|\cos \dfrac{\pi}{3}\right| - (-\ln |\cos 0|)$

$$= -\ln \frac{1}{2} + \ln 1$$

$$= -\ln \frac{1}{2} + 0$$

$$= -\ln \frac{1}{2}$$

or $\ln \left(\frac{1}{2}\right)^{-1} = \ln 2$

33. $\displaystyle\int_{\pi/2}^{2\pi/3} \cos x \, dx = \sin x \Big|_{\pi/2}^{2\pi/3}$

$$= \sin \frac{2\pi}{3} - \sin \frac{\pi}{2}$$

$$= \frac{\sqrt{3}}{2} - 1$$

35. $\displaystyle\int_0^\infty e^{-x} \sin x \, dx$

Use entry 47 of the table of integrals with $a = -1$, $b = 1$, and $u = x$.

$$\int e^{-x} \sin x \, dx = \frac{e^{-x}}{(-1)^2 + 1^2}(-\sin x - \cos x) + C$$

$$= -\frac{e^{-x}}{2}(\sin x + \cos x) + C$$

$$\int_0^\infty e^{-x} \sin x \, dx = \lim_{b\to\infty} \int_0^b e^{-x} \sin x \, dx$$

$$= \lim_{b\to\infty} \left[-\frac{e^{-x}}{2}(\sin x + \cos x) \Big|_0^b \right]$$

$$= \lim_{b\to\infty} \left\{ -\frac{e^{-b}}{2}(\sin b + \cos b) - \left[-\frac{1}{2}(0 + 1) \right] \right\}$$

$$= 0 + \frac{1}{2} = \frac{1}{2}$$

37. $S(t) = 500 + 500 \cos \left(\frac{\pi}{6}t\right)$

Total sales over a year's time are approximated by the area under the graph of S during any twelve-month period due to periodicity of S.

Total sales $\approx \displaystyle\int_0^{12} S(t)\,dt$

$$= \int_0^{12} \left[500 + 500 \cos\left(\tfrac{\pi}{6}t\right)\right] dt$$

$$= \int_0^{12} 500\,dt + 500 \int_0^{12} \cos\left(\tfrac{\pi}{6}t\right) dt$$

$$= 500t \Big|_0^{12} + 500\left(\tfrac{6}{\pi} \sin\left(\tfrac{\pi}{6}t\right)\Big|_0^{12}\right)$$

$$= 6000 + \frac{3000}{\pi}\left[\left(\sin\left(\tfrac{\pi}{6}\cdot 12\right) - \sin\left(\tfrac{\pi}{6}\cdot 0\right)\right)\right]$$

$$= 6000 + \frac{3000}{\pi}(\sin 2\pi - \sin 0)$$

$$= 6000 + \frac{3000}{\pi}(0 - 0)$$

$$= 6000$$

39. $V(t) = 170 \sin (120\,\pi t)$

Root mean square $= \sqrt{\dfrac{\int_0^T V^2(t)\,dt}{T}}$

(a) The period is $T = \dfrac{2\pi}{120\pi} = \dfrac{1}{60}$ sec.

(b) $\displaystyle\int_0^T V^2(t)\,dt = \int_0^T [170 \sin (120\,\pi t)]^2 \, dt$

$$= 170^2 \int_0^T \sin^2 (120\,\pi t)\,dt$$

$$= 170^2 \int_0^T \tfrac{1}{2}[1 - \cos (240\pi t)]\,dt$$

$$= \frac{170^2}{2} \int_0^T [1 - \cos (240\,\pi t)\,dt$$

$$= \frac{170^2}{2}\left[t - \frac{1}{240\pi} \sin (240\,\pi t)\right]\Big|_0^T$$

$$= \frac{170^2}{2}\left[T - \frac{1}{240\pi} \sin (240\,\pi T)\right] - \frac{170^2}{2}\cdot 0$$

Since $T = \frac{1}{60}$,

$$\int_0^T V^2(t)\,dt = \frac{170^2}{2}\left(\frac{1}{60} - \frac{1}{240\pi}\sin 4\pi\right)$$

$$= \frac{170^2}{120} \text{ and}$$

Root mean square $= \sqrt{\dfrac{\dfrac{170^2}{120}}{\dfrac{1}{60}}}$

$$= \sqrt{\frac{170^2}{2}}$$

$$= \frac{170}{\sqrt{2}}$$

$$\approx 120.21.$$

Thus, the voltage from a standard wall outlet is approximately 120 volts.

Section 11.4

1. $y = \sin^{-1}(-\sqrt{3}/2)$

By the definition of the inverse sine function,

$$\sin y = -\frac{\sqrt{3}}{2}.$$

Since $\sin(-\pi/3) = -\sqrt{3}/2$ and $-\pi/3$ is in $[-\pi/2, \pi/2]$,

$$y = \sin^{-1}\left(-\frac{\sqrt{3}}{2}\right) = -\frac{\pi}{3}.$$

3. $y = \tan^{-1} 1$

By the definition of the inverse tangent function,

$$\tan y = 1.$$

Since $\tan \pi/4 = 1$ and $\pi/4$ is in $(-\pi/2, \pi/2)$,

$$y = \tan^{-1} 1 = \frac{\pi}{4}.$$

5. $y = \sin^{-1}(-1)$

By the definition of the inverse sine function,

$$\sin y = -1.$$

Since $\sin(-\pi/2) = -1$ and $-\pi/2$ is in $[-\pi/2, \pi/2]$,

$$y = \sin^{-1}(-1) = -\frac{\pi}{2}.$$

7. $y = \cos^{-1}(1/2)$

By the definition of the inverse cosine function,

$$\cos y = -\frac{1}{2}.$$

Since $\cos \pi/3 = 1/2$ and $\pi/3$ is in $[0, \pi]$,

$$y = \cos^{-1}\left(\frac{1}{2}\right) = \frac{\pi}{3}.$$

9. $y = \cos^{-1}(-\sqrt{2}/2)$

By the definition of the inverse cosine function,

$$\cos y = -\frac{\sqrt{2}}{2}.$$

Since $\cos(3\pi/4) = -\sqrt{2}/2$ and $3\pi/4$ is in $(0, \pi)$,

$$y = \cos^{-1}\left(-\frac{\sqrt{2}}{2}\right) = \frac{3\pi}{4}.$$

11. $y = \tan^{-1}(-\sqrt{3})$

By the definition of the inverse tangent function,

$$\tan y = -\sqrt{3}.$$

Since $\tan(-\pi/3) = -\sqrt{3}$ and $-\pi/3$ is in $(-\pi/2, \pi/2)$,

$$y = \tan^{-1}(-\sqrt{3}) = -\frac{\pi}{3}.$$

For Exercises 13–23, set calculator in degree mode.

13. $\sin^{-1}(-.1392)$

Enter .1392 $\boxed{+/-}$ $\boxed{INV}$ $\boxed{sin}$

Display -8.001556351

$\sin^{-1}(-.1392) \approx -8°$

15. $\cos^{-1}(-.8988)$

Enter .8988 $\boxed{+/-}$ $\boxed{INV}$ $\boxed{cos}$

Display 154.0007782

$\cos^{-1}(-.8988) \approx 154°$

17. $\cos^{-1} .9272$

Enter .9272 $\boxed{INV}$ $\boxed{cos}$

Display 21.99753044

$\cos^{-1} .9272 \approx 22°$

19. $\tan^{-1} 1.111$

Enter 1.111 $\boxed{INV}$ $\boxed{tan}$

Display 48.00993839

$\tan^{-1} 1.111 \approx 48°$

21. $\tan^{-1}(-.9004)$

Enter .9004 $\boxed{+/-}$ $\boxed{INV}$ $\boxed{tan}$

Display -41.99987203

$\tan^{-1}(-.9004) \approx -42°$

23. $\sin^{-1} .9242$

Enter .9242 $\boxed{INV}$ $\boxed{sin}$

Display 67.54802935

$\sin^{-1} .9242 \approx 68°$

25. $y = \sin^{-1} 12x$

$$y' = \frac{1}{\sqrt{1-(12x)^2}} D_x(12x)$$

$$= \frac{12}{\sqrt{1-144x^2}}$$

27. $y = \tan^{-1} 3x$

$$y' = \frac{1}{1+(3x)^2} D_x(3x)$$

$$= \frac{3}{1+9x^2}$$

29. $y = \cos^{-1}\left(-\frac{2}{x}\right)$

$$y' = \frac{-1}{\sqrt{1-\left(-\frac{2}{x}\right)^2}} D_x\left(-\frac{2}{x}\right)$$

$$= \frac{-1}{\sqrt{1-\frac{4}{x^2}}} \cdot \frac{2}{x^2}$$

$$= \frac{-2}{x^2\sqrt{\frac{x^2-4}{x^2}}}$$

$$= \frac{-2}{x^2 \cdot \frac{\sqrt{x^2-4}}{\sqrt{x^2}}}$$

$$= \frac{-2}{x^2 \cdot \frac{\sqrt{x^2 - 4}}{|x|}}$$

$$= \frac{-2}{|x|\sqrt{x^2 - 4}}$$

31. $y = \tan^{-1}(\ln |7x|)$

$$y' = \frac{D_x (\ln |7x|)}{1 + (\ln |7x|)^2}$$

$$= \frac{\frac{7}{7x}}{1 + (\ln |7x|)^2}$$

$$= \frac{1}{x[1 + (\ln |7x|)^2]}$$

33. $y = \ln |\tan^{-1}(x + 1)|$

$$y' = \frac{D_x[\tan^{-1}(x + 1)]}{\tan^{-1}(x + 1)}$$

$$= \frac{\frac{1}{1 + (x + 1)^2} \cdot D_x(x + 1)}{(\tan^{-1}(x + 1)}$$

$$= \frac{\frac{1}{1 + (x + 1)^2}}{\tan^{-1}(x + 1)}$$

$$= \frac{1}{[1 + (x + 1)^2]\tan^{-1}(x + 1)}$$

35.   Let $x = \cos y$.   Then,

$$D_x (x) = D_x(\cos y)$$

$$1 = -\sin y \cdot \frac{dy}{dx}$$

$$\frac{dy}{dx} = -\frac{1}{\sin y}.$$

If $x = \cos y$, then $\sin y = \pm\sqrt{1 - x^2}$.
However, for $y = \cos^{-1} x$, $0 \le y \le \pi$
and $\sin y$ is positive so $\sin y = \sqrt{1 - x^2}$, and

$$\frac{dy}{dx} = -\frac{1}{\sqrt{1 - x^2}}.$$

Let $x = \tan y$.   Then,

$$D_x(x) = D_x(\tan y)$$

$$1 = \sec^2 y \cdot \frac{dy}{dx}$$

$$\frac{dy}{dx} = \frac{1}{\sec^2 y}$$

$$= \frac{1}{1 + \tan^2 y}$$

$$= \frac{1}{1 + x^2}$$

37. $\int \frac{x^3}{1 + x^8} \, du$

Let $u = x^4$.
Then $du = 4x^3 \, dx$.

$$\int \frac{x^3}{1 + x^8} \, dx = \int \frac{1}{1 + u^2}\left(\frac{1}{4} \, du\right)$$

$$= \frac{1}{4} \int \frac{1}{1 + u^2} \, du$$

$$= \frac{1}{4} \tan^{-1} u + C$$

$$= \frac{1}{4} \tan^{-1} x^4 + C$$

39. $\int \frac{e^x}{\sqrt{1 - e^{2x}}} \, dx$

Let $u = e^x$.
Then $du = e^x \, dx$.

$$\int \frac{e^x}{\sqrt{1 - e^{2x}}} \, dx = \int \frac{1}{\sqrt{1 - u^2}} \, du$$

$$= \sin^{-1} u + C$$

$$= \sin^{-1} e^x + C$$

41. $\int \dfrac{4}{\sqrt{1 - 9x^2}} \, dx$

Let $u = 3x$.

Then $du = 3 \, dx$

$\dfrac{1}{3} \, du = dx$.

$\int \dfrac{4}{\sqrt{1 - 9x^2}} \, dx = \int \dfrac{4}{\sqrt{1 - u^2}} \left(\dfrac{1}{3} \, du\right)$

$= \dfrac{4}{3} \int \dfrac{1}{\sqrt{1 - u^2}} \, du$

$= \dfrac{4}{3} \sin^{-1} u + C$

$= \dfrac{4}{3} \sin^{-1} 3x + C$

43. $\int \dfrac{\dfrac{1}{x}}{1 + (\ln x)^2} \, dx$

Let $u = \ln |x|$.

Then $du = \dfrac{1}{x} \, dx$.

$\int \dfrac{\dfrac{1}{x}}{1 + (\ln x)^2} \, dx$

$= \int \dfrac{\dfrac{1}{x}}{1 + (\ln |x|)^2} \, dx$

$= \int \dfrac{1}{1 + u^2} \, dx$

$= \tan^{-1} u + C$

$= \tan^{-1} \ln |x| + C$

45. $\int_0^1 \dfrac{1}{1 + x^2} \, dx = \tan^{-1} x \Big|_0^1$

$= \tan^{-1} 1 - \tan^{-1} 0$

$= \dfrac{\pi}{4} - 0$

$= \dfrac{\pi}{4}$

47. $\int_0^2 \dfrac{1}{\sqrt{16 - x^2}} \, dx = \int_0^2 \dfrac{\dfrac{1}{4} \, dx}{\dfrac{\sqrt{16 - x^2}}{4}}$

$= \int_0^2 \dfrac{\dfrac{1}{4} \, dx}{\sqrt{\dfrac{16 - x^2}{16}}}$

$= \int_0^2 \dfrac{\dfrac{1}{4} \, dx}{\sqrt{1 - \dfrac{x^2}{16}}}$

$= \int_0^2 \dfrac{\dfrac{1}{4} \, dx}{\sqrt{1 - \left(\dfrac{x}{4}\right)^2}}$

Let $u = \dfrac{x}{4}$. Then $du = \dfrac{1}{4} dx$.

If $x = 2$, $u = \dfrac{1}{2}$. If $x = 0$, $u = 0$.

$\int_0^2 \dfrac{1}{\sqrt{16 - x^2}} \, dx = \int_0^{1/2} \dfrac{du}{\sqrt{1 - u^2}}$

$= \sin^{-1} u \Big|_0^{1/2}$

$= \sin^{-1} \dfrac{1}{2} - \sin^{-1} 0$

$= \dfrac{\pi}{6} - 0$

$= \dfrac{\pi}{6}$

49. $\displaystyle\int_0^\infty \frac{1}{1 + x^2}\, dx = \lim_{b\to\infty} \int_0^b \frac{1}{1 + x^2}\, dx$

$$= \lim_{b\to\infty} \left[(\tan^{-1} x)\Big|_0^b \right]$$

$$= \lim_{b\to\infty} (\tan^{-1} b - \tan^{-1} 0)$$

$$= \lim_{b\to\infty} (\tan^{-1} b - 0)$$

$$= \lim_{b\to\infty} (\tan^{-1} b)$$

$$= \frac{\pi}{2}$$

51. $\theta = \tan^{-1}\left(\dfrac{4}{x}\right) - \tan^{-1}\left(\dfrac{1}{x}\right)$

   (a) If $x = 1$,

$$\theta = \tan^{-1} 4 - \tan^{-1} 1$$
$$\approx 76.0° - 45.0°$$
$$\approx 31°.$$

   (b) If $x = 2$,

$$\theta = \tan^{-1} 2 - \tan^{-1}\left(\frac{1}{2}\right)$$
$$\approx 63.4° - 26.6°$$
$$\approx 37°.$$

   (c) If $x = 3$,

$$\theta = \tan^{-1}\left(\frac{4}{3}\right) - \tan^{-1}\left(\frac{1}{3}\right)$$
$$\approx 53.1° - 18.4°$$
$$\approx 35°.$$

   (d) If $x = 4$,

$$\theta = \tan^{-1} 1 - \tan^{-1}\left(\frac{1}{4}\right)$$
$$\approx 45.0° - 14.0°$$
$$\approx 31°.$$

53. Refer to the figure below.

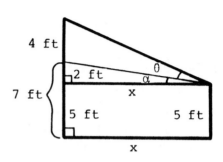

$\tan(\alpha + \theta) = \dfrac{6}{x}$ and $\tan \alpha = \dfrac{2}{x}$

$\alpha + \theta = \tan^{-1}\left(\dfrac{6}{x}\right)$ and $\alpha = \tan^{-1}\left(\dfrac{2}{x}\right)$

Thus,

$\tan^{-1}\left(\dfrac{2}{x}\right) + \theta = \tan^{-1}\left(\dfrac{6}{x}\right)$, and

$\theta = \tan^{-1}\left(\dfrac{6}{x}\right) - \tan^{-1}\left(\dfrac{2}{x}\right)$.

$$\frac{d\theta}{dv} = \frac{-\dfrac{6}{x^2}}{1 + \dfrac{36}{x^2}} - \frac{-\dfrac{2}{x^2}}{1 + \dfrac{4}{x^2}}$$

$$= \frac{-6}{x^2 + 36} + \frac{2}{x^2 + 4}$$

$$= \frac{-6x^2 - 24 + 2x^2 + 72}{(x^2 + 36)(x^2 + 4)}$$

$$= \frac{-4x^2 + 48}{(x^2 + 36)(x^2 + 4)}$$

If $\dfrac{d\theta}{dx} > 0$, then

$$-4x^2 + 48 = 0$$
$$x^2 = 12$$
$$x = \pm\sqrt{12}$$
$$x = \pm 2\sqrt{3}.$$

The value $x = -2\sqrt{3}$ is impossible in this problem.

If $x < 2\sqrt{3}$, $\dfrac{d\theta}{dx} > 0$. If $x > 2\sqrt{3}$,

$\dfrac{d\theta}{dx} < 0$.

Therefore, θ is maximum when

$x = 2\sqrt{3} \approx 3.46$.

To maximize θ, Patricia should stand $2\sqrt{3}$ ft ≈ 3.46 ft from the wall.

Section 11.5

1. Let $f(x) = \sin x$.

Then $f'(x) = \cos x$,

so

$$f'(0) = \cos 0 = 1.$$

The slope of the tangent line to the graph of $y = \sin x$ at $x = 0$ is 1.

3. Let $f(x) = \cos x$.

Then $f'(x) = -\sin x$,

so

$$f'\left(\frac{\pi}{2}\right) = -\sin \frac{\pi}{2} = -1.$$

The slope of the tangent line to the graph of $y = \cos x$ at $x = \pi/2$ is -1.

5. Let $f(x) = \tan x$.

Then $f'(x) = \sec^2 x$,

so

$$f'(0) = \sec^2 0 = \frac{1}{\cos^2 0} = 1.$$

The slope of the tangent line to the graph of $y = \tan x$ at $x = 0$ is 1.

7. $f(.2) = 1000e^{2 \sin(.2)}$

$\approx 1000e^{2(.1987)}$

$\approx 1000(1.4880)$

≈ 1490

9. $f(.5) = 1000e^{2 \sin(.5)}$

$\approx 1000e^{2(.4794)}$

$\approx 1000(2.6086)$

≈ 2610

11. $f(1) = 1000e^{2 \sin 1}$

$\approx 1000e^{2(.8415)}$

$\approx 1000(5.3817)$

≈ 5380

13. $f(1.8) = 1000e^{2 \sin(1.8)}$

$\approx 1000e^{2(.9738)}$

$\approx 1000(7.0118)$

≈ 7010

15. $f(3) = 1000e^{2 \sin 3}$

$\approx 1000e^{2(.1411)}$

$\approx 1000(1.3260)$

≈ 1330

17. Use the following table to plot points.

t	f(t)	
0	1000	
.2	1490	
.4	2180	
.5	2610	
.8	4200	
1	5380	
1.4	7180	
1.6	7380	Maximum Value
1.8	7010	
2.3	4440	
3	1330	
3.1	1090	
4.7	135	Minimum Value
7.9	7370	
11	135	

Note that additional values are
found by finding critical numbers
using the first derivative test.
See the answer graph in the back
of your textbook.

19. If $s(t) = A \cos (Bt + C)$, then

$$s'(t) = -A \sin (Bt + C) \cdot B$$
$$= -AB \sin (Bt + C), \text{ and }$$
$$s''(t) = -AB \cos (Bt + C) \cdot B$$
$$= -AB^2 \cos (Bt + C).$$

Thus,

$$s''(t) = -B^2 [A \cos (Bt + C)]$$
$$s''(t) = -B^2 s(t).$$

21. $u(x, t) = 16 + 11e^{-.00706x} \cos(wt - ax).$

Since the temperature must be main-
tained between $16 \pm 2°C$ and
$\cos (wt - ax)$ has a maximum and
minimum value of 1 and -1, it follows that $11e^{-.00706x} = 2$. Solving
gives

$$-.00706x = \ln \frac{2}{11} \quad \text{or} \quad x \approx 241 \text{ cm.}$$

23. $u = T_0 + A_0 e^{-ax} \cos (wt - ax)$

$$\frac{\partial u}{\partial t} = A_0 e^{-ax} [-\sin (wt - ax)](w)$$
$$= -wA_0 e^{-ax} \sin (wt - ax)$$

$$\frac{\partial u}{\partial x} = aA_0 e^{-ax} \cos (wt - ax)$$
$$\quad + A_0 e^{-ax} [-\sin (wt - ax)](-a)$$
$$= -aA_0 e^{-ax}$$
$$\quad \cdot [\cos (wt - ax) - \sin(wt - ax)]$$

$$\frac{\partial^2 u}{\partial x^2}$$
$$= a^2 A_0 e^{-ax} [\cos(wt - ax) - \sin(wt - ax)]$$
$$\quad -aA_0 e^{-x} [(-a)\sin(wt - ax) + a \cos(wt - ax)]$$
$$= a^2 A_0 e^{-ax} [\cos (wt - ax) - \sin(wt - ax)]$$
$$\quad - a^2 A_0 e^{-ax} [\sin(wt - ax) + \cos(wt - ax)]$$
$$= -2a^2 A_0 e^{-ax} \sin (wt - ax)$$

Now $a = \sqrt{\dfrac{w}{2k}}$ implies that

$$a^2 = \frac{w}{2k}$$

$$k = \frac{w}{2a^2}.$$

It follows that

$$k \frac{\partial^2 u}{\partial x^2}$$

$$= \frac{w}{2a^2}[-2a^2 A_0 e^{-ax} \sin (wt - ax)]$$
$$= -wA_0 e^{-ax} \sin (wt - ax)$$
$$= \frac{\partial u}{\partial t}.$$

Thus,

$$\frac{\partial u}{\partial t} = k \frac{\partial^2 u}{\partial x^2}.$$

25. $\sin \theta = \dfrac{s}{L_2}$

$$L_2 \sin \theta = s$$

$$L_2 = \frac{s}{\sin \theta}$$

27. $\cot \theta = \dfrac{L_0 - L_1}{s}$

$$s \cot \theta = L_0 - L_1$$
$$L_1 + s \cot \theta = L_0$$
$$L_1 = L_0 - s \cot \theta$$

29. $R_2 = k \cdot \dfrac{L_2}{r_2{}^4}$ where R_2 is the resistance along DC.

31. $R = k\left(\dfrac{L_1}{r_1{}^4} + \dfrac{L_2}{r_2{}^4}\right)$

$\quad = k\left(\dfrac{L_0 - s\cot\theta}{r_1{}^4} + \dfrac{s}{(\sin\theta)r_2{}^4}\right)$

33. Using Exercise 32 gives

$$\dfrac{ks}{\sin^2\theta}\left(\dfrac{1}{r_1{}^4} - \dfrac{\cos\theta}{r_2{}^4}\right) = 0.$$

or $\dfrac{k(s\csc^2\theta)}{r_1{}^4} + \dfrac{k(-s\cos\theta)}{\sin^2\theta\ r_2{}^4} = 0.$

35. Using Exercise 34 gives

$$k\left(\dfrac{1}{r_1{}^4} - \dfrac{\cos\theta}{r_2{}^4}\right) = 0$$
$$\dfrac{1}{r_1{}^4} - \dfrac{\cos\theta}{r_2{}^4} = 0 \quad k \neq 0$$
$$\dfrac{1}{r_1{}^4} = \dfrac{\cos\theta}{r_2{}^4}$$
$$\cos\theta = \dfrac{r_2{}^4}{r_1{}^4}$$

37. If $r_1 = 1.4$ and $r_2 = .8$, then

$$\cos\theta = \dfrac{(.8)^4}{(1.4)^4} \approx .1066.$$

Thus,

$$\theta \approx 84°.$$

Chapter 11 Review Exercises

5. $90° = 90\left(\dfrac{\pi}{180}\right) = \dfrac{90\pi}{180} = \dfrac{\pi}{2}$

7. $210° = 210\left(\dfrac{\pi}{180}\right) = \dfrac{210\pi}{180} = \dfrac{7\pi}{6}$

9. $360° = 2\pi$

11. $7\pi = 7\pi\left(\dfrac{180}{\pi}\right)^° = 1260°$

13. $\dfrac{9\pi}{20} = \dfrac{9\pi}{20}\left(\dfrac{180}{\pi}\right)^° = 81°$

15. $\dfrac{13\pi}{20} = \dfrac{13\pi}{20}\left(\dfrac{180}{\pi}\right)^° = 117°$

17. When an angle of 60° is drawn in standard position, one choice of a point on its terminal side is $(x, y) = (1, \sqrt{3})$. Then

$$r = \sqrt{x^2 + y^2} = \sqrt{1 + 3} = 2$$
$$\sin 60° = \dfrac{y}{r} = \dfrac{\sqrt{3}}{2}.$$

19. When an angle of $-30°$ is drawn in standard position, one choice of a point on its terminal side is $(x, y) = (\sqrt{3}, -1)$. Then

$$r = \sqrt{x^2 + y^2} = \sqrt{3 + 1} = 2.$$
$$\cos(-30°) = \dfrac{x}{r} = \dfrac{\sqrt{3}}{2}.$$

21. When an angle of 120° is drawn in standard position, one choice of a point on its terminal side is $(x, y) = (-1, \sqrt{3})$. Then

$$r = \sqrt{x^2 + y^2} = \sqrt{1 + 3} = 2$$

$$\csc 120° = \frac{r}{y} = \frac{2}{\sqrt{3}} = \frac{2\sqrt{3}}{3}.$$

23. When an angle of $\frac{\pi}{6}$ is drawn in standard position, on choice of a point on its terminal side is $(x, y) = (\sqrt{3}, 1)$. Then

$$r = \sqrt{x^2 + y^2} = \sqrt{3 + 1} = 2,$$

$$\sin \frac{\pi}{6} = \frac{y}{r} = \frac{1}{2}.$$

25. When an angle of $\frac{5\pi}{4}$ is drawn in standard position, one choice of a point on its terminal side is $(x, y) = (-1, -1)$. Then

$$r = \sqrt{x^2 + y^2} = \sqrt{1 + 1} = \sqrt{2}$$

$$\sec \frac{5\pi}{4} = \frac{r}{x} = \frac{\sqrt{2}}{-1} = -\sqrt{2}.$$

27. $\sin 47° \approx .7314$

29. $\tan 81° = 6.314$

31. $\sin 1.4661 = .9945$

33. $\cos .5934 = .8290$

37. Let $y = \tan^{-1}(-1)$.
By the definition of the inverse tangent function,

$$\tan y = -1.$$

Since $\tan(-\pi/4) = -1$ and $-\pi/4$ is in $(-\pi/2, \pi/2)$,

$$y = \tan^{-1}(-1) = -\frac{\pi}{4}.$$

39. Let $y = \sin^{-1}(\sqrt{3}/2)$.
By the definition of the inverse sine function,

$$\sin y = \frac{\sqrt{3}}{2}.$$

Since $\sin \pi/3 = \sqrt{3}/2$ and $\pi/3$ is in $[-\pi/2, \pi/2]$,

$$y = \sin^{-1} \frac{\sqrt{3}}{2} = \frac{\pi}{3}.$$

41. Let $y = \tan^{-1} \sqrt{3}$).
By the definition of the inverse tangent function,

$$\tan y = \sqrt{3}.$$

Since $\tan \pi/3 = \sqrt{3}$ and $\pi/3$ is in $(-\pi/2, \pi/2)$,

$$y = \tan^{-1} \sqrt{3} = \frac{\pi}{3}.$$

43. A graph of $y = \cos x$ is given in the textbook. To get $y = 3 \cos x$, each value of y in $y = \cos x$ must be multiplied by 3. This gives a graph going through $(0, 3)$, $(\pi, -3)$ and $(2\pi, 3)$. See the graph in the answer section of your textbook.

45. The graph of $y = \tan x$ appears in Figure 15 in Section 11.1. The difference between the graph of $y = \tan x$ and $y = -\tan x$ is that the y-values of points on the graph of $y = -\tan x$ are the opposites of the y-values of the corresponding points on the graph of $y = \tan x$.

A sample calculation: when $x = \frac{\pi}{4}$,

$$y = -\tan \frac{\pi}{4} = -1.$$

See the graph in the answer section of your textbook.

47. $y = -4 \sin 7x$
$y' = -4(\cos 7x) \cdot D_x(7x)$
$\quad = -4(\cos 7x)7$
$\quad = -28 \cos 7x$

49. $y = \tan(4x^2 + 3)$
$y' = \sec^2(4x^2 + 3) \cdot D_x(4x^2 + 3)$
$\quad = \sec^2(4x^2 + 3) \cdot (8x)$
$\quad = 8x \sec^2(4x^2 + 3)$

51. $y = 3 \cos^6 x$
$y' = 3 D_x(\cos x)^6$
$\quad = [3 \cdot 6(\cos x)^5] \cdot D_x(\cos x)$
$\quad = 18(\cos x)^5(-\sin x)$
$\quad = -18 \sin x \cos^5 x$

53. $y = \cot(4x^5)$
$y' = [-\csc^2(4x^5)] \cdot D_x(4x^5)$
$\quad = [-\csc^2(4x^5)] \cdot 20x^4$
$\quad = -20x^4 \csc^2(4x^5)$

55. $y = x^2 \cos x$
$y' = x^2 D_x(\cos x) + \cos x\, D_x(x^2)$
$\quad = x^2(-\sin x) + \cos x(2x)$
$\quad = -x^2 \sin x + 2x \cos x$

57. $y = \dfrac{\sin x - 1}{\sin x + 1}$
$y' = \dfrac{(\sin x + 1)(\cos x) - (\sin x - 1)(\cos x)}{(\sin x + 1)^2}$
$\quad = \dfrac{\sin x \cos x + \cos x - \sin x \cos x + \cos x}{(\sin x + 1)}$
$\quad = \dfrac{2 \cos x}{(\sin x + 1)^2}$

59. $y = \dfrac{x - 2}{\sin x}$
$y' = \dfrac{\sin x \cdot 1 - (x - 2)\cos x}{\sin^2 x}$
$\quad = \dfrac{\sin x - (x - 2)\cos x}{\sin^2 x}$
$\quad = \dfrac{\sin x - x \cos x + 2 \cos x}{\sin^2 x}$

61. $y = \ln |\cos x|$
$y' = \dfrac{D_x(\cos x)}{\cos x}$
$\quad = \dfrac{-\sin x}{\cos x}$
$\quad = -\tan x$

63. $y = \sin^{-1}(-3x)$
$y' = \dfrac{1}{\sqrt{1 - (-3x)^2}} \cdot D_x(-3x)$
$\quad = \dfrac{1}{\sqrt{1 - 9x^2}} \cdot (-3)$
$\quad = -\dfrac{3}{\sqrt{1 - 9x^2}}$

65. $y = \tan^{-1}(5x^2)$

$y' = \dfrac{D_x \ (5x^2)}{1 + (5x^2)^2}$

$= \dfrac{10x}{1 + 25x^4}$

67. $\displaystyle\int \sin 2x \ dx$

Let $u = 2x$.

Then $du = 2 \ dx$

$\dfrac{1}{2} \ du = dx.$

$\displaystyle\int \sin 2x \ dx = \int (\sin u)(\tfrac{1}{2} \ du)$

$= \dfrac{1}{2} \displaystyle\int \sin u \ du$

$= \dfrac{1}{2}(-\cos u) + C$

$= -\dfrac{1}{2} \cos 2x + C$

69. $\displaystyle\int \tan 9x \ dx$

Let $u = 9x$.

Then $\dfrac{1}{9} \ du = dx.$

$\displaystyle\int \tan 9x \ dx = \int (\tan u)(\tfrac{1}{9} \ du)$

$= \dfrac{1}{9} \displaystyle\int \tan u \ du$

$= \dfrac{1}{9}(-\ln |\cos u|) + C$

$= -\dfrac{1}{9} \ln |\cos 9x| + C$

71. $\displaystyle\int 5 \sec^2 x \ dx = 5 \int \sec^2 x \ dx$

$= 5 \tan x + C$

73. $\displaystyle\int x \sin 3x^2 \ dx$

Let $u = 3x^2$.

Then $\dfrac{1}{6} \ du = x \ dx.$

$\displaystyle\int x \sin 3x^2 \ dx = \int \sin u (\tfrac{1}{6} \ du)$

$= \dfrac{1}{6} \displaystyle\int \sin u \ du$

$= \dfrac{1}{6}(-\cos u) + C$

$= -\dfrac{1}{6} \cos 3x^2 + C$

75. $\displaystyle\int \sqrt{\cos x} \cdot \sin x \ dx$

Let $u = \cos x$.

Then $du = -\sin x \ dx$

$-du = \sin x \ dx.$

$\displaystyle\int \sqrt{\cos x} \cdot \sin x \ dx$

$= \displaystyle\int \sqrt{u}(-du)$

$= -\displaystyle\int u^{1/2} \ du$

$= -\dfrac{2}{3}u^{3/2} + C$

$= -\dfrac{2}{3}(\cos x)^{3/2} + C$

77. $\displaystyle\int x \tan 11x^2 \ dx$

Let $u = 11x^2$.

Then $du = 22x \ dx.$

$\displaystyle\int x \tan 11x^2 \ dx$

$= \dfrac{1}{22} \displaystyle\int (\tan 11x^2) \cdot (22x \ dx)$

$= \dfrac{1}{22} \displaystyle\int \tan u \ du$

$= \dfrac{1}{22}(-\ln |\cos u|) + C$

$= -\dfrac{1}{22} \ln |\cos 11x^2| + C$

79. $\int (\sin x)^{5/2} \cos x\, dx$

Let $u = \sin x$.

Then $du = \cos x\, dx$.

$\int (\sin x)^{5/2} \cos x\, dx$

$\quad = \int u^{5/2}\, du$

$\quad = \dfrac{y^{5/2+1}}{5/2 + 1} + C$

$\quad = \dfrac{u^{7/2}}{7/2} + C$

$\quad = \dfrac{2}{7}u^{7/2} + C$

$\quad = \dfrac{2}{7}(\sin x)^{7/2} + C$

81. $\int \dfrac{3}{\sqrt{1 - x^2}}\, dx = 3\int \dfrac{1}{\sqrt{1 - x^2}}\, dx$

$\qquad\qquad\qquad = 3\sin^{-1} x + C$

83. $\int \dfrac{-6x^4}{1 + x^{10}}\, dx$

Let $u = x^5$.

Then $du = 5x^4$.

$\int \dfrac{-6x^4}{1 + x^{10}}\, dx$

$\quad = \int \dfrac{-6x^4}{1 + (x^5)^2}\, dx$

$\quad = -\dfrac{6}{5}\int \dfrac{5x^4}{1 + (x^5)^2}\, dx$

$\quad = -\dfrac{6}{5}\int \dfrac{du}{1 + u^2}$

$\quad = -\dfrac{6}{5}\tan^{-1} u + C$

$\quad = -\dfrac{6}{5}\tan^{-1} x^5 + C$

85. $\int_0^{\pi/2} \cos x\, dx$

$\quad = \sin x \Big|_0^{\pi/2}$

$\quad = \sin \dfrac{\pi}{2} - \sin 0$

$\quad = 1 - 0 = 1$

87. $\int_0^{2\pi} (10 + 10\cos x)\, dx$

$\quad = \int_0^{2\pi} 10\, dx + \int_0^{2\pi} 10\cos x\, dx$

$\quad = 10(2\pi) + 10\int_0^{2\pi} \cos x\, dx$

$\quad = 20\pi + 10\left(\sin x \Big|_0^{2\pi}\right)$

$\quad = 20\pi + 10(\sin 2\pi - \sin 0)$

$\quad = 20\pi + 10 \cdot 0$

$\quad = 20\pi$

93. Refer to the figure below.

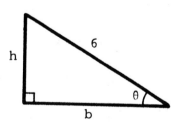

Let A be the area of the triangle then,

$$A = \frac{1}{2} bh.$$

However,

$$\sin \theta = \frac{h}{6} \qquad \text{and} \cos \theta = \frac{b}{6}$$

$$h = 6 \sin \theta \quad \text{and} \quad b = 6 \cos \theta.$$

Thus,

$$A = \frac{1}{2}(6 \cos \theta)(6 \sin \theta)$$

$$= 18 \sin \theta \cos \theta$$

$$= 9(2 \sin \theta \cos \theta)$$

$$A = 9 \sin (2\theta).$$

$$\frac{dA}{d\theta} = 9 \cos(2\theta) \cdot 2 = 18 \cos(2\theta)$$

If $\frac{dA}{d\theta} = 0$,

$$18 \cos(2\theta) = 0$$

$$\cos 2\theta = 0$$

$$2\theta = \frac{\pi}{2}$$

$$\theta = \frac{\pi}{4}.$$

If $\theta = \frac{\pi}{4}$, $\frac{dA}{d\theta} > 0$. If $\theta > \frac{\pi}{4}$, $\frac{dA}{d\theta} < 0$.

Thus, A is maximum when $\theta = \frac{\pi}{4}$ or 45°.

95. **(a)** $\frac{d}{dx} \ln |\sec x + \tan x|$

$$= \frac{\sec x \tan x + \sec^2 x}{\sec x + \tan x}$$

$$= \frac{\sec x(\tan x + \sec x)}{(\sec x + \tan x)}$$

$$= \sec x$$

(b) $\frac{d}{dx}(-\ln |\sec x - \tan x|)$

$$= -\frac{\sec x \tan x - \sec^2 x}{\sec x - \tan x}$$

$$= -\frac{\sec x(\tan x - \sec x)}{(\sec x - \tan x)}$$

$$= -(-\sec x)$$

$$= \sec x$$

(d) $D(\theta) = k \int_0^\theta \sec x \, dx$

$$= k \ln |\sec x + \tan x| \Big|_0^\theta$$

$$= k \ln |\sec \theta + \tan \theta|$$
$$\quad - k \ln |\sec 0 + \tan 0|$$
$$= k \ln |\sec \theta + \tan \theta|$$
$$\quad - k \ln |1 + 0|$$
$$= k \ln |\sec \theta + \tan \theta|$$
$$\quad - k \ln 1$$
$$= k \ln |\sec \theta + \tan \theta| - 0$$
$$= k \ln |\sec \theta + \tan \theta|$$

Since $D(34°03') = 7$,

$$7 = k \ln |\sec 34°03' + \tan 34°03'|$$

$$k = \frac{7}{\ln |\sec 34°03' + \tan 34°03'|}$$

$$k \approx 11.0635$$

$D(40°45')$

$$\approx 11.0635 \ln |\sec 40°45' + \tan 40°45'|$$

$$\approx 8.63.$$

New York city should be placed approximately 8.63 in from the equator.

(e)

$D(25°46')$

$$\approx 11.0635 \ln |\sec 25°46' + \tan 25°46'|$$

$$\approx 5.15$$

Miami should be placed approximately 5.15 in from the equator.

Extended Application

1. $f(\theta) = 6hs + \dfrac{3}{2}s^2\left(-\cot \theta + \dfrac{\sqrt{3}}{\sin \theta}\right)$

  $f'(\theta) = D_\theta \ (6hs) + \dfrac{3}{2}s^2 \ D_\theta \left(-\cot \theta + \dfrac{\sqrt{3}}{\sin \theta}\right)$

  $f'(\theta) = 0 + \dfrac{3}{2}s^2\left(\csc^2 \theta - \dfrac{\sqrt{3} \cos \theta}{\sin^2 \theta}\right.$

  $f'(\theta) = \dfrac{3}{2}s^2\left(\dfrac{1}{\sin^2 \theta} - \dfrac{\sqrt{3} \cos \theta}{\sin^2 \theta}\right)$

2. $f'(\theta)$ vanishes if $f'(\theta) = 0$.

  Therefore, $\dfrac{3}{2}s^2\left(\dfrac{1}{\sin^2 \theta} - \dfrac{\sqrt{3} \cos \theta}{\sin^2 \theta}\right) = 0$ or $\dfrac{1}{\sin^2 \theta} = \dfrac{\sqrt{3} \cos \theta}{\sin^2 \theta}$

  Therefore, $1 = \sqrt{3} \cos \theta$.

CHAPTER 11   TEST

[11.1]   1. Convert 72° to radian measure. Express your answer as a multiple of π.

[11.1]   2. Convert $\frac{7\pi}{20}$ to degree measure.

[11.1]   3. Evaluate $\tan \frac{11\pi}{6}$ without using a calculator or table.

[11.1]   4. Find the value of sin 2.986.

[11.1]   5. Graph one period of $y = -4 \sin x$,

[11.1]   6. Graph one period of $y = \frac{1}{2} \tan x$.

[11.2]   **Find the derivatives of each of the following functions.**

   7. $y = -3 \cos 4x$ 8. $y = \cot^2 \frac{x}{2}$ 9. $y = 3x^2 \sin x$

   10. $y = \frac{\sin^2 x}{\sin x + 1}$ 11. $y = \ln |3 \cos x|$

[11.3]   **Find each integral.**

   12. $\int 3 \sin 2x \, dx$ 13. $\int 4 \sec^2 x \, dx$

   14. $\int x \cos (2x^2) \, dx$ 15. $\int \tan 3x \, dx$

   16. $\int (\cos x)^2 \sin x \, dx$ 17. $\int_{\pi/6}^{\pi/3} \cos x \, dx$

[11.4]   **Give each value in radians.**

   18. (a) $\sin^{-1} \left(-\frac{\sqrt{3}}{2}\right)$ (b) $\tan^{-1} \left(-\frac{\sqrt{3}}{3}\right)$

[11.4]   **Find the derivative of each function.**

19. $y = \sin^{-1}(3x)$ 20. $y = \tan^{-1}(x + 2)$

[11.4]   **Find each integral.**

21. $\displaystyle\int \frac{-12}{\sqrt{1 - x^2}}\, dx$ 22. $\displaystyle\int \frac{1}{\sqrt{1 - 4x^2}}\, dx$

23. $\displaystyle\int \frac{x^5}{1 + x^{12}}\, dx$

[11.5]   24. Find the slope of the tangent to the graph of $y = \tan x$ at $x = \pi/3$.

[11.5]   25. Find the maximum and minimum values of y for the simple harmonic motion given by $y = 3 \sin(2x + 3)$.

[11.1]   26. The range of $y = \sin x$ is $[-1, 1]$ and $\csc x = 1/\sin x$. Use this to explain why the range at $y = \csc x$ is $(-\infty, -1] \cup [1, \infty)$.

[11.2]   27. Explain why the slope of $y = \tan x$ is never zero.

[11.3]   28. If b is a positive real number, explain why $\displaystyle\int_{-b}^{b} \sin x\, dx$ will always be zero.

CHAPTER 11   TEST ANSWERS

1. $\dfrac{2\pi}{5}$ 2. $63°$ 3. $-\dfrac{\sqrt{3}}{3}$ 4. $.1550$ 5.

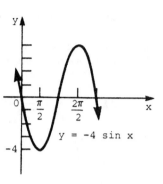

6.

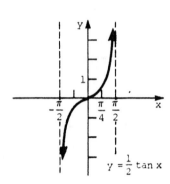

7. $12 \sin 4x$ 8. $-\cot \dfrac{x}{2} \csc^2 \dfrac{x}{2}$

9. $6x \sin x + 3x^2 \cos x$ 10. $\dfrac{\sin^2 x \cos x + 2 \sin x \cos x}{(\sin x + 1)^2}$

11. $-\tan x$ 12. $-\dfrac{3}{2} \cos 2x + C$ 13. $4 \tan x + C$ 14. $\dfrac{1}{4} \sin (2x^2) + C$

15. $-\dfrac{1}{3} \ln |\cos 3x| + C$ 16. $-\dfrac{1}{3} \cos^3 x + C$ 17. $\dfrac{\sqrt{3} - 1}{2}$ or $.366$

18. (a) $-\dfrac{\pi}{3}$ (b) $-\dfrac{\pi}{6}$ 19. $\dfrac{3}{\sqrt{1 - 9x^2}}$ 20. $\dfrac{1}{1 + (x + 2)^2}$

21. $12 \cos^{-1} x + C$ or $-12 \sin^{-1} x + C$ 22. $\dfrac{1}{2} \sin^{-1} 2x + C$

23. $\dfrac{1}{6} \tan^{-1} (x^6) + C$ 24. 4 25. $3, -3$

26. The reciprocals of numbers in $(0, 1]$ are $[1, \infty)$, and the reciprocals numbers in $[-1, 0)$ are $(-\infty, -1]$. Thus, the range of $y = \csc x$ is of $(-\infty, -1] \cup [1, \infty)$.

27. The slope of y = tan x is given by y' = sec² x. Since sec x is never zero, the slope of y = tan x is never zero.

28. $$\int_{-b}^{b} \sin x \, dx = -\cos x \Big|_{-b}^{b}$$

$$= -\cos b - [-\cos (-b)]$$

$$= -\cos b + \cos (-b)$$

For the cosine function, cos (−x) = cos x for all values of x. Thus,

−cos b + cos (−b) = 0 for all values of b, and $\int_{-b}^{b} \sin x \, dx$ for all

real numbers b.

NOTES

NOTES

NOTES

NOTES

NOTES

NOTES